World Soils Book Series

Series Editor

Alfred E. Hartemink
Department of Soil Science, FD Hole Soils Laboratory
University of Wisconsin–Madison
Madison, WI
USA

The World Soils Book Series publishes peer-reviewed books on the soils of a particular country. They include sections on soil research history, climate, geology, geomorphology, major soil types, soil maps, soil properties, soil classification, soil fertility, land use and vegetation, soil management, soils and humans, soils and industry, future soil issues. The books summarize what is known about the soils in a particular country in a concise and highly reader-friendly way. The series contains both single and multi-authored books as well as edited volumes. There is additional scope for regional studies within the series, particularly when covering large land masses (for example, The Soils of Texas, The Soils of California), however, these will be assessed on an individual basis.

More information about this series at http://www.springer.com/series/8915

Bipin B. Mishra
Editor

The Soils of India

Springer

Editor
Bipin B. Mishra
Department of Soil Science
Bihar Agricultural University
Sabour, Bhagalpur, Bihar, India

ISSN 2211-1255 ISSN 2211-1263 (electronic)
World Soils Book Series
ISBN 978-3-030-31080-6 ISBN 978-3-030-31082-0 (eBook)
https://doi.org/10.1007/978-3-030-31082-0

This Springer imprint is published by the registered company Springer Nature Switzerland AG
The registered company address is: Gewerbestrasse 11, 6330 Cham, Switzerland

The Soil

That enjoys the shifting coarse of the river Ganga

That never allows to construct a house on river bank

That does change with periodically occurring floods

But, never says for shelter elsewhere

Every year, new "Alluvium" appears that grows the crops

But, never says to wait until "Alluvium turns into Soil".

It is the Ganga with aquifer, while riverbed ecosystem with rhizosphere.

Such a Holy Soil of ***Gokhale Nagar Bishnupur*** *is my trait,*

That is strange, full of wisdom; though rests on trust of
Vasudeva kutumbkam.

Thus, I see the soils from Arunachal to Saurastra stretching Kerala,

Kashmir to Kanyakumari extending to Indira Point of Nicobar Islands;

O My Soil ! The Land of My India!
—Bipin B. Mishra

Foreword

India is uniquely a diversified country with a complex mosaic of landscapes covering with variegated thin layer of a variety of soil types. The book "The Soils of India" under the World Soils Book Series provides, in a reader-friendly manner, an overview of the distribution, properties, functions and emerging challenges of soils in Indian agriculture following the future strategic land use planning processes, soil governance and soil research opportunities. It provides a complete synthesis of the soils of India with special emphasis on glimpses of soil research, climate and climate change, geology and geomorphology, mineralogy and micromorphology, biogeochemistry, major soil types and classification, soil health and emerging threats to soil safety, with logical and conceptual action plans towards sustainable management options.

The continued pressing demand for land on way to find the most remunerative land use options and subsequent challenges with finite soil resources would have prompted the United Nations Organization to declare 5 December as World Soil Day as a mark to connect people with soils and raise awareness on their critical importance in our lives in order to assure the food security, food safety and food nutrition. The mission of the International Union of Soil Sciences with the World Soils Book Series to prepare books for each country is a powerful aid towards UNO's commitment. I am pleased that "The Soils of India" in this series would be a milestone towards the sustainable Indian agriculture.

Professor Bipin B. Mishra, as Editor of this book, is a frank and seasoned soil scientist with an acclaimed credibility and has contributed commendably in shaping this book as a treasure for farmer-friendly land use planning options in order to benefit the Indian agriculture in different facets including climate change mitigation and adaptation and organic farming approaches with a positive vision towards second green revolution in India following the system approaches in soil resilience.

Most of the chapters are authored by experienced scientists from across the country (both ICAR and SAUs) on emerging issues, opportunities, threats and potential mitigation options in order to move Indian agriculture towards sustainable development under assured economic growth. This book is intended for students, scientists, land use planners, policy-makers and corporate sectors including farming communities at large who are indulged to know the ground truth about the state of the art in soil sciences following the land evaluation for reliable planning process in India.

April 2017

Dr. Trilochan Mohapatra
Secretary, Department of Agricultural Research and Education

Director General, ICAR
New Delhi, India

Preface

The nature, properties and qualities of soils today in Indian subcontinent are definitely not of the same or even similar as they were in the past, although the soil resource has been deeply rooted to human's existence because food, fibre, fuel, fodder, water and air are directly under the influence of soil resource in all pedogenic, edaphologic and environmental perceptions. Change is no doubt the natural law of the environment, wherein humans look for adaptation to such changing situations or exploring mitigation options. But, if the change is because of anthropologic ill-interferences, it leads gradually to distortion of the natural equilibrium causing a disastrous consequence in time and space. The soils in India have been made victimized with such consequences as a result of continuing unplanned and imbalanced activities since long past. Within the wheel of rapidly growing human population as well as shrinking trend of agriculturally viable land area, the demand for soil and land has become pressing. Such pressing demand has led to intensification of land uses without caring for soil health and sustainability, and as a result, the soil and land resources in India are virtually now the threatened scarce resources that also suffer from varying nature of conflicts. The book covers a wide range of issues for inventorying the soil resources for systematic evaluation following the strategic planning for improvement to desired extent. Accordingly, the book is aimed at synthesizing the geologic, mineralogic, micromorphometric, climatic, pedologic, edaphologic, anthropologic, biotic, abiotic, biogeochemistry and ecologic aspects of these soils in relation to emerging threats to soil health, stability, safety, biodiversity, productivity, profitability and overall sustainability on the landscape. Efforts are made to project detailed information on challenges and opportunities of soil resources in priority areas. Chapter 1 is an introduction highlighting the background as well as emerging values of soils in everyone's lifestyle, threats to soils, opportunities for restoration of soil health and quality, relevance of land use planning and second green revolution. Subsequently, education and research in soil science need to be upgraded with changing scenario under the umbrella of ICAR.

The history of soil research in India is merely at youthful stage and still needs advancement in terms of principles, laws, theories and technologies in order to meet the pressing demands for healthy soil resources in India. The ancient history witnesses the importance of soils as the "Mother Earth". Soil as a natural resource exists on landscape in continuum beyond the man-made boundary, and so India believes on "*Vasudheva Kutumbkam*". Chapter 2 is a systematic projection of complete historical spectrum of growth and development in soil research in order to establish the priorities for assurance of sound soil health and sustainable production.

We sow the seeds in soil, but grow the plant on land. So, soil works under the umbrella of climate and collectively called the land, i.e. soil and climate. Chapter 3 highlights the climate and their possible variation in time and space with emphasis on trends of various weather parameters over the year across the country besides their impacts on soil parameters. This chapter is also devoted to the changing scenario of climate and its impacts on soils that is the lowest boundary of the entire earth's atmosphere excluding ocean, water bodies and rock outcrop. The current changing scenario of climate is becoming a threat even to soil qualities necessitating the adaption as well as mitigation approach with emphasis on carbon sequestration in soil environment.

Soils being the thin skin of the earth's crust develop primarily on rocks as conditioned by geological orientation and physiography of the landscape. Chapter 4 covers both geology and geomorphology in shaping the diversified soil types. The soil is dynamic as a three-dimensional natural body in an open ecosystem and diversified particularly in differing geology, climate, geomorphology, relief, hydrology and vegetation. Chapter 5 thus covers different major soil types and their taxonomic equivalence with World Reference Base (WRB) and USDA soil taxonomy following their relevance in land use options and transfer of technology. Importantly, mineral fractions including clays and clay minerals in soils are the driving force for almost all the production functions of the soils. Chapter 6 is accordingly designed to highlight the scope and distribution of these minerals in differing soil groups of India and their vital interactions relevant to different soil functions. The edaphology is always substantiated by pedology, and for soil and land evaluation, pedology plays a vital role. Accordingly, soil micromorphology, being the study of soil as well as sediment in thin section, can only disclose whether the soil is formed *in situ* or through colluvium or alluvium. Chapter 7 discusses on use of micromorphology in Indian soils and to answer some researchable problems related especially to pedology and edaphology.

Soil is a living (biotic) and non-living (abiotic) natural system, wherein elemental cycling as mediated by complex interactions between biotic and abiotic soil components is by and large usual. However, the human activities have substantially altered biogeochemical cycling of several key elements including carbon, nitrogen, phosphorus, potassium and other secondary and minor nutrients over the past few decades, which, in turn, had serious environmental consequences. Chapter 8 presents the soil biogeochemistry of Indian soils in both natural ecosystems and managed agricultural systems. A better understanding of soil biogeochemical cycling is needed as it controls ecosystem functioning of both natural and managed (i.e. agricultural) ecosystems, productivity and sustainability.

Chapter 9 is devoted to the benchmark soils in major agro-ecological regions. It is a representative of the most extensive soils in major land resource area or agro-ecological zone. Such efforts would be helpful in identifying the use of soil resource inventory and classification in optimal land use and production system. Chapter 10 is accordingly intended to land evaluation and land use planning being the powerful means of making decision for specific use of land uses in a given land unit for the most remunerative return on sustainable basis, since the land in India is not only scarce but threatened too. This chapter is devoted to issues, opportunities and different procedures followed in India for land evaluation and land use planning.

Soil as a natural resource has capability to produce, and so the soil must be healthy for optimal production, safe and resilient with minimum or zero level of degradation, profitable and sustainable to meet the community's requirements without environmental abuse following the adaptation and mitigation options against the emerging challenges due to climate change. Chapter 11 opens the most burning issues with the health and fertility of soils in four major spectra, viz. physical, chemical, biological and pollution, and critically identifies the opportunities for managing the soil health following the policy initiatives taken by the Government of India. It is apparent that the Indian Council of Agricultural Research, besides promoting the soil science and pedology, is constantly encouraging for assurance of healthy soils in each parcel of land in collaboration with other departments, state governments and central and state agricultural universities.

Chapter 12 covers the burning issues on quantification of soil resilience "*the capacity of a soil to recover its functional and structural integrity after a disturbance*". This chapter helps to understand the declining trend of the partial factor productivity of inputs like fertilizers in India. Contrary to ever-increasing food demand in India, the partial factor productivity of inputs like fertilizers decreased from 20 kg grain/kg NPK applied in 1985 to 14 in 1996 and to 9.2 in 2010 due to deterioration in physical, chemical and biological quality of soils. Since the soil resilience is the soil's inherent ability to restore its quality after disturbance that influences the soil health and quality, the sustainability thus depends on resilience and resistance of soil by conservation and improvement measures in time and space.

Since historic past, soil has been serving the societies with its commercial and industrial values also. Chapter 13 discusses various soil forms of industrial importance. Type-specific soils and their fractions including parent materials are already used differently in trades and industries across the country. However, success in adoption of conservation agriculture will open opportunity for soil-based carbon trading in days to come. Besides, by linking land use planning with supply chain process may strengthen the scope of soil in commercialization through the farming communities and ensure the alleviation of poverty.

Chapter 14 is devoted to "soil and future" and covers a wide range of issues including governance, restructuring the national soil and land policies and modernization of smart soil-based education, research and development aspects, land use-based soil classification scheme, effective land use planning process, legal restriction on diversion of agricultural land, soil health insurance and evaluation as well as auditing of project recommendations for property rights and so on.

The book, as a whole, may be an aid to future line of action to ensure a healthy soil in each parcel of land for optimal return with time and space in order to meet the over-pressing demands towards food security in terms of quantity, quality, nutrition and nourishment for the growing population of India.

June, 2017

Bipin B. Mishra
Haramaya University, Dire Dawa, Ethiopia

Acknowledgements

I would like to thank the United Nations Organization for declaring 5 December as the "World Soil Day" and subsequently 2015 as the "International Year of Soils" following the "International Decade of Soil (2015-2024)" that have strengthened the everlasting foundation of the International Union of Soil Sciences (IUSS) with its vision, mission and challenging goals, Late Dr. Jon Hempel, Coordinator of the Working Groups in formulizing the "Universal Soil Classification" under IUSS, who inducted me to lead one of the USC Working Groups, Professor Alfred E. Hartemink, University of Wisconsin, Madison, and the Series Editor of the "World Soils Book Series" for assigning this task.

Dr. Sudhanshu K. Ghosh, Former Head, Division of Soil Science & Agricultural Chemistry, IARI, New Delhi, and Past President, The Clay Minerals Society of India; Dr. M. Velayutham, Former Director, ICAR-NBSS & LUP, Nagpur; Dr. A. Subba Rao, Former Director, ICAR-IISS, Bhopal; Dr. Suresh K. Chaudhari, Assistant Director General (Soil and Water Management), Indian Council of Agricultural Research, New Delhi, and President, Indian Society of Soil Science; Dr. S. K. Singh, Director, ICAR-NBSS & LUP, Nagpur, and President, The Clay Minerals Society of India; Dr. S. C. Datta, Emeritus Scientist, Division of Soil Science and Agricultural Chemistry, ICAR-IARI, New Delhi; Dr. H. Pathak, Director, ICAR-NRRI, Cuttack; Dr. Kamaljit Ray, Regional Coordinator, IMD, New Delhi; Dr. Sanjay Kr Choudhary, T.N.B. College, Bhagalpur; Dr. K S Anil Kumar and Dr. V. Ramamurthy, ICAR-NBSS & LUP, Regional Centre, Bangalore; Dr. Sayantani Neogi, University of Cambridge, UK; Dr. K. Sammi Reddy, Head, CRIDA, Hyderabad; Dr. Sean Taylor, McDonald Institute for Archaeological Research, University of Cambridge, UK; Dr Debjani Sihi, USA; and Dr Biswanath Dari, USA, together with other authors and co-authors of respective chapters are thankfully acknowledged.

I sincerely acknowledge Dr. Rabi N. Sahoo and Dr. Prameela Krishnan, Division of Agricultural Physics, IARI, New Delhi, for valuable cooperation at planning stage. Dr. J. Somasundaram, IISS, Bhopal; Dr Debjani Sihi, USA; Dr Biswanath Dari, USA; and Dr. Sayantani Neogi, University of Cambridge, UK, for technical support in reviewing some chapters with constructive comments are also acknowledged.

I thankfully acknowledge Dr. IP Abrol, Former DDG (Soils), ICAR; Dr. KV Raman, Former Director, NAARM, Hyderabad; Dr. G Narayanasamy and Dr. R. K. Rattan, Former Presidents, Indian Society of Soil Science; and Dr NG Raghumohan, Former Pedologist, IARI, New Delhi, for encouragement.

Sincere thanks are extended to Dr. Ajoy K. Singh, Vice Chancellor, Bihar Agricultural University, Sabour, Bhagalpur, for encouragement. Dr.Anshuman Kohli (Soil Science) along with Navin Kumar Singh (College Store), Sanjeev Kumar (College Establishment) and Shiv Shankar Poddar (Computer Operator) of the Bihar Agricultural University, Sabour, Bhagalpur, are acknowledged for technical and computer access. Thanks also to Subhash Kumar, Digital Techno, Bhagalpur, for providing computer input.

I am highly thankful to Sanjeet Kumar Sharma (Department of Computer Applications, T. N.B. College, Bhagalpur) along with Dr. Anand Kumar Pandey, Dr.Rakesh Kumar Pandey and Er. Ravi Bhusan Yadav for computer support including e-proofing and online correction.

Sincere thanks are extended to Prof. Awadh Kishore Roy, Vice Chancellor, B.N.M.U., Madhepura, and T.M.B.U., Bhagalpur, for encouragement and technical support.

I would sincerely acknowledge Dr. Ayyanadar Arunachalam, Assistant Director General (International Relations), ICAR, New Delhi, for constant support and encouragement.

Thanks are due to Suresh Rettagunta, Chennai, for constant support, sincere advice and everlasting encouragement on behalf of Springer International Publishing AG Switzerland.

I feel gratefully thankful to Professor Alfred E. Hartemink for constant technical support, valuable comments and constructive directives in finalizing the manuscript to its present shape with encouragement at every stage of communication.

I am gratefully indebted to Dr. Trilochan Mohapatra, Secretary, Government of India, Department of Agricultural Research & Education, and Director General, Indian Council of Agricultural Research (ICAR), Ministry of Agriculture and Farmer's Welfare, Krishi Bhavan, New Delhi, for writing a scholarly but visionary foreword of this book "The Soils of India" in the World Soils Book Series.

I want to thank everyone who ever said anything positive during completion of this book including my grandson Himnish (Hani).

Contents

1 **Introduction** 1
Bipin B. Mishra

2 **History of Soil Research** 17
A. Subba Rao and J. Somasundaram

3 **Climate** 41
Kamaljit Ray, S. D. Attri, H. Pathak, Anjani Kumar, and Dibyendu Chaterjee

4 **Geology and Geomorphology** 57
K. S. Anil Kumar, S. Thayalan, R. S. Reddy, M. Lalitha, B. Kalaiselvi, S. Parvathy, K. Sujatha, Rajendra Hegde, S. K. Singh, and Bipin B. Mishra

5 **Major Soil Types and Classification** 81
K. S. Anil Kumar, M. Lalitha, Shivanand, K. Sujatha, K. M. Nair, R. Hegde, S. K. Singh, and Bipin B. Mishra

6 **Soil Mineralogy and Clay Minerals** 109
S. C. Datta, Sudhansu K. Ghosh, and Debarup Das

7 **Soil Micromorphology** 129
Sayantani Neogi and Sean Taylor

8 **Soil Biogeochemistry** 143
Debjani Sihi and Biswanath Dari

9 **Benchmark Soils in Agro-ecological Regions** 159
K. S. Anil Kumar, K. S. Karthika, M. Lalitha, R. Srinivasan, Shivanand, K. Sujatha, K. M. Nair, R. Hegde, S. K. Singh, and Bipin B. Mishra

10 **Land Evaluation and Land Use Planning** 191
V. Ramamurthy, O. Challa, L. G. K. Naidu, K. S. Anil Kumar, S. K. Singh, D. Mamatha, K. Ranjitha, and Bipin B. Mishra

11 **Soil Health and Fertility** 215
Suresh K. Chaudhari, P. P. Biswas, and Hemlata Kapil

12 **Soil Resilience** 233
K. Sammi Reddy, K. L. Sharma, K. Srinivas, A. K. Indoria, M. Vassanda Coumar, Pushpanjali, and V. Girija Veni

13 **Soil and Industry** 243
Bipin B. Mishra, Sanjay K. Choudhary, and Richa Roy

14 Soil and Future 261
Mariappan Velayutham and Bipin B. Mishra

Index 273

Contributors

S. D. Attri Agro Advisory Services Division (AASD), Meteorological Department India, New Delhi, India

P. P. Biswas Indian Council of Agricultural Research, New Delhi, India

O. Challa ICAR-National Bureau of Soil Survey and Land Use Planning, Nagpur, India

Dibyendu Chaterjee ICAR-National Rice Research Institute, Cuttack, Odisha, India

Suresh K. Chaudhari Indian Council of Agricultural Research, New Delhi, India;
Indian Society of Soil Science, New Delhi, India

Sanjay K. Choudhary TNB College, TM Bhagalpur University, Bhagalpur, India

M. Vassanda Coumar ICAR-Central Research Institute for Dryland Agriculture, Hyderabad, Telangana, India

Biswanath Dari Aberdeen Research and Extension Center, University of Idaho, Aberdeen, ID, USA

Debarup Das Division of Soil Science and Agricultural Chemistry, ICAR-Indian Agricultural Research Institute, Pusa, New Delhi, India;
The Clay Minerals Society of India, New Delhi, India

S. C. Datta Division of Soil Science and Agricultural Chemistry, ICAR-Indian Agricultural Research Institute, Pusa, New Delhi, India;
The Clay Minerals Society of India, New Delhi, India

Sudhansu K. Ghosh Division of Soil Science and Agricultural Chemistry, ICAR-Indian Agricultural Research Institute, Pusa, New Delhi, India;
The Clay Minerals Society of India, New Delhi, India

R. Hegde ICAR-National Bureau of Soil Survey and Land Use Planning, Hebbal, Bangalore, India

A. K. Indoria ICAR-Central Research Institute for Dryland Agriculture, Hyderabad, Telangana, India

B. Kalaiselvi ICAR-National Bureau of Soil Survey and Land Use Planning, Bangalore, India

Hemlata Kapil Indian Council of Agricultural Research, New Delhi, India

K. S. Karthika ICAR-National Bureau of Soil Survey and Land Use Planning, Bangalore, India

Anjani Kumar ICAR-National Rice Research Institute, Cuttack, Odisha, India

K. S. Anil Kumar ICAR-National Bureau of Soil Survey and Land Use Planning, Hebbal, Bangalore, India

M. Lalitha ICAR-National Bureau of Soil Survey and Land Use Planning, Hebbal, Bangalore, India

D. Mamatha ICAR-National Bureau of Soil Survey and Land Use Planning, Bangalore, India

Bipin B. Mishra Bihar Agricultural University, Sabour, Bhagalpur, India;
Pedology and Land Use Planning, School of Natural Resources Management and Environmental Sciences, Haramaya University, Dire Dawa, Ethiopia;
International Union of Soil Sciences, Vienna, Austria

L. G. K. Naidu ICAR-National Bureau of Soil Survey and Land Use Planning, Bangalore, India

K. M. Nair ICAR-National Bureau of Soil Survey and Land Use Planning, Hebbal, Bangalore, India

Sayantani Neogi McDonald Institute for Archaeological Research, University of Cambridge, Cambridge, UK

S. Parvathy ICAR-National Bureau of Soil Survey and Land Use Planning, Bangalore, India

H. Pathak ICAR-National Rice Research Institute, Cuttack, Odisha, India

Pushpanjali ICAR-Central Research Institute for Dryland Agriculture, Hyderabad, Telangana, India

V. Ramamurthy ICAR-National Bureau of Soil Survey and Land Use Planning, Bangalore, India

K. Ranjitha ICAR-National Bureau of Soil Survey and Land Use Planning, Bangalore, India

Kamaljit Ray Ministry of Earth Sciences, Lodhi Road, New Delhi, India

K. Sammi Reddy ICAR-Central Research Institute for Dryland Agriculture, Hyderabad, Telangana, India

R. S. Reddy ICAR-National Bureau of Soil Survey and Land Use Planning, Bangalore, India

Richa Roy Department of Biotechnology, TNB College, TM Bhagalpur University, Bhagalpur, India

K. L. Sharma ICAR-Central Research Institute for Dryland Agriculture, Hyderabad, Telangana, India

Shivanand ICAR-National Bureau of Soil Survey and Land Use Planning, Bangalore, India

Debjani Sihi Oak Ridge National Laboratory, Environmental Sciences Division, Climate Change Science Institute, Oak Ridge, TN, USA

S. K. Singh ICAR-National Bureau of Soil Survey and Land Use Planning, Nagpur, India

J. Somasundaram ICAR-Indian Institute of Soil Science, Bhopal, Madhya Pradesh, India

K. Srinivas ICAR-Central Research Institute for Dryland Agriculture, Hyderabad, Telangana, India

R. Srinivasan ICAR-National Bureau of Soil Survey and Land Use Planning, Bangalore, India

A. Subba Rao ICAR-Indian Institute of Soil Science, Bhopal, Madhya Pradesh, India

K. Sujatha ICAR-National Bureau of Soil Survey and Land Use Planning, Hebbal, Bangalore, India

Sean Taylor McDonald Institute for Archaeological Research, University of Cambridge, Cambridge, UK

S. Thayalan ICAR-National Bureau of Soil Survey and Land Use Planning, Bangalore, India

Mariappan Velayutham ICAR-National Bureau of Soil Survey and Land Use Planning, Nagpur, India;
Indian Society of Soil Science & Indian Society of Soil Survey & Land Use Planning, MS Swaminathan Research Foundation (MSSRF), Chennai, India

V. Girija Veni ICAR-Central Research Institute for Dryland Agriculture, Hyderabad, Telangana, India

Abbreviations

%	Per cent
°C	Degree centigrade
A&N	Andaman & Nicobar
AA	Amorphous aluminosilicate
AAS	Atomic absorption spectrophotometer
AC	Ammonium chloride
ACR	Agro-climatic region
ADMB	Aravalli–Delhi Mobile Belt
AE	Agronomic efficiency
AER	Agro-ecological region
AEZ	Agro-ecological Zones
AFAS	Amorphous ferri-aluminosilicate
AICARP	All India Coordinated Agronomic Research Project
AICP	All India Coordinated Project
AICRP	All India Coordinated Research Project
AICRPDA	All India Coordinated Project on Dryland Agriculture
AICRP-LTFE	All India Coordinated Research Project on Long-Term Fertilizer Experiments
AICRP-STCR	All India Coordinated Research Project on Soil Test Crop Response
AINPB	All India Network Project on Biofertilizer
AIS&LUS	All India Soil and Land Use Survey
AnMC	Carbon mineralized under anaerobic condition
AP	Andhra Pradesh
ARP	Arunachal Pradesh
AS	Ammonium sulphate
As	Arsenic
Av. K	Available potassium
Av. N	Available nitrogen
Av. P	Available phosphorus
AWC	Available water capacity
BCE	Before Common Era (Before Christian Era)
BCR	Benefit–cost ratio
BD	Bulk density
BG	β-glucosidase
BGA	Blue-green algae
BMC	Biomass carbon
BMN	Biomass nitrogen
BNF	Biological nitrogen fixation
BP	Before present
BPHO	Phosphodiesterase
BRDF	Bidirectional reflectance distribution function

BSP	Benchmark Soils Project
BSR	Basal respiration
C	Carbon
C:N	Carbon-to-nitrogen ratio
CA	Conservation agriculture
$CaCO_3$	Calcium carbonate
CAN	Calcium ammonium nitrate
CAZRI	Central Arid Zone Research Institute
CBI	Coir Board of India
CCRM	Community climate risk manager
Cd	Cadmium
CEC	Cation-exchange capacity
CEC_R	Reduced cation-exchange capacity
CEC_{UT}	Cation-exchange capacity of untreated sample or original cation-exchange capacity
CH	Colloidal (<0.2 μm) clay-humus complex
CH_4	Methane
CHN	Carbon, hydrogen and nitrogen (CHN) analyser
CI	Coefficient of improvement
C_L	Labile carbon
C_{NL}	Non-labile carbon
CO_2	Carbon dioxide
COLE	Coefficient of linear extensibility
CQI	Climatic quality index
Cr	Chromium
CS	Carbon sequestration
CSIR	Council of Scientific & Industrial Research
CSSRI	Central Soil Salinity Research Institute
CSWCRTI	Central Soil and Water Conservation Research and Training Institute
CTCZ	Continental tropical convergence zone
CTW	Conventional tillage wheat
Cu	Copper
DAC	Department of Agriculture and Cooperation
DAP	Diammonium phosphate
DCD	Dicyandiamide
DH	Dehydrogenase
DLDD	Digitized Legacy Data Development
DLR	Department of Land Resources
DNA	Deoxyribonucleic acid
DNDC	DeNitrification-DeComposition
DOC	Dissolved organic carbon
DoS	Department of Space
DRONE	Dynamic Remotely Operated Navigation Equipment
DSR	Direct seeding of rice
DSR	Ditrect seeded rice
DSS	Decision support system
DSSAT	Decision Support System for Agrotechnology Transfer
DST	Department of Science and Technology
E	East
ECEC	Effective cation-exchange capacity
Ed., ed., Eds	Editor(s)

El Nino	Complex weather patterns resulting from variations in ocean temperatures
ENE	East-north-east
ENSO	El Niño–Southern Oscillation
EPA	Environment (Protection) Act, 1986
ESP	Exchangeable sodium percentage
FA	Fulvic acid
FAI	Fertiliser Association of India
FAO	Food and Agriculture Organization
FCA	Forest (Conservation) Act, 1980
FCC	False colour composite
FCC	Fertility Capability Classification
FCI	Fertilizer Corporation of India (Sindri)
FCO	Fertilizer Control Order
FDHA	Fluorescein di-acetate hydrolysis
FMM	Flagship mission mode
FRR	Fertilizer response ratio
FYM	Farm yard manure
GDP	Gross domestic product
GHG	Greenhouse gas
GIS	Geographic information system
GM	Green manure
GOI	Government of India
GPS	Global Positioning System
GS	Green Seeker
GWP	Global warming potential
ha	Hectare
HA	Humic acid
HC	Hydraulic conductivity
HCH	Hexachlorocyclohexane
HCS	High-charge smectite
Hg	Mercury
HIS	Hydroxy-interlayered smectite
HIV	Hydroxy-interlayered vermiculite
HP	Himachal Pradesh
HRS	Hyperspectral remote sensing
HWSC	Hot-water-soluble carbon
HySI	Hyperspectral Imager
HYV	High-yielding variety
I & B	Information and Broadcasting
IADP	Intensive Agricultural District Programme
IARI	Imperial Agricultural Research Institute
IARI	Indian Agricultural Research Institute
IBM	Indian Bureau of Mines
IBSNAT	International Benchmark Sites Network for Agrotechnology Transfer
ICAR	Indian Council of Agricultural Research
ICP-OES	Inductively Coupled Plasma–Optical Emission Spectrophotometer
ICRISAT	International Crops Research Institute for the Semi-Arid Tropics
ICT	Information and communication technology
IGB	Indus, Ganges, and Brahmaputra
IGP	Indo-Gangetic Plain

IISS	Indian Institute of Soil Science
IISWC	Indian Institute of Soil and Water Conservation
IMD	India Meteorological Department
IMY	Indian Minerals Yearbook
INM	Integrated Nutrient Management
IPCC	Intergovernmental Panel on Climate Change
IPL	Indian Potash Limited
IPNS	Integrated plant nutrient supply system
IRMS	Isotopic Ratio Mass Spectroscopy
ISO	Indian Standards Organization
ISRO	Indian Space Research Organisation
ISSS	Indian Society of Soil Science
ISSSLUP	Indian Society of Soil Survey and Land Use Planning
IUSS	International Union of Soil Sciences
IWMP	Integrated Watershed Management Programme
J&K	Jammu and Kashmir
K	Potassium
Kao	Kaolin or kaolinite
Km	Kilometre
KVK	*Krishi Vigyan Kendra*
LCC	Land Capability Classification
LCC	Leaf colour chart
LCS	Low-charge smectite
LE	Linear extensibility
LED	Light-emitting diode
LGP	Length of growing period
LP	Lorentz–polarization
LP factor	Lorentz–polarization factor
LQI	Land quality index
LRA	Local and regional authorities
LTFE	Long-term fertilizer experiment
LUCII	Land use change intensity index
LUP	Land use planning
LUPAS	Land use planning and analysis system
LUR	Land use requirement
LUT	Land utilization type
M	Metre
Ma	Millions of years
mA	Milliampere
MBC	Microbial biomass carbon
MBN	Microbial biomass nitrogen
MBT	Main Boundary Thrust
MCT	Main Central Thrust
MDS	Minimum Data Set
MFC	Microbial fuel cell
MoEFCC	Ministry of Environment, Forest and Climate Change
MFT	Main Frontal Thrust
MGLP	Multiple goal linear programming
Mha	Million hectare
Mi	Miles

MIR	Mid-infrared
MIS	Marine isotope stage
ML	Marginal land
MLRA	Major Land Resource Area
MMA	Macro Management of Agriculture
MoA	Ministry of Agriculture
MoEF	Ministry of Environment & Forests
MOP	Muriate of potash
MoPR	Ministry of Panchayati Raj
MoRD	Ministry of Rural Development
MoTA	Ministry of Tribal Affairs
MOVCDNER	Mission Organic Value Chain Development for North Eastern Region
MoWR	Ministry of Water Resources
MP	Madhya Pradesh
MRT	Mean residence time
MS	Maharashtra
MSNP	Micro- and Secondary Nutrients and Pollutant Elements
MSSRF	MS Swaminathan Research Foundation
mW	Milliwatts
N	Nitrogen
N	North
N_2O	Nitrous oxide
NAAS	National Academy of Agricultural Sciences
NAP	National Agricultural Policy, 2000
NREGA	National Rural Employment Guarantee Act, 2005
NARES	National Agricultural Research and Education System
NASA	National Aeronautics and Space Administration (USA)
NBS	Nutrient-based subsidy
NBSS&LUP	National Bureau of Soil Survey and Land Use Planning
NBSS	Nutrient-based subsidy scheme
NCH	Non-colloidal (2–0.2 μm) clay-humus complex
NDVI	Normalized difference vegetation index
NE	North-eastern
NE	North-east
NEP	National Environmental Policy, 2006
NER	North Eastern Region
NE-SW	North-east–south-west
NFP	National Forest Policy, 1988
NFSM	National Food Security Mission
Ni	Nickel
NIR	Near infrared
Nm	Nanometre
NMSA	National Mission for Sustainable Agriculture
NMSHC	National Mission on Soil Health Card
NNW	North–north-west
NNW-SSE	North–north-west–south–south-east
NOCU	Neem oil coated urea
NPC	Non-pedogenic $CaCO_3$
NPF	National Policy for Farmers
NPK	Nitrogen, phosphorus, potassium

NPMSHF	National Project on Management of Soil Health and Fertility
NPP	Net primary productivity
NRAA	National Rainfed Area Authority
NRSA	National Remote Sensing Agency
NRSC	National Remote Sensing Centre
N-S	North–south
NW	North-west
NWDPRA	National Watershed Development Programme for Rainfed Areas
NWP	National Water Policy 2012
OC	Organic carbon
OTC	Open top chamber
P&S	Phosphorus and sulphur
P	Phosphorus
PI	Productivity index
PAH	Polycyclic aromatic hydrocarbon
Pb	Lead
PBB	Polybrominated biphenyl
PC	Pedogenic $CaCO_3$
PCB	Polychlorinated biphenyl
PCDF	Polychlorinated dibenzofuran
PCh	Pseudo-chlorite
PFP	Partial factor productivity
Pg	Picogram
PGPR	Plant growth-promoting rhizobacteria
PGS	Participatory Guarantee System
pH	Negative log of hydrogen ion activity
PHO	Phosphomonoesterase
PHO	Phosphatase
PIL	Prime Irrigated Land
PKVY	P*aramparagat Krishi Vikas Yojana*
POC	Particulate organic carbon
PON	Particulate organic nitrogen
POP	Persistent organic pollutant
Ppm	Part per million
PPP	Public–private partnership
PRL	Prime Rainfed Land
qCO_2	Metabolic quotient
qCO_2	Microbial metabolic quotient
R&D	Research and development
RADAS	Reclamation and Development of Alkali and Acid Soils
RAS	Royal Agricultural Society
RCT	Resource conservation technology
RKVY	*Rashtriya Krishi Vikas Yojana*
RNA	Ribonucleic acid
ROS	Reactive oxygen species
RP	Rock phosphate
RSG	Reference Soil Group (WRB)
S	South
S1	Highly suitable
S2	Moderately suitable
S3	Marginally suitable

SA	Sustainable agriculture
SAARC	South Asian Association for Regional Cooperation
SAARC STORM	South Asian Association for Regional Cooperation: Severe Thunderstorm Observations and Regional Modelling
SAC	Space Applications Centre
SAPZ	Special agricultural production zone
SAR	Synthetic-aperture radar
SAUs	State Agricultural Universities
SAZ	Special agricultural zone
SBT butanoate	S-benzylisothiouronium butanoate
SBT furoate	S-benzylisothiouronium furoate
SE	South-eastern
SFM	Sustainable forest management
SI	Storie index
SIC	Soil inorganic carbon
SIR	Storie index rating
SIS	Soil Information System
SLM	Sustainable land management
Sm	Smectite
SMB	Soil microbial biomass
SOC	Soil organic carbon
SOIL	Soul Of Infinite Life
SOM	Soil organic matter
SON	Soil organic nitrogen
SQI	Soil quality index
SRI	System of Rice Intensification
SRM	Soil resource map
SRO	Short-range order
SS&LUPS	Soil Survey and Land Use Planning Scheme (at Sabour)
SSE	South-south-east
SSNM	Site-specific nutrient management
SSP	Single super phosphate
SSSA	Soil Science Society of America
STCR	Soil Test Crop Response
STD	South Tibetan Detachment
STFR	Soil Test and Fertilizer Recommendation
STL	Soil testing laboratories
STORM	Severe Thunderstorm Observations and Regional Modelling
SW	South-west
SWOT	Strength, weakness, opportunities and thrust
SYI	Sustainable yield index
SysNet	Systems research network
TC	Total carbon
TC	Tropical cyclone
TCM	Technical Cooperation Mission (USA)
TDS	Total dissolved solid
TERI	The Energy and Resources Institute
TGA	Total geographical area
TM-FCC	Thematic mapper-false colour composite
TN	Tamil Nadu

TN	Total nitrogen
TOC	Total organic carbon
TPR	Transplanted puddled rice
TSP	Triple super phosphate
UN	United Nations
UNDP	United Nations Development Programme
UNECE	United Nations Economic Commission for Europe
UNESCO	United Nations Educational, Scientific and Cultural Organization
UNMDTF	United Nations Millennium Development Task Force
UP	Uttar Pradesh
UPASI	United Planters' Association of Southern India
US$	American dollar
USA	United States of America
USDA	United States Department of Agriculture
USSR	United Soviet Socialist Republic
UV	Ultraviolet
V	Volt (electric potential difference)
VAM	Vesicular-arbuscular mycorrhiza
VisNIR DRS	Visible near-infrared diffuse reflectance spectroscopy
VKC	Village Knowledge Centre
WB	West Bengal
WCSS	World Congress of Soil Science
WDC-PMKSY	Watershed Development Component of the "*Pradhan Mantri Krishi Sinchayee Yojana*"
WFPS	Water-filled pore space
WHC	Water holding capacity
WHO	World Health Organization
WMO	World Meteorological Organization
WRB	World Reference Base (for Soil Resources)
WSA	Water-stable aggregate
WSC	Water-soluble carbon
WSW	West-south-west
WUE	Water-use efficiency
XRD	X-ray diffraction
Zn	Zinc
ZT	Zero tillage
ZTW	Zero-tillage wheat

1 Introduction

Bipin B. Mishra

Abstract

India constitutes a part of South Asian Subcontinent and is situated between the latitudes of 08° 04′ and 37° 06′ N and longitudes of 68° 07′ and extended to 87° 25′ E in the north of equator with a geographical area of 329 M ha (million hectare). In three broad physiographic regions, viz. Peninsula (a triangular plateau in the Deccan and south of the Vindhyayan), Extra Peninsula (a mountain region of the Himalayas) and Indo-Gangetic plain followed by distinct geological setup and climatic arrangement, India is gifted to have diversified types of soil groups within specific agroclimatic regions, viz. alluvial soils, black soils, red soils, laterite and lateritic soils, forest and mountain soils, arid and desert soils, acid soils, salt-affected soils, and peaty and marshy soils.

Indian soils have rich history with most significant contributions to the lifestyle, culture, civilization and overall livelihood besides supporting for survival and nourishment of south Asian population. The continued pressing demand for land and soil due to increasing population has made these soils highly threatened and scarce resource as a result of soil health deterioration, soil degradation, loss of biodiversity, nutrient mining, emergence of heavy metal toxicity, declining soil organic matter and impact of climate change. The chapter highlights on generalized but conceptual action plan for successful implementation of land use planning on the ground-based interventions as covered in subsequent chapters. In addition, education and research in soil science deserve strengthening through a breakthrough in the National Agricultural Research and Education System (NARES). Soils by and large possess ample scope for improvement in order to meet the growing demands of the nation. This chapter is thus a generalized overview on vision, mission and goal of soil resources towards Second Green Revolution in the twenty-first century.

Keywords

Indian soils • Challenges • Soil evaluation • Education • Land use planning • Second Green Revolution

1.1 Introduction/Background

India is a land of soil museum covered with diversified types of climate, vegetation, geology, geomorphology and hydrology following a complex system of river network. Soil is a fingerprint of human health, instinct, culture and humanity. The relationship between landforms and associated soils is indicative of precise recognition as well as mapping of homogeneous soil units (Sharma and Raychoudhury 1988; Sharma et al. 2001). Soil as a natural resource exists in continuum on a landscape and preaches for *Vasudevakutumbkam* (meaning "earth is a family"), since a soil may adapt to any part of the globe. Starting from Arunachal Pradesh to Kerala welcoming world with its flavour of spices and the richest soil museum (Fig. 1.1), as well as Kashmir to Kanyakumari extended to Rameswaram and Indira Point of Nicobar Islands, the soils of this great land of golden history with competitive present and challenging future stand as collective symbol of unity within unique diversity in their functions. Agriculture being the backbone of nation's prosperity and development is however facing acute challenges due to rapidly rising population besides diversion of agricultural lands for other sectors. Such continued increasing trend in human population in India has resulted in a very high

B. B. Mishra (✉)
Bihar Agricultural University, Sabour, Bhagalpur, 813210, India
e-mail: bbmsoil@rediffmail.com

B. B. Mishra
Pedology and Land Use Planning, School of Natural Resources Management and Environmental Sciences, Haramaya University, P O Box 138 Dire Dawa, Ethiopia

B. B. Mishra
International Union of Soil Sciences, Vienna, Austria

B. B. Mishra (ed.), *The Soils of India*, World Soils Book Series,
https://doi.org/10.1007/978-3-030-31082-0_1

Fig. 1.1 Kerala soil museum. *Source* https://journeyviakerala.wordpress.com/2017/05/15/kerala-soil-museum

demand of agricultural land and so the soil, while area of such land is not only restricted but shrinking with the emergence of multidimensional threats and conflicts. As a result, the pressure on the same land and soil through land use intensification has made the existing soils threatened. Thus, soils of today need a very systematic as well as reliable means of verification through inventorying the components of soils, climate, surrounding soil-site characteristics, aquifer, biodiversity and so on through evaluation. The chapter is aimed at one-to-one win-win enlisting of all possible attributes that may facilitate in translating a tangible planning approach that can be reliably acceptable and adoptable for restoration and enhancement of productivity, safety, profitability and sustainability of soil as an integral component of the land resource without degradation and environmental threats (Kesavan and Swaminathan 2008; Bhaduri et al. 2015). There can be no greater issue than that of conservation of soil in this country. Just as we conserve men, women and children, we must conserve the resources of the land on which they live. Obviously, it is vital to conserve the soil so that the next generation would have a land that is more but never less fertile than ever before.

1.2 Soil as a Strange, Full of Wisdom

Soil, as a natural resource, is still a strange, full of wisdom that needs to be captured for its enormous uses in infinite purposes to meet the human need and livelihood. As early as 5000 B.C, the *Vedas* and the *Upanishads* in Indian history witnessed the soil as synonymous with land, the symbol of Mother, for survival and nourishment of all lives on the earth. Soil is accordingly worshiped as Mother Earth, while, on the other hand, soil is a huge storehouse for variety of waste materials on way to keep our environment clean. The idol of Goddess Durga is made of soil (Fig. 1.2), while *Deep* or *Diya* for *Diwali* lightening is traditionally prepared by *Kumhar* with soils (Fig. 1.3). *Kumhar* (potter) is a group of people in India meant for making materials with type-specific soils and they are the true soil technocrats in the rural societies. Also, different structures made of soils are used in the Hindu marriage celebration. In railway journey, tea in soil-made *kullhar* is special attraction to the common passengers. Poor men in villages make their house with mudwalls, whereas rich men use soil-made bricks and tiles for construction of buildings. Potters use soils for earthenwares and utensils including *Surahi*, *Metia*, *Ghara*,

Fig. 1.2 Soil made idols (Goddess *Durga*), Kolkata. *Source* https://www.scoopwhoop.com/soil-from-brothels-durga-puja-idols/

Fig. 1.3 *Deep* or *Diya* made of soil for lightening. *Source* https://commons.wikimedia.org/wiki/File:Diya.jpg

Berua, etc., while sculptures for models. Similarly, different types of containers used for cooking food, collecting milk, storing grains and other farm produces are made of soils. In rural areas, type-specific soils are used in plastering of walls. If mudwalls are made of soils of vertic features (cracking), the wall plastering is done with red soils. Soil is not only important for agriculture, but engineers, hydrologists and environmentalists do also have their important options for type-specific soils in respective profession. Even in medical treatments and applications, the type-specific soils or clays are gaining recognition (Mishra and Roy 2015; Gaétan et al. 2012).

Soil has its origin, growth and development and its end also, albeit as a natural resource, it cannot be a waste and does not need any rest, if it is managed systematically in a planned framework. It has its past, present and future too, but its present is surrounded with threats, degradation, limitations and uncertainties while future is full of challenges. Its age may be recent, young or even old in terms of maturity subject to alteration. Since soil cannot be produced or manufactured within a night or day or month and even in years, it is virtually highly precious and accepted as a non-renewable resource. It behaves like the organism, but it is not mobile and cannot reproduce. "Soil" is often said to be dead when organic cycle is inoperative. Clay in a soil has a very important bearing on the genesis, characteristics and physical, chemical as well as biological properties of soil (Pal et al. 2000).

Among different soil groups like alluvial soils, black soils, red soils, laterite and lateritic soils, forest and mountain soils, arid and desert soils, acid soils, salt-affected soils, and peaty and marshy soils, the alluvial soils have certain traits that make these soils easily manageable even in active floodplains. The silt dominant active floodplain

Table 1.1 Salient features of profile pit in a floodplain soil of Supaul, Bihar (BAU 2016)

Depth (cm)	Soil texture (%)			Salient features				
	Sand	Silt	Clay	EC (dS m^{-1}) (1:2.5)	pH (1:2.5)	N (kg/ha)	P_2O_5 (kg/ha)	K_2O (kg/ha)
0–30	28	62.4	7.54	0.12	7.42	74	14	112
30–50	27	64.2	6.94	0.07	7.45	65	12	98
50–80	32	65.3	0.94	0.09	7.89	61	7	110
80–100	34	62.5	0.45	0.11	7.64	42	10	84
>100	34.3	61.5	1.84	0.06	7.41	38	8	74

Fig. 1.4 Deep roots of *Saccharum spontaneum* characteristic to sand/silt dominant Kosi floodplain soils. *Source* BAU (2016)

Fig. 1.5 Planting of cucurbits in silt dominant floodplain soil showing emergence of deep rooted *Saccharum spontaneum*. *Source* BAU (2016)

Fig. 1.6 Tomato cultivation in silt dominant floodplain soil covered with mulch. *Source* BAU (2016)

soils (Table 1.1) could be utilized after emergence of the deep rooted *Saccharum spontaneum* (Fig. 1.4) for successful cultivation of cucurbits and vegetables in Kosi floodplain (Figs. 1.5 and 1.6) as reported by the Bihar Agricultural University, Sabour (BAU 2016).

Being the uppermost thin layer of the earth's crust, it is the lowest boundary of the earth's atmosphere and, so it is subject to interactions with incoming radiation (Mishra 2016). Soil is influenced by nuclear and electromagnetic radiations, gravitation, electricity, magnetism and sound in different forms impacting the livelihood even. In fact, the soil lies beneath our feet where we have our home and shelter to live in and work. But, the soil often does not get the credit it deserves for its life-supporting functions. Even the old people in India prefer to walk daily on soil without using shoes and often work in kitchen garden (https://www.healthline.com/health/walking-barefoot). Connection of human body with the earth's electrons has been found to promote intriguing physiological changes. Earthing or grounding refers to the discovery of benefits including better sleep and reduced pain from walking barefoot outside or sitting, working, or sleeping indoors connected to conductive systems that transfer the earth's electrons from the ground into the body (Gaétan et al. 2012). They have established that electrons from antioxidant molecules neutralize reactive oxygen species (ROS, or in popular terms, free radicals) involved in the body's immune and inflammatory responses (Gaétan et al. 2012).

Soil obeys the classical laws of science, but that too in an open system, and so its state of existence is highly dynamic in nature. Like other physical entities, soil too has its unit, "A Soil" or "Pedon" in three-dimensional framework and forms through pedogenesis as the outcome of soil forming factors and processes, a natural but continuous process taking thousands of years. The history of systematic knowledge on soils begins virtually with the work of Russian Pedologist V. V. Dokuchaev around 1880 (Dokuchaev 1883), though sporadic reports on soil fertility prior to Dokuchaev are well documented (von Liebig 1840). Obviously, the history of soil science is by and large at youthful stage with immense scope for research and development for scientific advancement under the pressing demand of land and land uses in agriculture, pasture, forest, wildlife, tourism, recreation, urbanization, construction of buildings, roads and highways. Such demand is becoming more pressing due to population pressure, land shrinkage as a result of non-farming uses as well as soil degradation (physical, chemical, biological and environmental). The extent of horizontal expansion of lands is by and large reaching its climax following the rising trend of challenges in respect of overall deterioration of soil health, safety and sustainability (Kesavan and Swaminathan 2008; Sihi et al. 2017; Mondal et al. 2015; Katyal et al. 2001). The existing soil-related scenario in different aspects clearly indicates the nutrient mining and fertility imbalance leading to decline in partial factor productivity (Muralidharudu et al. 2012) as well as loss of biodiversity with emergence of toxic levels of heavy and undesirable materials, incidence of soil sickness, multi-nutrient deficiencies, soil water depletion, compaction, erosion and lowering of soil organic matter. These adverse signs lead to make it more scarce that simultaneously cause conflicts on so-called scarce land resource in India. Soil preservation and sustainable land management have, therefore, become essential for reversing the trend of soil degradation and ensuring productivity, safety, profitability and overall sustainability of soil and land resources. As per FAO projection, the demands

Fig. 1.7 Bank erosion of alluvium in Ganga basin active floodplain soils. *Source* Mishra (2015)

of a growing population for food, feed and fibre are estimated to result in a 60% increase by 2050 (http://www.fao.org/soils-2015/en/). Thus, soils cannot be ignored in ecosystem mechanisms as it was before. Even public awareness is now growing. More importantly, the catastrophic events like countywide recent floods, tsunami, earthquakes and droughts have made common people increasingly aware of the need for a more rational use and a more effective protection of soil and land resources. Riverbank erosion (Fig. 1.7) is common in many river basins like the Brahmaputra, Kosi, Ganga, Gandak, Mahananda, Yamuna (Shyampura et al. 1992; Mishra 2015). Importantly, soil cannot be substituted by so-called hydroponics even, since hydroponics is currently not better than a showpiece with a very limited production outcome at high cost under sophisticated technical supervision and expensive plant protection measures. By and large, many disciplines are involved in soil research such as soil science, agronomy, mineralogy, hydrology, geology, geography, botany, zoology, microbiology, biotechnology, anthropology and medical science, but they lack mutual coordination and integration in order to achieve a sound and truthful outcome and thus remain partly hidden and strange.

1.3 Soil and Land Resource

Even if the reality is ignored, one cannot ignore its consequences. India has the second largest human population in the world with limited and scarce land resources. India's population until 2018 has reached almost 1.36 billion, i.e. around 17% of the world's population (Population density, 457 person/km^2). The growing population results in growing urbanization, which causes speedy diversion of productive agricultural land irreversibly for non-agricultural uses. Reports indicate that the urban population in 1981 was 23.4% of total, which approached 25.7% in 1991, 27.8% in 2001 and 31.2% in 2011, and thus, a total of 7.8% in urban population is increased within the last four decades (GoI 2011). In India, the net sown area was about 0.12 ha per capita [140 M ha (million hectare)] in 2012–13, which was less than even half of the world average of 0.23 ha per capita (GoI 2015). Surprisingly, a total agricultural land of around 3.16 M ha was lost to other sectors for non-farming uses in the years between 1991–92 and 2012–13 (GoI 2015). A more recent estimate puts the extent of degraded land in India at 120 M ha, which is about 38% of our total geographical area (NAAS 2010). In a report of the Indian Institute of Management, Sharma (2015) critically opined that the current economic growth in India is directly linked with subsequent growth in road, rail, playground, stadium, port, infrastructure and market that require enough area of land. Sharma (2015) further stated that the total road length has increased by about 45%, from about 3.37 million km in 2001 to nearly 4.87 million km in 2012, wherein the land acquisition is almost neglected. The cultivable land has shrunk marginally by 0.43% to 182.39 M ha in the last five years. This is due to shift in area for non-agricultural purposes such as buildings, road and railways among others. The total agricultural land in 2003–04 was 183.19 M ha against 182.39 M ha in 2008–09, a fall of 0.80 M ha, according to the government data (PTI 2017). This signifies that our soil and land resources suffer from ignorance of fundamental soil policies particularly for agriculture and livestock. There are aspiring soil development programme launched by the Indian Government, but for want of sound policies and governance, the development programme often faces threats of completeness. During the past few decades, India has commendably made significant strides in achieving self-sufficiency in food production. However, in order to meet the ever-increasing food requirement of the growing population, it is but everlasting compulsion to keep soil in work in order to produce more and more from each unit area of land.

Besides, in practical sense, diversion of agricultural land or land area shrinkage seems to be unavoidable because of rapidly growing population, high rate of urbanization followed by emerging trend of industrialization and infrastructure development. Sharma (2015) critically projected that the area under non-agricultural uses increased by about 23% (21.3–26.3 M ha) during the last two decades. Importantly while states like Uttar Pradesh, Andhra Pradesh, Odisha, Madhya Pradesh, Bihar and Tamil Nadu showed the higher rate of increase, Gujarat, and some north-eastern states showed the lower rate of increase in land under non-agricultural uses during the last two decades (Sharma 2015). This is one of the burning issues on the way to restore soil resources for agriculture, livestock, forestry and other life-supporting enterprises (Kesavan and Swaminathan 2008) by generating strategic models for sustainability. In order to stop conversion of agriculture land for non-agriculture purposes, the government has formulated the National Policy for Farmers and the National Rehabilitation and Resettlement Policy. The National Rehabilitation and Resettlement Policy 2007 envisages that the projects should be set up on wasteland, degraded land or un-irrigated land. Besides this, the policy has stated that the acquisition of cultivable land for non-agriculture purposes should be kept to the minimum (PTI 2017).

If self-sufficiency in terms of quantity and quality of food in India is to be restored and maintained, there is need to look into many more issues concerning decline in soil quality and risk of soil degradation as already warned earlier (Abrol and Sehgal 1972; Chhonkar 2008; Mondal et al. 2015). The capacity of a soil to produce is limited, and the limits to production are set by its intrinsic characteristics and agro-ecological settings followed by sustainable use and integrated management strategies (Sehgal 1995). Since there is great variability in Indian soils, it is vital to treat each parcel of land differently but scientifically after systematic soil evaluation (Riquier et al. 1970; Sihi et al. 2017). Until now, we have been producing what we need by hook and crook, by getting hold of the best land, best inputs and best of everything. But, now there is a need to diagnose the nature, properties, potential and problems or limitations of soils in each parcel of land. It is thus imperative to map the potentially viable soils for developing rational land use, which can be obtained through soil resource mapping program (Sehgal 1995).

1.4 Climatic Variability

Soil and climate are the two vital physical units of the natural environment that constitute the land as a natural resource for the purpose of land use options such as arable, forestry, plantation, forage, pasture, wildlife, tourism, recreation or even infrastructures including roads and buildings. Hilgard (1892) in USA emphasized the relationship between soils and climate, which is known as the climatic zone concept. However, India has world's most-pronounced monsoonic climate, wherein wet and dry seasons with changing annual temperatures produce three distinct climatic events in a year, viz. (i) hot wet weather from mid-June to the end of September; (ii) cool dry weather from early October to February; and (iii) hot dry weather normally with high atmospheric humidity from March to mid-June (IMD 2014, 2015). However, the spatial variations with associated regional differences like elevation, relief, geomorphology, river basin, vegetative succession, proximity to waterbodies including latitude and longitude often cause changes in their occurrence. In India, the wet season called the south-west monsoon normally appears from mid-June to early October, when winds, from the Indian Ocean, carry moisture-laden air across the subcontinent causing heavy rainfall as well as flood. Importantly, around three-fourths of the total annual precipitation falls during these months. Virtually, the monsoon is characterized by a seasonal reversal of existing wind directions followed by alternating the wet–dry seasons (IMD 2014, 2015). Chapter 3 is devoted to climate and its impacts on soils.

However, for practical purposes in the soil management, agroclimatic as well as agro-ecological zones are frequently used. The agroclimatic zone is a land unit in terms of major climate, superimposed on length of growing period, when soil moisture availability is within permissible limits (FAO 1983). The agro-ecological zone, on the other hand, is the land unit carved out of agroclimatic zone and superimposed on landform that may act as the modifier to climate and length of growing period (Velayutham et al. 1999). India possesses varying landforms and differing climatic conditions such as the mountainous landscapes, the river basins, peninsular plateaus, altitudinal differing forests, complexity in geological formations besides thermal variation from Kashmir to Kanyakumari, and rainfall extremes from aridity with a rainfall of even less than ten centimetres to as high as 1120 cm (world's maximum rainfall). Such spatial variations in environmental situations have resulted in widely differing soil types compared to most parts of the world. Therefore, the systematic appraisal of agro-ecological regions encompassing relatively homogenous regions in terms of soil, climate, physiography and conducive moisture availability periods, i.e. length of growing period (LGP), may help in planning of land use options (Sehgal et al. 1992, 1995; Velayutham et al. 1999; Mandal et al. 1999; Natarajan et al. 2002). By and large, soil carbon sequestration by itself impacts on climate change and food security globally (Lal 2004).

India is also known to have common occurrence of droughts in areas like Andhra Pradesh, Maharashtra, Rajasthan, Bihar and Uttar Pradesh caused by insufficient rainfall resulting in the deficiency of soil moisture in the soil profiles affecting agricultural productivity adversely. The National Commission on Agriculture (1976) has classified drought into meteorological, hydrological and agricultural drought. Drought is usually a period of at least 15 consecutive days, none of which had rainfall of 0.25 mm or more (Air Ministry 1936), whereas Ramdas (1960) considers drought as a situation when actual seasonal rainfall is deficient by more than twice the mean deviation. Subrahmanyam (1964) defined drought in terms of moisture deficiency or aridity index as the ratio of annual moisture deficiency to annual water demand. Reddy et al. (2001) attempted to identify and categorize drought-prone zones of Andhra Pradesh. However, the frequency of occurrence of both drought and flood is currently of serious concern in India.

1.5 Pedology and Soil Classification

India owes diversified types of landform, climate, geology and hydrology within a very compact boundary of the Himalayas in the north, Arabian Sea in the south-west, Indian Sea in the south and Bay of Bengal in the east excluding neighbouring territories with Nepal, Bangladesh, China and Pakistan. The accumulated influences of site-specific diversified types of the soil forming factors led to the formation and development of infinite number of soil types with major groupings of the Himalayan foothill and tarai soils, forest soils, podzols, alluvial soils, red soils, laterite and lateritic soils, floodplain soils (*Diara* and *Tal* land), black soils, mountain and hill soils, desert (arid) soils, high altitude soils, salt-affected soils, dune soils, peat and marshy soils, etc. Buchanan in 1807 used the term *Laterite* for the highly ferruginous, vesicular and non-stratified material observed in Malabar hills in Kerala (Buchanan 1807; Shivarajasingham et al. 1962), and the soils developed are known as laterite and lateritic soils. Jha (1972) was one of the pioneer pedologists after independence, who started soil survey work in Bihar (including Jharkhand) in 1954 and grouped the soils in 23 soil associations. Jha's original work in USA on fragipan with Cline is commendable (Jha and Cline 1963). Agarwal and Yadav (1954), Abrol and Bhumbla (1971), Karale et al. (1974), Murthy et al. (1982) and Gawande et al. (1980) initiated the excellent teamwork on survey, classification and distribution of soils particularly the salt-affected soils in India. Mishra et al. (1994) and Mishra (2015) attempted to establish the pedogenic development of *Diara* and *Tal* land soils in the Ganga floodplains of Bihar and proposed the 13th Order (Fluvisols) in India, Soil Taxonomy is popularly known as USDA Soil Taxonomy but in some countries, US Soil Taxonomy is often used.

The criteria being applied to classify the Indian soils were based mainly on geology, fertility, relief, chemical composition and physical structure, but with inherent drawback. Raychaudhuri and Govindarajan (1971) reviewed the efforts made on soil genesis and classification in India. Murthy et al. (1982) edited the exhaustive works being done on benchmark soils of India. Sehgal (1989) in his commendable efforts presented the soil resource map of India (1:1 M scale), which has subsequently been improved by NBSS&LUP in 2000. Virtually, Sehgal (1989) outlined the soil resource mapping project of India. Under this reconnaissance soil survey project, the soil resource map of different states were brought out in 1:250,000 scale during the 1990s and the soil map of India was prepared after necessary abstraction, printed and released by the NBSS&LUP, ICAR, in 2000. Velayutham et al. (1999) at NBSS&LUP developed the Agro-Ecological Sub-regions of India for Planning and Development. However, India does not owe its own national scheme of soil classification. Sehgal (2003) reviewed on soil classification in India. Bhattacharyya et al. (2009) compiled and classified Indian soils at family level using USDA Taxonomy (Soil Survey Staff 1999). Subsequently, Bhattacharyya et al. (2013) made efforts to develop logical relations of differing soil groups in taxonomic terms on different zones of India. Recently, efforts are being made for classifying the Indian soils on their relevance to land use options so that it would be acceptable at grass-root level (Mishra 2016). However, this requires a cumulative effort under the umbrella of National Bureau (ICAR-NBSS&LUP) in India.

1.6 Threats to Soils

The soil, being so valuable in its functions and responsibilities, now suffers from a number of risks, uncertainties and fears like degradation by mismanagement, over exploitation, natural erosion (water, wind and riverbank), pollution, deforestation, drainage, groundwater recession, salinity, sodicity, calcareousness, acidity, toxicity, contamination, imbalanced fertilization, nutrient mining, organic carbon depletion besides soil sealing, loss of soil biodiversity and climate extremes. An area of around 6.74 M ha suffers from salt accumulation out of which 3.78 M ha are alkaline or sodic and 2.96 M ha saline (Mandal et al. 2010). It is projected that by 2025, the area under salt-affected soils would be 13 M ha (Sharma and Chaudhari 2012). Similarly, out of about 26.62 M ha of Vertisols (shrink–swell soils), almost 5.6 M ha mostly in basaltic terrain in Maharashtra indicate sign of sub-soil sodicity (Lingade et al. 2008). Soil in India is thus not only limited and scarce, but it is a threatened resource too. Due to increasing population, soil is under pressure not only of intensification, but also of competing uses for cropping, fruit and tree plantation, agro-forestry, pasture, wildlife, recreation, tourism and urbanization. The unplanned land uses to meet the people's immediate requirements result in unsustainable land resource development. Degradation of land in general and the soil systems in particular have been highlighted earlier too (Bhumbla and Khare 1984; NRSA 1985; Sehgal and Abrol 1994).

Soil as a system must have a balance among its physical, chemical and biological environments, since soils are the integral part of the landscape, wherein their characteristics are largely governed by landforms on which they have developed (Yadav et al. 1998; Sahoo et al. 2003; Nogiya et al. 2017). The physical environment is the one that drives most of the soil responsibilities. Soil texture, for example, works as a skeleton for soil system in such a complex manner in which, through the process of aggregation, the aggregates are binding the soil particles together forming the pore space, wherein the micropores are to hold water tightly, but inhibit water percolation, normally unavailable to plants, and macropores too hold water that can be extracted and utilized by plants. When soil aggregates are in water, they hold both micro- and micropores together. However, if aggregates get broken, the finer particles (silt and clay) shift downward in the profile through percolating water leaving sand behind in the upper soil layers. These fine particles in the bottom profile fill the macropores, thus inhibiting water percolation further. There is thus need to improve soil aggregate stability, so that available water holding capacity of a soil could be protected and restored above water table. This is a vital issue to be due attended through a system approach. Besides, technologic manipulation is further required to minimize the evaporation loss of soil water above the surface. Once the soil physical environment is congenial, manipulation to soil biological environment needs to be prioritized in order to compromise with chemical and nutritional manipulation in a planned manner to suit with land use options (http://www.fao.org/soils-2015/en/). How to insure "per drop for more crop" depends not only on congenial soil physical environment, but on systematic uses of water resources too. Linking of rivers may be one possibility, but water trading from water excess area (flood prone) to water-deficient area (drought-prone) should be a priority besides encouraging other means to promote soil water balance (Mishra 2015). Rainfall–runoff are the important components contributing significantly to the hydrological cycle, wherein runoff estimation is crucial in hydrologic design of soil water conservation structures (Balvanshi and Tiwari 2014; Nandgude et al. 2017).

1.6.1 Soil Health Decline: Overview

The soil health decline is evident from stagnating or declining productivity growth rate in the production of food grain crops as well as emergence of multiple nutrient deficiencies in major crop production systems. Thus, emphasis on soil health involving soil physical, chemical and biological parameters is required instead of focus on a few major plant nutrients. Regular monitoring of soil health is equally important. The major strategy to improve soil health should be based on integrated plant nutrient supply system (IPNS), promotion of balanced fertilization and finding ways to incorporate organic matter into the soil. There is a need to develop the rapid soil quality test kits and identify the key soil indicators, particularly biological as well as mineralogical indicators, for an easy monitoring of soil health at the farm as well as landscape scale.

Recent reports indicate that continuous traditional cultivation of rice–maize–wheat cropping systems has resulted in many problems. Therefore, soybean–wheat–maize–gram have emerged as a good alternative both for crop diversification and for maintaining the sustainable soil health (Chander et al. 2013). Both these cropping systems require better nutrient, water and soil management practices to express their full potential. Soybean is a versatile crop, which fits very well in many soybean-based cropping systems and is one of the dominant crops in *kharif* season in many parts of central India. Yield gap in soybean–wheat system is owing to mineral deficiencies in the biophysical crop growth environment that are not properly addressed by the prevalent soil, nutrient, water and crop management practices. Neither inorganic fertilizers nor organic manures alone can sustain productivity (Prasad et al. 1999). So judicious uses of organic manures and inorganic fertilizers are essential to safeguard the soil health and augment productivity and input use efficiency (Bandyopadhyay et al. 2010). Sihi et al. (2017) attempted to evaluate the soil health in organic versus conventional farming of basmati rice in North India, while Sunanda Biswas (2016) developed an innovative approach to recover degraded soil (soil resilience).

1.6.2 Dwindling Factor Productivity

The dwindle in partial factor productivity (PFP) and compound growth rates of major crops under intensive cropping systems and poor nutrient use efficiency are owing to deterioration in soil quality. Continuous traditional crop management leads to a decline in organic C levels by 50–70% to equilibrium levels governed by climate and precipitation. The primary reasons identified for soil quality deterioration include the wide nutrient gap between nutrient demand and supply, high nutrient turn over in soil–plant system coupled with low and imbalanced fertilizer use, emerging (multi-nutrient) deficiencies in soils, soil acidity, nutrient leaching in sandy soils, nutrient fixation in red, laterite and clayey soils, impeded drainage in swell–shrink soils, soil salinization and sodification or sodiumization (IISS Vision 2030 2009). However, the total factor productivity is used as an important measure to evaluate the performance of production system and sustainability of its growth pattern. The partial factor productivity of fertilizers is declining under intensive cropping systems in India and other developing countries. SubbaRao (2009) under the future planning for soils in India indicated that the partial factor productivity of fertilizers during the last three and half decades showed a declining trend from 15 kg food grains/kg NPK fertilizer in 1970 to 5 kg food grains/kg NPK fertilizer in 2005. In urgency for higher production, no serious attention was given to the long-term soil quality and sustained high productivity. As a consequence, the annual compound growth rate of major crops has declined from 3.36% in 1981–85 to 0.11% in 2001–05. Such gloomy trend was also registered in case of pulses and oilseeds, while cotton exhibited even negative growth rate. The inputs mainly include nutrient supply, irrigation, energy, plant protection measures and cropland. The current status of nutrient use efficiency is quite low in case of P (15–20%), N (30–50%), S (8–12%), Zn (2–5%), Fe (1–2%) and Cu (1–2%). There is thus need to re-work with type-specific soils and surrounding environments integrating the defined set of inputs in order to achieve the target. Biswas and Sharma (2008) however forwarded an approach for estimating fertiliser response ratio under the Indian scenario.

1.6.3 Soil Biodiversity

Functioning of terrestrial ecosystems, plant biodiversity, productivity, variability and stability is directly dependent on soil microbial diversity (Kaur et al. 2005; Laxminarayana 2006). Soil biodiversity is an abstract aggregated property of species in the context of communities or ecosystems. Functional diversity rather than taxonomic diversity (community structure) or species richness per se is the major determinant of ecosystem functioning. It may be thus more important to understand the linkages between the actions of a key species or the functional groups and ecological functions of different ecosystems than to search for the diversity index or the species richness and try to relate the same to ecosystem or community functioning. Characterization of functional communities of soil organisms (flora and fauna) and soil biological activities under different soil crop situations for enhancing nutrients availability and also characterization of microbial biodiversity and functional communities (particularly N-fixers, P&S solubilizers, Lignin-2 cellulolytic organisms), testing of mixed biofertilizer formulations and diversity of biofertilizers in agriculture should receive top priority. Since the organic materials are a scarce commodity, biofertilizers particularly plant growth-promoting rhizobacteria (PGPR) and mycorrhiza (VAM) are increasingly being deployed and good responses are being obtained showing that increasing the soil microbial diversity through such inoculation is benefiting the soil health in a similar way as addition of organics which also promote the proliferation of native soil health-promoting microorganisms (IISS Vision 2030 2009). Because Indian soils are by and large deficient to very deficient in organic matter and further its declining trend due to climate change could result on loss of biodiversity (Koshy 2018).

1.6.4 Organic Waste Management and Recycling

In order to minimize the entry of heavy metals and other organic pollutants through solid and liquid wastes into agricultural land/food chain, several countries have formulated regulatory mechanisms, like maximum cumulative loading limit, maximum annual loading limit, maximum concentration limit of metal in soil to receive further input, maximum concentration in soil amendment materials. The different approaches for such formulations have resulted widely differing numerical limits for the same metal. India has not put forward any soil protection policy to restrict heavy metal build-up during their inadvertent addition through different amendments. Future researches should focus on following important areas in order to protect our soil resources from metal contamination, viz. determination of baseline concentration as well as maximum threshold limits of heavy metals in different soil types, and environmental risk and impact assessment on use of solid and liquid wastes under diverse agro-ecosystems (IISS Vision 2030 2009). Obviously, more emphasis is desired to use soil biogeochemistry to understand the cycling processes under organic solid waste management (Bhaduri et al. 2015; Mondal et al. 2015).

1.7 Soil Evaluation

Indian Council of Agricultural Research through its Institutes, National Bureaus, Centres including SAUs and other agencies is well enriched with soil resource maps of each state and union territory. These maps are subject to interpretation by expert people for being further used to determine the agricultural values of the soils and their susceptibility for improvement following further ground verification through soil evaluation. The simple procedure as outlined by Riquier et al. (1970) and revised as well as improved in India (Natarajan et al. 2002; Velayutham 2012; Naidu et al. 2017) needs further refinement after logical modification based on site-specific pedologic as well as edaphologic limitations (physical, chemical and biological) as associated with type-specific degradation, pollution, deficiency or toxicity including impacts induced by climate change. Such comprehensive approach covering all possible limiting parameters could be fixed in the form of a mathematical function for soil evaluation in a given land unit or area in order to assess actual and potential productivities. Productivity (actual productivity) is employed simply as the soil capability to produce a certain amount of crop or land use per hectare per year and expressed as the percentage of optimum yield per hectare of the same crop grown on the soil following the recommended simple farming practices. However, potential productivity signifies the productivity of a soil after all possible improvement measures are employed with least or no limitation. The change in improvement could parametrically/quantitatively be expressed by the ratio of the two indices called the coefficient of improvement (CI), i.e. potential productivity/actual productivity (Riquier et al. 1970). This index indicates that the actual productivity of a soil could be multiplied by this ratio after application of all suitable management measures to soil.

The chapters in this book cover different physical, chemical and biological aspects including multidirectional threats covering degradation, pollution, vulnerability, sickness, deterioration in biogeochemical environment, nutrient mining, loss of biodiversity, emergence of heavy metals, etc., to soils. How these limiting traits in soil could be integrated and translated in a logical framework for soil evaluation seems to be of scientific interest. However, it is apparent that soils in India must undergo for systematic evaluation on location-specific basis in order to fix and enlist limitations or threats, which could possibly be minimized through effective but reliable improvement measures so that CI values could be appreciated to the extent of the potentiality of a soil (Choudhary et al. 2009). Such exercise bears direct alignment to any smart land use planning approach. Detailed emphasis is made in the respective chapter.

1.8 Land Use Planning

Land is merely an area of the earth's surface comprising all physical, chemical and biological environments that influence the land use as outlined by FAO (1976). Obviously, land refers not only to soil but also to climate, landforms, hydrology, geology, natural vegetation including fauna and flora. The land use simply means for management of a land for specific uses to meet the human needs in all rural, urban and industrial context, whereas the land use plan refers to a coherent set of decisions about the ways to achieve the desired level preferably on economic background. Thus, the land use planning is the systematic evaluation and assessment of a land and associated attributes for the purpose of identifying the best land use options beneficial to landowners or users without degrading the land resources or environment. It includes not only the landowner or users, but also the policy or decision makers as well as economic experts for financial measures from planning stage or logistic planning to execution. Let the concept of LUP be translated into land economics and subsequently into agri-business, so that the essence of corporatization could be integrated with LUP. It may take the shape of cooperative farming or even contractual farming following the provision of inputs by contiguous farmers, but in no way, landownership is challenged or even disturbed (Mishra 2017). Naidu and Hunsigi (2001) concluded that the combination of soil potential rating approach was found to be a better approach to assess the land suitability for crops compared to FAO approach (FAO 1976).

Since soil is basically a resource intended for conservation and management by the farmers, who care for business, there is logistic to shift this business towards insured profitability, since more than 70% of Indian population is dependent on this profession. The key factor for success in this giant business lies on best agricultural practices in line with land use planning followed by the best business strategies. So, there must be a clear vision to empower the poor farmers with current reliable technological updating in order to compete with the emerging business environment and its processes, from the point of production (soil/land) to the point of consumption (market) through well-planned farming system approaches preferably in line with conservation agriculture (Somasundaram et al. 2017).

1.9 Education and Research in Soil Science

Soil as a science by itself has split up into different branches of science, and there is a fear of losing track of the basic principles of soils that possess attributes, viz. flexibility (adaptability to changing environment); resilience (ability to recover against disturbance); resistance (ability to buffer the change); detoxification (ability for adsorption of toxic ions); residence time (exchange capacity to store and release ions); productivity (capability to produce the yield through plant growth); responsiveness (capacity to benefit plants); plasticity (ability to deform); stickiness (capacity to adhere); and sustainability (restoring equilibrium of interactions). If students have learned the right things, they ought to be able to perform right approach with the acquired knowledge and keep the soil linked with plant, animal, human health, environment and society. A generalized linkage must include soil geology, geomorphology, soil mineralogy, soil micromorphology, pedology and classification, soil biogeochemistry, soil biodiversity, soil biotechnology, land evaluation, soil health and edaphology besides industrial, commercial and medicinal relevance of the soils.

Soil science education in India thus needs a systematic, but comprehensive base in course curricula in order to impart learning on conceptual, theoretical, practical, experiential, analytical and above all interpretational view points. It may require a smart class, but more importantly, there must be a field/farm and laboratory-based interactive environment, wherein creative attitude of a student in soil science would find satisfaction towards completeness of the course curriculum as would be prescribed under the umbrella of the Indian Council of Agricultural Research (ICAR). India wants soil scientists as the true practitioners to restore soils for their optimal functions and consequently for optimal production without any scope of existing limitations or risks. The ICAR with sincere cooperation of the State Agricultural Universities (SAU) has been maintaining the excellence in upgrading the agricultural education.

Soil science research accordingly needs to be upgraded through training, demonstration and soil museum following the concept with bullock ploughing. The simple definition of any scientific research is to perform a methodical study in order to establish and prove a hypothesis or answer for a specific but relevant question. Such answer would be the central goal of an experiment. In other words, research is a process of relating a theory (in classroom) to fact (in laboratory or experimental plot). The research must be systematic following a set of steps. Soil-based research works are almost in open system and so it is full of risk, fear, uncertainties, doubts and limitations. Management strategies for agricultural production must be based on a reliable skill for input integration through step-by-step evaluation of resources involved. Soil, for example, is a basic resource contributing as the foundation to sustainable production and its evaluation for actual as well as potential productivity may be undertaken well before suitability identification of land use choice. More often research projects suffer from misinterpretation of basic principles, methodology and evaluation on timescale, viz. logframe matrix. Under the present scenario when soils in India are by and large threatened, research, either for degree or for project must yield reliable recommendations.

Technology is simply the assemblage of scientific laws and principles in the form of a tool or device or machine or even procedure to attain the goal targeted. So, the technology works in a system that is developed within a set of inputs or components or may be interacting set of sub-systems too in order to yield or generate outputs of interest. Reliable recommendations of a research programme may often facilitate in developing a technology. Soil science in this respect is currently more emerging in generating technologies for keeping the soil resource healthy, productive and sustainable.

It is of high relevance that ICAR should promote a soil science-based medical education in view of the fact that type-specific soil or clay behaves like a protective medical tool or even treatment, but need scientific validation critically. There is further need to promote relevance of industrial uses of soil in selected areas including soil battery and earth generator in order to enrich opportunities for generating the clean energy. The Indian Institute of Soil Science (IISS) is the pioneer national institute under ICAR providing scientific base for enhancing and sustaining productivity of soil resource with minimal environmental degradation (IISS 2018) under the following key objectives: (i) to carry out basic and strategic research on soils especially physical, chemical and biological processes related to management of nutrients, water and energy; (ii) to develop advanced technology for sustainable systems of input management in soils those are most efficient and least environmental polluting; (iii) to develop expertise and backstop other organizations engaged in research on agriculture, forestry, fishery and various environmental concerns; (iv) to exchange information with scientists engaged in similar pursuits through group discussions, symposia, conferences and publications; (v) to collaborate with State Agricultural Universities, National, International and other Research Organizations in the fulfilment of the above objectives; (vi) develop database repository of information on soils in relation to quality and productivity (IISS 2018).

Importantly, education system in soil science must not be static as in basic sciences. It must be dynamic and smart based on reliable findings out of research and proven technology. So the relevant research findings within the University must be well proven and subsequently translated to course curriculum at least within three years. Besides, education may control both research and technology, if such practice is made more relevant and burning through ICAR. Rattan (2013), Sanyal (2015) and Katyal (2015) have discussed the need and priority of smart Soil Science Education in India.

Even to sustain the agricultural production, it is first priority to look for a prime land that rests on healthy soil and its functions. The comparative visibility of agriculture in relation to medical and engineering sciences (Table 1.2) may show how soil resource is the wheel of entire agricultural development of a nation. Sanyal (2015) stressed that despite good schemes of exercises being achieved through ICAR/SAU and the targeted results being achieved in

Table 1.2 Comparative visibility of agriculture in relation to medical science and engineering

Distinguishing traits	Agriculture	Medical science	Engineering
Origin/genesis	Close to nature	Synthetic and natural	Artificial
Importance and use	Nourishment and survival of life	Maintenance, cure and repairing for life	Comfort/luxury and energy support to life
In case of failure	Existence of life ends	Sick person at risk or die	Life style may be hard and uncomfortable
Alternative	It cannot be substituted	Medicinal plants, *Yoga* and even soil in some cases	Indigenous and traditional means
Education	Student–teacher–farmer	Student–teacher–patient	Student–teacher–industry
Academic goal	Farming as self-business–entrepreneur–employment	Medical practice–private clinic–employment	Corporate sector, self-industry–employment
Bottleneck in profession	Lack of confidence in farming practice and entrepreneurship	Almost nil due to professional satisfaction	Only in some cases due to shortage of placement
Choice/option of student for admission	Poor or last option for admission in agriculture	First option (biology group)	First option (mathematics group)
Road map of agriculture	Soil-based strategic planning to be formulated	Integration with soil, clay and medicinal plants	Mechanization and processing tools
What needs to be done to make agriculture more realistic?	Reliable and well-proven approach towards productivity, profitability and sustainability in agriculture through land use planning	Linking medical to food, water, soil, clay, medicinal and aromatic plants	Linking engineering to agricultural mechanization and processing options
Linking soil to consumers via supply chain	Point of produce (soil) to point of consumption (market)	Produce must be medically sound	Produce preferably processed/value added
Food quality versus soil quality	Food chain directly linked to soil	Soil being protective medical treatment is seldom appreciated	Physical manipulation of soil and food processing
Climate change mitigation	Soil has immense potential to mitigate and adapt	Virtually nil	May help in technology generation

Table 1.3 Shifts in approaches between the First and Second Green Revolutions in India

First Green Revolution	Second Green Revolution (proposed)
High yielding varieties demanding high water and nutrients coupled with intensive tillage, and sufficient crop protection measures to meet the quantitative requirements of growing population	Quantitative requirements of food to be met by maintaining the productivity, quality, safety, profitability and overall sustainability of land/soil besides climate change mitigation
Least or no care of changing soil health on whole soil basis (pedon)	Both edaphological and pedological interventions in managing the soil health mandatory
Horizontal and vertical production approach based on availability of land to get more and more yield	Horizontal, vertical, mixed farming, value addition and food processing followed by agri-business approach through demand-supply chain framework
Attraction for organic farming is confined to integrated nutrient/pest management options	Efforts to be made to grow the crops organically in order to minimize degradation, pollution/health risks
Rice, wheat and maize to have been the main attraction	Improved cultivars in differing soils for cereals, oilseed, pulses, millets, cucurbits, vegetables and others besides value addition and marketing
Conservation agriculture (CA) has been least appreciated	Conservation agriculture with mixed/integrated farming as the main thrust area including milk, fishery, honeybee and related agri-business based activities
Land use planning (LUP) practically insignificant	Land use planning be the prerequisite for the most remunerative land use choice in each land unit

(continued)

Table 1.3 (continued)

First Green Revolution	Second Green Revolution (proposed)
Labour plight due to mechanization in farming process	Labour and woman empowerment due to adoption of CA, organic/integrated farming approach and value addition following the demand-supply chain framework
Sustainability undermined at the actual farming level	Sustainability is priority towards technology generation under LUP
Maximum tillage practice on priority	Minimum or zero tillage/no till as mandatory in conservation agriculture and organic farming
Social, economic, environmental and soil health-related issues were least cared	All these issues are integral to LUP, besides promotion to soil biodiversity, soil carbon stock followed by carbon sequestration to minimize C-emission, emerging threats of heavy metals, partial factor productivity, potassium depletion in soils, etc.
Social and regional imbalance causing price differentials	Social harmony, peace and overall prosperity expected in a balanced manner due to emerging opportunities in farming system under LUP
Life-sustaining approach	Business-oriented approach with the involvement of information technology, corporate sector and related investment agencies/marketing at farmer's door
Second agricultural succession (ploughing to harvesting) to get the produce	Third agricultural succession (ploughing–harvesting–threshing–processing–marketing to enhance farmer's income) for economic growth and livelihood security
Poverty alleviation not assured	Poverty alleviation may be assured, if all the above practices are ensured with LUP

respect of desired quality education of soil science, one has still remained largely handicapped to implement such smart recommendations at the grass-root level. Katyal (2015) emphasized that the education in soil science must include topics linking and strengthening the public policies on resource management by providing professional inputs. He further stressed that the development of soft skills on creative thinking, strategic planning, problem-solving and public communication is the topics that may inspire convergence of science.

Logical perception on comparative linkage of agriculture with medical science and engineering under relevant distinguishing traits (Table 1.2) clearly indicates the relevance of soil resource for healthy, prosperous and quality food secured India and there is thus a logistic to shift Indian agriculture towards Second Green Revolution. However, even with scarce and threatened land resources, India must look for a golden twenty-first century in agriculture, provided that the soils should be made responsive by restoring the soil health and resilience, and accordingly, the soil science education and research need to be strengthened. The possible shifts in approaches between the First and the projected Second Green Revolutions are given in Table 1.3.

1.10 Conclusions

Soils in India occur with wide diversities. Both the human nutrition and food security depend on our efforts to protect and restore the qualities of soils within their diversities. The health and protection of soils should also be the part of our public health agenda. Soil may have disease promoting quality like soil-borne pathogens, or it may have health-promoting quality. The land use planning process following the adoption of conservation agriculture through land evaluation would lead to poverty alleviation, steadily carbon sequestration, organic farming options and quality food production by restoring the overall soil health. Let India be the "employment provider" through strong employment base under land use planning programme. The soils in India are by and large deficient in soil organic carbon and surely need high-tech approaches for enhancement of soil carbon stock and thus make India a "Buffer Nation" in slowing down the impact of climate change, since these soils need more and more carbon stock. If the Second Green Revolution is the mission for the twenty-first century, each chapter in the book will be the pillar to construct "Ever Green India".

References

Abrol IP, Bhumbla DR (1971) Saline and alkali soils in India: their occurrence and management. World soil resources report 41. FAO, Rome

Abrol IP, Sehgal J (1972) Degraded lands and their rehabilitation in India. In: Greenland DJ, Szaboles I (eds) Soil resilience and sustainable land use. CAB International, UK, pp 129–144

Agarwal RR, Yadav JSP (1954) Saline and alkali soils of the Indo-Gangetic alluvium in Uttar Pradesh. J Soil Sci 5:300–306

Air Ministry (1936) British rainfall. London Meteorological Office, London

Balvanshi A, Tiwari HL (2014) A comprehensive review of runoff estimation by the curve number method. Int J Innov Res Sci Eng Technol 3:17480–17485

Bandyopadhyay PK, Saha S, Mani PK, Mandal B (2010) Effect of organic inputs on aggregate associated organic carbon concentration under long-term rice–wheat cropping system. Geoderma 154:379–386. https://doi.org/10.1016/j.geoderma.2009.11.011

BAU (2016) Soil and land use report on silt dominant sand dunes in Supaul district of Bihar 2016. Bihar Agricultural University, Sabour, Bhagalpur

Bhaduri D, Pal S, Purakayastha TJ, Chakraborty K, Yadav RS, Akhtar MS (2015) Soil quality and plant-microbe interactions in the rhizosphere. In: Lichtfouse E (ed) Sustainable agriculture reviews. Springer International Publishing, Heidelberg, Germany, pp 307–335

Bhattacharyya T, Sarkar D, Sehgal J, Velaytham M, Gajbhiye KS, Nagar AP, Nimkhedkar SS (2009) Soil taxonomic database of India and the states (1:250,000 scale), vol 143. NBSS&LUP Publication, p. 266

Bhattacharyya T, Pal DK, Mandal C, Chandran C, Ray SK, Sarkar D, Velmourougane K, Srivastava A, Sidhu GS, Singh RS, Sahoo AK, Dutta D, Nair KM, Srivastava R, Tiwary P, Nagar AP, Nimkhedkar SS (2013) Soils of India: historical perspective, classification and recent advances. Curr Sci 104(10):1308–1323

Bhumbla DR, Khare A (1984) Estimate of wastelands in India. Society for Promotion of Wastelands Development, New Delhi

Biswas PP, Sharma PD (2008) A new approach for estimating fertiliser response ratio—the Indian scenario. Indian J Fertil 4:59–62

Buchanan F (1807) A journey from Malabar through the countries of Mysore, Canara and Malabar. Geol Mag 2:436–460

Chander G, Wani SP, Kanwar LS, Prasad JK, Pal CK, Pal DK, Mathur TP (2013) Balanced and integrated nutrient management for enhanced and economic food production: case study from rainfed semi-arid tropics in India. Arch Agron Soil Sci 59(12):1643–1658. https://doi.org/10.1080/03650340.2012.761336

Chhonkar PK (2008) Organic farming and its relevance in India. In: Tarafdar JC, Tripathi KP, Mahesh K (eds) Organic agriculture. Scientific Publishers, Jodhpur, India, pp 5–33

Choudhary BC, Prasad SN, Mohan K, Mishra BB (2009) 25 years perspective plan of land use planning of Bihar, 2000–2025. Rajendra Agricultural University, Pusa, Bihar

Dokuchaev VV (1883) Russian chernozems (RusskiiChernozems). Israel Prog Sci Trans, Jerusalem (1967) (trans from Russian: Kaner N). Available form U.S. Dept. of Commerce, Springfield, VA

FAO (1976) A framework for land evaluation. Soils Bull 32

FAO (1983) Land evaluation of rainfed agriculture. Soil Bull 52

Gaétan C, Stephen TS, James LO, Karol S, Pawel S (2012) Earthing: health implications of reconnecting the human body to the earth's surface electrons. J Environ Public Health (Article ID 291541). http://dx.doi.org/10.1155/2012/291541

Gawande SP, Joshi OP, Singh P (1980) The genesis and classification of saline-sodic soils of the Bevar distributor command area in the Etah district, Uttar Pradesh, India. In: Proceedings international symposium on salt affected soils. Central Soil Salinity Research Institute, Karnal, 18–21 Feb 1980, pp 102–109

GoI (2011) Census of India 2011, population totals paper 1 of 2011, India series 1. Registrar General and Census Commissioner, Govt. of India, New Delhi

GoI (2015) Land use statistics at a glance 2003–04 to 2012–13. Directorate of Economics & Statistics, Department of Agriculture & Cooperation, Ministry of Agriculture, Govt. of India, New Delhi, May 2015

Hilgard EW (1892) A report on the relations of soil to climate. Weather Bull 3:1–59

http://www.CO2.earth/

https://www.healthline.com/health/walking-barefoot

https://www.fao.org/soils-2015/en/

https://journeyviakerala.wordpress.com/2017/05/15/kerala-soil-museum

https://www.scoopwhoop.com/soil-from-brothels-durga-puja-idols/

https://commons.wikimedia.org/wiki/File:Diya.jpg

IISS (2018) Annual report 2017–18. ICAR-Indian Institute of Soil Science, Bhopal

IISS, Vision 2030 (2009) Document, published by Director, ICAR-Indian Institute of Soil Science, Bhopal, India, pp 1–41

IMD (2014) Annual climate summary. Indian Meteorological Department, National Climate Centre, Pune

IMD (2015) Annual climate summary. Indian Meteorological Department, National Climate Centre, Pune

Jha PP (1972) Soils of Bihar. In: Soils of India. The Fertiliser Association of India Publication, New Delhi

Jha PP, Cline MG (1963) Morphology and genesis of a sol brun acide with fragipan in uniform silty material. Soil Sci Soc Am Proc 27:339–344

Karale RL, Bisdam EBA, Jongerus A (1974) Micromorphological studies on diagnostic subsurface horizons of some alluvial soils of Meerut district of Uttar Pradesh. J Indian Soc Soil Sci 22:70–76

Katyal JC (2015) Withering soil science education-revival and resurgence boosters. J Indian Soc Soil Sci 63(1):1–13

Katyal JC, Rao NH, Reddy MN (2001) Critical aspects of organic matter management in the tropics: the example of India. In: Martius C, Tiessen H, Vlek PLG (eds) Managing organic matter in tropical soils: scope and limitations. Proceedings of a workshop organized by the Center for Development Research at the University of Bonn (ZEF Bonn), Germany. Springer, Netherlands, Dordrecht, The Netherlands, 07–10 June 1999, pp 77–88

Kaur K, Kapoor KK, Gupta AP (2005) Impact of organic manures with and without mineral fertilizers on soil chemical and biological properties under tropical conditions. J Plant Nutr Soil Sci 168:117–122

Kesavan PC, Swaminathan MS (2008) Strategies and models for agricultural sustainability in developing Asian countries. Philos Trans Roy Soc B 363:877–891

Koshy J (2018) India among nations that face grave danger to soil biodiversity: world wide fund for nature. The Hindu, New Delhi, 31 Oct 2018

Lal R (2004) Soil carbon sequestration impacts on global climate change and food security. Science 304:1623–1627

Laxminarayana KP (2006) Effect of integrated use of inorganic, biological and organic manures on rice productivity and soil fertility in ultisols of Mizoram. J Indian Soc Soil Sci 54:213–220

Lingade SR, Srivastava R, Prasad J, Saxena RK (2008) Occurrence of sodic vertisols in Nagpur district, Maharashtra. J Indian Soc Soil Sci 56:231–232

Mandal C, Mandal DK, Srinivas CV, Sehgal J, Velayutham M (1999) Soil climatic database for crop planning in India. Tech Bull no 53. NBSS and LUP, 1014 pp

Mandal AK, Sharma RC, Singh G, Dagar JC (2010) Computerized database of salt-affected soils in India. Technical bulletin. CSSRI, Karnal, p 28

Mishra BB (2015) Soil based integrated management inputs for flood and flood plain soils of Bihar, India. EC Agric 1.2(2015):109–123

Mishra BB (2016) Indian system of soil classification: a way forward. Agric Res Technol Open Access J 3(2):555606. https://doi.org/10.19080/artoaj.2016.03.555606

Mishra BB (2017) Land economics vs land use planning. In: Sheraz Mahdi S, Kumar S, Dhawan AK (eds) Proceedings national conference on climate change & agricultural production: adapting crops to climate variability and uncertainty. Bihar Agricultural University, Sabour, 6–8 Apr 2017, pp 42–50

Mishra BB, Roy R (2015) Soil science vs science for medicine. EC Agric 2.5(2015):454–461

Mishra BB, Choudhary CD, Singh NK, Mall J, Singh VN (1994) "Fluvisols" as a proposed Order in soil taxonomy for active flood plain soils. In: Transactions XVth world congress of soil science, Acapulco, Mexico, vol 6B, pp 55–56

Mondal S, Singh RD, Patra AK, Dwivedi BS (2015) Changes in soil quality in response to short-term application of municipal sewage sludge in a typic haplustept under cowpea-wheat cropping system. Environ Nanotechnol Monit Manage 4:37–41

Muralidharudu Y, Subba Rao A, Sammi Reddy K (eds) (2012) District-wise soil test based fertilizer and manure recommendations for balanced nutrition of crops. Indian Institute of Soil Science, Bhopal, pp 1–270

Murthy RS, Hirekerur LR, Deshpande SB, Venkata Rao BV, Shankaranarayana HS (eds) (1982) Benchmark soils of India: morphology, characteristics and classification for resource management. NBSS & LUP, Nagpur, 374 p

NAAS (2010) Degraded and wastelands of India; status and spatial distribution. Directorate of Information and Publications of Agriculture, ICAR, Pusa, New Delhi

National Commission on Agriculture (1976) Final report - National Commission of Agriculture (PDF). Indian Government, 1976, New Delhi

Naidu LGK, Hunsigi G (2001) Application and validation of FAO-framework and soil potential ratings for land suitability evaluation of sugarcane soils of Karnataka. Agropedology 11:91–100

Naidu LGK, Ramamurthy V, Challa O, Velayutham M (2017) Land evaluation for sustainable agriculture. Brillion Publishing Co., New Delhi, p 356

Nandgude SB, Jadhav GS, Shinde SS, Mahale DM, Bhattacharyya T, Wandre SS (2017) Computation of runoff by SCS-CN method from microwatersheds of Urmodi basin in Maharashtra state using RS and GIS. Agropedology 27:145–152

Natarajan A, Krishnan P, Velayutham M, Gajbhiye KS (2002) Land resources of Kudangulam, Vijayapatti and Erukkandurai villages, Radhapuram Taluk, Tirunelveli district, methodology at 1:10,000 scale using cadastral map as base map. NBSS&LUP Pub no 36, p 277

NRSA (1985) Wasteland mapping of India using landsat false colour composite on 1.1 million scale. Project report, National Remote Sensing Agency (NRSA), Department of Space, Government of India

Nogiya M, Verma TP, Moharana PC, Singh RS, Tailor BL, Singh R, Meena RL, Kumar S, Singh SK (2017) Influence of landform on the spatial variability of soil fertility in central state farm, Jetsar, Sri Ganganagar district of Rajasthan. Agropedology 27(2):125–130

Pal DK, Bhattacharyya T, Deshpande SB, Sharma VAK, Velayutham M (2000) Significance of minerals in soil environment of India. NBSS review series 1. NBSS&LUP, Nagpur, p 19

Prasad R, Gangaiah B, Aipe KC (1999) Effect of crop residue management in rice-wheat cropping system on growth and yield of crops and on soil fertility. Exp Agric 35:427–435

Press Trust of India (PTI) (2017) Cultivable land in India shrinking. The Hindu Business Line, 11 Nov 2017

Ramdas LA (1960) Crops and weather in India. Indian Council of Agricultural Research, New Delhi, p 127

Rattan RK (2013) Soil, science and society. Presidential address delivered at the 78th annual convention of the Indian society of soil science, CAZRI, Jodhpur

Raychaudhuri SP, Govindarajan SV (1971) Soil genesis and classification. In: Review of soil research in India. Indian Society of Soil Science, New Delhi, pp 107–135

Reddy RS, Naidu LGK, Srinivas S, Niranjana KV, Ramesh M, Harindranath CS, Shivaprasad CR, Thayalan S (2001) Identification and categorization of agricultural drought-prone zones of Andhra Pradesh. Agropedology 11:101–109

Riquier J, Bramao DL, Cornet J (1970) A new system of soil appraisal in terms of actual and potential productivity (first approximation). In: FAO, soil resources, development and conservation service, land and water development division. Approaches to land classification. Soil Bull 22. FAO, Rome, 120 pp

Sahoo AK, Chattopadhyay T, Singh RS, Shyampura RL (2003) Characterization of soils and vertical distribution of available micronutrients under different landforms of Malwa plateau, Rajasthan. Indian Agric 47:217–223

Sanyal SK (2015) Quality soil science education in India. J Indian Soc Soil Sci 63(1):14–15

Sehgal JL (1989) Soil resources map of India (1:1 M Scale): meeting the challenges. J Soil Surv Land Eval (UK) 7(2):123–125

Sehgal J (1995) Land resources appraisal for land use planning to meet the challenges of 21st century. J Indian Soc Soil Sci 43(4):504–528

Sehgal JL (2003) Soil classification. In: Fundamentals of soil science. Indian Society of Soil Science, New Delhi

Sehgal J, Abrol IP (1994) Soil degradation in India: status and impact. Indian Council of Agricultural Research, New Delhi, Oxford, and IBH Publishing Co., PVI, Ltd., p 80

Sehgal J, Mandal DK, Mandal C, Vadivelu S (1992) Agro-ecological regions of India, 2nd edn. Tech Bull no 24, NBSS and LUP, 130 pp

Sehgal J, Mandal DK, Mandal C (1995) Agro-ecological sub-regions of India (Map). NBSS and LUP, Nagpur

Sharma VP (2015) Dynamics of land use competition in India: perceptions and realities. W.P. no 2015-06-02, working paper series IIMA, June 2015. Research and Publications, Indian Institute of Management, Ahmadabad

Sharma DK, Chaudhari SK (2012) Agronomic research in salt affected soils of India: an overview. Indian J Agron 57(3rd IAC Special Issue):175–185

Sharma JP, Raychoudhury C (1988) Soil-landform relationship in a basaltic terrain. J Indian Soc Soil Sci 36:755–760

Sharma JP, Landey RJ, Kalbande AR, Mandal C (2001) Characteristics and classification of soils of Kathiawar region of Gujarat as influenced by topography. Agropedology 11:83–90

Shivarajasingham S, Alexander LT, Cady JG, Cline MG (1962) Laterite. Adv Agron 14:1–60

Shyampura RL, Krishna NDR, Gopal Ram, Giri JD (1992) Genesis and characterization of the soils across Ghaggar flood plain in Rajasthan. Ann Arid Zone 3:1199–1204

Sihi D, Dari B, Sharma DK, Pathak H, Lata N, Sharma OP (2017) Evaluation of soil health in organic vs. conventional farming of basmati rice in North India. J Plant Nutr Soil Sci 180:389–406. https://doi.org/10.1002/jpln.201700128

Soil Survey Staff (1999) Soil taxonomy, 2nd edn. USDA-SCS Publication, Washington DC

Somasundaram J, Reeves SH, Wang WJ, Heenan MA, Dalal RC (2017) Impact of 47 years of no-tillage and stubble retention on soil aggregation and carbon distribution in a vertisol. Land Degrad Dev. https://doi.org/10.1002/ldr.2689

SubbaRao A (2009) Implications of soil fertility to meet the future demands: the Indian scenario. In: Future planning for soils in India. Proceedings IPI-OUAT-IPNI international symposium on role and benefits of potassium in improving AT, Bhubaneswar, Odisha. Nutrient management for food production, quality and reduced environmental damage, 5–7 Nov 2009, OU

Subrahmanyam VP (1964) Climatic water balance of Indian arid zone. In: Proceedings of symposium on problems of Indian arid zone, Jodhpur, pp 405–411

Sunanda Biswas (2016) Soil resilience: a new approach to recover degraded soil. Biotech articles; views 1333; www.biotecharticles.com/Agriculture-Article/Soil

Velayutham M (2012) National soil information system (NASIS) and land resource mapping for perspective land use planning and pragmatic farm level planning. In: 4th Dr. S. V. Govindarajan memorial lecture. Madras Agric J 99(4–6):147–154. http://sites.google.com/site/majmasu/archive/99-4-6

Velayutham M, Mandal DK, Mandal C, Sehgal J (1999) Agro-ecological subregions of india for planning and development. NBSS and LUP, Publ no 35, 372 pp

von Liebig J (1840) Chemistry in its application to agriculture and physiology. Playfaeir, Taylor and Waton, London

Yadav SS, Jagdish P, Gaikwad ST, Thyalan S (1998) Landform-soil relationship and its impact on soil properties. J Indian Soc Soil Sci 46:332–335

History of Soil Research

2

A. Subba Rao and J. Somasundaram

Abstract
History enables to plan with creative thinking skills for sustained better present and brighter future. Soil resource is deeply rooted to human civilization as being the key to sustainable agriculture as well as other functions. The Hindu *Vedic* literature as well as Bible supported the vital role of soils. This chapter focuses on the history of soil research under (i) ancient era (ii) medieval era and (iii) recent era including pre- and post-independence periods. Researches on colloidal chemistry and acidity during early 1930s have set a platform for in-depth understanding of different facets of soil research including pedology and edaphology. The All India Coordinated Research Projects (AICRP) such as long-term fertilizer experiments (LTFE), soil genesis, survey, mapping and classification have contributed significantly in strategic land use planning processes and their implementation. Soil test crop response (STCR) correlation and micro-nutrient and secondary elements have played crucial role in strengthening the soil research. Although the soil research in India is relatively at youthful stage, however, it has taken rapid strides in terms of development as a single major subject and capable of accommodating and fostering other allied subjects under its umbrella.

Keywords
Ancient • Medieval and pre-independence • Post-independence soil research priorities

A. Subba Rao (✉) · J. Somasundaram
ICAR-Indian Institute of Soil Science, Nabibagh, Berasia Road, Bhopal, Madhya Pradesh 462038, India
e-mail: asrao110@gmail.com

B. B. Mishra (ed.), *The Soils of India*, World Soils Book Series,
https://doi.org/10.1007/978-3-030-31082-0_2

2.1 Introduction

Everything around humans has a history and a past. The human race invented agricultural practices around 8000–10,000 years ago in the area between Nile Valley in Egypt and Indus Valley in India (Krishna 2002). In the *Ramayana*, it has been stated that all the dead things and rotten garbage returned to earth are transformed into wholesome things that nourish life. "Upon this handful of soil our survival depends. Husband it and it will grow our food, our fuel and our shelter and surround us with beauty. Abuse it and the soil will collapse and die taking man with it"—*1500 BC; Sanskrit proverb*. In India, during Neolithic period (third to second centuries BC), farming communities were found in areas with fertile black (*regur*) soils (Vertisols) on the Deccan Plateau. Literature from the fourth century AD did mention about irrigation of fields and fines or penalties for those who allowed breaches in the irrigation system (Krupenikov 1992). These farming communities spread at an early date to include the fertile floodplains of major rivers like the Ganges. Ancient documents in India indicated ploughing of soils as a common practice at least as early as 5000 years ago (Lal 2007a, b). In about 1400 BC, the Bible depicted Moses as he realized that fertile soil was essential to the well-being of his people (Brevik and Hartemink 2010).

By the fourteenth century AD, irrigation, fertilization through manure application, fallow periods to restore fertility and selection of crops based on soils were largely adopted in India (Krupenikov 1992). Soil research in India has a history of less than 150 years, although the reference to the importance of soil, or more strictly land, finds a mention in the *Vedas*, the *Upanishads* and other ancient scriptures which date back to 5000 BC. The initiation of soil research was made in the last decade of the nineteenth century, as a part of research in agricultural chemistry, with the appointment of an agricultural chemist to the Government of India. However, it received a great impetus only in 1915 with the establishment of the Imperial (now Indian) Agricultural

Research Institute, the agricultural colleges and research laboratories in the State Departments of Agriculture.

2.2 The Ancient Era

Archaeological records and literatures indicated that agricultural crop production was regularly practiced (wheat, barley, jujube) during the Indus Valley civilization, which began about 2500–2000 BC (Botterro et al. 1965). In the period between 2000 BC and 500 AD, the economy of the *Rigvedic* Indian civilization centred around well-developed agriculture (Kenoyer 1995; Gangopadhyay 1932). According to fertility, soils were mainly divided into two classes—*Urvara* (fertile) and *Anurvara* or *Usara* (sterile). *Urvara mrittika* is again sub-classified into different kinds according to their fitness to the cultivation of different kinds of crops, for example, the soil fit for barley—*vavya*; for sesame—*tila* and for rice—*vraiheyasaleya*. *Anurvara mrittikawa* is subdivided into *usara* (salt ground) and *marn* or desert. Two broad divisions of soils are found to occur in the *Rigveda*: fertile (*apnaswati*) and barren (*astana*). In the later Vedic literature, they are expressed as alkaline (*usara*) and non-alkaline (*anusara*). Fields with better fertility were classified as *urvara*.

About 15,000–10,000 years before present (ybp), when agriculture developed in the Fertile Crescent region that extended from Israel through northern Syria to Western Iran, there was another eastward wave of human migration (Cavalli-Sforza and Piazza 1994; Renfrew 1996), a part of which also appeared to have entered India. As per one postulation, this has possibly brought the Dravidian languages into India (Renfrew 1996). Subsequently, the Indo-European (Aryan) language family was introduced into India about 4000 ybp. The value of manure in cultivation was appreciated in India as early as the time of *Rigveda*. In addition to bones, animal flesh, fish washings, and vegetable and animal products, the manure they mainly used consisted of the excreta of various animals mixed with litter (Abrol and Nambiar 2008). The exact chemical composition of different kinds of soils was not known to the ancient authors but from the following typical maxims of Khana, the wife of *Mihira*, the famous ancient astronomer; it appeared that they made extensive experimental observations and obtained a masterly knowledge regarding the characteristic suitability of soils for cultivation of different kinds of crops. "Sandy soil is suitable for cultivation of *Aus* paddy and clayey soil for that of jute" (Majumdar 1935). "Your expectation will be fulfilled if you cultivate *patrol (Trichosanthes dioica)* in sandy alluvial soil" (Majumdar 1935). The earliest reference of manuring the soil can be traced as far back as a verse in *Atharva Veda* (1500–500 BC). But more elaborate instructions are given in the *Brhat Samhita* (500 AD), the *Agnipurana* (500–700 AD), the *Krishi Sangraha* of Parasara (500–1000 AD; 400 BC) and Sukraniti (Abrol and Nambiar 2008). Use of organic manures is mentioned in *Rig Veda* (2500–1500 BC), while use of green manure is mentioned in *Atharva Veda* (1000 BC). In *Sukra*, it is mentioned that, to cause healthy growth, the plant should be nourished by dung of goat, sheep, cow and water. In *Vrkshayurveda*, by Surpala, use of manures and manuring and remains of animals were considered beneficial for plant growth. Use of oilcakes and animal excreta had been reported in Kautilya's *Arthashastra* (300 BC). The directive of *Parasara* to keep the dung heap undisturbed up to the month of *Magha*, i.e. for ten months of a year is very crucial for fermentation (decomposition). This process reduces the active ammonia which could otherwise be injurious to the seed and tender plant roots. Some examples from Majumdar's (1935) translation of *Upavana-Vinoda (1300 AD)* are reproduced below:

"The yield of coconut is sure to increase if salt, *Kapala*, husk and sugar are poured on its roots".

"The areca nut tree would give rich fruit if every year its basin is filled with ordure in the rainy season".

"In the month of *Magha*, a dung heap is raised by a spade, dried in the sun and made into smaller balls. In the month of *Phalguna*, these are placed into the holes dug for the purpose and afterwards scattered in the field at the time of sowing. The paddy plant only grows without manure, and it does not bear fruit". Kasyapa (800 AD), an ancient Indian sage, dealt with the cultivation practices of some of the important individual crops (Raychaudhuri et al. 1964). Some examples are quoted below:

"A wise husbandman should start cultivation suitably, keeping in view the fertility of the soil and after having assured himself of necessary water supply from rainfall or from a big water reservoir or from canal".

2.3 The Middle Era

It is often noticed by almost all foreign travellers who have travelled to India about the natural fertility of soils prior to the twentieth century (1200–1899). According to French traveller Jean de Thevenot who travelled to India in the seventeenth century, fish manure was used in Gujarat for sugarcane cultivation (Abrol and Nambiar 2008). The writings of Francis Buchanan who toured several southern states in 1801 indicated the use of green leaf as manure as a common practice (Raychaudhury 1984). He recorded that "the leaves or shoots used by farmers here as manure are the Handur; Cango or *Rabinia mitis*; the Yacada or *Aspedias gigantean*; the Calli or *Euphorbium tirucalli*; the Devadarum or *Erythroxylon sideroxyloides*. A report during June 1805 to the Board of Directors of the East India Company states that '*It is an undisputed fact that the produce of the soil is definitely below what it is capable of yielding under proper management*'. Subsequent paragraphs reflect the interest of the company. '*Substantial advantage must be derived by the State from the increased wealth and prosperity of the people*'. The government will be enriched by the taxes on additional produce from the land without putting burden on the people". The report ended with a proposal to set up experimental farms for the introduction of better farming system, soil management and improvement of cattle herd. A scientific study on soils during British colonial rule was initiated in 1807 by Buchanan who termed the red honey-combed mass of earth in the Malabar Coast of India "Laterite", a term based on local terminology of brick making. Figure 2.1 is the monument of laterite brickstones at Angadipuram, Kerala, which commemorates where laterite was first described and discussed by Buchanan-Hamilton in 1807 (Thurston 1913).

In India, the first proposal for a special department of agriculture originated with a commission appointed following the great famine in Bengal and Orissa in 1866 (Randhawa 1979). One of the recommendations was that "the first class expert who should make a general enquiry into the character of the soils and agricultural conditions in the country which was immediately required" (Randhawa 1979). In 1889, Dr. J. V. Voelcker, consulting chemist to the Royal Agricultural Society, was invited to advice upon the best course to be adopted in order to apply agricultural chemistry to improve Indian agriculture. Amongst other recommendations, he also advised the appointment of a "really first class man" as agricultural chemist. Dr. Voelcker observed that Indian soils were generally low in nitrogen and lack of availability of synthetic nitrogen compounds prevented large-scale experimentation with chemical fertilizers. The appointment of Dr. J. W. Leather in 1892 marked the beginning of research in agricultural chemistry and soil science. Considerable credit goes to Dr. Leather for initiating permanent manure experiments for critically evaluating soil productivity on a long-term basis and also identified the four major soil regions, i.e. alluvial soils of the Indo-Gangetic plains, black cotton or regur soils of the Deccan plateau and the red and laterite soils of the peninsular and east India. The first permanent plot experiment was established at Kanpur (Uttar Pradesh) in 1885 based on the Rothamsted model. Two more such experiments were initiated at Pusa (Bihar) in 1905 and Coimbatore (Tamil Nadu) in 1908. These experiments showed that crop yields could be improved remarkably with the balanced use of chemical NPK fertilizers (Leather 1900; Kalamkar and Singh 1935).

Fig. 2.1 Monument of laterite brickstones. *Source* https://en.wikipedia.org/wiki/Laterite

A fundamental department of agriculture named "Department of Revenue, Agriculture and Commerce" started in the year 1871 remained mainly as revenue, and there was no work on agricultural development, but this did mark a commencement and appreciation of the agriculture sector by the colonial government. The credit for the modest foundation goes to Lord Mayo, who was the fourth Viceroy of India, and to A. O. Hume who was a civilian of the Bengal Civil Service and one of the founders of the Indian National Congress. Primarily, the department was established by the government with a view to supply cotton to the textile industries of Manchester, and not to feed the famine-ravished India. Actually, the inherent desire of the colonial rulers was to adjust the development of Indian agriculture for the production of raw materials to the economic and industrial necessities of British capitalism. The early development of agricultural research in India was lined with the occurrence of famines. The repeated occurrence of famines and starvation death of peasants in the second half of the nineteenth century opened the eyes of the imperial rulers. During this period, 24 famines, big and small, took a life toll of more than 28 million souls (Chandy 2002). However, at that time, little precedence was accorded to agricultural research and development in the country (Borthakur and Singh 2013).

2.4 Recent History—Pre-independence Era

The history of the Division of Soil Science and Agricultural Chemistry is traced back to the Chemical Section headed by Imperial Agricultural Chemist, which came into existence in 1905 with the establishment of the Imperial Agricultural Research Institute at Pusa in Bihar (Figs. 2.2 and 2.3) housed in the Phipps Laboratory, grand edifice named after the American Philanthropist Mr. Henry Phipps. The Great Bihar Earthquake of 1934 led to the shifting of the Imperial Institute to New Delhi where the building of the erstwhile Chemical Section continues to carry the name of Phipps Laboratory. With the country's independence in 1947, the Chemical Section was renamed as the Division of Soil Science and Agricultural Chemistry. Dr. Leather left India in 1916, and in the following decade or so, there was no scientific leadership in this area.

The Royal Commission on Agriculture appointed India in 1926 to examine and report on the present conditions of agriculture and rural economy in British India and to make recommendations for the improvement of agriculture and submitted its report in 1928. Some of the significant recommendations of the commission (Abrol and Nambiar 2008) which had an impact on soil management strategies in the country were (i) A soil survey of the whole India at the present time was not recommended, (ii) No sensible diminution in the fertility of long cultivated soil was to be anticipated, (iii) Manure experiments on unirrigated land were especially important, (iv) Experimental work was required to discover the green manure crops which could best be included in the cultivator's rotations, (v) No further investigation under government auspices of the possibilities of manufacturing synthetic nitrogen in India would be required, and (vi) Steps were to be taken to promote better preservation of farmyard manure. Royal Commission on Agriculture set up in the early twenties realized the importance of holistic agriculture research system and subsequently established ICAR system.

Fig. 2.3 Logo of the Imperial Agricultural Research Institute. *Source* https://en.wikipedia.org/wiki/Indian_Agricultural_Research_Institute

2.4.1 Establishing of Research Institutes and Organizations

A beginning in soil research was made in the last decade of the nineteenth century, as a part of research in agricultural chemistry, with the appointment of an agricultural chemist (Dr. J. A. Voelcker) to the Government of India. Dr. J. A. Voelcker, a consulting Chemist to the Royal Agricultural Society, happened to be the first scientist deputed to India by the British Government as early as 1889, to advise upon the best course of action to apply the teaching of agricultural

Fig. 2.2 Phipps Laboratory, Imperial Agricultural Research Institute, Pusa, Samastipur, Bihar. *Source* https://en.wikipedia.org/wiki/Indian_Agricultural_Research_Institute#/media/File:IARI_Pusa_Bihar.jpg

chemistry. Further, on the recommendations of the Famine Enquiry Commission, the Government of India established in 1905 the Imperial (now Indian) Agricultural Research Institute at Pusa Farm in Bihar. Subsequently, the government took steps for the development of agricultural research and education in different provinces by establishing agricultural colleges and research institutes. Agricultural colleges were established or reorganized at Pune, Kanpur, Nagpur, Coimbatore, Sabour (Fig. 2.4) and Lyallpur (now Faisalabad, in Pakistan).

Based on the reports of the Famine Commissions of 1880, 1898 and 1900, the Government of India (GoI) was determined to set up a central "Department of Agriculture" controlled by the Imperial Secretariat, and the agriculture departments were to be set up in the provinces to primarily look after agricultural activities, agricultural development and famine relief in the country. In 1892, an agricultural chemist and an assistant chemist were allotted to look after research and teaching in India in the Department of Revenue and Agriculture. Eventually, in 1901, an Inspector General of Agriculture was appointed to advise the imperial and the provincial governments on agricultural matters. During the severe famines of 1899–1900, Lord Curzon, the then Viceroy of India, was convinced that the GoI must urgently concentrate on the agricultural sector to overcome the destructions caused by frequent famines. As a consequence, Imperial Agriculture Research Institute now the Indian Agricultural Research Institute (IARI) and a college for advanced agriculture training were together established at Pusa in Bihar in the year 1905, and its Director was the agriculture adviser to the GoI till 1929. Meanwhile, the Indian Council of Agricultural Research (ICAR), formerly known as Imperial Council of Agricultural Research, was established on 16 July 1929 as a registered society under the Societies Registration Act, 1860 in pursuance of the report of the Royal Commission on Agriculture. The ICAR has its headquarters at New Delhi. India's agriculture growth during the past more than 100 years is closely connected with the researches done and technologies generated by the IARI/ICAR Institutes. The Green Revolution emerged from the fields of IARI, which has been instrumental in the development of high-yielding varieties of major crops including wheat, maize and rice in the country and taking major initiatives in integrated soil–water–nutrient management practices.

Similarly, United Planters' Association of Southern India (UPASI) dates back to late nineteenth century, even though tea has been introduced and cultivated in certain pockets of Southern India. In pre-independent era, until 1924, UPASI depended on the financial assistance provided by the then State Governments of Madras and Mysore Durbar for agricultural research on plantation crops. With the auspices of government's aid, the Tea Experimental Station at Peermade, the Rubber Experimental Station at Thenmalai and the Coffee Research Station in Coorg were established; however, the experimental station at Mundakayam was administered entirely by UPASI. The first milestone in the history of UPASI was the establishment of a separate Tea Experimental Station at Davershola in 1927 where Dr. W. S. Shaw was appointed as the First Scientific Officer followed by the appointment of an Assistant Scientific Officer in the same year. To meet the tea research expenses, UPASI levied a cess of eight annas per acre on its members. Between 1926 and 1932, the area under tea in South India expanded by 14,000 ha from 34,000 to 48,000 ha. UPASI appointed a Soil Chemist in 1935 to work on composts which evoked considerable interest at that time.

The Agricultural School at Saidapet, Chennai, which was established as early as 1868, was later relocated to Coimbatore during 1906. Also, a branch for teaching agriculture in the College of Science at Pune (established in 1879) was subsequently developed into a separate College of Agriculture in 1907. Similar agricultural colleges were established at Kanpur, Sabour, Nagpur and Lyallpur (now in Pakistan) between 1901 and 1905. An All-India Board of Agriculture was established in 1905 with a view to bring the provincial governments more in touch with one another and making suitable recommendations to the GoI. The Indian Agriculture Service was constituted in 1906.

(i) *International Union of Soil Sciences (IUSS)*

First international meeting of pedologists was held in 1901 which led to the foundation of International Society of Soil Science in 1924, currently renamed as International

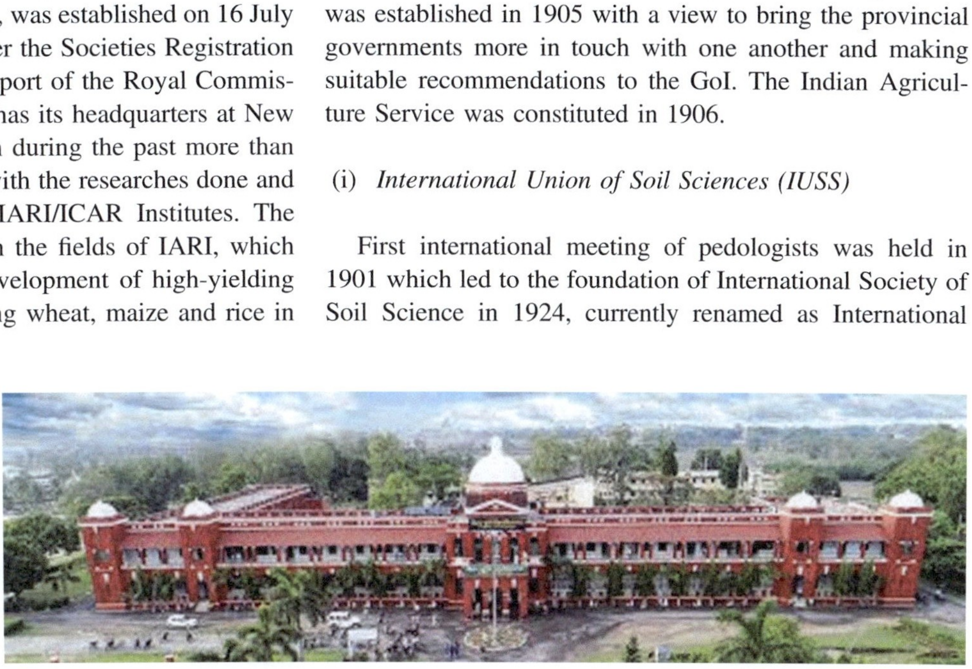

Fig. 2.4 Bihar Agricultural College, Sabour (established in 1908). *Source* http://www.bausabour.ac.in/college-homepage.aspx?CollegeCode=001

Union of Soil Sciences. The IUSS has organized 20 World Congresses of Soil Science, thousands of smaller conferences, meetings and workshops in the past 90 years (Hartemink 2015). The IUSS has 60,000 members and is organized into Divisions, Commissions and Working Groups that deal with all aspects of soil research. Participation of Indian Soil Scientists in different World Congresses has been commendable. The 12th International Congress of Soil Science was held in 1982 at New Delhi and Dr. J. S. Kanwar (Fig. 2.5) was the President.

(ii) *Indian Society of Soil Science (ISSS)*

Ten years later, similar to Global Society (IUSS), a Society called the Indian Society of Soil Science (ISSS) came into existence in India on 22 December 1934 in Calcutta under Act XXI of 1860 (Registration No. 8164/252 of 1934–35) with 28 members. It was formally inaugurated on 3 January 1935 at the Presidency College, Calcutta. Sir Bryce C. Burt, the then Agricultural Commissioner in the Imperial (now Indian) Council of Agricultural Research, was the first President of the Society and Prof. J. N. Mukherjee (Fig. 2.6) its first Secretary. Now, this ISSS expanded with more than 1700 members (both life and annual members) consisting of more than 45 chapters across the country. During the last 84 years (1934–2018) of its existence, it has made commendable contributions to fostering of soil science research in the country, which has earned the national and international appreciations and laurels. This society has organized 81 Annual Conventions periodically in various locations across India. Besides this, the society is also organizing various memorial lectures on soil science/research through its different chapters. It played key roles in organizing an International Symposium on Soil Fertility Evaluation in 1971 and the 12th International Congress on Soil Science in 1982. Besides, two important soil science societies, namely the Clay Minerals Society of India and the Indian Society of Soil Survey and Land Use Planning were subsequently established in 1982 and 1986, respectively.

Fig. 2.5 Dr. J. S. Kanwar. *Source* http://insaindia.res.in/detail/N75-0357 or https://www.icrisat.org/end-of-an-era-in-agricultural-science-a-trail-to-follow-dr-jaswant-singh-kanwar/

Fig. 2.6 Prof. J. N. Mukherjee. *Source* https://en.wikipedia.org/wiki/Jnanendra_Nath_Mukherjee

(iii) *Old and New Long-Term Experiments*

Sir Pheroza Kharegal, the then Vice Chairman of the Imperial Council of Agricultural Research (1939–44) prepared a memorandum on the development of agriculture and animal husbandry in India (Randhawa 1979). In line with "Rothamsted Classical Experiments", Indian long-term experiments were established at Kanpur, Uttar Pradesh (1885); Pusa, Bihar (1908) and at Coimbatore, Tamil Nadu (1909). During the same period, three permanent manure trials were established with more heavily fertilized crops such as sugarcane at Shahjahanpur (Uttar Pradesh) in 1935 and Padegaon (Maharashtra) in 1939. Later on some more long-term experiments were also established at Indore (Madhya Pradesh) 1947, Muzaffarnagar (Uttar Pradesh) in 1949 and at Anakapalle (Andhra Pradesh) in 1950 and on cereal-based cropping system at Ranchi (Jharkhand) in 1956. Many of them were discontinued later. However, the experiments at Coimbatore and Ranchi are still being continued. The effects of these inorganic fertilizers alone and in various combinations were compared with those of FYM and rape cake in most of the experiments (Singh et al. 2004). These permanent experimental plots showed the importance of fertilizers, manure and sequential cropping in maintaining crop productivity and fertility of the soils (Iyer and Pande 1954; Patwardhan et al. 1956).

The food crisis that broke out during the Second World War, and the Bengal famine in 1943, prompted the British Government to start a "Grow more food" campaign in 1943. Dr. D. W. Burns, the then Agriculture Commissioner, was assessed to prepare a note on the technological possibilities of agricultural development in India and also to review the production of all major crops since 1920. In his report, Dr. Burns indicated that the average yield of rice could be increased by 50%, 10% with the use of good variety and 40% through manuring. Similar increases were also mentioned for other crops such as wheat, sorghum, pearl millet, jute, etc. (Randhawa 1979).

2.4.2 History of Scientific Composting

The invention of modern composting is attributed to Sir Albert Howard, a British agronomist, who came to India in 1905 and spent almost 30 years experimenting with organic gardening and farming. Ultimately, about 1000 cart loads of compost were made at Indore each year with the collection and admixture of vegetables and animal wastes of the area farmed into pits. Howard had acknowledged the lessons he had learnt from the centuries-old experience of the Indian farmers, whom he named his "Professors". In 1931, he published a book "The Waste Products of Agriculture: their utilization as Humus", co-authored with Wad (Howard and Wad 1931). Howard (1940) believed that the introduction of improved crops and better soil management could improve the total agricultural production of India. The work of composting including both town and village wastes was initiated on a nation-wide scale by the GOI under Dr. C. N. Acharya after 1931. The Indore process for the manufacture of humus from vegetable to animal wastes was devised at the Institute of Plant Industry, Indore during 1924–1931. Howard has been called the "*father of modern composting*", for his refinement of a traditional Indian composting system into what is now known as the ***Indore method***.

2.4.3 Fertilizer Manufacture and Fertilizer Research

The first fertilizer factory (single super phosphate) was opened at Ranipet (Tamil Nadu) in 1906 creating history of the Indian fertilizer industry. In India, the fertilizer industry is in the core sector and second to steel in terms of investment. Similarly, first synthetic ammonia production factory was set up in Belagual at Mysore in 1941. Subsequently in 1946, the Fertilizers and Chemicals Travancore Ltd. at Udyogmandal (near Cochin) produced ammonia using wood as feedstock. In the wake of the great famine, which swept Bengal in 1943, the government decided to make all efforts to increase the food production. The plans to set up the first large ammonium sulphate plant at Sindri were thus conceived.

After the technology of ammonia reactors was set up (Sekhon, source http://www.ipipotash.org/), Imperial Chemical Industries assisted to conduct field experiments, mostly in the plant industry in the government research stations. In 1945, A. B. Stewart of Macaulay Institute of Soil Research, Aberdeen was invited by the then British Government to report on soil fertility investigations in India. His report concluded that Indian soils were generally low in nitrogen status but they did not require phosphate. But legumes, particularly Egyptian clover and shallow-rooted wheat crop responded to fertilizer phosphorus in certain parts of the Punjab and also in light-textured red soils.

2.4.4 Salt-Affected Soils and Reclamation

The Indian *Reh* Committee deliberated on the origin of salts in soils during 1877. The committee identified two important sources of salts in soils, i.e. weathering of soil constituents and role of canal water. Subsequently, there were evidences of existence of salt-affected soils in the then United Provinces of Agra and Oudh quoted by Moreland and Tamhane (1901). He also reported that cultivators used indigenous methods for reclamation of sodic soils in the United Provinces at that time, much before any scientific research on this grey subject. The Director of Land Records and Agriculture in his letter number 1247, dated 28 July 1888 informed the Secretary Board of Revenue that the extent of affected area of "*usar*" in North-Western Provinces and Oudh was 3,129,053 acres (1,271,977 ha). The same figure was reported in later estimates of alkali lands in Uttar Pradesh (Tiwari et al. 1989; Sharma and Chaudhuri 2012). The Punjab part of the Indus plains also suffered from soil salinity before any large-scale canal irrigation was introduced in the province. The affected belt was reported to be 16 km × 4.8 km (Jameson 1852). In 1906, the GoI called for reports from experienced local irrigation and revenue officers regarding the prevalence of *kallar* or alkali in Sind.

Research on reclamation of sodic soils India was started before independence (Dhar 1934; Talati 1947). There were two broad approaches in reclamation and management of salt-affected soils: (i) reclamation of sodic soils by chemical amendments and (ii) reclamation of saline and waterlogged saline soils by surface and sub-surface drainage. Moreover, various chemical compounds with appreciable soluble calcium content were also tried in varying agro-ecological situations (Kanwar and Bhumbla 1957; Yadav and Agarwal 1959). In 1939, the ICAR sanctioned a scheme for the survey of coastal soils in Bakharganj district (now in Bangladesh) for the occurrence, among others, of a hard subsoil layer, called clay pan.

2.4.5 Research on Soil Colloids and Soil Acidity

The theory of soil acidity was published in Nature as early as 1922, and subsequent papers on the nature of reactions responsible for soil acidity were published in the Indian Journal of Agricultural Science and also in the Journal of Colloidal Science and Nature starting from 1931 to 1975 (Panda 1987). Prof. Hans Jenny in his classical review paper on "*Reflections on the Soil acidity Merry-go-round*" had discussed thoroughly the humic acids and hydroxide adsorption theories, early aluminium theories, the H-clay theory and the newer aluminium theories of soil acidity (Jenny 1961). Soil research in India during the 1930s focused mainly on soil colloids. Other research schemes pertained to clay minerals and soil colloids which led to the concept of the role of aluminium in clay complexes in determining the availability of nutrients, particularly phosphates. The physical chemists were active in investigating soil colloids and their relationship to the soil fertility problem (Randhawa 1979).

Prof. J. N. Mukherjee and his co-workers in a series of papers published from 1942 recognized liberations of H, Al and Fe from prepared H clays. During 1955, P. F. Low's paper on the role of aluminium and hydrogen in the titration of bentonites helped the acceptance of Al theory in the USA. (Panda 1987). Parallel to Western literature, Prof. J. N. Mukherjee's contribution to colloidal chemistry and electrokinetic double layer gained lots of appreciation and recognition. He joined the Imperial College of Science and Technology in London in 1919 to carry out research in the Physical Chemistry laboratory of F. G. Donnan. Prof. J. N. Mukherjee published a paper in Nature in 1921 on the origin of charge on colloids and its neutralization, read earlier at one of the sessions of a seminar organized jointly by the Faraday Society, London and the Physical Society, was greatly appreciated by critics for his new ideas (Sanyal 1997). Two of Mukherjee's other contributions in the field of colloids refer to that of an accurate method, known as the moving boundary method of measuring the cataphoretic speed of colloidal particles which are electrically charged, by the passage of an electric current. This work he started while at Donnan's laboratory and published a paper in 1928 in the Proceedings of the Royal Society, London. This method was an improvement of an earlier method by Burton, but unfortunately Mukherjee's method is not referred to in the British textbooks, whereas other European authors recognize the superiority of Mukherjee's method. A similar incidence of injustice was observed when the rule of coagulation of colloids by electrolytes, discovered by Mukherjee and Sen and communicated in the Journal of the Chemical Society, London is credited to Burton and Bishop, the inadequacy of whose work was specifically discussed in Mukherjee and Sen's paper. The nature of soil acidity published in Nature (1921) was suggested to be a significance of colloid chemical reactions, which, apparently, did not explain all the aspects of the complex soil colloid system. This has encouraged J. N. Mukherjee to submit a large scheme to the then Imperial (now Indian) Council of Agricultural Research (ICAR) to be continued for a number of years. This project has provided the basic information on the changes in behaviour of colloidal systems with changes in the composition of the intermicellary medium. Those who contributed significantly to these basic studies were B. Chatterjee, S. Mukherjee and N. P. Datta (Sanyal 1997).

2.4.6 Research on Crop Rotation and Soil Organic Matter

It was established that the humus reserves of well-drained Indian soils vary more than a hundred fold, from 0.010% nitrogen to over 1.0% carbon. So, planning was formulated jointly by representatives of the Indian Agricultural Research Institute (IARI) and the United States Technical Cooperations Mission (TCM) to elucidate the intricacies of nitrogen and carbon dynamics in Indian soils. About 500 soil samples were collected in India by Dr. N. C. Mehta and his team of IARI with district agricultural chemists during December 1954 to April 1955. Half of each sample was deposited at IARI in New Delhi, and the other half was shipped to Berkeley, California, where all samples were analysed in the laboratory. Special contracts with the International Cooperation Administration and a grant from the Rockefeller Foundation facilitated to carry out analyses and computations. This has further helped in furnishing valuable information on soil management and cropping histories on soil organic carbon (Jenny and Raychaudhuri 1960). In their classical work, they also highlighted the importance of soil sampling in accordance with soil-forming factors. They concluded from 500 soil samples collected from cultivated fields and forest in accordance with a climatic grid that the low level of organic matter in many Indian soils is primarily caused by environment, and only secondarily by cultural practices. Moreover, hot climates accelerate loss of soil organic matter, and the solution of India's nitrogen economy must be based on nitrogen fixation by annual agricultural crops/plants (Jenny and Raychaudhuri 1960).

2.4.7 Research on Soil Erosion

Soil conservation research in India started as early as 1923 with the establishment of dry farming scheme at Manjri near Pune (Kanitkar et al. 1960). The experiments carried out at

Manjri from 1929 onwards to determine the quantity of rainwater lost by run-off and quantity of soil lost by erosion were laid out on the same plan as was followed by Duley and Miller at Missouri in the USA. However, the soil and water conservation research gained much interest during the first and second Five Year Plans at the same time, a chain of soil conservation research, demonstration and training centres were established at Dehradun, Kota, Bellary, Udhagamandalam, Vasad, Agra and Chandigarh by Government of India in 1950s and later transferred to ICAR on 1967 (Singh et al. 1992).

2.4.8 Soil Fertility

In India, Professor N. R. Dhar (1892–1986), Father of Indian Physical Chemistry, discovered the process of thermal and photochemical fixation of atmospheric nitrogen in the soil. He was the Founder-Director of the Sheila Dhar Institute of Soil Science, Allahabad University (1935–1986). Harrison and Das (1921) were among the first to observe that alkaline earths were extremely active for the retention of soluble phosphates in the calcareous soils. The real pace in this direction was achieved in India only after 1940 when Raychaudhuri and Mukherjee (1941) reported that Indian red and laterite soils fixed considerable amount of phosphates. In latter studies, Patel and Vishwanath (1946) reported that black soils of Madhya Pradesh, Bombay (now Mumbai), and Deccan had the high phosphate-fixation capacity.

2.4.9 Soil Classification and Mapping

The original investigations on Indian soils were carried out by Voelcker dates back to 1893 and by Leather to 1898. They classified the soils of the country into four major groups, namely the Indo-Gangetic alluvium, the black cotton soil or regur, red soil and laterite soil. Schokalskaya (1932) has published a soil map of India based on the Russian concept which describes 16 soil groups such as climate, vegetation, soil-forming materials, salinity, alkalinity, swamps and peats (Bhattacharyya et al. 2013). In 1935, Wadia and his co-workers collected a soil map of India with significance on geological formations and classified the soils as red, black (*regur*), laterite and lateritic soils of Peninsular India and also Indo-Gangetic Plains (IGP) which includes delta, desert, *bhabar*, *terai* and alkali soils. Vishwanath and Ukil (1943) published a soil map of India by placing the soils in different climatic zones for the first time (Fig. 2.7). This was prepared by Imperial Agricultural Research Institute, New Delhi, and published in the year 1943. There were 17 kinds of soils identified, viz. deep black soils, black clay soils, black loamy soils, black sandy soils, red sandy soils, brown sandy soils, red and yellow soils, dark reddish brown soils, mixed soils (white and yellow), alluvium (Indus, Gangetic, Brahmaputra), coarse alluvium, soils of swamp lands, calcareous soils, usar soils, kallar soils, coastal alluvium soils and soils of Great Himalayan Ranges.

2.5 Recent History—Post-independence Era

In India, the importance of soil science has been realized during pre-independence era, i.e. after establishing the then Imperial Council of Agricultural Research for "*grow more food campaign*" coupled with the establishment of Chemical Section during 1905 as Phipps laboratory at Pusa, Bihar. However, major developments were noticeable in the organization of research on soils which took place in the post-independence era, when soil science emerged as a distinct discipline in agricultural universities' curriculum. Research on soil fertility and soil testing, soil genesis, soil clay mineralogy, soil physics and soil chemistry started receiving attention in fifties. Soil survey works were started in 1954, in most of the states as a part of their agricultural development programme, and Bihar (Soil Survey and Land use Planning Scheme, Sabour) was pioneer. In 1954, the soil survey work was started under the leadership of Dr. P. P. Jha (Fig. 2.8) following the conventional procedures in Bihar (including Jharkhand) to delineate the soil associations based on geology, relief, physical and chemical characteristics of soils for the first time in India by the then Soil Survey Scheme at Sabour (Mishra et al. 2001; Mishra 2015, 2016). Meanwhile, the National Atlas Organization, Kolkata, prepared a soil map of India in 1957 classifying Indian soils into six major groups and 11 broad types. Murthy and Pandey (1983) applied the working principles of the USDA 7th Approximation in classifying the Indian soils with available data on soil classes and thus used the US system for the first time at great soil group level, which was transformed into a soil map of India on 1:6.3 million scale. Mishra et al. (1994) attempted to propose "Fluvisols" as the 13th Order in USDA Soil Taxonomy for active floodplain soils frequently occurring in different parts of India and elsewhere. Mishra (2015, 2016) proposed a framework of Indian system of soil classification mainly based on land use options.

The Indian work on different aspects of soil had been reviewed in 1971, and it was published by the Indian Society of Soil Science. Since then, a series of bulletins on special topics such as acid soils, soil potassium, clay minerals, soil phosphorus and soil nitrogen have also been published by the society. During 1982, Dr. N. S. Randhawa and his team have brought a comprehensive review of soil science during the 12th International Congress of Soil Science covering various recognized sub-disciplines of soil science Similarly, Indian Society of Soil Science has produced numerous publications on Soil Science, viz. Potassium in Soils, Crops and Fertilizers

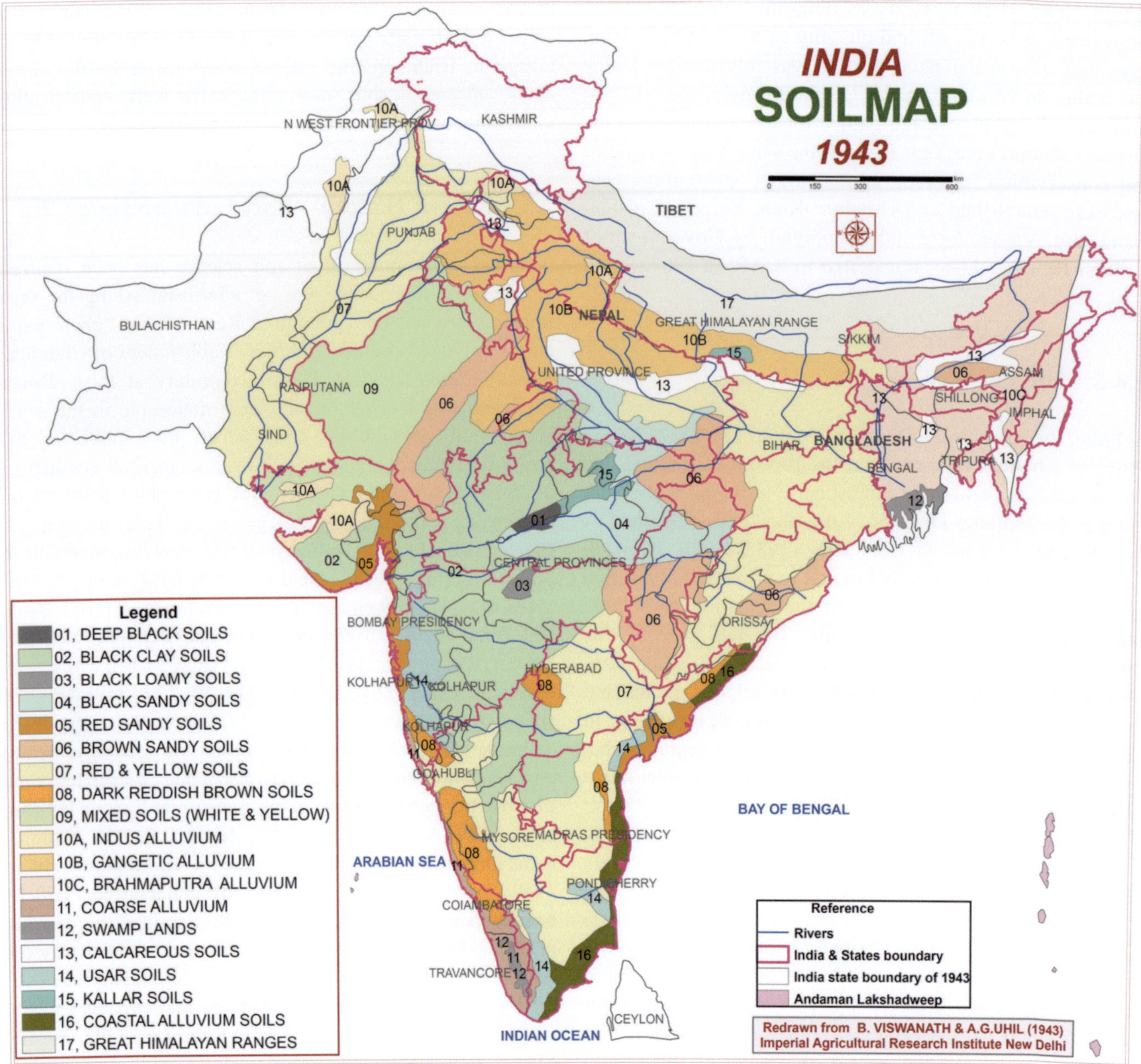

Fig. 2.7 Different types of soils of India prepared in the pre-independence period. *Source* Regional Centre, ICAR-NBSS&LUP, Bangalore, 2018

(ISSS 1975); Acid Soils of India: Their Genesis, Characteristics and Management (ISSS 1976); Phosphorus in Soils, Crops and Fertilizers (ISSS 1979); Nitrogen in Soils, Crops and Fertilizers (1984); Soil Science in India (ISSS 1984); Soil Related Constraints in Crop Production (ISSS 1991); Soil Management for Sustainable Agriculture in Dry land Areas (ISSS 1994); Soil Management in Relation to Land Degradation and Environment (ISSS 1996); Sustainable Soil Productivity under Rice-Wheat System (1997); Organic Matter and Organic Residue Management for Sustainable Productivity (1998), etc.

2.5.1 Establishment of Agricultural Research Institutes and Universities

The ICAR was established during pre-independence era (i.e. 1929). However, it has greatly expanded in post-independence (www.icar.org.in). The ICAR has played a pioneering role in ushering Green Revolution and changing the scenario of "*ship to mouth*" existence dramatically into food sufficiency and subsequent development of agriculture in India through its research and technology development and could help the country to increase the production of food grains by 5 times,

Fig. 2.8 Dr. P. P. Jha

horticultural crops by 9.5 times, fish by 12.5 times, milk 7.8 times and eggs 39 times since 1951–2014, thus making a visible impact on the national food and nutritional security. It has played a crucial role in promoting excellence in higher education and research in agriculture (www.icar.org.in).

The concept of "rural universities" conceived by the University Education Commission (1949) under the chairmanship of Dr. S. Radhakrishnan had ultimately led to the establishment of agricultural universities along the lines of Land Grant Universities in the USA. Subsequently, the Kothari Commission recommended as a policy on higher education that each state should have at least one Agricultural University. The concept of agricultural universities received a special stimulus from the reports of the joint Indo-US review committee of 1955 and 1960 on agricultural research and education. The G.B. Pant University of Agriculture and Technology at Pantnagar (Uttarakhand) was the first agricultural university set upon this pattern in India in 1960 (Fig. 2.9). It was followed by subsequent creation of more than 70 state agricultural universities and central agricultural universities besides different ICAR institutes across the country (www.icar.org.in).

2.5.2 Coordinated Research Projects on Soil Productivity and Fertilizer Use

Several All India Coordinated Research Projects related to nutrient management were initiated, which laid a strong foundation for soil fertility research.

(i) *Coordinated Project on Correlation of Soil Test with Crop Response Correlation (STCR)*

Dr. N. P. Datta at the Division of Soil Science and Agricultural Chemistry initiated research on soil-testing methods, soil test crop response correlation studies and the use of radio tracers in soil fertility research. He also initiated studies with ^{15}N and ^{32}P for the balance sheet of nitrogen and phosphorus in the soil–plant system. His work was followed by Dr. B. Ramamoorthy et al. (1967) who developed the concept of soil test-based fertilizer recommendations for targeted yields of crops. Taking a clue from Truog regarding the basis for fertilizer application for targeted yields, Ramamoorthy and co-workers in (1967) established the theoretical basis and experimental proof for the fact that Liebig's law of the minimum operates equally well for N, P and K. Among the various methods of fertilizer recommendation, yield targeting is a unique method which not only indicates soil test-based fertilizer dose but also the higher level of yield the farmer can hope to achieve if good agronomic practices are followed in raising the crop. The essential basic data required for formulating fertilizer recommendation for targeted yield are (i) nutrient requirement in kg q^{-1} of produce, grain or other economic produce (ii) the per cent contribution from the soil available nutrients (iii) the per cent contribution from the applied fertilizer nutrients. Simultaneously, the STCR project was initiated in 1967–68 with eight centres at IARI, New Delhi, by Dr. B. Ramamurthy. During 1970–71, five more centres were added. Presently, STCR project is working with seventeen cooperating centres and lead centre located at ICAR-IISS, Bhopal. Apart from developing fertilizer prescription equation, this AICRP has generated commendable work on soil fertility maps for major and micro-nutrients for 173 districts covering across every parts of the country.

(ii) *Fertilizer Recommendations and Geo-referenced Soil Fertility Maps*

Scientists at Indian Institute of Soil Science and AICRP on soil test crop response correlation project generated information on targeted yield equations for different crops on different soils, soil fertility index data of different states, cropping seasons and suitable targeted yield for different crops. Based on this information, on-line software and soil fertility maps of different states and fertilizer recommendation system for targeted yields of crops were developed. The soil fertility data on N, P, and K index values at district level for the states of Andhra Pradesh, Maharashtra, Chhattisgarh, West Bengal, Haryana, Orissa, HP, Karnataka, Punjab,

Fig. 2.9 GB Pant University of Agriculture and Technology, Pantnagar

Tamil Nadu and Bihar have been developed. From the attribute database, the different thematic layers have been reclassified to generate various thematic maps on N, P and K index values using ARC GIS platform (Anonymous 2005). The system works as a ready beckoner to give prescription in the form of fertilizer available (e.g. Urea, SSP, MOP, etc.). In case of known fertility status, one can give the known values for N, P and K and submit for calculation. This system provides real use of fertility maps to the users/farmers (SubbaRao et al. 2009). The calculated soil test values were incorporated into the fertility maps to prescribe nutrients, if soil test values are not available with farmers, for prescribing fertilizers for targeted yields of crops (http://www.iiss.nic.in; SubbaRao et al. 2009). This user-friendly system has the facility to input actual soil test values of the farmers' fields to obtain optimum dose. Besides this, soil test calibration was developed for calculating fertilizer doses after adjusting nutrient supplied through farmyard manure, green manure and bio-fertilizers at some centres. Critical limits of nutrients were determined in variety of soils to diagnose their sufficiency or deficiency in the soils and to predict the crop response to applied nutrients (Chaudhary et al. 2008).

(iii) *Old and New Long-Term Experiments*

From 1957 onwards, several British classical experiments were modified to evaluate the residual effects of the annually repeated dressings of different combinations of nutrients. The British long-term experiments were continued for more than 160 years continuously and yielded most valuable information, which were used to develop an efficient approach for managing the crops and cropping systems and also helpful in refining the strategy and policy to enhance productivity without adverse effect on the environment.

Research on the use of chemical fertilizer was started in 1945, and the first scheme sanctioned on the use of ammonium sulphate on rice lands in West Bengal. The other projects compared organic and inorganic manures, oilcakes and phosphates including the method of applying phosphatic fertilizers. Research on rapid soil tests was taken up at the Imperial Agricultural Research Institute which paved the way for the development of soil-testing activities in the country (Randhawa 1979). After achieving independence, the long-term experiments were started in Muzaffarnagar, Uttar Pradesh (1949); Chinsura and Suri, West Bengal (1948); Berhampur, West Bengal (1949); Cuttack, Orissa (1948); Anakapalle, Andhra Pradesh (1950) and Ranchi, Bihar (1956). However, these experiments were discontinued except ones located at Coimbatore and Ranchi. Permanent manurial experiment at Coimbatore completed 100 years of existence. During late sixties, high-yielding varieties (HYV) were introduced which later on proved to be the main pillar of Green Revolution. Intensification of agriculture under irrigated condition resulted in acceleration of nutrient mining from soil to harness the potential of the HYV for long term. To sustain the productivity, a need for interventions in soil fertility maintenance programme was felt (Singh and Wanjari 2013). In view of these emerging compulsions, the ICAR decided to launch the "*All India Coordinated Research Project on Long-Term Fertilizer Experiments (AICRP-LTFE)*" in September 1970. Dr. A. B. Ghosh was the first Principal Investigator of the project. Initially, the project experiments were started at 11 centres in irrigated and intensively cultivated areas representing different agro-climatic regions and subsequently extended at other centres across the country in different soil orders. The long-term meteorological data coupled with soil and crop data set offer good opportunity for measuring the influence of climatic aberrations on soil processes and crop performance (LTFE, Annual Report 2012–13).

(iv) *Micro-nutrients in Soils and Plants*

The significance of micro-nutrients in Indian agriculture was realized only in the late sixties. Dr. J. S. Kanwar together with Dr. N. S. Randhawa did pioneering research on micro-nutrients in plants and soils in Punjab and established a sound foundation for micro-nutrient research in India. His work developed into a coordinated micro-nutrient project with Dr. Randhawa as its first coordinator in 1967 with an aim to delineate the micro-nutrient deficient areas and to alleviate the nutrition stresses in crops. The ICAR initiated the All India Coordinated Scheme of Micro-nutrient in Soils and Plants in 1967 with its National Headquarter at the Punjab Agricultural University, Hisar. Subsequently, it was shifted to Punjab Agricultural University, Ludhiana, in 1970 and finally shifted to ICAR-IISS, Bhopal, during 1988/89 and six coordinating centre located at Lucknow, Hisar, Jabalpur, Ranchi, Anand and Coimbatore. Later Ludhiana, Delhi and Hyderabad centres were also developed. Realizing the need for strengthening micro-nutrient research, three more centres, viz. Akola for Maharashtra, Bhubaneswar for Orissa and Pantnagar for Uttar Pradesh were added in the year 1996. Two bulletins on micro-nutrient research on soils and plants in India are prepared by Kanwar and Randhawa (1967) and Katyal and Randhawa (1983). Takkar and co-workers (1989) have consolidated the scattered information on micro- and secondary-nutrient status in soils generated during twenty years (1967–1987) by various cooperating centres. In that report, various research aspects of micro-nutrients were covered, namely (i) diagnosis of micro- and secondary-nutrient deficiencies, (ii) delineation of micro- and secondary-nutrient deficient areas, (iii) responses of crops to micro- and secondary-nutrients, (iv) amelioration of micro- and secondary-nutrient deficient areas, (v) nutrient interactions and (vi) depletion of micro-nutrients from soils (Takkar et al. 1989). In addition to this, Singh and Behera (2011) have also consolidated various information of different states and centre such as critical concentration in plants and soils, diagnosis of micro-nutrient disorder in plant and soils, nutrient indexing for forecasting emerging micro-nutrient deficiencies in soils, soil fertility mapping through delineation of micro- and secondary-nutrients deficient areas, etc. (Singh and Behera 2011; Khamparia et al. 2010). They have also reported that analysis of 2.5 lakhs surface soil samples collected from different parts of the country revealed the predominance of zinc deficiency in divergent soils. Of these samples, 49, 12, 4, 3, 33 and 41% soil are tested to be deficient in available zinc (Zn), iron (Fe), manganese (Mn), copper (Cu), boron (B) and sulphur (S), respectively. The magnitude of zinc deficiency varied widely among soil types and within the various states (Singh and Behera 2011). In recent days, boron and molybdenum deficiencies are also being observed in some crops (Shukla and Behera 2011; Behera et al. 2011). Similarly, state wise recommendation of micro- and secondary-nutrients has been consolidated by Shukla et al. (2012).

(v) *Coordinated Agronomic Research*

The concept of coordinated and multidisciplinary agronomic research was conceived by Dr. A. B. Stewart (1947) of the Macaulay Soil Research Institute, Aberdeen. He had been invited by the Government of India in 1946 as a consultant from British just before Independence. He proposed simple fertilizer trials on cultivators' fields, since research on soil fertility at research farm did not match with the farmers' fields. He also advised model agronomic experiments in order to develop a package of practices especially on fertilizer use for different agro-climatic and soil conditions.

2.5.3 Soil Fertility and Soil Health Improvement

During 1950s, much emphasis has been given to the establishment of permanent manure plot experiments (Abrol and Nambiar 1997). Parker's (1951) method of calculating nutrient index (NI) values was used to indicate fertility status of soils for the purpose of mapping. Even a minimum of 500 soil samples were taken for working out nutrient index for a given area since the total number of soil analyses data was rather small at the country level. Ramamurthy and Bajaj (1969) prepared a soil fertility maps showing available nitrogen, phosphorus and potassium status of Indian soils. These maps were based on 1.3 million soil samples analysed by the end of 1967 by 31 soil-testing laboratories.

(i) *Fertilizer Manufacture and Fertilizer Research*

Prior to 1960/61, India manufactured only straight nitrogenous fertilizers [ammonium sulphate (AS), urea, calcium ammonium nitrate (CAN), ammonium chloride and single superphosphate (SSP)]. The production of NP complex fertilizers commenced in 1960/61 (Tables 2.1). Currently, India produces a large number of grades of NP/NPK complex fertilizers. These include 16–20–20, 20–20–0, 28–28–0, 15–15–15, 17–17–17, 19–19–19, 10–26–26,

Table 2.1 Chronology of fertilizer production in India

Year of manufacture	Fertilizer product	Total number of units
1906	SSP	65
1933	AS	10
1959	Ammonium sulphate nitrate	No longer manufactured
1959	Urea	29
1959	Ammonium chloride	1
1960	Ammonium phosphate	3
1961	CAN	3
1965	Nitro phosphate	3
1967	DAP	11
1968	TSP	No longer manufactured
1968	Urea ammonium phosphate	2
1968	NPK complex fertilizers	6
1970	Diammonium phosphate	Not available
1980	DAP and urea	
1990	DAP, urea and muriate of potash (MOP)	Not available
2016	Neem-coated urea, DAP and MOP	Not available

Source Adapted from http://www.fao.org/docrep/009/a0257e/a0257e03.htm

12–32–16, 14–28–14, 14–35–14 and 19–19–19. In addition, India produces various grades of simple and granulated mixtures (Table 2.1).

The domestic production of N and P_2O_5 was 29,000 and 10,000 tonnes, respectively, in 1951/52. By 1973/74, this had increased to 1.05 million tonnes N and 0.325 million tonnes P_2O_5. As a result of the oil crisis in the mid-1970s and the sharp increase in the international prices of fertilizers, consequently the Government of India encouraged to invest in domestic fertilizer production plants in order to reduce dependence on imports.

(ii) *Fertilizer Association of India*: The Fertilizer Association of India (FAI) came into existence in 1955. Its aims were to promote the most productive use of balanced fertilizers for increasing production and productivity and ensuring the availability of adequate domestic supplies of fertilizers by assisting industry and increasing operating efficiency. The FAI maintained close contact with various departments and agencies of the Government of India, State Governments, research bodies, etc. The FAI today plays a significant role in promoting Indian agriculture through balanced and efficient use of fertilizer in soils.

(iii) *Nitrogen Research*: In 1928, Report of Royal Commission on Agriculture in India submitted to Parliament clearly indicated the poor fertility of the Indian soils. The report recognized the existence of deficiencies of nitrogen (N), phosphorus (P_2O_5) and potash (K_2O). Nitrogen is invariably deficient in Indian soils with around 99% of soils responding to N application. In this regard, Government of India undertook number of steps for improving the fertility of the Indian soils as a key to the sustainable agricultural development. Production of ammonium sulphate as a by-product of steel industry in 1933 by Tata Iron and Steel Company Limited, Jamshedpur, marked the beginning of manufacture of nitrogenous fertilisers in the country. Production of urea was first commenced in 1959 by FCI, Sindri, revolutionized the fertiliser N production with number of urea manufacturing units currently being thirty. Diammonium phosphate (DAP) production started in 1967 by Gujarat State Fertilizers and Chemicals Limited was a landmark in itself as N and P carrier. Calcium ammonium nitrate (CAN), nitrophosphate and subsequently NPK complex fertilisers offer an array of N carriers available to the farmers. Over the years, urea became the main source for N application. From 1950 to 51, fertiliser N use grew from a measly amount of about 0.45 kg ha^{-1} to nearly 87 kg ha^{-1} in 2014–15. Country became self-sufficient in food grains in 1990–91 and has marched on to have over-flowing warehouses during the beginning of twenty-first century. Bulletin on Nitrogen in Soils, Crops and Fertilizers (1984) was published by the Indian Society of Soil Science. Similarly, Indian Institute of Soil Science (IISS), Bhopal, has also made valuable contribution in N management in major soil types in India (SubbaRao et al. 2009).

(iv) *Phosphorus Research*: Phosphorus (P) is one of the major plant nutrients of limited availability in Indian soils. Kanwar and Grewal (1960) found that Punjab soils fixed appreciable quantities of phosphates. Mukherjee (1943) and Chatterjee and Datta (1951) studied phosphate fixation in clay minerals. De (1963) briefly reviewed the India's contribution to the study of phosphate fixation by soils, clay minerals, hydrous oxides and lime. Kanwar and Grewal (1974) reviewed on P status, fixation in Indian soil, forms of P in Indian soils and factors affecting P availability. Similarly, Tandon (1987) published a book on Phosphorus Research and Agricultural Production in India. Phosphorus fertility status in India based on 9.6 million soil tests, 49.3% of districts and union territories are low in available P, 48.8% are medium, and 1.9% are high (Hasan 1996). In comparison with an earlier compilation by Ghosh and Hassan (1979), this present survey indicates the low P fertility class has increased by 3.0% while medium and high categories have decreased by 2.7 and 0.3%, respectively. Both surveys highlight the need for P fertilizer application for proper crop growth in nearly 98% of India's districts. Tandon (1987) consolidated all the information available on phosphorus in his publication on "Phosphorus research and agricultural production in India". After that many ICAR institutes have taken up research projects in addressing phosphorus (P), improving P use efficiency, using low-grade indigenous rock phosphate (RP), mobilization of P from RPs, some best agronomic practices to effectively and economically utilize RPs and P recovery from solid to liquid wastes (Subbarao et al. 2015). Recently, ICAR launched a project on DAC (Division of INM) Sponsored Project on "GIS & GPS Based Soil Fertility Maps for Precise Fertilizer Recommendations for the Farmers of the Country" during 2010–12. Similarly, Indian Institute of Soil Science (IISS), Bhopal, has also made valuable contribution in P management in soils (SubbaRao et al. 2009).

(v) *Potassium Research*: According to Sekhon's article (www.ipipotash.org), the most notable change in the landscape of potassium (K) research in India has been a very substantial expansion of work during the last five decades. Pioneering work of field trials by potash scheme and subsequently by Indian Potash Limited supplemented the efforts of the ICAR and state agricultural research system and established the need for potassium in large areas for many crops. Stewart's report was the foundation of work since mid-fifties in simple fertilizer trials in different districts of Indian states and a chain of multi-factorial agronomic experiments at various research stations throughout the country. The establishment of Potash Research Institute of India in 1977 was a major initiative by the Indian Potash Limited (IPL). Bulletin on Potassium (K) in soils, crops and fertilisers published by Indian Society of Soil Science (1975) has brought consolidated information on various facets of potassium such as status of K in Indian soils and its role in crop production, plantation and cash crops and soil test for K recommendation. Ghosh and Hasan (1975) reported that 20, 42 and 38% of the districts were low, medium and high in K status, respectively, based on 4.5 million samples analysed during 1968–74 in the soil-testing laboratories all over countries.

The earlier estimates of soil fertility for K based on data generated from soil-testing laboratories in the country (Ghosh 1976; Sekhon et al. 1992; SubbaRao et al. 1993; Subbarao et al.1996; Datta 1996; Swarup and SrinivasaRao 1999) indicated discrepancies in the percentage of samples testing high, though the overall soil K fertility of soils declined. Due to larger contribution of non-exchangeable K to plant K needs, lack of crop responses to applied K has been reported even in soils with low exchangeable K. The major sources of non-exchangeable K in soils are K rich 2:1 clay minerals such as micas and vermiculite (Mengel and Uhlenbecker 1993; Srinivasarao et al. 2006). Good research work done on potassium in India led to building up some information on mineralogy of soil potassium, forms of potassium and their content in Indian soils (SubbaRao et al. 2009). Potassium dynamics in mica-rich soils (both biotite and muscovite) of Jharkhand and Bihar were earlier studied by Mishra and Ghosh (1995), Mishra et al. (1992, 1995a, b, c) and Upadhyay and Mishra (1993, 1994, 1997). Similarly, Datta (2011) studied potassium release and fixation patterns in Indian soils. Basak and Biswas (2009) studied the dynamics of K release from waste mica inoculated with potassium solubilizing microorganism (*Bacillus mucilaginosus*) and also evaluated its effectiveness as potassic-fertilizer using sudan grass (*Sorghum vulgare* Pers.) var *Sudanensis* as test crop grown under two Alfisols. At the Nuclear Research Laboratory, IARI, New Delhi, Mishra et al. (1992) studied the self diffusion of ^{86}Rb in four surface soils derived from mica-rich parent materials at field capacity and reported that the values ranged from 0.54×10^{-10} to 1.674×10^{-10} $cm^2\ s^{-1}$ at zero level of K and from 1.788×10^{-10} to 7.02×10^{-10} $cm^2\ s^{-1}$ at 50 ppm level K. These values were found much lower than the values reported elsewhere earlier. However, enrichment of these soils with 50 ppm carrier K resulted in a tremendous increase in the values for self diffusion (^{86}Rb).

2.5.4 Soil and Water Conservation

India was among the first few countries to have taken timely cognizance of the enormity of problems of soil erosion due to rainwater. Since the establishment of Soil Conservation Research, Demonstration and Training Centres during the 1950s in different problem areas of the country and their reorganization into the present set-up of Indian Institute of Soil and Water Conservation (erstwhile known as Central Soil and Water Conservation Research and Training Institute (CSWCRTI)) in 1974, the R&D activities of the institute and its centres have continuously focused on evolving strategies/technologies for soil and water conservation and watershed management (Vision 2050 2015a). Out of an estimated 175 million hectares of degraded and cultivated lands (Narayana 1993), nearly all of them are subject to serious erosion hazards of water and wind, shifting cultivation, water logging, salinity and alkalinity and changing water courses. Initially, the research work was mainly concentrated with constructional aspects of check dams, terraces, etc., and these required only minor modifications from the available knowledge (Narayana 1993). Central Soil and Water Conservation Research and Training Institute (CSWCRTI) pioneered in operationalizing the watershed concept through four Operational Research Projects at Sukhomajri (Haryana), Nada (Chandigarh), Fakot (Tehri-Garhwal in Uttarakhand) and G. R. Halli (Chitradurga, Karnataka). In addition to these research centres, a Desert Afforestation Research Station was established at Jodhpur in 1952, which was later redesignated in 1957 as Desert Afforestation and Soil Conservation Station, and now it is Central Arid Zone Research Institute (CAZRI), Jodhpur. After realizing tremendous tangible and intangible benefits from these watersheds, the ICAR developed 47 model watersheds in sixteen states in collaboration with State Agricultural Universities and State Departments during 1983. Encouraged with the success of the model watersheds, the Ministry of Agriculture conceived of a massive National Watershed Development Programme for Rainfed Areas (NWDPRA) for conservation and sustainable development in 29 states during 1990–91. Narayana and Babu (1983) estimated soil erosion rate in India based on first approximation; soil erosion is taking place at the rate of 16.35 ton/ha/annum which is more than the permissible value of 4.5–11.2 ton/ha. About 29% of the total eroded soil is lost permanently to the sea. Ten per cent of it deposited in reservoirs. The remaining 61% is dislocated from one place to the other. An iso-erodent map of India, based on distribution of erosion index values as developed during 1970s (Babu et al. 1978) shows potential erosivity of rainfall. Also, an erosion map was prepared in the 1970s and later modified (Ahmad 1973; Das 1977). Government of India has also taken up a massive programme for conservation of the soil in the country and proposed to spend Rs 6400 million during sixth Five Year Plan (1980–85). Moreover, during 12th International Congress on Soil Science, it was felt that there was a great need to formulate a World Soil Policy to combat desertification, to reduce erosion hazards to ensure rational utilization of the world's land and water resources (Kanwar 1982). In India, an estimated area of 187 Mha has been categorized as degraded lands by the National Bureau of Soil Survey and Land Use Planning (NBSS&LUP) in 1994 and revised it to 147 Mha in 2004, which includes water erosion (94 Mha), acidification (16 Mha), flooding (14 mha), wind erosion (9 mha), salinity (6 Mha), combination of factors (7 mha). However, the latest estimate made by the National Wasteland Development Board and NBSS&LUP was around 123 and 120 Mha area, respectively, under degraded lands (Maji et al. 2010).

2.5.5 Basic and Strategic Soil Research

The establishment of Indian Institute of Soil Science on 16 April 1988 was a follow-up of the recommendation by Working Group of the ICAR and Task Force Committee. The committee also realized the need for doing basic and strategic research per se relating to soil and nutrient management. The institute is only national centre in the country which is carrying out the research in all aspects of soil science comprising soil physics, soil chemistry and fertility, environmental soil science and soil biology (SubbaRao et al. 2009, 2014; Manna et al. 2014; Rao and Patra 2009). Down the lane, IISS has completed its silver jubilee (1988–2013) and has leapfrogged in soil research in areas such as nutrient management, fertility improvement, management of soil physical and biological components, soil quality for sustainable productivity and minimizing environmental pollution. This institute has strong research focus on four major flagship programmes such as (i) Soil health management and input use efficiency for food and environmental security, (ii) Soil microbial diversity and its role in nutrient use efficiency, climate mitigation and Bio-remediation for sustainable agriculture, (iii) Conservation Agriculture and Carbon sequestration and (iv) Soil pollution and remediation. A phospho-compost/N-enriched phospho-compost technology was developed using phosphate solubilizing microorganisms, namely *Aspergillus awamori*, *Pseudomonas straita* and *Bacillus megaterium*, phosphate rock, pyrite and bio-solids to increase the manurial value compared to ordinary FYM and compost. In addition to this, rapid composting techniques have been developed by the Institute (Singh 2004; SubbaRao et al. 2009; Manna et al. 2015). The IISS has also been attempting to use advanced modern research like modified N fertilizer using pine oleoresin-coated urea for slow release of nitrogen, nano-technology for enhancing nutrient use efficiency (e.g. nano-rock phosphate, zinc fertilizers, etc., urea granules coated with nano-particles of Zn, Fe, Cu and Si using oleoresin), use of biochar for carbon sequestration and conservation agriculture research for enhancing soil health and crop productivity in black soils.

2.5.6 Improvement of Soil Physical Conditions

It is estimated that out of the 328 mha of the total geographical area in India, 173.65 mha are degraded, producing less than 20% of its potential capacity (GOI 1990) and out of this, 89.52 mha suffers from one or the other form of physical constraints, viz. shallow depth, soil hardening, slow and high permeability, sub-surface compacted layer, surface crusting, soil cracking and temporary water logging, etc. (Anonymous 1982; Painuli and Yadav 1998). These soil physical constraints can adversely affect the crop production. A bulletin on soil-related constraints in crop production was published by the Indian Society of Soil Science (Biswas et al. 1991). An another bulletin on "Soil management for sustainable agriculture in Dryland areas" (Biswas et al 1994) addresses the important issues, strategies and management of soils of dryland areas. Similarly, a bulletin on "Soil management in relation to land degradation and environment" edited by Biswas and Narayansamy (1996a, b) vividly covered land degradation, water and wind erosion, management for soil physical constraints and its ameliorative measures.

2.5.7 Amelioration of Problem Soils

(i) *Salt-Affected Soils*

The Govt. of India constituted an Indo-American team to assist the Indian Council of Agricultural Research to develop a comprehensive water management programme for the country. As a follow-up of these recommendations, it was decided to establish the Central Soil Salinity Research Institute in Fourth Plan period. The Institute started functioning at Hisar (Haryana) on 1 March 1969. Later on, it was decided to shift this Institute to Karnal during October 1969. In February 1970, the Central Rice Research Station, Canning Town, West Bengal, was transferred to CSSRI, Karnal, to conduct research on problems of coastal salinity. Another Regional Research Station for carrying out research on problems of inland salinity prevailing in the black soil region of western parts of the country started functioning at Anand (Gujarat) from February 1989. Keeping in view the need of undertaking research for situations under surface drainage congestion, high water table conditions, relatively heavy textured soils and indurate pan for managing alkali soils of Central and Eastern Gangetic Plains, another Regional Station was established during October 1999 at Lucknow. The coordinating unit of AICRP on management of salt-affected soils and use of saline water in agriculture is located at the Institute with a network of eight research centres located in different agro-ecological regions of the country (Agra, Bapatla, Bikaner, Gangawati, Hisar, Indore, Kanpur and Tiruchirappalli). The coordinating unit of AICRP on water management functioned at the institute from early seventies till it was shifted to Rahuri (Maharashtra) in 1990. A pilot scheme for reclamation of 1000 ha of land was undertaken in 1954 with the cooperation of the Government of Netherlands. The area is located 32 km north to the city of Bhavnagar, Gujarat. The observations on the salt content of soil indicated that June–October (rainy season) is the period of desalinization and October–May the period of resalination.

A book on "Saline and alkali soils of India" by Agarwal et al. (1968) highlighted the up-to-date information of problems of salinity and alkalinity in soils and their effect on crop production. Various drainage and leaching techniques were followed as amelioration techniques during late 1960s such as physical and hydro-technical amelioration, biological (bulky organic manures) amelioration and chemical amelioration (Agarwal et al. 1968). During 1970s, soil amendments were extensively used to reclaim the problematic soils (Singh et al. 1978). Application of gypsum technology for sodic land reclamation has drawn attention of farmers and policy planners (Abrol et al. 1978; Sharma and Chaudhuri 2012). Besides this, pyrite, calcium chloride, press mud, spent wash, acids, acid formers, fly ash and organic materials were extensively evaluated as soil amendments for reclamation of alkali soils (Chhabra and Abrol 1977; Abrol et al. 1988; Mishra 1977). Sharma and Chaudhuri (2012) reviewed on reclamation of salt-affected soils of India. The tentative estimate indicates that area under salt-affected soils in India may increase from 6.74 to 16.2 million ha by 2050. The problems of water quality especially polluted waters are also likely to increase exponentially due to increasing urbanization and industrialization. Management of these degraded land and water resources is likely to pose formidable problems because many new pollutants are emerging (Vision 2050 2015a).

(ii) *Acid Soil*

Out of 49 million hectares of acid soils in India, 26 million hectares have a pH below 5.6 and 23 million hectares between 5.6 and 6.5. Most of these soils are classified under the soil order Oxisols, Ultisols, Spodosols, Alfisols, Mollisols, Inceptisols and Entisols and are concentrated in the Eastern India. Among various north-eastern states, Nagaland having the highest area under acid soils (100% to total geographical area) followed by Meghalaya (99.9%), Tripura (99.5%), Manipur (98.1%) and Arunachal Pradesh (81.1%). Similarly, Kerala and Goa are having the area under acid soils of 94.7 and 83.3%, respectively, compared to other states of India.

Application of lime to neutralize the exchangeable Al has been found effective for soils of Sikkim (Pradhan and Khera 1976) and Meghalaya (Prasad et al. 1982). Acid soils of Sikkim responded well with lime application (Pradhan and Khera 1976). In the soils of Khasi hills of Meghalaya, Woodruff's buffer method gave the highest correlation coefficient ($r = 0.93^{**}$). For the soils of East Khasi hills, addition of lime equivalent to 25% lime requirement is good enough to neutralize exchangeable aluminium. Lime requirement by Woodruff's buffer (1948) method ranges between 1.35 and 4.95 t/ha for the soils of Tripura (Lasker and Dadhwal 1981). In Bihar (present Jharkhand), Mandal et al. (1955) obtained significant effect of lime on wheat and maize but not in paddy even though soil pH was 5.2. Encouraged by these results, Mandal et al. (1966) conducted extensive field and laboratory studies on varieties of crops which yielded excellent results of practical use of lime in acid soils. Another study reported the application of only 10% of lime on the basis of lime requirement every year like fertilizer in furrows recorded higher crop yield of soybean, ground-nut, etc than the application of full dose of lime. In acid soils of Orissa, 25% of the lime requirement is good enough when lime application is done during alternative years. It has been shown that 25% lime requirement just neutralizes the exchangeable Al of Alfisols, Inceptisols and Entisols studied, and this amount of lime is equal to 2–3 times higher than the exchangeable Al present in the soil (Misra et al. 1989).

2.5.8 Crop Rotations and Green Manuring

The practice of green manuring is as old as art of manuring crops. Literature evidence on the value of green manures can be found as early as 500–400 BC in the writings of *Varahamihira* in India (Kadke 1965). In India, leguminous green manure crops had been known as an important source of N for wetland rice much before the advent of modern agricultural technology (Singh et al. 1991). Their use in crop production is recorded as early as 1134 BC in China (Palaniappan 1992; Tandon 1992). Similarly, value of green manure was recognized by farmers in India for thousands of years, as mentioned in treatises like *Vrikshayurveda*. The type of green manures, their nutritive value and extent of residues of principal crops in India have been worked out by Palaniappan et al. (1990). Green manure crops can be leguminous as well as non-leguminous and can be grown *in situ* or brought from outside as cuttings of trees and shrubs. The latter practice is called green leaf manuring (Singh et al. 1991). Some studies have shown that up to 100 kg N/ha or more can be accumulated by *Sesbania aculeate*, a green manure crop, in a 7–8 week fallow period available before transplanting of the wet season rice crop (Beri and Meelu 1981; Bhardwaj and Dev 1985; Meelu et al. 1994). Kalidurai and Kannaiyan (1989) reported substantial yield increases of rice after incorporation of *Sesbania rostrata*. Similarly, Wani et al. (1994) found that adopting green manures and organic amendments in crop rotations provided a quantifiable increase in soil organic matter quality and other soil quality parameters as compared to continuous monoculture of cereal systems. Tandon (1992) has consolidated the available information relevant to fertilizers, organic manures, green manuring, biological and industrial waste and bio-fertilizers.

2.5.9 Soil Classification and Mapping

Subsequent to the recognition of soil survey as a National Priority in 1947, a need was felt for creating a centralized information warehouse to assimilate, verify and disseminate information on the nature, extent and distribution of soils in the country. Consequently, the government launched All-India Soil Survey Scheme in 1956, which expanded in 1959 as the All India Soil and Land Use Survey Organization (AIS&LUS). Integrating the effects of climate, vegetation and topography, 16 major and 108 minor soil regions were identified and brought under 27 units by Raychaudhari et al. (1963). Later, a revised soil map of India was generated with 23 major soil groups under FAO/UNESCO's scheme on World Soil Map project, and subsequently this map was refined with 25 broad soil classes represented on a 1:7 million scale map (Bhattacharyya et al. 2013). In 1969, the AIS&LUS was divided into two wings, one being under the Indian Council of Agricultural Research (ICAR). It was later reconstituted as a Directorate through a Presidential Notification in 1973. The Directorate was accorded the status of a Bureau in 1976 and was renamed as National Bureau of Soil Survey and Land Use Planning (NBSS&LUP) with its Hqrs. at Nagpur, Maharashtra. Subsequently, the soil resources have been identified and classified through soil survey by various agencies such as the department of agriculture, soil conservation, forestry and irrigation, the All India Land Use and Soil Survey Organization, Government of India, Department of Agriculture and the National Bureau of Soil Survey and Land Use Planning of the ICAR. Dr. S. P. Raychaudhuri, Chief Soil Survey Officer carried out pioneering work in the collection and collation of published and unpublished information on soils and their classification. It was later published in the "Final Report of the All-India Soil Survey Scheme". More than 2000 soil series were identified (Raychaudhari et al. 1963). The National Bureau also brought out a book on "*Benchmark Soils of India*" to coincide with the 12th Congress of the International Society of Soil Science in 1982. Since then, National bureau (NBSS&LUP) used to bring out state, district and taluk level maps with the scale of 1:250,000, 1:50,000 and 1:10,000/1:5000, respectively, for efficient land use planning.

2.5.10 Environmental/Contaminated Soil Research

Pollution in soil is associated with geogenic pollution, indiscriminate use of fertilizers/pesticides/herbicides, excess of salts and dumping of large quantities of solid waste, etc. (SubbaRao and Panwar 2010). Le Riche (1968) first reported metal contamination of soil resulting from the long-term use of sewage sludge. The use of such materials without ascertaining their composition could be hazardous. However, studies have been largely directed to understand the waste characteristics of urban solid wastes in relation to its suitability as a manure (Gupta et al. 1984; Mutatkar 1985). Apart from other organization, the ICAR-IISS has made commendable work on recycling of distillery effluents in agriculture, developing database on soil and water pollution in India (Panwar et al. 2010), bio-remediation/phytoremediation of contaminated soils, evaluated sink capacity of soils for pollutants, developed compost production and quality of standards as well as wastewater assessment and recycling (SubbaRao and Reddy 2010; SubbaRao et al. 2014, Vision 2050 2015b). Sewage sludge from the textile industry contains high concentrations of elements such as boron, lead, cadmium, copper and chromium that have a potential cause of environmental degradation (Maiti et al. 1992). Elevated levels of potentially toxic elements have been reported in soils treated with urban solid wastes (Jeevan Rao and Shantaram 1999).

2.5.11 Soil Testing, Health Enhancement and Monitoring

The soil-testing programme was started in India during the year 1955–56 with the setting up of 16 soil-testing laboratories (STL) under the Indo-US Operational Agreement for "Determination of Soil Fertility and Fertilizer Use". In 1965, five of the existing laboratories were strengthened, and nine new laboratories were established with a view to serve the Intensive Agricultural District Programme (IADP) in selected districts. To meet the increasing requirement of soil-testing facilities, the STL has increased progressively to 177 by 1971, 330 by 1980 and 533 by 2000 indicating an annual growth rate of 6.94% over a period of thirty years (SubbaRao and Srivastava 2005). Continuing efforts have been made to modernize, strengthen and intensify such mission for restoring the soil health as well as fertility (Vision 2050 2015b).

2.6 Conclusions

History of soil research is a part of research approaches already made in past. It tells us how and why research efforts moved and happened in the past. Consequently, it helps to prevent from repeating past mistakes and cautions against fears in order to understand the present. Soil science research in India has taken rapid strides in terms of development as a single major subject and able to accommodate and foster other allied subjects under its umbrella. For achieving sustained food production and plant–animal–human nutrition as well as energy security, improving environment, enhancing water availability and quality and mitigating the adverse impact of the climate change, it is the "SOIL" which holds the key. It is the soil research in general and soil scientist in particular, who has to take pragmatic approach to safeguard the soil resource for posterity. Moreover, soil scientists can no longer work in isolation, and a holistic multidisciplinary approach is essential to handle our soil resources with care for *clean-green-healthy soil* as well as *environment*.

References

Abrol IP, Nambiar KKM (1997) Fertility management of Indian soils-A historical perspective. Adv Geoecology 29:293–310

Abrol IP, Nambiar KKM (2008) Fertility management of Indian soils—a historical perspective. Asian Agri-Hist 12(1):3–18

Abrol IP, Saha AK, Acharya CL (1978) Effect of exchangeable sodium on some soil physical properties. J Indian Soc Soil Sci 26:98–105

Abrol IP, Yadav JSP, Massoud FI (1988) Salt affected soils and their management. FAO Soils Bull 39:131

Agarwal RR, Yadav JSP, Gupta RN (1968) Saline and alkali soils of India. Indian Council of Agricultural Research, New Delhi, pp 1–127

Ahmad S (1973) Soil erosion in India. Asia Publ. House, Bombay, India

Anonymous (1982) 20 Point programme. Field Manual on Dry land Agriculture, published by ICAR and Department of Agriculture and Cooperation, Ministry of Agriculture, New Delhi, pp 1–87

Anonymous (2005) Invited papers of the national seminar on soil testing, Indian Institute of Soil Science, Bhopal, Madhya Pradesh, India, 21–22 Jan 2005, pp 1–64

Babu R, Tejwani KG, Agarwal MC, Bhushan LS (1978) Distribution of erosion index and iso-erodent map of India. J Soil Water Conserv 6 (1):1–14

Basak BB, Biswas DR (2009) Influence of potassium solubilizing microorganism (*Bacillus mucilaginosus*) and waste mica on potassium uptake dynamics by sudan grass (*Sorghum vulgare* Pers.) grown under two Alfisols. Plant Soil 317:235–255. https://doi.org/10.1007/s11104-008-9805-z

Behera SK, Lakaria BL, Singh MV, Somasundaram J (2011) Molybdenum in soils, crops and fertilizers. Indian J Fertil 7(5):52–57

Beri V, Meelu OP (1981) Substitution of nitrogen through green manure in rice. Indian Farming 31(5):6–7

Bhardwaj KKR, Dev SP (1985) Production and decomposition of *Sesbania cannabina* (Retz.) Pers. in relation to its effects on yields of wetland rice. Trop Agric Trin 62:233–236

Bhattacharyya T, Pal DK, Mandal C, Chandran P, Ray SK, Sarkar D, Velmourougane K, Srivastava A, Sidhu GS, Singh RS, Sahoo AK, Dutta D, Nair KM, Srivastava R, Tiwary P, Nagar AP, Nimkhedka SS (2013) Soils of India: historical perspective, classification and recent advances. Curr Sci 104(10):1308–1323

Biswas TD, Narayanasamy G (eds) (1996a) Soil-related constraints in crop production. Bulletin No 15. Indian Council of Agricultural Research, New Delhi, pp 1–176

Biswas TD, Narayanasamy G (eds) (1996b) Soil Management in relation to land degradation and environment. Bulletin No 17. Indian Council of Agricultural Research, New Delhi, pp 1–167

Biswas TD, Narayanasamy G, Yadav JSP, Dev G, Katyal JC, Sidhu PS (eds) (1991) Soil management for sustainable agriculture in dryland areas. Bulletin No 16. Indian Council of Agricultural Research, New Delhi, pp 1–143

Biswas TD, Narayanasamy G, Yadav JSP, Dev G, Katyal JC, Sidhu PS (1994) Soil management for sustainable agriculture in dryland areas, published Indian society of soil science, New Delhi, pp 1-143

Borthakur A, Singh P (2013) History of agricultural research in India. Curr Sci 105(5):582–592

Botterro I, Elena C, Vercontter J (1965) The nearest: the early civilization. Weidinfield and Nicolson, London, p 782

Brevik EC, Hartemink AE (2010) Early soil knowledge and the birth and development of soil science. Catena 83(1):23–33

Cavalli-Sforza LL, Piazza M (1994) The history and geography of human genes

Chandy KT (2002) Agricultural research and education in India. Booklet No. 435. Agricultural Extension Education, New Delhi

Chatterjee B, Datta S (1951) Phosphate fixation by clay minerals, montmorillonite and kaonilite. J Indian Soc Soil Sci 2(2):224–233

Chhabra R, Abrol IP (1977) Reclaiming effect of rice grown in a sodic soil. Soil Sci 124:49–55

Chaudhary RS, Damodar Reddy D, Singh AB, Ramana S, Lakaria BL (2008) IISS—a profile. Indian Institute of Soil Science, Nabibagh, Bhopal, Madhya Pradesh, India, pp 1–42

Das DC (1977) Soil conservation practices and erosion control in India—a case study in soil conservation and management in developing countries. Soil Bulletin No. 33. Food and Agricultural Organization, United Nations, Rome, Italy, pp 11–50

Datta SC (1996) Modification of a kinetic model interpreting threshold concept of potassium release and fixation. J Indian Soc Soil Sci 44:644–648

Datta SC (2011) Potassium dynamics and status in Indian soils. Karnataka J Agric Sci 24(1):7–11

De SK (1963) India's contribution to the study of phosphate fixation by soils, clay minerals, hydrous oxides and lime. J Indian Soc Soil Sci 11:195–201

Dhar NR (1934) Molasses and press-mud in alkali land reclamation. J Indian Chem Soc 2:105–111

Gangopadhyay R (1932) Some materials for the study of agriculture and agriculturists in ancient India. N.C. Mukherjee & Co., Serampore

Ghosh AB (1976) Non-conventional sources of fertilizer potassium. Bull Indian Soc Soil Sci 10:317–323

Ghosh AB, Hasan R (1975) Available potassium status in Indian soils. Bull Indian Soc Soil Sci 10:164–169

Ghosh AB, Hassan R (1979) Phosphorus fertility status of soils of India. Bulletin No. 12. ISSS, pp 1–8

Government of India (GOI) (1990) Indian agriculture in brief, 23rd edn. Ministry of Agriculture & Cooperation, New Delhi, p 160

Gupta MD, Chatterjee N, Gupta SK (1984) Utilization of Calcutta City wastes as manures. In: Seminar on soil resources and productivity management, Indian Agricultural Research Institute, New Delhi

Harrison WH, Das S (1921) The retention of soluble phosphates in calcareous and non-calcareous soils. Imperial Department of Agriculture in India

Hartemink AE (2015) 90 years IUSS and global soil science. Soil Sci Plant Nutr 61:579–586

Hasan R (1996) Phosphorus status of soils in India. Better Crops Int 10 (2):4–5

Howard A (1940) An agricultural testament. London, p 262

Howard A, Wad Y (1931) The waste products of agriculture: their utilization as humus. Humphrey Milford & Oxford University Press, Oxford. Retrieved 9 Aug 2010

http://www.fao.org/docrep/009/a0257e/a0257e03.htm

https://en.wikipedia.org/wiki/Laterite

ISSS (1975) Potassium in soils, crops and fertilizers. Indian Society of Soil Science, New Delhi

ISSS (1976) Acid soils of India: their genesis, characteristics and management. Indian Society of Soil Science, New Delhi

ISSS (1979) Phosphorus in soils, crops and fertilizers. Indian Society of Soil Science, New Delhi

ISSS (1984) Nitrogen in soils, crops and fertilizers. Indian Society of Soil Science, New Delhi

ISSS (1991) Soil related constraints in crop production. Indian Society of Soil Science, New Delhi

ISSS (1994) Soil management for sustainable agriculture in dryland areas. Indian Society of Soil Science, New Delhi

ISSS (1996) Soil management in relation to land degradation and environment. Indian Society of Soil Science, New Delhi

ISSS (1997) Sustainable soil productivity under rice-wheat system. Indian Society of Soil Science, New Delhi

ISSS (1998) Soil organic matter and organic residue management for sustainable productivity. Indian Society of Soil Science, New Delhi

Iyer SS, Pande RK (1954) A statistical study of the long term NPK experiment at Shahjahanpur. In: Proceedings of 2nd biennial conference of sugarcane research and development workers, Part II, pp 479–486

Jameson DR (1852) On the physical aspect of Punjab: its agriculture and botony. J Agric Hortic Soc 8:132–138

Jeevan Rao K, Shantaram MV (1999) Potentially toxic elements in soils treated with urban solid wastes. Indian J Environ Health 40(4):364–368

Jenny H (1961) Reflections on the acidity merry-go-round. Soil Sci Soc Am J 25:428–432

Jenny H, Raychaudhuri SP (1960) Effect of climate and cultivation on nitrogen and organic matter reserves in Indian soils. Indian Council of Agricultural Research, New Delhi, pp 1–127

Kadke JR (1965) Hastening decomposition of incorporated green manure. Indian J Agron 10:443–446

Kalamkar RJ, Singh S (1935) A statistical examination of yield of wheat at the Kanpur Agricultural College farm. Indian J Agric Sci 5:346

Kalidurai M, Kannaiyan S (1989) Effect of *Sesbania rostrata* on N uptake and yield of upland rice. J Agron Crop Sci 163:284–288

Kanitkar NV, Surur SS, Gokhale DN (1960) Dryfarming research in India. ICAR, New Delhi

Kanwar JS (1982) Managing soil resources to meet the challenges to mankind: presidential address. In: 12th international congress on soil science, New Delhi, India, 8–16 Feb, pp 1–32

Kanwar JS, Bhumbla DR (1957) Reclamation of alkaline and saline soils of the Punjab. Report of Indian Council of Agricultural Research Scheme

Kanwar JS, Grewal JS (1960) Phosphate fixation in Punjab soils. J Indian Soc Soil Sci 8:211–218

Kanwar JS, Grewal JS (1974) A review—phosphorus fixation in Indian soils. ICAR technical bulletin (Agric) No. 29. Indian Council of Agricultural Research, New Delhi, pp 1–48

Kanwar JS, Randhawa NS (1967) Micronutrient research in soils and plants in India: a review. Indian Council of Agricultural Research, New Delhi, India

Katyal JC, Randhawa NS (1983) Micronutrients. FAO fertilizer and plant nutrition bulletin 7. Rome, p 80

Kenoyer JM (1995) Interaction systems, specialized crafts and culture change: the Indus valley tradition and the Indo-Gangetic tradition in south Asia. In: The indo-aryans of ancient South Asia: language, material culture and ethnicity, pp 213–257

Khamparia RS, Singh MV, Sharma BL, Kulhare PS, Sharma GD (2010) Four decades (1970–2010) of research on micro- and secondary- nutrients and pollutant elements in crops and soils of Madhya Pradesh. ICAR-Indian Institute of Soil Science, Bhopal, Madhya Pradesh, India, pp 1–113

Krishna KR (2002) Soil fertility and crop production. Oxford & IBH Pub. Co., New Delhi

Krupenikov IA (1992) History of soil science, from its inception to the present. Amerind Publishing Co., New Delhi, India, p 352

Lal R (2007a) Evolution of the plow over 10,000 years and the rationale for no-till farming. Soil Till Res 93:1–12

Lal R (2007b) Ancient history of agriculture and the plow. IUSS Comm Hist Philos Sociol Soil Sci Newsl 14:27–30

Lasker S, Dadhwal KS (1981) Annual report. ICAR Research Complex, Shillong (Tripura Centre)

Le Riche NH (1968) Metal contamination of soil in the Woburn Market Garden experiment resulting from the application of sewage sludge. J Agric Sci 71:205–208

Leather JW (1900) Land records and agriculture, North-West Provinces, and Oudh. Bull no. 9

Maiti PS, Sah KD, Gupta SK, Banerjee SK (1992) Evaluation of sewage sludge as source of irrigation and manure. J Indian Soc Soil Sci 40:168–172

Maji AK, Reddy GO, Sarkar D (2010) Degraded and wastelands of India: status and spatial distribution. Directorate of Information and Publications of Agriculture, Indian Council of Agricultural Research, New Delhi, pp 1–167

Majumdar GP (1935) Upavana vinoda. Indian popular science series No. 1. Indian Research Institute Publication

Mandal SC, Das SK, Mukherjee HN (1955) Lime requirement of an acid sandy soil. J Indian Soc Soil Sci 3:71–75

Mandal SC, Sinha H, Prasad CR, Ali MA (1966) Studies on liming of acid red loam soils. J Indian Soc Soil Sci 14:127–131

Manna MC, Sahu A, Singh AB, SubbaRao A, Khanna SS (2014) Quality of compost production from solid urban waste for enhancing crop productivity. Research bulletin No. 02/IISS/2014. ICAR-Indian Institute of Soil Science, Bhopal, pp 1–52

Manna MC, Sahu A, Patra AK, Khanna SS, Chaudhari SK, Sikka AK (2015) Rapid composting technique-ways to enhance soil organic carbon, productivity and soil health. ICAR-Indian Institute of Soil Science, Bhopal, pp 1–8

Meelu OP, Singh Y, Bijay-Singh (1994) Green manuring for soil productivity improvement. World roil resources reports 76. Food and Agriculture Organization of the United Nation, pp 1–123

Mengel K, Uhlenbecker K (1993) Determination of available interlayer potassium and its uptake by ryegrass. Soil Sci Soc Am J 57:761–766

Mishra BB (1977) Possibility of pyrite to use in soil in monsoon to reclaim its alkalinity. Food Farm Agric 9:136–137

Mishra BB (2015) Soil classification: issues and opportunities for Indian system. The 26th Dr. S.P. Raychaudhuri memorial lecture, 19th October, 2015, ICAR-National Bureau Soil Survey & Land Use Planning, Nagpur. J Indian Soc Soil Sci 63(Supplement):S41–S52

Mishra BB (2016) Indian system of soil classification: a way forward. Agric Res Technol Open Access J 3(2):555–606. https://doi.org/10.19080/ARTOAJ.2016.03.555606

Mishra BB, Ghosh SK (1995) Characterization of soils derived from mica rich parent materials in two toposequences. J Indian Soc Soil Sci 43:92–98

Mishra BB, Ghosh SK, Sachdev P, Sachdev MS, Deb DL, Arora RP (1992) Self diffusion coefficient of ^{86}Rb in mica-rich soils. Curr Sci 63:681–682

Mishra BB, Choudhary CD, Singh NK, Mall J, Singh VN (1994) Fluvisol as a proposed order in soil taxonomy for active flood plain soils. In: Transactions of XVth world congress of soil science, 6B, Acapulco, Mexico, pp 55–56

Mishra BB, Ghosh SK, Ghosh G (1995a) Studies on potassium in soils derived from mica-rich parent materials of Bihar: I. Forms, distribution and fixation of potassium in soils. J Potassium Res 11:249–256

Mishra BB, Ghosh SK, Ghosh G (1995b) Studies on potassium in soils derived from mica-rich parent materials of Bihar: II. Quantity—intensity relationship of potassium. J Potassium Res 11:257–261

Mishra BB, Ghosh SK, Ghosh G (1995c) Studies on potassium in soils derived from mica-rich parent materials of Bihar: III. Potassium release from mica species using chemical extractants. J Potassium Res 11:262–270

Mishra BB, Mall J, Choudhary J, Singh RA (2001) Nutrient mining in different agro-climatic zones of Bihar. Fertil News 46(11):21–43

Misra UK, Satapathy S, Panda N (1989) Characterization of some acid soils of Orissa: I—nature of soil acidity. J Indian Soc Soil Sci 37 (1):22–28

Moreland WH, Tamhane UA (1901) Reh. Agric Ledger 13:416–463

Mukherjee MK (1943) The role of sesquioxide in the fixation of phosphate. Bull Indian Soc Soil Sci 5:25–26

Murthy RS, Pandey S (1983) Soil map of India (1:6.3 m). National Bureau of Soil Survey & Land Use Planning, ICAR, Nagpur

Mutatkar VK (1985) Prospects of organo-inorganic fertilizers. Fertil News 12:60–67

Narayana DV (1993) Soil and water conservation research in India. Indian Council of Agricultural Research, ICAR, KAB, Pusa, New Delhi, pp 1–454

Narayana DV, Babu R (1983) Estimation of soil erosion in India. J Irrig Drain Eng 109(4):419–434

Painuli DK, Yadav RP (1998) Tillage requirements of Indian soils. In: 50 years of natural resource management research, pp 245–262

Palaniappan SP (1992) Green manuring: nutrient potentials and management. In: Tandon HLS (ed) Fertilizers, organic manures, recyclable wastes and biofertilizers. Fertilizer Development and Consultation Organization, New Delhi, pp 52–71

Palaniappan SP, Siddeswaran K, Srinivasalu Reddy D (1990) Evaluation of green manures for the sustainable rice farming systems. In: International symposium on natural resource management for sustainable agriculture, New Delhi, p 220

Panda N (1987) Acid soils of Eastern India—their chemistry and management. Directorate of Agriculture and Food Production, Orissa, Bhubaneswar, pp 1–23

Panwar NR, Saha JK, Adikari T, Kundu S, Biswas AK, Rathore A, Ramana S, Srivastava S, SubbaRao A (2010). Soil and water pollution India: some case studies. IISS Technical Bulletin, Indian Institute of Soil Science, Bhopal, Madhya Pradesh, pp 1–40

Parker FW, Nelson WL, Winter E, Miller IE (1951) The broad interpretation of soil test informations. Agron J 43:105–112

Patel DK, Vishwanath D (1946) Comparative studies on Indian soils VI. Phosphate fixation capacities of soils. Indian J Agric Sci 16:428–434

Patwardhan GK, Kadrekar SB, Arakeri HR (1956) Role of compost in sugarcane culture. In: Proceedings of 9th Congress International Society of Sugar Cane Technologists, India, 1, pp 150–156

Pradhan HR, Khera MS (1976) Lime needs of soils of Sikkim based upon exchangeable aluminum test. Bull Indian Soc Soil Sci 11:291–299

Prasad RN, Dhawal RS, Munna Ram (1982) J Res Assam Agric Univ 313:131

Ramamoorthy B, Narsimhan RL, Dinesh RS (1967) Fertilizer application for specific yield targets of sonora-64. Indian Farm 25(5):43

Ramamurthy B, Bajaj JC (1969) Soil fertility map of India. Indian Agricultural Research Institute, New Delhi

Randhawa NS (1979) A history of the Indian Council of Agricultural Research (1929–1979). Indian Council of Agricultural Research, New Delhi, India

Rao DLN, Patra AK (2009) Soil microbial diversity and sustainable agriculture. J Indian Soc Soil Sci 57(4):513–530

Raychaudhari SP, Aggarwal RR, Datta Biswas NR, Gupta SP, Thomas PR (1963) Soils of India. Indian Council of Agricultural Research, New Delhi, India

Raychaudhuri SP, Mukherjee MK (1941) Studies on Indian red soils II. Fixation of phosphates. Indian J Agric Sci 11:205–209

Raychaudhuri SP, Kaw RK, Raghavan D (1964) Agriculture in ancient India

Raychaudhury SP (1984) Development of soil productivity in ancient and medieval India. Soil Science in India, Indian Society of Soil Science, New Delhi, India, pp 1–13

Renfrew (1996) Language families and the spread of farming. In: Harris DR (ed) The origins and spread of agriculture and pastoralism in Eurasia, pp 70–92

Sanyal SK (ed) (1997) 62nd annual convention & national seminar on developments in soil science, Oct 18–21 1997. Univerity College of Agriculture, Calcutta Unviersity, West Bengal, India, pp 1–73

Schokalskaya ZJ (1932) Contribution to the knowledge of Soils Asia. 2

Sekhon GS, Brar MS, SubbaRao A (1992) Potassium in some benchmark soils of India. Potash Research Institute of India, Gurgaon, Haryana, India, pp 1–82

Sharma DK, Chaudhuri SK (2012) Agronomic research in salt affected soils of India: an overview. Indian J Agron 37(3rd IAC special issue):175–185

Shukla AK, Behera SK (2011) Zinc management in Indian agriculture: past, present and future. Indian J Fertil 7(10):14–33

Shukla AK, Behera SK, SubbaRao A, Singh AK (2012) State-wise micro and secondary nutrients recommendation for different crops and cropping systems. Research bulletin No. 1/2012. In: All India coordinated research project of micro and secondary nutrients and pollutant elements in soils and plants. IISS, Bhopal, pp 1–77

Singh MV, Behera SK (2011) All India coordinated research project of micro- and secondary-nutrients and pollutant elements in soils and plants—a profile. Research bulletin No. 10. ICAR-Indian Institute of Soil Science, Bhopal, pp 1–57

Singh M (ed) (2004) Proceedings of the national seminar on standards and technology for rural/urban compost held at Indian Society of Soil Science, Bhopal during 17 Jan 2004

Singh NT, Hothi BS, Chhibha IM, Thind SS (1978) Indian Farm 28:7

Singh Y, Khind CS, Singh B (1991) Efficient management of leguminous green manures in wetland rice. Adv Agron 45:135–189

Singh G, Babu R, Narain P, Bhushan LS, Abrol IP (1992) Soil erosion rates in India. J Soil Water Conserv 47(1):97–99

Singh MV, Wanjari RH, Adhikari T (2004) Nutrient dynamics, crop productivity and sustainability under long-term fertilizer experiments in India. ICAR-Indian Institute of Soil Science, Bhopal, Madhya Pradesh, India, pp 1–119

Singh M, Wanjari RH (2013) Annual report—2012–13. All India coordinated research project on longterm fertilizer experiments to study changes in soil quality, crop productivity and sustainability. AICRP—LTFE, Indian Institute of Soil Science (ICAR), Nabibagh, Bhopal, pp 1–142

Srinivasarao C, Rupa TR, SubbaRao A, Singh SP, Bansal SK (2006) Kinetics of nonexchangeable potassium release from important benchmark soils of India: effects of mineralogy, soil depth and extraction media. Commun Soil Sci Plant Anal 36(3, 4):1–19

Stewart AB (1947) Report on soil fertility investigations in India with special reference to manuring. Army Press, New Delhi, India

SubbaRao A, Panwar N (2010) Soil pollution: Indian scenario. In: Kundu et al (eds) Souvenir 75th annual convention of Indian Society of Soil Science, Bhopal during 14–17 Nov 2010, pp 47–66

SubbaRao A, Reddy KS (2010) Indian Institute of Soil Science at a glance. In: Kundu et al (eds) Souvenir 75th annual convention of Indian Society of Soil Science, Bhopal during 14–17 Nov 2010, pp 1–7

SubbaRao A, Srivastava S (2005) Retrospect and prospect of soil testing service in India. In: Anonymous (ed) Invited papers of the national seminar on soil testing, Indian Institute of Soil Science, Bhopal, Madhya Pradesh, India, 21–22 Jan 2005, pp 1–64

SubbaRao A, SeshaSai MVR, Pal SK (1993) Non-exchanageable potassium reserves and their categorization in some soils of India. J Indian Soc Soil Sci 41:667–673

SubbaRao A, Biswas AK, Sammi Reddy K, Hati KM, Ramana S (eds) (2009) IISS: two decades of soil research. Indian Institute of Soil Science, Nabibagh, Bhopal, Madhya Pradesh, pp 1–132

SubbaRao A, Singh AB, Wanjari RH, Ramesh K, VassandaCoumar M, Shinogi KC (2014) Glimpses of IISS contribution in technology generation and dissemination. Indian Institute of Soil Science, Bhopal, pp 1–124

SubbaRao A, Srivastava S, Ganeshamurty AN (2015) Phosphorus supply may dictate food security prospects in India. Curr Sci 108 (7):1253–1261

Swarup A, SrinivasaRao C (1999) Current status of crop responses to nutrients. Experiences from all India coordinated research project on long-term fertilizer experiments. Fertil News 44(4):27–43

Subba Rao A, Srinivasarao Ch, Srivastava S (1996) Potassium status and crop response to potassium on the soils of agroecological regions of India. IPI Research Topics No. 20, International Potash Institute, Horgen, Switzerland, pp 1–185

Takkar PN, Chhibba IM, Mehta SK (1989) Twenty years of co-ordinated research on micronutrients in soils and plants (1967–87), IISS bulletin I. ICAR-Indian Institute of Soil Science, Bhopal, Madhya Pradesh, India, pp 1–314

Talati RR (1947) Field experiments on the reclamation of salt-affected lands in Baramati of Bombay, Deccan. Indian J Agric Sci 17:153–174

Tandon HLS (1987) Phosphorus research and agricultural production in India. Fertilizer Development and Consultation Organization, New Delhi, India, pp 1–160

Tandon HLS (1992) Fertilizers, organic manures, recyclable wastes and biofertilizers. Fertilizer Development and Consultation Organization, New Delhi, pp 1–148

Thurston E (1913) The Madras Presidency, with Mysore, Coorg and the associated states, provincial geographies of India. Cambridge University Press. Retrieved 6 Apr 2010

Tiwari KN, Sharma DN, Tripathi SK (1989) Salt affected soils of Uttar Pradesh, their reclamation and management. CS Azad University of Agriculture and Technology, Kanpur, India, p 34

Upadhyay R, Mishra BB (1993) Evaluation of waste mica as a source of potassic fertilizer and changes in the forms of soil potassium due to mica incorporation. J Potassium Res 9:325–332

Upadhyay R, Mishra BB (1994) Release of potassium from mica as affected by different factors. J Potassium Res 10:49–55

Upadhyay R, Mishra BB (1997) Potassium release from mica by some chemical and biological treatments. J Potassium Res 13:101–104

Vishwanath B, Ukil AC (1943) Soil map of India. Indian Agricultural Research Institute, New Delhi

Vision 2050 (2015a) Indian Institute of Soil and Water Conservation (Indian Council of Agricultural Research). Director, Dehradun Uttarakhand

Vision 2050 (2015b) Central Soil Salinity Research Institute, (Indian Council of Agricultural Research). Director, Karnal, Haryana, pp 1–48

Wani SP, McGill WB, Haugen-Kozyra KL, Robertson JA, Thurston JJ (1994) Improved soil quality and barley yields with fababeans, manure, forages and crop rotation on a gray luvisol. Can J Soil Sci 74:75–84

Woodruff CM (1948) Testing soils for lime requirement by means of a buffered solution and the glass electrode. Soil Sci 66(1):53–64

Yadav JSP, Agarwal RR (1959) Dynamics of soil changes in reclamation of saline alkali soils of Indo-Gangetic alluvium. J Indian Soc Soil Sci 7:317–322

Climate

3

Kamaljit Ray, S. D. Attri, H. Pathak, Anjani Kumar, and Dibyendu Chaterjee

Abstract

Climate is one of the five soil-forming factors and has a significant influence on the properties of soil. India has a great geographic and geomorphologic diversity with a number of different climate regimes. We sow the seeds in soil, but grow the plants on land that also constitutes the climate. Agriculture depends on the mean climate of a particular region. Indian agriculture is primarily rain-fed, with a majority of farmers dependent on monsoon for irrigating their crops. Drought, floods, variation in onset dates, wet and dry spells further pose detrimental effects on agriculture. The chapter highlights the effects of climate on agriculture and summarizes the climate in various seasons in different parts of the country. This also covers the trends of various weather parameters over the years. The trends indicate that annual mean surface air temperatures have shown significant warming of about 0.63 °C/100 year during the past 115 years (1901–2015). Frequency of depressions and cyclonic storms formed over the north Indian Ocean (1951–2015) during the monsoon and post-monsoon seasons have shown decreasing trends. Through the effects of precipitation and temperature, climate affects the rates of biological, chemical and physical processes involved in soil formation. The development of soil profile and its physical and chemical composition are greatly affected by weather and also have a role in crop performance. The chapter also covers the impacts of climate change on soil quality, nutrient dynamics, organic matter turnover and soil erosion. Possible adaptation and mitigation approaches are highlighted. Opportunities for carbon sequestration as a tool to mitigate climate change have been pinpointed.

Keywords

Climate • Monsoon • Climate change • Soil qualities • Carbon sequestration

3.1 Introduction

The balance between the soils and the climate influences the nature and distribution of the ecosystems as well as growing medium. As climate changes, so too will the soil's function to support ecosystems. One of the main reasons for poor productivity in agriculture is uncertain weather and its associated natural disasters. Though the average annual rainfall of India is considerably high, the crop productivity is not satisfactory in several agroclimatic zones. The main reason for this is the uneven distribution of rainfall and extreme climatic events. Food security in India is severely challenged by natural disasters such as floods, cyclones and drought. India is prone to flood in about 49.8 million hectares which accounts for 12.3% of the geographical area. Out of the total geographical area of India, about one-sixth area with 12% of the population is found to be susceptible to drought and 8% of the total area of India is cyclone prone (Kumar et al. 2014). Climate and soil are the integral components of land and any change in climate may intend to alter the land capability followed by its suitability for the preferred land-use options (Mishra 2016). Soil may be managed for improvement, but climate can be looked into for adaptation within the existing changing situations. As such, information about climate of an area is vital for soil and land management for sustainable production. The most important change in soils expected as a result of climate change would be a gradual improvement in fertility and physical conditions of soils in humid and sub-humid climate, change from one major soil-forming process to another in

K. Ray (✉)
Ministry of Earth Sciences, Lodhi Road, New Delhi 110003, India
e-mail: raykamaljit@gmail.com

S. D. Attri
Agro Advisory Services Division (AASD), Meteorological Department India, Lodhi Road, New Delhi 110003, India

H. Pathak · A. Kumar · D. Chaterjee
ICAR-National Rice Research Institute, Cuttack, Odisha, India

B. B. Mishra (ed.), *The Soils of India*, World Soils Book Series,
https://doi.org/10.1007/978-3-030-31082-0_3

certain fragile tropical soil and changes in soil property due to poleward retreat of the permafrost boundary. Mishra (2016) reviewed the issues of global climate change and opined for integrated mitigation options primarily based on soil and land-use planning approaches in differing soil and agroclimatic zones of India.

The climate at any place is governed by well known atmospheric circulation patterns of pressure and winds over the globe. These pressure and wind belts are shifting north and southwards as per the revolution of the earth around the sun and thus govern the climate of that place. India lies in the tropical region, bounded on the north by subtropical high region (above 30° N) and equator in the south. India's climate is strongly influenced by the Himalayas and the *Thar* Desert. The Himalayas act as a barrier to the cold katabatic winds flowing down from Central Asia, keeping the bulk of the Indian subcontinent warmer than most locations at similar latitudes. As such, land areas in the north of the country have a continental climate with severe summer conditions and with very cold winters, when the temperatures plunge to freezing point.

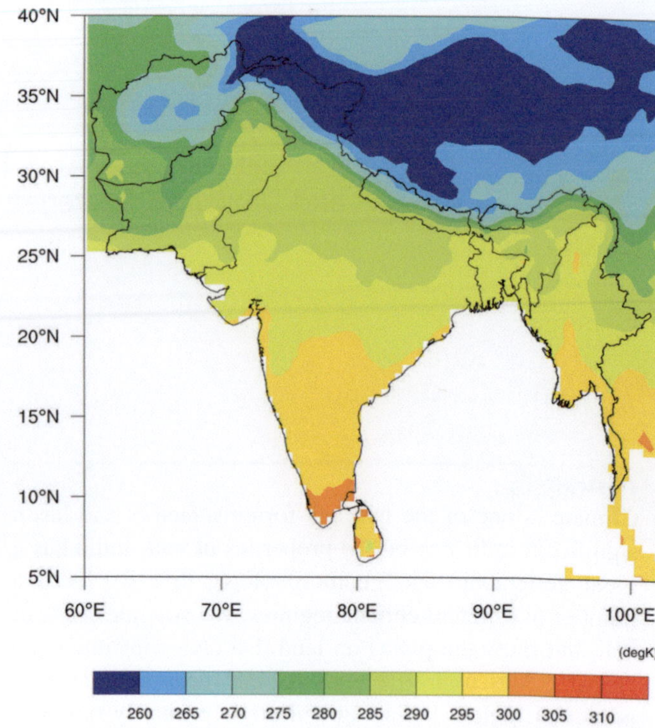

Fig. 3.1 Mean surface temperature in January (1981–2010). *Source* Regional Climate Centre (RCC-RA-II Region), India Meteorological Department (IMD)

3.2 Major Tropical Systems During Different Seasons

The climate of India is controlled by some major tropical systems which are prevalent during the various seasons. The seasons are (a) Southwest monsoons (June–September), (b) Northeast Monsoons/Post-Monsoon season (October–December), (c) Winter season (January–March), (d) Premonsoon season (March–May). Details of weather along with associated systems during different seasons are presented as under.

3.2.1 Winter Season/Cold Weather Season (January and February)

India Meteorological Department (IMD) has categorized the months of January and February in winter season; however, December can be included in this season for north-western parts of the country. This season starts in early December associated with clear skies, fine weather, light northerly winds, low humidity and temperatures, and large diurnal variations of temperature. The cold air mass extending from the Siberian region has profound influence on the Indian subcontinent (at least all of the north and most of central India) during these months. The mean surface temperatures vary from −10 to 32 °C during January (Fig. 3.1). Kargil or Dras (in Jammu & Kashmir) experiences biting cold because the night temperatures drop to −40 °C, while Thiruvananthapuram experience temperatures around 27 °C. The rains during this season generally occur over the western Himalayas, the extreme north-eastern parts, Tamil Nadu and Kerala (Attri and Tyagi 2010). Western disturbances and associated trough in westerlies are main rain-bearing system in northern and eastern parts of the country. These disturbances also cause precipitation over the plains and adjoining mountain ranges of north India. During winter, at about 8 km above sea level, a westerly jet stream blows at very high speed over the subtropical zone. This jet stream is bifurcated by the Himalayan ranges. The northern branch of this jet stream blows along the northern edge of this barrier. The southern branch blows eastwards south of the Himalayan ranges along 25° N latitude. This branch of jet stream exercises a significant influence on the winter weather conditions over India. This jet stream is responsible for bringing western disturbances from the Mediterranean region into Indian subcontinent. Winter rain and hail storms in north-western plains and occasional heavy snowfall in hilly regions are caused by these disturbances. The passage of the westerly trough associated with these western disturbances is generally followed by cold waves in whole of northern plains. These cold waves are detrimental to crops as they cause stress and frost due to sudden fall in temperature. The low temperatures as a result of cold waves may impede the intake of nutrients and also the soil moisture intake by plants may stop when temperatures fall below 1 °C. There are threshold/cardinal Maximum and optimum temperatures for winter season crops, below which the crop growth and hence yield is affected. Heavy snowfall followed by strong winds, breaks off the leaves, twigs and branches of the plants and can reduce the growth for a number of years. The glazing or coating of leaves and stems causes a lot of damage to the plants by way of suffocation, accumulation of toxic material and oxygen deficiency. Ice, in contact with the crowns and roots, also inhibits carbon dioxide diffusion.

3.2.2 Pre-monsoon Season/Summer Season/Hot Weather Season (March, April and May)

The region of maximum solar radiation covers most of north-western India, extending southwards to central India in pre-monsoon season. The region experiences the hottest temperatures during this season with an average of 35–45 °C, with occasional highs of about 50 °C (Pant and Rupa Kumar 1997). As a result of high temperatures over the north Indian plains, an intense thermal low-pressure area develops over north-western India during this season. In addition to the high temperature, the general wind flow is westerly to north-westerly. The westerly continental airflow brings very little moisture throughout the continental region of north India during pre-monsoon season. Therefore, rainfall episodes over most parts of northern India occur in association with synoptic scale westerly troughs that bring moisture to this region (Srinivasan et al. 1973). During May, the north-western plains experience high temperature around 45 °C when areas of Rajasthan desert record day temperatures around 55 °C, while the temperatures around Gulmarg or Pahalgam in Kashmir are hardly around 20 °C. However, weather remains mild in coastal areas of the country owing to the influence of land and sea breezes. Weather over land areas is influenced by thunderstorms associated with rain and sometimes with hail in this season. During the pre-monsoon season of April and May, entire India gets severe thunderstorms. In this season, Gangetic West Bengal and surrounding areas get severe thunderstorms called Nor'westers, which is locally called as '*Kal-Baisakhi*'. The northwest India gets convective dust-storms called locally as '*Andhi*'. Severe thunderstorms create lot of damage to property and crops and human and animal fatalities through the strong surface wind squalls, large hail and occasional tornadoes accompanying them. Damage by hail differs across regions and with the stage of the growing plants. Philip and Daniel (1976) observed that hail storms occur with an average frequency of 10 hailstorms per year over extreme northeast Assam, six over west Uttar Pradesh, Agartala and portions of east Bihar and about two to three over east Madhya Pradesh and west and north Bihar. April and May were identified as the period with highest Hailstorm activity over Northwest and northern parts of the country. Over central and north-western parts, March was observed to be the month of hailstorm activity. A hailstorm can decimate foliage, break branches and shatter the fruit. Cereal crops suffer more damage due to lodging. Lesions are caused on the plants owing to mechanical rupture of the mesophyll cells.

In 1994–95, 0.46 mha cropped area, 1.2 mha in 1997–98 and 2.9 mha in 1998–99 was damaged in Haryana, Punjab, Himachal Pradesh, Rajasthan, Uttar Pradesh, Maharashtra and Andhra Pradesh due to hail storm activity. During 5–8 January 2002, many parts of Karnataka state were affected by hailstorm and the estimated loss suffered by the farming community was around Rs. 27.5 crores. In erstwhile Andhra Pradesh, hailstorm caused huge damage to 77,000 ha of agricultural fields in 2005–06. The state of Madhya Pradesh was badly hit during 2–14 March 2006 by heavy hailstorm causing widespread damage to standing winter crops. During February 2007, most parts of Rajasthan were badly hit by hailstorm. In March 2007, heavy rains accompanied by hailstorm damaged wheat, sugarcane and oilseed crops in thousands of hectares in Punjab and Haryana (Bhardwaj et al. 2007). The SAARC STORM-2015 carried out in all the SAARC countries including India gave some interesting statistics. It was observed that the highest hailstorm events were recorded over Northwest India (50) followed by Southern Peninsula (19). They both accounted for about 70% of the total hailstorm events over the country. Jammu & Kashmir recorded maximum number of hailstorms (23) followed by Karnataka (17) and Himachal Pradesh recorded 12 hailstorms during the Pre-monsoon period in 2015 (Ray et al. 2015).

The season is also characterzed by cyclonic storms, which are intense low-pressure systems spread over hundreds to thousands of kilometres associated with surface winds more than 33 knots over the Indian seas viz. Bay of Bengal and the Arabian Sea. These systems generally move towards a north-westerly direction and some of them recurve to northerly or northeasterly path. Storms forming over the Bay of Bengal are more frequent than the ones originating over the Arabian Sea. On an average, frequency of these storms is about 2.3 per year. The frequent occurrence over major areas of incidence of cyclones and floods, significantly limits the agricultural productivity in India which in turn affects the socio-economic prospects of farmers and food security in these areas. The cyclone and its associated flood incidence cause severe damage to agriculture in several ways. The crops get affected both in terms of establishment and productivity. The stagnation of water inside the crop fields result in crop damage. The pulses and oil seeds and vegetable crops are highly susceptible to flood. In case of paddy, the duration of flood will decide the extent of damage to crop. The average annual flood damage was found to be about 3.57 mha of cultivated area based on the survey of

47 years between 1953 and 1999 (http://www.mapsofindia.com/top-ten/geography/india-flood.html). Cyclone associated flood water also brings sand and silt along with it, to the crop fields and thereby resulting in change in soil physical condition. The topsoil was found to be severely altered for a longer period due to deposition of clay which was witnessed in case of flash flood damage in Uttarakhand in 2013. In case of Cyclones the soils of the affected land area became saline due to intrusion of seawater.

3.2.3 Southwest Monsoon/Summer Monsoon (June, July, August and September)

The summers and winter monsoons of South Asia constitute most spectacular manifestations of regional anomalies in the general circulations of the atmosphere resulting from land–sea contrasts and orographical features. The SW monsoon is the most significant feature of the Indian climate. Indian agriculture is mostly rain-fed, with majority of farmers dependent on monsoon for irrigating their crops. Over-reliance on the monsoon poses uncertainties for food production and rural incomes as droughts, floods and variation in rainfall pattern greatly affects the agriculture sector (MOEF Govt of India 2015). Approximately 30% of the cultivated area in Uttarakhand was severely affected resulting in crop loss and several water conservation and harvesting structures were damaged due to flood event in June 2013. Media reported enormous damage to the agriculture sector, resulting in crop loss of Rs. 948 crore in 2.38 lakh hectares. The horticulture sector was also the worst hit with the loss estimated at Rs. 1339.2 crore (87,984 ha).

The Indian Ocean is landlocked to the north by the Asian continents. The geographical features give rise to the extreme thermal contrast between land and the sea in summers and in winters which is crucial factor in the development of the most pronounced monsoon circulations over this part of globe. The summer monsoon circulations get established toward the end of May and continue till the end of September. Consequent to the intense heat of the summer months, the northern Indian landmass becomes hot and draws moist winds over the oceans causing a reversal of the winds over the region which is called the summer or the southwest monsoon. This is most important feature controlling the Indian climate because about 75% of the annual rainfall is received during a short span of four months (June–September). While large-scale features of monsoons are repetitive from year to year, anomalies from one year to another year and the associated variations of rainfall in space and time can be appreciable, leading to floods and droughts over the large areas. The season is spread over four months, but the actual period at a particular place depends on onset and withdrawal dates. It varies from less than 75 days over West Rajasthan to more than 120 days over the south-western regions of the country contributing to about 75% of the annual rainfall. The onset of the SW monsoon normally starts over the Kerala coast, the southern tip of the country by 1 June, advances along the Konkan coast in early June and covers the whole country by middle of July. However, onset occurs about a week earlier over islands in the Bay of Bengal. Normal rainfall during Southwest monsoon is 890 mm for the country as a whole and 611 mm in NW India, 994 mm in central India, 723 mm in southern India and 1427 mm in NE India. Out of the 890 mm of the normal rainfall in the country, June month accounts for 160 mm followed by 293 mm in July, 262 mm in August and 175 mm in September. The mean Southwest monsoon (June, July, August and September) rainfall contributes 74.2% of the annual rainfall. Variability in the onset, withdrawal and quantum of rainfall during the monsoon season has profound impacts on water resources, power generation, agriculture, economics and ecosystems in the country. The monsoon is a special phenomenon exhibiting regularity in onset and distribution within the country, but inter-annual and intra-annual variations are observed. The monsoon is influenced by global and local phenomenon like El-Nino, northern hemispheric temperatures, sea surface temperatures, snow cover, etc. The monsoonal rainfall oscillates between active spells associated with widespread rains over most parts of the country and breaks with little rainfall activity over the plains and heavy rains across the foothills of the Himalayas. Heavy rainfall in the mountainous catchments under 'break' conditions results in flooding over the plains. However, very uncomfortable weather due to high humidity and temperatures is the feature associated with the Breaks.

Droughts are also one of the vagaries of monsoon, which affect agriculture, livestock and soils. On an average, severe drought occurs once in five years in most of the tropical countries, though often they occur on successive years causing misery to human life and livestock. About two-third of the geographic area of India receives low rainfall (<1000 mm) which is also characterized by uneven and erratic distribution. Out of the net sown area of about 142 million hectares, two-third is reported to be vulnerable to drought conditions (Table 3.1). In the last decade, 2009–10 and 2013–14 have been drought-affected years with huge impacts on agricultural production and livestock (Sesha Sai et al. 2016).

Rajeevan and Pai (2006) indicated that most of the severe droughts over India are associated with El-Nino events. But all the El-Nino years have not resulted in drought occurrence. About 70% of the annual rainfall over India is contributed by Southwest Monsoon, which commences in the month of the May from the southernmost tip. Rainfall in terms of the amount and distribution is the most important single factor influencing the incidence of drought. Droughts

Table 3.1 Probability of occurrence of drought in different meteorological subdivisions of India

Meteorological sub-division	Frequency of deficient rainfall
Assam	Very rare, once in 15 years
West Bengal, Madhya Pradesh, Konkan, Bihar and Odisha	Once in 5 years
South Interior Karnataka, Eastern Uttar Pradesh and Vidharba	Once in 4 years
Gujarat, East Rajasthan, Western Uttar Pradesh	Once in 3 years
Tamil Nadu, East Rajasthan, Western Uttar Pradesh	Once in 2.5 years
West Rajasthan	Once in 2 years

Source Rao and Rao (2016)

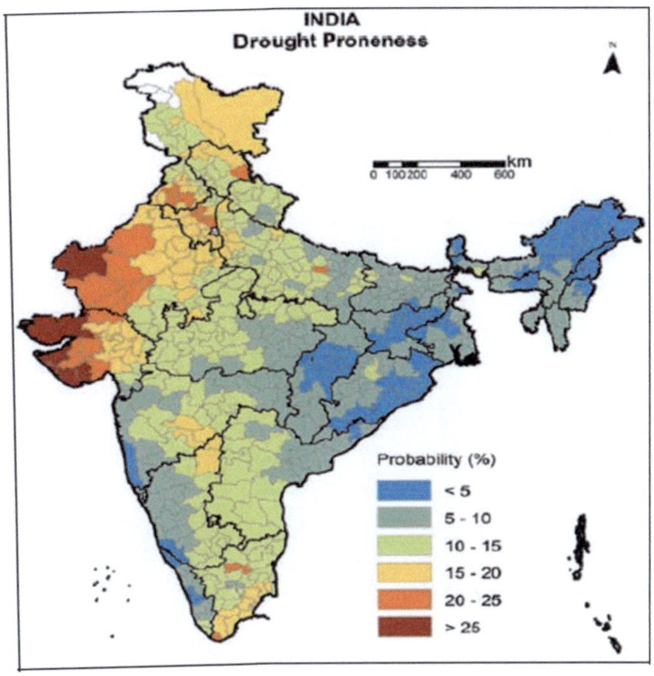

Fig. 3.2 Probability (percent) of occurrence of drought across India. *Source* Rama Rao et al. (2013)

in the Indian region are mainly due to various kinds of failure of rains from SW Monsoon. Based on the analysis of long series of yearly data, probability of occurrence of drought in different meteorological subdivisions is given in Table 3.1. Most of the drought-prone areas lie in the arid, semi-arid and sub-humid regions. The spatial probability of occurrence of severe drought in India has been worked out by Rao and Rao (2016) and shown in Fig. 3.2.

3.2.4 Post-Monsoon/NE Monsoon/Retreating Monsoon (October, November, December)

During the withdrawal phase of the summer monsoon, lower level winds over south Asia reverse their direction from southwest to northeast. This change is associated with the southward movement of the continental tropical convergence zone (CTCZ) and the subtropical anticyclone. IMD refers to the October to December period as the northeast monsoon, which is a part of the northeast trades. During the northeast monsoon season, the country receives about 11% of its annual rainfall, while many districts over the south peninsula receive 30–60% of their annual rainfall. October to December is the major period of rainfall in south Peninsula, particularly the eastern half, comprising of the meteorological subdivisions of coastal AP, Rayalaseema, and Tamil Nadu. In Tamil Nadu, this is the rainy season accounting for nearly 60% of the annual rainfall in the coastal districts and about 40–50% in the interior districts of the state. Though the principal rainy season for south Interior Karnataka, Kerala, and Arabian Island is the SW monsoon, the period October to December contributes 20% of the annual total. In Rayalaseema, September and October contribute 36% of the annual total.

The weather conditions are influenced by high pressure developed over North-western part of the subcontinent. This results in the blowing of cold dry winds from this region towards southern low-pressure areas lying over water bodies surrounding peninsular India. Since these winds are cold and dry, they do not cause rainfall and weather conditions, however; wherever these Northeast monsoon winds collect moisture while passing over the Bay of Bengal, they bring rain to Peninsular India.

This season is also dominated by Tropical Cyclones (TC). As per the climatology, TCs occur in the Indian Ocean prominently during pre-monsoon season and post-monsoon season. In Bay of Bengal the maximum frequency is in the two post-monsoon months of October and November. The Bay of Bengal TCs more often strike Orissa-West Bengal coast in October, Andhra coast in November and the Tamil Nadu coast in December. Over 60% of the TCs in the Bay of Bengal strike different parts of the east coast of India, 30% strike coasts of Bangladesh and Myanmar and about 10% dissipate over the sea itself (Tyagi et al. 2010). Nearly 7% of the global TCs form in the North Indian Ocean. About 5–6 TCs occur in the North Indian Ocean annually, including 4 over the Bay of Bengal and 1 over the Arabian Sea. The monthwise distribution of TCs over the north Indian Ocean is shown in Fig. 3.3. The frequencies of TCs over the Bay of Bengal and Arabian Sea are shown in Figs. 3.4 and 3.5

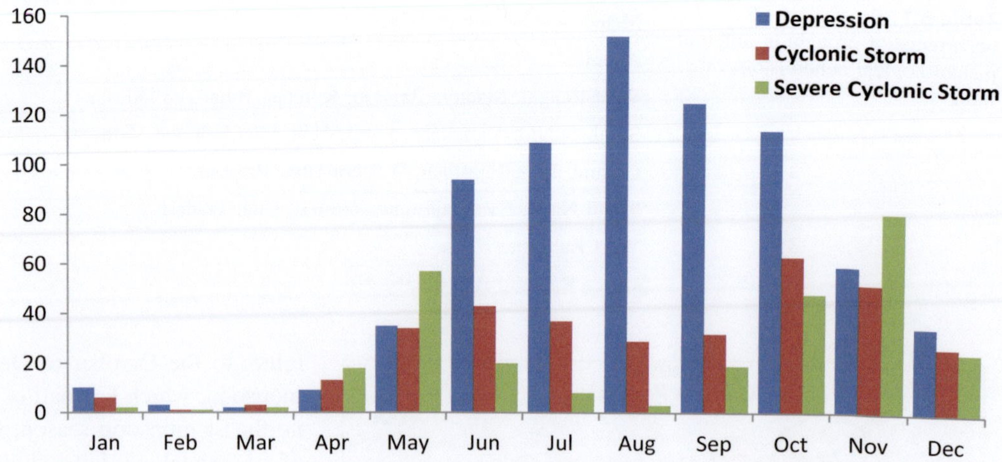

Fig. 3.3 Monthly frequency of cyclonic disturbances during 1891–2008 over North Indian Ocean. *Source* Regional Specialised Meteorological Centre (RSMC), New Delhi, IMD

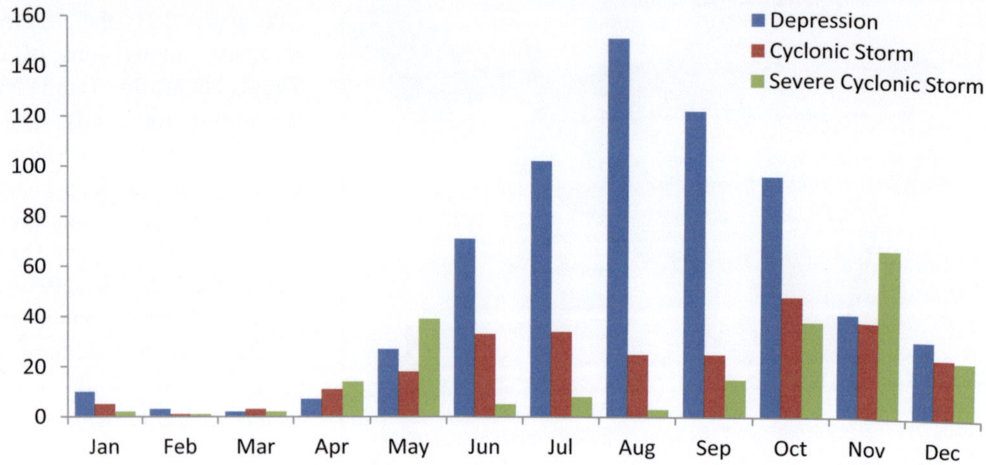

Fig. 3.4 Monthly frequency of cyclonic disturbances during 1891–2008 over Bay of Bengal. *Source* Regional Specialised Meteorological Centre (RSMC), New Delhi, IMD

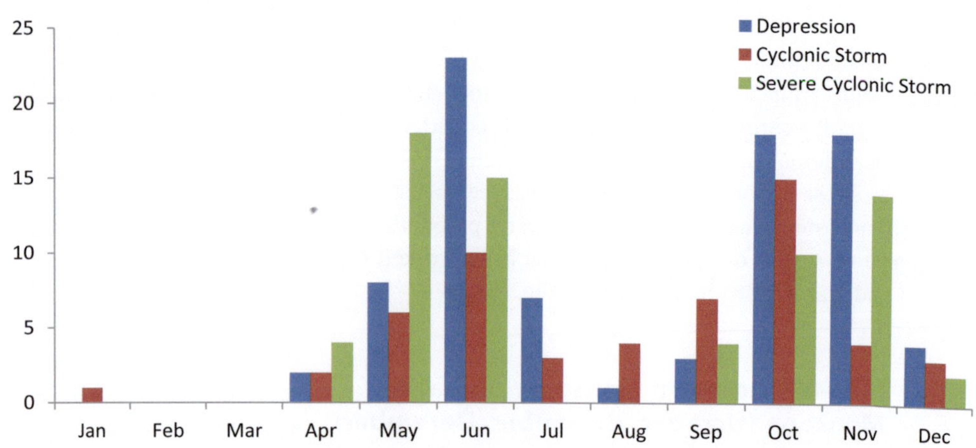

Fig. 3.5 Monthly frequency of cyclonic disturbances during 1891–2008 over Arabian Sea. *Source* Regional Specialised Meteorological Centre (RSMC), New Delhi, IMD

respectively. The cyclones crossing different coastal states are shown in Fig. 3.6. About 50% of the cyclones over the Arabian Sea dissipate over the sea itself. Maximum landfall takes place over Gujarat coast. The disastrous impacts of cyclones are due to heavy rain, strong winds and storm surge. Mohapatra et al. (2012) have developed the cyclone-prone districts mapping based on the frequency of tropical cyclones, associated mean surface wind, heavy

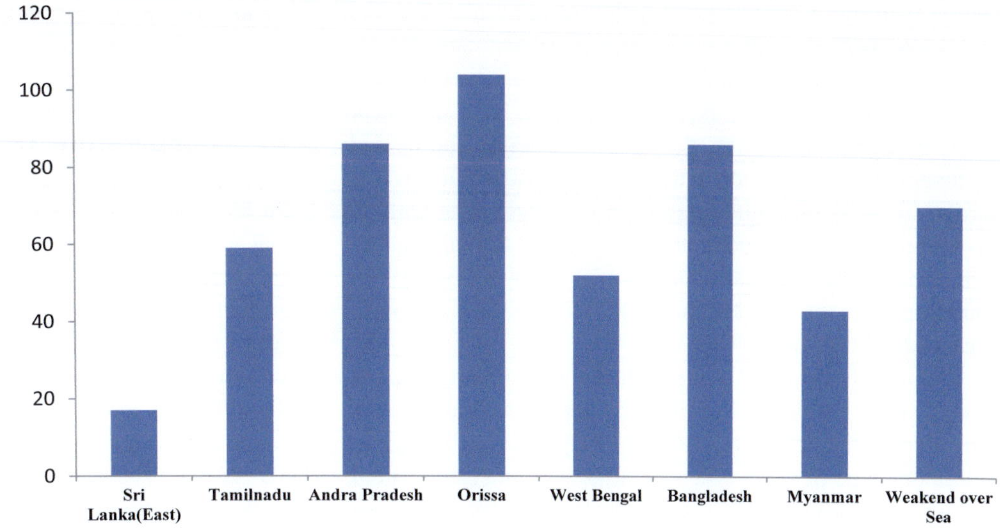

Fig. 3.6 Monthly frequency of cyclone/sever cyclone over Bay of Bengal landfalling over different coastal states during 1891–2008. *Source* Regional Meteorological Specialised Centre (RSMC), New Delhi, IMD

rainfall and probable maximum storm surge. It indicates that north Odisha and West Bengal coastal districts are very highly prone to cyclonic activity.

The tropical Cyclones cause irreparable damage to agriculture by destruction of vegetation, crops, orchards, livestock and long term loss of soil fertility, from the saline deposits over land, flooded by seawater. The effects of strong winds in coastal areas are seen in stunted and often much-sculpted trees. In addition to the battering effect of winds, there is the additional damage caused by airborne sea salt which occurs within a few hundred metres of the coast (Das et al. 2003).

During Cyclone Phailin, the heavy rainfall and the storm surge of 2 m along the coast led to flooding in coastal Odisha districts. The water could not recede back due to swollen sea and thus led to damage of standing rice crop. Ray et al. (2016) assessed the rice crop inundation due to flood using microwave remote sensing data from Indian SAR (Synthetic Aperture Radar) satellite RISAT-1. The maximum rice area inundated due to floods was in Balasore, Bhadrak, Jajpur, Jagatsinghpur and Kendrapara districts. The total rice inundated area was 132.4 thousand ha.

3.3 Soils and Climate

Soil forms the lowest boundary of earth's atmosphere and is subject to interactions with incoming radiation (Mishra 1996). The atmosphere is largely unheated by the passage of solar radiation through it; therefore, the soil surface unless fully covered by vegetation plays a significant role in the energy exchange between atmosphere and earth. Soil climate is generally characterized by soil air, soil temperature and soil moisture.

Climate affects the soil formation as it determines the amount of water that is available for processes such as the weathering of minerals, the transportation of minerals and the release of elements. Climate also influences the temperature of the soil, which determines the rate of chemical weathering. Climates that are warm and moist encourage rapid growth of plants. This leads to a high production of organic matter. The decomposition of organic matter is accelerated in type-specific climate and break down of such processes may be possible under freezing, thawing, wetting and drying conditions. Through the effects of precipitation and temperature, climate affects the rates of biological, chemical and physical processes involved in soil formation. High precipitation and low-temperature increase organic matter in soil. Leaching of soluble materials increases with increasing precipitation. Movement of clay in soil profile increases, with increasing precipitation. Silicate clay and Al and Fe oxide formation increase with increasing temperature. The profiles of a soil bear imprint of the passage of climate over very long periods of time. Ghosh (1965) has cited the presence of heavy segregation of calcium, formation of pan at deeper levels and presence of conglomerates in Rajasthan soils and the occurrences of rounded hollows and potholes on mountain soil are indication of prevalence of wet climate over Rajasthan earlier. Parent material influences soil characteristics under low to moderate rainfall conditions but under high rainfall conditions, the effect of parent rocks is obliterated leading to formation of soils with characteristics governed mainly by climatic factors.

Tripathi and Mishra (1969) found that the clay percentage, exchange capacity and $CaCO_3$ increased with decreasing rainfall and that soils of the central arid regions had a higher Na content and a poorer physical condition than soils of western or eastern regions. Regions with mean annual precipitation below 75 cm had mostly calcareous soils and region with mean annual precipitation of more than 200 cm and more had no calcareous soils. Tamhane and Karale (1967) found that in Maharashtra, basaltic soils of low rainfall region had high alkalinity and were base saturated, while in high rainfall zone, the soils were acidic and tended to develop laterite characteristics. Climate also influences the fertility of soils through variation in C:N ratio. The soils in cool climates were several times richer in organic matter than those in hot climates. Jenny and Raychoudhuri (1960) found an increase in C:N ratio of 8:5 in semi-arid region to 10:5 in humid region.

Climate of an area is a major factor in soil formation. It tends to reduce differences caused by parent material. That is why two different parent materials may develop the same soil in one type of climate regime. Likewise, the same parent material may produce different types of soil in different types of climates. Hot summer and low rainfall develop black soil, as found in some districts of Gujarat and Tamil Nadu, irrespective of the parent rock. Principal source of soil water is rainfall (precipitation). Some of the water which falls during rainfall is lost as run-off, which goes back to the source, i.e. sea, ocean and some is lost during evaporation and the remaining percolates into the ground.

3.3.1 Climate Change Trends

Agricultural production is highly vulnerable to climate variability and change. The changing pattern in temperature, precipitation and extreme events by the end of this century are likely to have significant implications for rural poverty and for both rural and urban food security. At the same time, agriculture also presents new adaptation and mitigation opportunities as large land area falls under crops and rangeland, and the additional mitigation potential of aquaculture. As such, we need to characterize the climate in a holistic manner. Salient feature of climate change trends are presented as under:

The latest analysis of observations by the World Meteorological Organisation (WMO) shows that atmospheric levels of carbon dioxide (CO_2), methane (CH_4), and nitrous oxide (N_2O) reached new highs in 2014. Globally-averaged atmospheric levels of CO_2, CH_4 and N_2O were of the order of 397.7 ppm, 1833 ppb and 327.1 ppb which were ~143, 254 and 121% of the pre-industrial level, respectively (WMO 2015). Climate has shown warming of 0.89 [0.69–1.08] °C over the period 1901–2012 which is mainly attributed to anthropogenic activities (IPCC 2013). The newer findings indicate that warming is more pronounced than expected. Further, it has been projected that the global mean surface temperature and sea level may increase by 0.3–1.7 °C and 0.26–0.54 m for RCP 2.6, 1.1–2.6 °C and 0.32–0.62 m for RCP 4.5, 1.4–3.1 °C and 0.33–0.62 m for RCP 6.0 and 2.6–4.8 °C and 0.45–0.81 m for RCP 8.5, respectively by 2181–2100 (IPCC 2013).

3.3.2 Temperature Trends

India has also shown warming trends in conjunction with global scenario. Annual mean surface air temperatures show a significant warming of about 0.63 °C/100 year during the 115 years (1901–2015) (Fig. 3.7a). It may be mentioned that 12 out of the 15 warmest years in India were during the recent past fifteen years (2001–2015). Also the past decade (2001–2010 or 2006–2015) was the warmest decade on record with decadal mean temperature anomaly of 0.49 °C (IMD 2015). The year 2015 was significantly warm in respect of temperature in India also. The annual mean temperature for the country was +0.67 °C above the 1961–1990 average, thus making the year 2015 as the third-warmest year on record since the nation-wide records commenced in 1901. The other 9 warmest years on record in order were: 2009 (0.77 °C), 2010 (0.75 °C), 2003 (0.61 °C), 2002 (0.59 °C), 2014 (0.53 °C), 1998 (0.49 °C), 2012 (0.48 °C), 2006 (0.43 °C) and 2007 (0.41 °C). Much of this increase has taken place since 1975 (IMD 2014). It further suggests that this warming is primarily due to rise in maximum temperature across the country, over larger parts of the data set. However, since 1990, minimum temperature is steadily rising and rate of its rise is slightly more than that of maximum temperature. Warming has been observed in all seasons in the country (Fig. 3.7b–e). The mean surface air temperature anomaly was 0.86 °C during post-monsoon, 0.70 °C during winter, 0.56 °C during pre-monsoon and 0.48 °C during monsoon season during the period 1901–2015 (IMD 2015). The mean temperature for the post-monsoon season (with anomaly +1.1 °C above average) in 2015 was the highest since 1901, thus making it the warmest post-monsoon season. The other five warmest post-monsoon years in order were 2011 (with anomaly +0.73 °C), 2008 (0.728 °C), 2009 (0.72 °C), 1979 (0.66 °C) and 2006 (0.63 °C). Also all the three individual months of the 2015 post-monsoon season viz. October (with anomaly 1.2 °C), November (1.3 °C) and December (1.2 °C) were warmest since 1901.

Mean temperature anomaly during 2015 Monsoon season was +0.72 °C making it the fourth warmest since 1901. The other five warmest monsoon years on record in order were 2014 (0.77 °C), 2009 (0.75 °C), 1987 (0.74 °C), 2003 (0.59 °C) and 1998 (0.56 °C). Considering the individual months of monsoon season, July was the third-warmest (with anomaly 0.9 °C, 1987 (0.98 °C) and 2002 (0.96 °C)), August was second warmest (0.93 °C, 2009 (0.97 °C)) and September (with anomaly +1.0 °C) was the warmest since 1901. Warmer temperature during the monsoon season

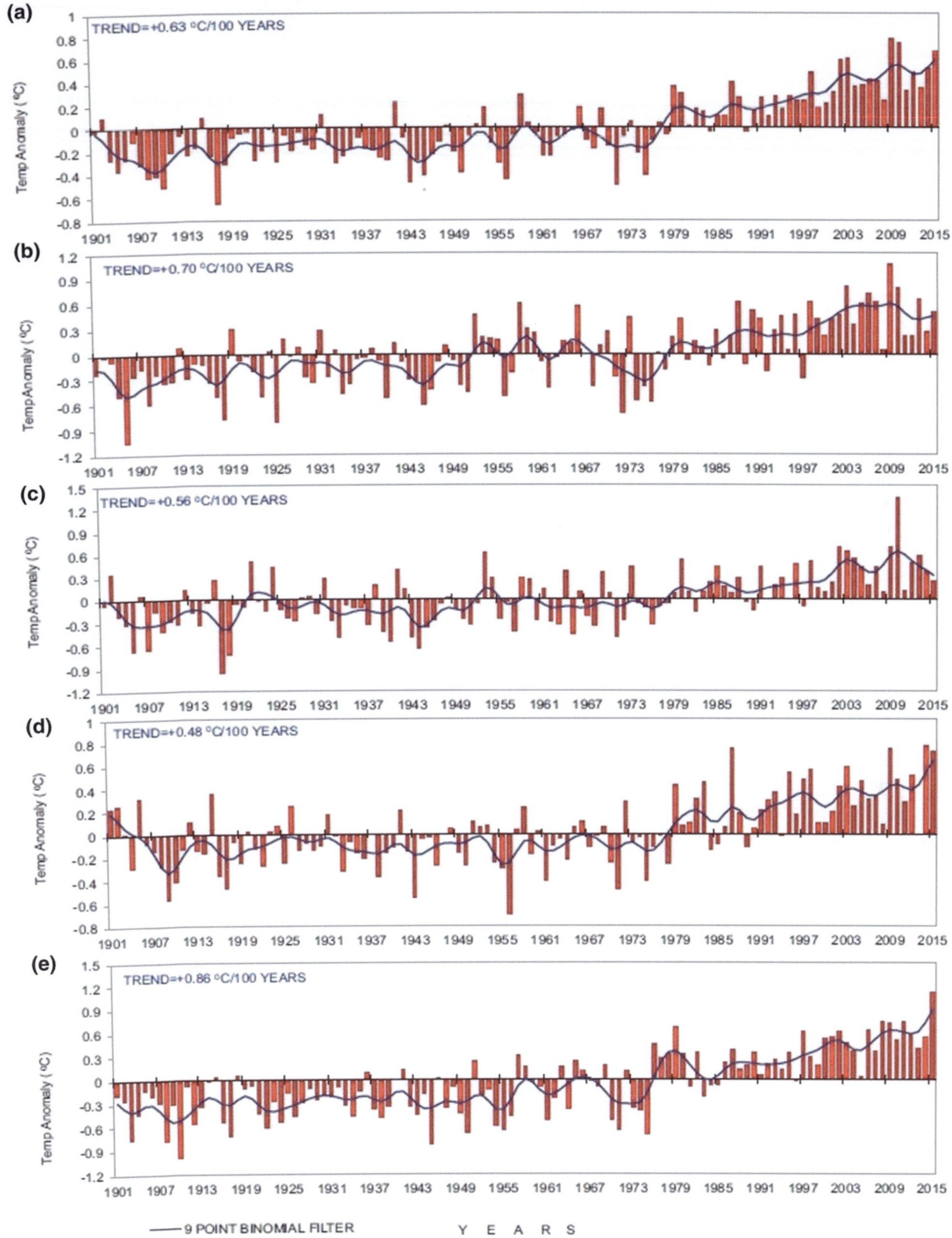

Fig. 3.7 All Indian mean temperature annually **a** annual, **b** winter, **c** pre-monsoon, **d** monsoon and **e** post-monsoon season during 1901–2015. *Source* IMD, 2015

(June–September, +0.72 °C above average) and the post-monsoon season (October–December, +1.1 °C above average) mainly contributed to the warmer annual temperature.

The climate of India was significantly warmer than normal during the year 2016 till October in line with the warmer than normal global climate observed during the period. The annual mean land surface air temperature averaged over the country during 2016 till October was +0.90 °C above the 1961–1990 average. The country averaged seasonal mean temperatures during the winter season (January–February, with anomaly +01.360 °C, warmest since 1901) and the pre-monsoon season (March–May, with anomaly +1.250 °C, second warmest ever since 1901) mainly accounted for the above normal January–October temperature for the year till date. The warm anomaly during 2016 can be attributed to the strong El Nino conditions that peaked during the 2016 winter season and weakened to weak level but persisted during the subsequent spring season.

Spatial pattern of mean annual temperature during last 115 years indicates positive (increasing) trends over most parts of the country except some parts of Rajasthan, Gujarat and Bihar, where negative (cooling) trends were observed (Fig. 3.8). During last 60 years (1951–2010), mean temperatures show a significant decreasing trends over Punjab, Uttarakhand and Jammu & Kashmir, no trends over Chhattisgarh, Haryana, Meghalaya, Orissa, Uttar Pradesh and West Bengal, while significantly increasing trends were observed remaining States in India (Rathore et al. 2013).

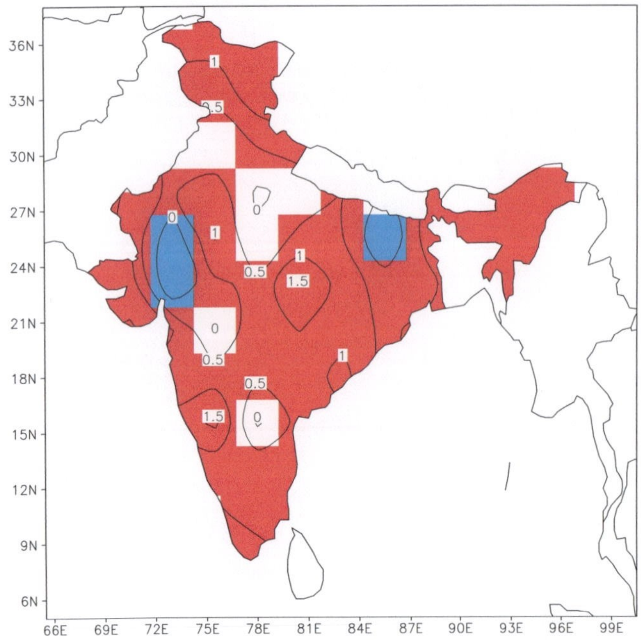

Fig. 3.8 Annual mean temperature trends (°C/100 years) during 1901–2015 (positive trends; red, negative trends: blue). *Source* IMD, 2015

3.3.3 Rainfall Trends

The monsoon rainfall at all India level does not show any trend but there are some regional patterns. Seasonal and annual rainfall trends during 1901–2015 have been presented in Fig. 3.9. However, decadal epoch has been observed. Similarly rainfall for the country as whole for the same period for individual monsoon months also does not show any significant trend. During the season, three subdivisions viz. Jharkhand, Chhattisgarh, Kerala show significant decreasing trend and eight subdivisions viz. Gangetic West Bengal, West Uttar Pradesh, Jammu & Kashmir, Konkan and Goa, Madhya Maharashtra, Rayalaseema, Coastal Andhra Pradesh and North Interior Karnataka show significant increasing trends. The alternating sequence of multi-decadal periods of thirty years having frequent droughts and flood years are observed in the all India monsoon rainfall data (Attri and Tyagi 2010). The decades 1961–70, 1971–80 and 1981–90 were dry periods. The first decade (1991–2000) in the next 30 years period already experienced wet period. However, during the winter season, rainfall is decreasing in almost all the subdivisions except for the subdivisions Himachal Pradesh, Jharkhand and Nagaland, Manipur, Mizoram and Tripura. Rainfall is decreasing over most parts of the central India during the pre-monsoon season. However, during the post-monsoon season, rainfall is increasing for almost all the subdivisions except for the nine subdivisions. Annual rainfall showed decreasing trends over 16 States and increasing trends over 15 States including Islands during 1951–2010 (Rathore et al. 2013).

3.3.4 Extreme Events

A number of studies point out an increasing trend in the observed frequency of heavy precipitation events (Rajeevan et al. 2008; Pattanaik and Rajeevan 2010), and a decreasing trend in the light rainfall events (Goswami et al. 2006) and moderate to heavy rainfall events especially over the Western Ghats (Krishnan et al. 2013). Frequency of depressions and cyclonic storms formed over the north Indian Ocean (1951–2015) during the monsoon and post-monsoon season have shown decreasing trends (IMD 2015). A significant increase was noticed in the frequency, persistency and spatial coverage of both of these high-frequency temperature extreme events (heat and cold wave) during the decade (1991–2000) (Pai et al. 2004).

Both total and low cloud cover over Arabian Sea and the equatorial Indian Ocean are observed to decrease during the ENSO events. However, cloud cover over Bay of Bengal is not modulated by the ENSO events. On inter-decadal scale, low cloud cover shifted from a 'low regime' to a 'high regime' after 1980 which may be associated with the

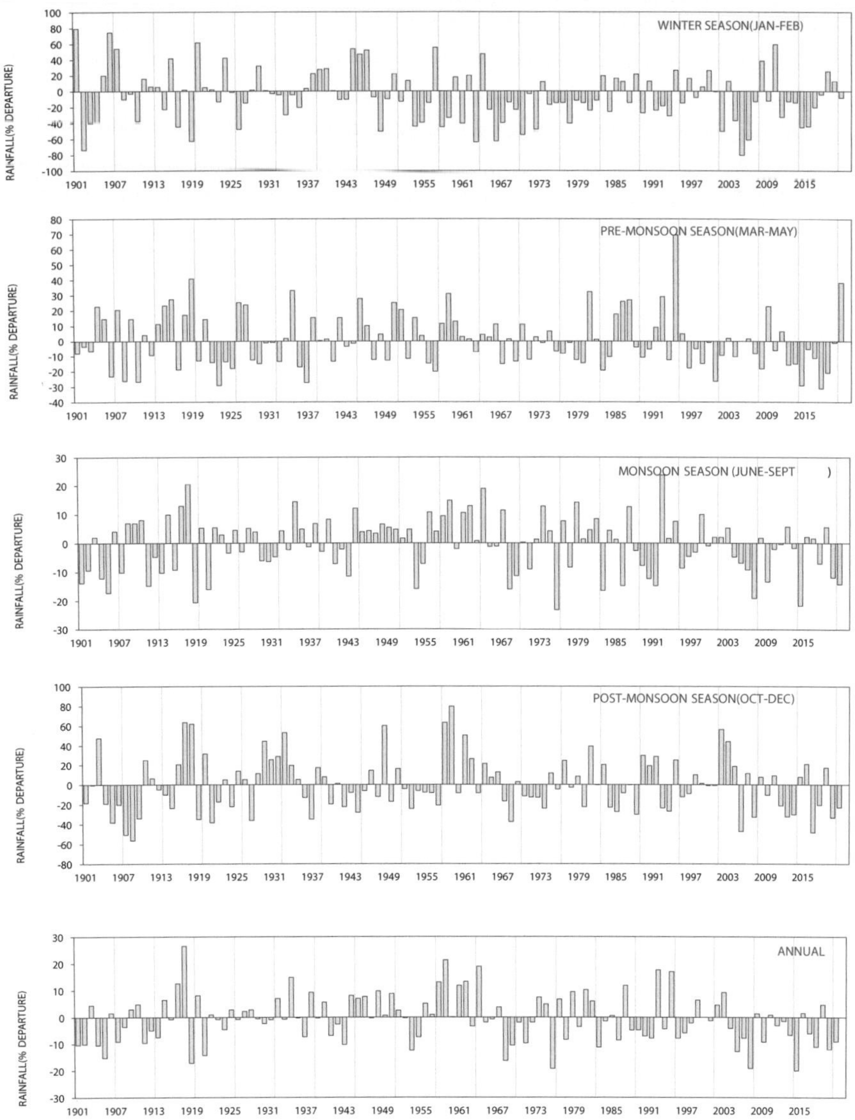

Fig. 3.9 Percentage departure of area-weighted seasonal and annual rainfall over the country (1901–2015). *Source* IMD, 2015

corresponding inter-decadal changes of sea surface temperatures over north Indian Ocean observed during the late 1970s (Rajeevan et al. 2000). Decrease in low and medium rainfall events and increase in heavy rainfall events have been observed over the country (Goswami et al. 2006). Based on rainfall data during 1901–2005, annual normal rainy days varied from 10 over extreme western parts of Rajasthan to the high frequency of 130 days over north-eastern parts of the country (IMD 2010).

Extreme weather events always had an impact on agriculture, particularly thunderstorms (Hails), Cyclones, Floods, Cold Waves and Heat Waves. Several research studies are able to establish the adverse effects of weather extremes especially rainfall and temperature on crop productivity and yield (Chakraborty et al. 2018), however, an adequate database for assessing the damage in quantitative terms is still uncertain. Agriculture risk assessments are necessary for building resilience to Climate Change and for sustenance of global agriculture and food systems. Sustainability of agriculture depends upon the availability of natural resources that support agriculture presently and also in future. The management of these resources in spite of the disruptions due to climate change and associated weather extremes would be a challenge. Development of new adaptation techniques and strategies for agriculture and surveillance of resources would be needed to tackle the adverse affects of climate change. As per FAO report on 'The Impact of Disasters on agriculture and food security' (www.fao.org/3/a-i5128e.pdf), the full impact of disasters on agriculture sector due to weather extremes is not well understood but still 25% of the economic impact caused by climate-related disasters falls in the agriculture sector. The first and the most basic requirement in agro meteorological hazard assessment for extreme events is an adequate database. If sufficient quality data are available, it may be feasible to estimate the risk of the extreme events and their damage in quantitative terms (Das et al. 2003).

3.4 Soil and Climate Change: Impacts and Mitigation

Agriculture contributes to greenhouse effect mainly through the emission of greenhouse gases (GHGs) such as methane (CH_4), nitrous oxide (N_2O) and carbon dioxide (CO_2). Submerged rice fields are the potential source of methane (CH_4) production through microbial decomposition of organic matter under anaerobic conditions. Long-time submergence, higher natural C substance and utilization of farmyard manure (FYM) in puddled soil encourage CH_4 to emit at larger quantity (Kumar et al. 2016). The enteric maturation in ruminants is another source of CH_4 emanation (IPCC 2014).

Nitrous oxide is generated in soils through the procedures of nitrification and denitrification. Nitrification is the aerobic microbial oxidation of ammonium to nitrate, and denitrification is the anaerobic microbial reduction of nitrate to nitrogen gas (N_2). Nitrous oxide is an intermediary product released in the process of denitrification and a by-product of nitrification that releases from microbial cells into the soil environment. One of the fundamental controlling elements in this reaction is the accessibility of inorganic N in soil through addition of manufactured or organic fertilizer, crop residue sewage or mineralization of N in soil organic matter. Water management practices like alternate wet and dry reduces emission of CH_4 as compared to flooded rice, however, this practice increases N_2O and CO_2 emissions (Kumar et al. 2016).

The main source of CO_2 in agriculture is the soil management practices such as tillage, which triggers emission of this gas through biological decomposition of soil organic matter (SOM). Tillage breaks the soil aggregates increases the oxygen supply and exposes the surface area of organic material promoting the decomposition of organic matter. Use of fuel for various agricultural operations and burning of crop residues are the other sources of CO_2. An off-site source of CO_2 is the manufacturing of farm implements, fertilizers and pesticides. Smouldering of post-harvest refuse build-ups likewise adds to the worldwide carbon dioxide budget. Utilization of fuel for different operations and blazing of residues are alternate wellsprings of CO_2 emanation.

Global warming, caused by the increase in concentrations of GHGs in the atmosphere, alters the energy balance of the atmospheric system, which leads to subsequent climate change. As a result, the phenomenon of global warming is coming in the picture due to which, snow and ice are melting fast and sea level is rising. Climate change may have a considerable impact on agriculture through direct and indirect effects on the crops, soils, livestock and pests. Development of technologies for adaptation and mitigation and their uptake at speedy rate by the farmers are essential for climate change management.

Soil plays a dual role, i.e. it acts as both a source and sink of GHGs. It is intricately linked to the atmospheric-climate system through carbon, nitrogen, and hydrologic cycles. In 2010, the worldwide emanation of GHGs was around 50 Bt CO_2 eq. in which India contributed around 2.34 Bt CO_2 eq., i.e., around 5% of the total emanation. Agricultural farming contributed more than 11% of the aggregate GHGs emanation globally (Pathak 2015). The share of Indian agriculture was around 7% of worldwide emanation from agriculture. The majority of GHG emission is from the energy sector (65%), followed by agriculture (18%) and industry (16%) (Fig. 3.10). Out of the total emission of 421 Mt of CO_2 eq. (Table 3.2) by Indian agricultural sectors (including crop and animal husbandry), enteric fermentation contributed the highest (56%), followed by agricultural soil (23%) and rice fields (18%). Burning of crop residues in field contributed 2% and manure management contributed 1% of the emission (Pathak 2015).

Among the various rice ecosystems, the highest emission was from the irrigated continuously flooded rice (34%), followed by rainfed flood-prone rice and irrigated single aeration rice (18%) (Bhatia et al. 2012; 2013). Rainfed drought-prone, deep water and irrigated multiple aeration rice ecosystems contributed 16, 8 and 6% of methane, respectively (Pathak et al. 2014). Fertilizer contributed 77% to the total nitrous oxide emission.

Effect of global warming on soil processes is affected by several factors. Increased precipitation increases the surface runoff in sloppy lands; increase infiltration and water storage within the soil in plain lands; enhance groundwater recharge in the highly permeable and well-drained soils; increase the evaporation on soil having low infiltration and transpiration in the case of well-developed plant canopies (Varallyay 2007). Increase in temperature will upsurge the potential evapotranspiration and decline the surface runoff, infiltration, water storage and groundwater recharge, especially if accompanied by low rainfall leading to drought particularly

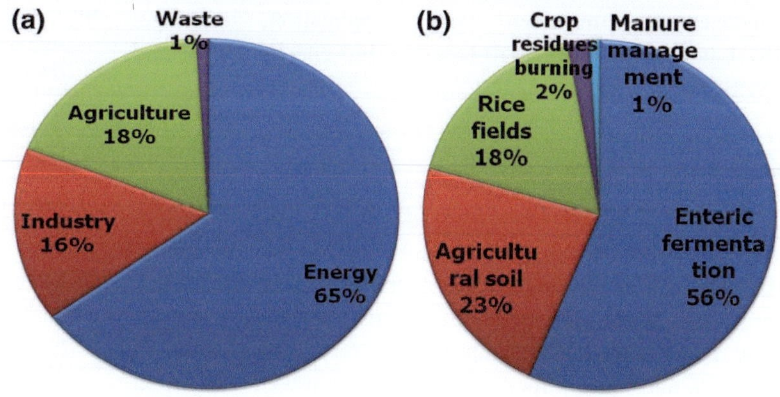

Fig. 3.10 Emission of greenhouse gases from **a** various sectors of Indian economy and **b** sub-sectors of agriculture in 2010. *Source* Adapted from Pathak (2015)

Table 3.2 Greenhouse gas emissions from Indian agriculture (2012–13)

Source	Methane	Nitrous oxide (Mt)	GWP (CO_2 eq.)
Enteric fermentation	10.9	–	228.9
Manure management	0.13	0.08	27.53
Rice cultivation	3.46	–	72.66
Agricultural soil	–	0.26	80.6
Crop residue burning	0.4	0.01	11.5
Total	14.89	0.35	421.19

Source Pathak (2015)

in drought-prone area (Table 3.3). There is a drastic reduction in total rainfall in drought years compared to normal years (Reddy et al. 2001). The main impacts of climate change on soil are listed below.

- Increase in sea level may lead to salt-water incursion in the coastal lands transforming them less suitable for conventional agriculture.
- Reduced amount and quality of organic matter content, which is as of now very low in Indian soil.
- Under elevated CO_2 concentration, the crop residues with higher C:N ratio may diminish their rate of decomposition and nutrient supply.
- Change in precipitation volume and recurrence, and wind intensity may alter the severity, frequency and extent of soil erosion.
- Increase in soil temperature increase N mineralization, but its availability may decrease due to volatilization and denitrification losses.

To minimize the negative impacts of climate change, IPCC (2014) has identified the following mitigation options in agriculture: (1) Plant management: improved variety, rotation, cropping system; (2) Nutrient management: type of fertilizer, urease/nitrification inhibitor; (3) Tillage/residue management: reduced tillage, residue retention; (4) Water

Table 3.3 Rainfall variation in normal versus drought years in drought-prone area

Type of drought-prone area	Normal year (NY) or drought year (DY)	Rainfall distribution (mm)				Annual rainfall (mm)
		June	July	August	September	
Mild	NY	105.0	179.0	171.0	139.0	799.0
	DY	72.4	95.5	96.5	65.3	481.0
Moderate	NY	56.6	82.6	99.7	157.2	733.2
	DY	55.3	56.2	36.8	81.3	452.3
Severe	NY	77.0	87.8	111.6	177.1	700.2
	DY	52.1	62.8	71.7	88.6	411.1
Chronic	NY	68.0	82.2	100.3	160.3	613.3
	DY	52.3	55.4	93.1	84.1	394.3

Source Reddy et al. (2001)

management: improved water application, drainage; (5) Land-use change: agro-forestry, bio-energy crops, and (6) Bio char application.

3.5 Conclusions

Soil is capable of acting both as a source and sink of carbon. The soil also helps to regulate other greenhouse gases such as nitrous oxide and methane. The overall climate change may have considerable impacts on the soil functions to perform, and more importantly, the future use of soils often needs significant adaptations in order to meet the challenges of changing climate. Changes in temperature and rainfall patterns can have a great impact on the organic matter and processes that take place in our soils, as well as the plants and crops that grow from them. By restoring degraded soils and adopting soil conservation practices, there is major potential to decrease the emission of greenhouse gases from agriculture, enhance carbon sequestration and build resilience to climate change. Soil carbon sequestration increases the ability of soils to hold soil moisture, withstand erosion and enrich ecosystem biodiversity, which helps cropping systems to better withstand droughts and floods.

References

Attri SD, Tyagi A (2010) Climate profile of India. IMD met monograph no. environment meteorology, 1/2010

Bhardwaj J, Singh S, Singh D (2007) Hailstorm induced crop losses in India: some case studies. In: 4th European conference on severe storms, Trieste, Italy, 10–14 Sept 2007

Bhatia A, Aggarwal PK, Jain N, Pathak H (2012) Greenhouse gas emission from rice and wheat-growing areas in India: spatial analysis and upscaling. Greenh Gas Sci Technol 2:115–125

Bhatia A, Jain N, Pathak H (2013) Methane and nitrous oxide emissions from Indian rice paddies, agricultural soil and crop residue burning. Greenh Gas Sci Technol 3:196–211

Chakraborty D, Sehgal VK, Dhakar R (2018) Changes in daily maximum temperature extremes across India over 1951–2014 and their relation with cereal crop productivity. Stoch Environ Res Risk Assess. https://doi.org/10.1007/s00477-018-1604

Das HP, Adamenko TI, Anaman KA, Gommes RG, Johnson G (2003) Agrometeorology related to extreme events. WMO technical note no. 201 (CAgM-XI working group on agrometeorology related to extreme events)

Ghosh B (1965) The genesis of desert plains in the central Luni basin of western Rajasthan. J Indian Soc Soil Sci 13:123–126

Goswami BN, Venugopal V, Sengupta D, Madhusoodanan MS, Xavier PK (2006) Increasing trend of extreme rain events over India in a warming environment. Science 314(5804):1442–1445

IMD (2010) Trends in precipitation extremes over India. NCC research report no. 3/2010

IMD (2014) Annual climate summary. National Climate Centre, Pune

IMD (2015) Annual climate summary. National Climate Centre, Pune

IPCC (2013) Climate change 2013—the physical science basis: summary for policymakers. A report of working group I of the IPCC

IPCC (2014) Climate change: mitigation of climate change. contribution of working group III to the fifth assessment report of the intergovernmental panel on climate change [Edenhofer O, Pichs-Madruga R, Sokona Y, Farahani E, Kadner S, Seyboth K, Adler A, Baum I, Brunner S, Eickemeier P, Kriemann B, Savolainen J]

Jenny H, Raychoudhuri SP (1960) Effect of climate and cultivation on the nitrogen & organic matter reserves in Indian soils. Indian Council of Agricultural Research (ICAR), New Delhi

Krishnan R, Sabin TP, Ayantika DC, Kitoh A, Sugi M, Murakami H, Turner AG, Slingo JM, Rajendran K (2013) Will the South Asian monsoon overturning circulation stabilize any further? Clim Dyn 40:187–211

Kumar A, Brahmanand PS, Nayak AK (2014) Management of cyclone disaster in agriculture sector in coastal areas. Directorate of Water Management, NRM Division (ICAR), Chandrasekharpur, Bhubaneswar

Kumar A, Nayak AK, Mohanty S, Das BS (2016) Greenhouse gas emission from direct seeded paddy fields under different soil water potentials in Eastern India. Agric Ecosyst Environ 228:111–123

Ministry of Environment, Forest and Climate Change (MOEF), Government of India (2015) India first biennial update report to the United Nations framework convention on climate

Mishra BB (1996) Theory of photopedology. J Indian Soc Soil Sci 44 (3):541–543

Mishra BB (2016) Climate change vs soil management: challenges, opportunities and policies. Agric Res Technol 2. https://doi.org/10.19080/artoaj.2016.02.555598

Mohapatra MS, Mandal, GK, Bandyopadhyay B, Tyagi A, Mohanty UC (2012) Classification of cyclone hazard prone districts of India. Nat Hazards 63. https://doi.org/10.1007/s11069-011-9891-8

Pai DS, Thapliyal V, Kokate PD (2004) Decadal variation in the heat and cold waves over India during 1971–2000. Mausam 55(2):281–292

Pant GB, Rupa Kumar K (1997) Climates of South Asia. Wiley, Baffins Lane, Chichester, pp 219–224

Pathak H (2015) Greenhouse gas emission from Indian agriculture: trends, drivers and mitigation strategies. Proc Indian Natl Acad Sci (In press)

Pathak H, Bhatia A, Jain N (2014) Greenhouse gas emission from Indian agriculture: trends, mitigation and policy needs. Indian Agricultural Research Institute, New Delhi, xvi + 39 pp

Pattanaik DR, Rajeevan M (2010) Variability of extreme rainfall events over India during southwest monsoon season. Meteorol Appl 17:88–104

Philip NM, Daniel CEF (1976) Hailstorms over India. IMD meteorological monograph, climatology no. 10

Rajeevan M, Pai DS (2006) On El-Nino-Indian monsoon predictive relationships. Research report no. 4/2006. National Climate Centre, IMD, Pune

Rajeevan M, Prasad RK, De US (2000) Cloud climatology of the Indian Ocean based on ship observations. Mausam 52:527–540

Rajeevan M, Bhate J, Jaiswal AK (2008) Analysis of variability and trends of extreme rainfall events over India using 104 years of gridded daily rainfall data. Geophys Res Lett. https://doi.org/10.1029/2008gl035143

Rama Rao CA, Raju BMK, SubhaRao AVM, Rao KV, Rao VUM, Kausaliya R, Venkateswarlu B Sikka AK (2013) Atlas on vulnerability of Indian agriculture to climate change. Central Research Institute for Dryland Agriculture, Hyderabad, p 116

Rao VUM, Rao BB (2016) Coping strategies for extreme weather in dryland agriculture. Mausam 67(1):5–14

Rathore LS, Attri SD, Jaswal AK (2013) State level climate change trends in India. Meteorological monograph no. ESSO/IMD/EMRC/02/2013. India Meteorological Department, New Delhi, India

Ray K, Bandopadhyay BK, Sen B, Sharma P, Warsi AH, Mohapatra M, Yadav BP, Debnath GC, Stella S, Das S, Duraisamy M, Rajeev VK, Barapatre V, Paul S, Shukla P, Madan R, Goyal S, Das AK, Bhan SC, Chakravarthy K, Rathore LS (2015) Pre-monsoon thunderstorms 2015: a report. IMD report no. ESSO/IMD/SAARC STORM PROJECT-2015/(01)(2015)/4

Ray K, Mohapatra M, Chakravarthy K, Ray SS, Singh SK, Das AK, Kannan BAM, Bandopadhyay BK (2016) Rainfall associated with tropical cyclone 'Phailin' in Bay of Bengal (NIO) in October 2013 and the assessment of rice inundation due to flooding. In: Tropical cyclone activity over the north Indian Ocean. Capital and Springer Publication, New Delhi

Reddy RS, Naidu LGK, Srinivas S, Niranjana KV, Ramesh M, Harindranath CS, Shivaprasad CR, Thayalan S (2001) Identification and categorization of agricultural drought-prone zones of Andhra Pradesh. Agropedology 11:101–109

Sesha Sai MVR, Murthy CS, Chandrasekar K, Jeyaseelan AT, Diwakar PG, Dadhwal VK (2016) Agricultural drought: assessment and monitoring. Mausam 67(1):131–142

Srinivasan V, Ramamurthy K, Nene YR (1973) IMD forecasting manual part III-discussion of typical synoptic situations. FMU rep no. III-2.2, Dec 1973, Office of DDGM (forecasting), Pune

Tamhane RV, Karale RL (1967) Studies on the basaltic soils of the Bombay Deccan. J Indian Soc Soil Sci 15:269–279

Tripathi BR, Mishra B (1969) Comparative study of salt affected soils of Uttar Pradesh developed under different agro-climatic conditions. Indian J Agric Sci 39:652–659

Tyagi AK, Bandyopadhyay B, Mohapatra M (2010) Monitoring and prediction of cyclonic disturbances over north Indian Ocean by regional specialised meteorological centre, New Delhi (India): problems and prospective. https://doi.org/10.1007/978-90-481-3109-9_13

Varallyay G (2007) Potential impacts of climate change on agro-ecosystems. Rev Agric Cons Sci 72:1–8

WMO (2015) WMO statement on the status of the global climate in 2015. WMO-1167

Geology and Geomorphology

4

K. S. Anil Kumar, S. Thayalan, R. S. Reddy, M. Lalitha, B. Kalaiselvi, S. Parvathy, K. Sujatha, Rajendra Hegde, S. K. Singh, and Bipin B. Mishra

Abstract

Parent material is a critical component in any pedogenic system. In order to interpret the soil data for specific land uses, all geological, sedimentological and geomorphological observations are of prime importance. Indian geology is diverse and contains rocks belonging to different geologic periods with reference to Deccan Traps, Gondwana and Vindhyayan. Plate tectonics, tectonic evolution, and proterozoic orogen with archaean era marking the drift of the Gondwana supercontinent and Mesozoic period, with greatest volcanic eruptions on peninsular India over Archaean gneiss and schists. This chapter also highlights the Cambrian formations being occurred in the salt range in Punjab and central Himalayas. Mesozoic, with the Deccan lava flows occurring in Peninsular India, consists of archean gneisses and schists, which are the oldest rocks. Cambrian Period is found in the salt range in Punjab and the Spiti area in central Himalayas. The Indian sub-continent comprises of three major geomorphological components, namely the Himalayas, the Great Plains and the Peninsular India. The Indian shield consists of archean gneisses and schists and Assam-Burma geological province which was a part of *Tethys* sea. Central Uplands comprise of the Malwa Plateau, Aravalli ranges, Madhya Bharat Pathar, east Rajasthan uplands, the Vindhyayan, the Bundelkhand uplands and the Narmada valley. The Deccan Plateau consists of Satpura range and Maharashtra plateau in the north and Karnataka and Telangana plateaus in the south Eastern plateau and is represented by Baghelkhand plateau, Chotanagpur plateau, Garhjat hills, Mahanadi basin and Dandakaranya upland. The Western Hills consist of North Sahyadri, Central Sahyadri, Nilgiris and South Sahyadri, running parallel to the west coast. The highly dissected Eastern Hills running in semi-circular fashion border the east coast of India and consist of Eastern Ghats and the Tamil Nadu Upland. There are two coastal plains, one along the Arabian Sea (west coast) and the other along the Bay of Bengal. There are two groups of Islands namely the Arabian Sea Islands and the Bay of Bengal Islands.

Keywords

Parent materials • Geology • Geomorphology • Physiographic divisions • Pedogenesis

4.1 Introduction

Soils are the integral part of a landscape, which is largely controlled by the landform on which the soils develop from parent materials (Chamuah et al. 1996; Yadav et al. 1998; Sahoo et al. 2003; Nogiya et al. 2017). The geology of an area controls the parent material being an integral part for pedogenesis. Parent material can be represented by a wide range of rock types with varying geochemical compositions. The soil is essentially the weathering rind of the earth's surface and provides insights into the processes and characteristics of the lithosphere, hydrosphere, biosphere, atmosphere and anthroposphere (human activities). Latin name for man is "homo", which is derived from humus and

K. S. A. Kumar (✉) · S. Thayalan · R. S. Reddy · M. Lalitha · B. Kalaiselvi · S. Parvathy · K. Sujatha · R. Hegde
ICAR-National Bureau of Soil Survey and Land Use Planning, Hebbal, Bangalore 560024, India
e-mail: anilsoils@yahoo.co.in

S. K. Singh
ICAR-National Bureau of Soil Survey and Land Use Planning, Amravati Road, University P.O., Nagpur 440033, India

B. B. Mishra
Bihar Agricultural University, Sabour, Bhagalpur 813210, India

B. B. Mishra
Pedology and Land Use Planning, School of Natural Resources Management and Environmental Sciences, Haramaya University, Dire Dawa, P.O. Box 138, Ethiopia

B. B. Mishra
Commission 1.4 (Soil Classification), International Union of Soil Sciences (2018–22), Vienna, Austria

B. B. Mishra (ed.), *The Soils of India*, World Soils Book Series,
https://doi.org/10.1007/978-3-030-31082-0_4

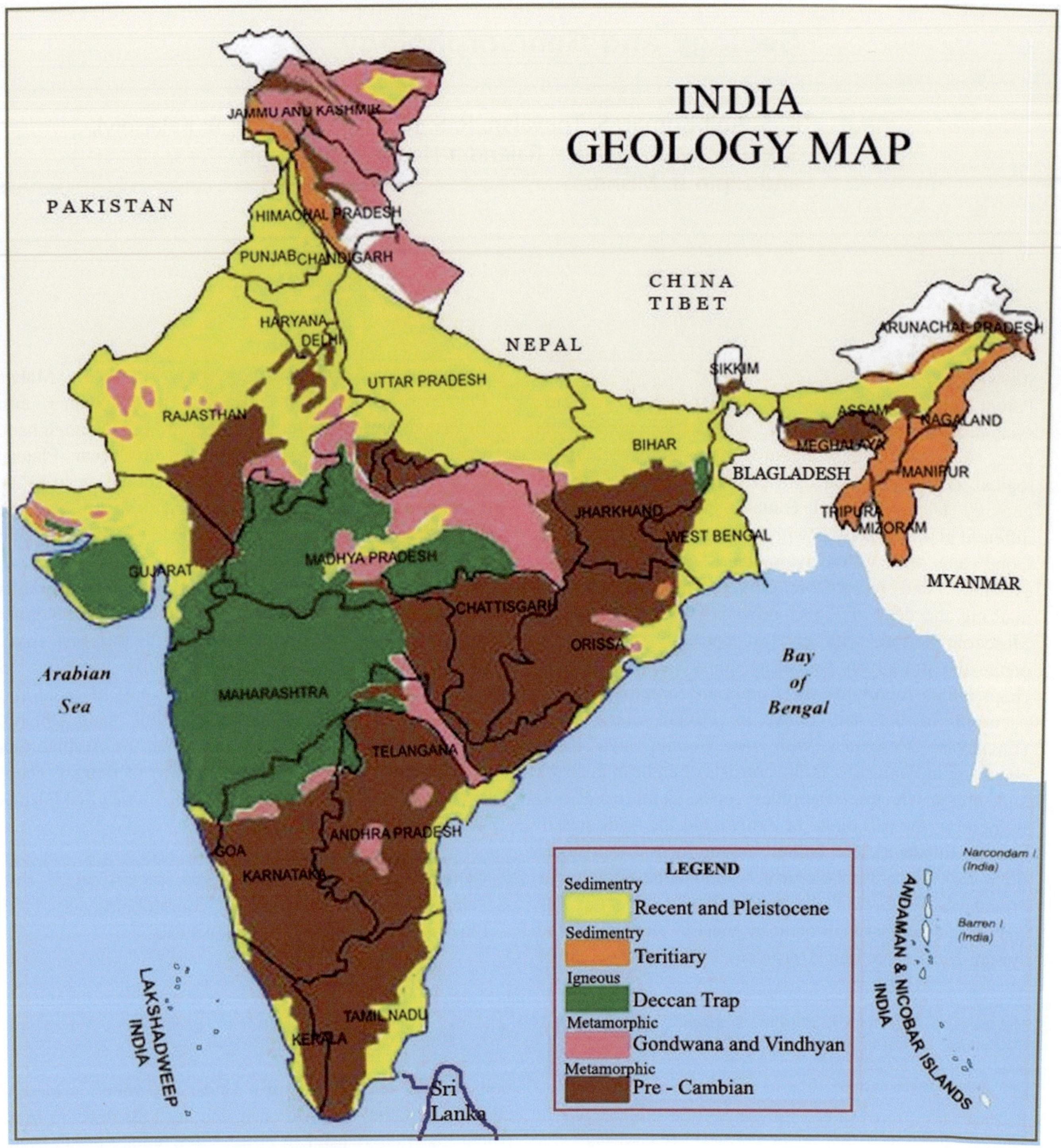

Fig. 4.1 Geology map of India. *Source* Limaye (2017)

humus is the tangible substance or stuff that goes into the makeup of a physical object called "the soil" with regolith following some specific process. Regolith or parent material is the weathering product of rock at zero time of soil formation. Soils are the integral part of the landscape and their characteristics are largely controlled by landforms on which soils have developed (Sawhney et al. 2005, 1992; Sharma et al. 1999; Wadia 1976).

India has great geomorphological diversity due to occurrence of varying ranges of mountains, hills and gullies, plateaus, deserts, floodplains, coastlands, wetlands, plains and deltas. Such diversity coupled with natural as well as human-induced climate change is steadily accelerating and altering the existing geomorphic processes which by and large lead to increase the frequency and magnitude of geomorphic hazards viz. landslide, wetland, flood, drought, desertification, acidity, sodicity, salinisation, changing river coarse, siltation, soil sealing, bank erosion, melting of glaciers, increasing the frequency of flood events, rising of sea level, depletion of soil moisture storage as well as levels of groundwater tables. Geomorphology is thus an area of geology with the study of landforms and processes that have shaped the landforms with the description and classification of physical features on landscapes. It is the systematic assessment of landforms and their interactions as records of geologic history, wherein the distribution of soils on earth's surface employs geologic concepts (Daniels et al. 1971). Soil geomorphology examines how the soil properties vary in a landscape as a function of local environmental conditions. It was earlier opined that in computation of even the Universal soil loss equation (USLE), the associated parent materials of the soils would preferably be included for better and reliable database for effective conservation measures (Gebrekidan et al. 2006). Soils are generally heterogeneous due to mass movement, irregular seepage and non-uniform deposition of sediments as a result of periodic flood and associated stream channels (Sawhney et al. 2005). Periodic erosion as well as deposition due to seasonal flooding may limit the pedogenic processes and thus delaying the evolution of diagnostic horizons (Sidhu et al. 1976).

The soils occurring in different parts of India are also diverse in characters depending upon climate, geology, geomorphology, parent materials, slope, altitude, biotic factors including biodiversity, vegetation, coastal environment, radiation inputs and management besides anthropogenic interventions and natural disasters. India occupies the greater part of South Asia. It is known from archaeological evidence that a highly developed culture, the Indus civilization, dominated the northwestern part of the subcontinent from about 2600 to 2000 BCE. Since the pre-historic past, India has been functioning virtually as a self-contained spiritual and cultural arena, which gave rise to a distinctive well defined tradition based primarily on The Hinduism, the roots of which can largely be traced to the Indus civilization and its vital linkage with soil, water, landforms and geology as well as geomorphology. It is bounded to the northwest by Pakistan, to the north by Nepal, China, and Bhutan and to the east by Myanmar (Burma). Bangladesh to the east is surrounded by India to the north, east, and west (Medlictt 1965). The island nation of Sri Lanka is situated about 65 km off the southeast coast of India. The land of India together with Bangladesh and most parts of Pakistan forms a well-defined subcontinent, set off from the rest of Asia by imposing the northern mountain rampart of the Himalayas and by adjoining mountain ranges to the east and west. The geology as well as geomorphology of India are complex and deserve systematic descriptions as both are related to soil formation and development. Any change in geomorphic processes influences the pedogenic processes (Gerrad 1981; Hall 1983; Sawhney et al. 1996).

4.2 Geology of India

India has a very unique geological and structural conditions of almost all ages of the geological time scale (Balasubramanian 2017) even from the Eoarchean Era (Krishnan 1960). According to Balasubramanian (2017) and others (Wadia 1961, 1976), all kinds of rock masses, mineral deposits, mineral fuels including coal and oil resources occur in India. About 100 years of geological investigations and exploration of various earth's materials have yield a tremendous amount of data to know about the geology of India pertaining to the ages viz. (i) Archaean Era (which includes the systems up to 2500 Million years) (ii) Proterozoic Era (which includes the systems between 2500 and 570 Million years) (iii) Palaeozoic Era (which includes the systems between 570 and 245 Million years) (iv) Mesozoic Era and (which includes the systems between 245 and 66 Million years), and (v) Cenozoic Era (which includes the systems between 66 and 0.01 Million years). Besides, the stratigraphy of India can be divided into several divisions such as Archean System, Dharwar System, Cudappah system, Vindhyan system, Paleozoic, Mesozoic, Gondwana, Deccan Trap, Tertiary and Alluvial (Balasubramanian 2017). The rocks are very deformed and altered with recently deposited alluvium, varied mineral deposits and fossil records in Indian subcontinent. India possesses an impressive fossil collection consisting of stromatolites, invertebrates, vertebrates and plant fossils. The Deccan traps, Gondwana and Vindhyan form the major geographical area of the country (Wadia 1961, 1976). The map showing the geology of India (Limaye 2017) is presented in Fig. 4.1. The states covering Maharashtra, parts of Gujarat, Madhya Pradesh and Andhra Pradesh have Deccan trap formations. Deccan trap is formed as a result of an extensive melting event underneath the Indian Craton. The Indian Plate broke off from the Gondwana and proceeded further northward, when an excessive melting event happened due to a geologic hotspot, called the Reunion hotspot. Such melting resulted in the formation of Deccan trap as well as separation of Madagascar from the Indian craton. The states of Andhra Pradesh, Maharashtra, Madhya Pradesh, Odisha, Bihar, Jharkhand, Chhattisgarh,

Fig. 4.2 Sheeting on coarse-grained granite. *Source* NBSS&LUP ARCHIVES 2017

West Bengal, Jammu and Kashmir, Punjab, Himachal Pradesh, Rajasthan and Uttarakhand are from the Gondwanas and Vindhyans. The fluvatile rocks of Permo-carboniferous period possess a unique sequence out of the Gondwana sediments (Van dorsser 1976). These rocks are observed in the Rajmahal Hills in eastern India and in the valleys of Damodar and Sone rivers. Harindranath and Shivaprasad (2009) attempted to establish relationship between rock types and expected soils with taxonomic grouping (Table 4.1).

4.2.1 Plate Tectonics

In geology, Pangaea was a supercontinent of which India was also a part of Indian subcontinent and was then known by the name Indian craton. The Indian craton had Madagascar and Southern Africa sharing the southwest coast and Australia in its east coast. Pangaea was later split to Gondwana to the south and Laurasia to the north as a result of the rifting in the Jurassic period about 160 million years ago. However, the Indian craton was with the supercontinent Gondwana till its separation that took place in the early Cretaceous period about 120 million years ago. The Indian plate later moved in a quicker pace towards north to the Eurasian Plate retaining the connection between Madagascar and Africa, marked as the fastest known movement of any plate. There are evidences that confirm this movement at about 50 million years ago, though the common belief is that the Indian plate got separated from Madagascar about 100 million years ago. The closure of Tethys Ocean has resulted in this orogenic event, which is still active. The Caucasus range in western Asia, the Alps in Europe as well as the Himalayan mountains and the Tibetan Plateau in South Asia are formed as a result of the closure of Tethys ocean. Asian continent gets deformed westward and eastward directions as a result of the current orogenic event (Mishra and Ravi Kumar 2016). The Indo-Australian Plate is formed as a result of the collision of the Indian Plate to that of adjacent Australian Plate (Chatterjee and Scotese 1999).

4.2.2 Tectonic Evolution

Indian craton possesses granites and gneisses in its core which got exposed during the initial phases of tectonic evolution as a result of cooling and solidification of the earth's upper crust (Yin and Harrison 1996, 2000). This tectonic evolution happened during the Archaean Era. Peninsula has exposed granites and gneisses. The Aravalli range, extending for about 500 km from northern region till its end at Delhi has several deformations of faulting, folding, minor igneous intrusions, indicating the main phase of orogenesis, is the remnant of an early Proterozoic orogen called the Aravalli-Delhi Orogen. The second phase of orogenesis is marked by the erosion of mountains and further deformation of the sediments of the Dharwarian group. The sediments consist of records of volcanic activities and intrusions in the second phase. The sediments in the Early to Late Proterozoic period are present in their original stratification and found undeformed as arenaceous and calcareous deposits in the humid and semi-arid climatic regimes in the basins of Cuddapah and Vindhyans (Fig. 4.2).

In the Jurassic period, large grabens were created in Central India with conglomerates and sandstones from the Upper Jurassic and Lower Cretaceous period. India began to move towards Asia, separating from Australia and Africa during the Late Cretaceous period. During this time, uplift in southern India happened which resulted in Indian Ocean sedimentation before the Deccan eruptions. The rocks are found along the coasts in Tamil Nadu and Pondicherry. The Deccan lava flows as a result of the greatest volcanic eruption happened in the Mesozoic era, over more than 500,000 km^2 area, marking the final rift from Gondwana. In the early Tertiary, the first phase of the Himalayan orogeny, the Karakoram phase occurred (Medlictt 1965). The Himalayan orogeny has continued to the present day.

Table 4.1 Relationship between Lithology and Soil

Type of rock	Kind of soil formed	Soil orders
Granite coarse-grained	Red gravelly loamy and clayey soils	Alfisols, Inceptisols, Entisols
Granite porphyry	Red gravelly loamy and clayey soils	Alfisols, Inceptisols, Entisols
Granite fine-grained, andesite and rhyolite	Red clayey soils	Alfisols, Inceptisols
Granodiorite	Red clayey soils, black clayey soils	Alfisols, Inceptisols, Vertisols
Diorite	Red clayey soils, black clayey soils	Alfisols, Inceptisols, Vertisols
Dolerite	Red clayey soils with iron concretions	Alfisols, Inceptisols
Basalt	Shallow to deep black cracking clayey calcareous soils	Inceptisols, Entisols, Vertisols
Breccia, conglomerate	Red stony and gravelly loamy and clayey soils	Alfisols, Inceptisols, Entisols
Sand stone	Shallow to deep, red sandy, loamy soils	Inceptisols, Entisols, Alfisols
Shale	Red clayey calcareous and non-calcareous soils, black cracking clayey calcareous soils	Alfisols, Inceptisols, Vertisols
Lime stone, dolomite	Red clayey calcareous soils with kankar black cracking clayey calcareous soils with kankar	Alfisols, Inceptisols, Vertisols
Gneisses coarse-grained	Red gravelly loamy and clayey soils	Alfisols, Inceptisols, Entisols
Gneisses fine-grained	Red clayey soils	Alfisols, Inceptisols
Schists (chlorite, biotite, amphibolite, phyllite, pyroxene)	Red clayey calcareous and non-calcareous soils, black cracking clayey calcareous soils	Alfisols, Inceptisols, Vertisols
Marble	Red clayey calcareous soils with kankar black cracking clayey calcareous soils with kankar	Alfisols, Inceptisols, Vertisols
Quatrzite	Shallow to deep, red sandy, loamy soils	Inceptisols, Entisols, Alfisols
Alluvium	Deep, sandy, loamy and clayey, calcareous and non-calcareous alluvial soils	Entisols, Inceptisols, Alfisols, Vertisols
All the above rocks under temperate climate of Himalayas	Shallow to deep, sandy, loamy and clayey, calcareous and non-calcareous hilly soils	Entisols, Mollisols, Inceptisols
All the above rocks under humid climate of north-east states, Western Ghats, Eastern Ghats and coastal area	Deep, lateritie and red acidic clayey soils	Ultisols, Mollisols, Inceptisols, Alfisols
Arid climate with some of the above rocks	Shallow to deep, sandy, loamy, clayey, gravelly red soils, saline sodic soils. Patches of black soils	Aridisols, Vertisols

Source Harindranath and Shivaprasad (2009)

Fig. 4.3 Geomorphology of India. *Source* NBSS&LUP (2002)

4.2.3 Precambrian

The Precambrian era is the earliest phase of tectonic evolution representing solidification and cooling of surface crust which is prior to 2.5 billion years. These rocks are classified into the Archaean and Proterozoic system. Archaean system has mostly igneous origin and also consists of metamorphosed granitic and basaltic rocks together with some amount of sedimentary rocks. The Archaean formations occupy most of the southern and eastern India and parts of Assam, Jharkhand, Madhya Pradesh and Rajasthan. This group includes Dharwar System, Khondalites Charnockites and unclassified crystalline rocks. After the Dharwar formation, there was a period of volcanic eruption and igneous intrusions which has granitic masses, charnockites, dykes and other rocks such as unclassified crystalline rocks (Balasubramanian 2017).

4.2.4 Paleozoic

The Palaeozoic belongs to the period ranging from 542 to 251 million years and are called as Dravidian systems in India. The geological formations include cambrian system, silurian system, devonian system, lower carboniferous system, lower gondwana system, upper carboniferous and permian systems (Elias 2013). The Cambrian period is the period in which plenty of fossil evidences have been obtained in India. Spiti in central Himalayas and the Salt range in Punjab possess a thick sequence of fossiliferous sediments in its rocks marking the evidences of Cambrian period. The Silurian rocks of Kashmir are exposed in the Lidar valley. The Devonian systems of rocks are represented by the Muth Quartzites of Spiti, Kumaon and Kashmir. The Carboniferous systems of rocks in India are distributed only in a few places in the Himalayan region in Kashmir. They contain fossiliferous limstones and shales. Gondwana System of formations in India are mainly sedimentary sequences and are distributed in Godavari Valley to the Rajmahal hills, Domodar, Sone and Narmada Valleys, along Mahanadi Valley and parts of the Himalayan foot hills spread over Nepal, Bhutan, Assam and in Kashmir (Balasubramanian 2017).

4.2.5 Mesozoic

Mesozoic era ranged from 251 to 65 million years and it includes formations from Triassic, Jurassic, Cretaceous systems and Deccan traps. The Triassic system has ammonite ceratite made of arenaceous limestones, calcareous sandstones and marls (Elias 2013). The jurassic and cretaceous systems are known for their marine transgression and exposed in Himalayan ranges, Central Tibet, Kashmir, Ahmednagar, Kutch, Narmada Valley, Tiruchinopoly, Ariyalur and Rajamahendri areas in north, west and south India (Balasubramanian 2017). The Mesozoic era witnessed fissure eruptions which lead to the development of Deccan traps in India spread over an area of 300,000 km^2 in India. By the end of Mesozoic era, there were enormous lava flows resulting in the development of basaltic rocks and laterite and bauxite cappings are also present in many parts of these formations (Sen 2001).

4.2.6 Cenozoic

The Cenozoic era belongs to the period ranging from 65 million years to the present day and the geological formations belong to this group are tertiary systems, Eocene system, Oligocene system, lower and the middle Miocene, Pliocene system and Pleistocene system. The tertiary period marked the beginning of Himalayan orogenic movements and the volcanism associated with the Deccan trap. Lignite, fire clays, ball clays, terra cotta clay, sandstones, shell-limestone, Kaolin and petroleum are the major economic natural resources of these formations. In addition, geological formation of Quaternary period also exists in India belongs to the period from 1.6 million years till the recent past which includes all the recent alluvial deposits of Indo-Gangetic plains in the northern parts of India (Balasubramanian 2017).

4.2.7 Quaternary Period

Quaternary period represents the past two million years in the geological time scale. It represents the late Pleistocene and Holocene times where the marine transgression and regression are common (Bloom 1983). The sea-level rise and fall have left their signatures not only in the sedimentaries of the coastal plains but also in inland areas (Pandarinath et al. 1998). Sea level change causes formation of variety of coastal land forms (tidal flats, beaches, marshy lands swamps, etc.), extensive sedimentation and geomorphological changes in the coastal areas. The Konkan coastal plain and the miliolite lime stone beds in Saurashtra are believed to have uplifted in recent times (Nair and Hashimi 1980). In inland, the changes are mostly alluvial deposits, consisting of sand, silt and clay of Indo-Gangetic plains, banks of Brahmaputra River, deltaic regions of Cauvery, Krishna and Mahanadi belong to this era which has formed as a result of erosion from the Himalayas.

4.3 Geomorphology of India

Geomorphology is the science of landforms, their origin, evolution and distribution within the physical landscape. An understanding of such geomorphological processes is important to know the physical geography of the landscapes including mountains, hills, volcanoes, islands, valleys, plateau, flood and coastal plains. Kanchan (2018) rightly said that the landscape, which looks so constant, is on the move and one has to explore and study it. The Indian sub-continent

Fig. 4.4 Folded quartzite mound, Nagpur. *Source* NBSS&LUP Archives 2017

comprises three major geomorphological components (Fig. 4.3), namely the Himalayas, the Great Plains and the Peninsula (Spate and Lear Mouth 1967). Geomorphic heterogeneity is normal in the sub-continent where land building and erosional forces have been active since geological times. Stratigraphic and tectonic history, relief and aggradational and degradational processes are responsible for the formation of different landforms assemblages viz., the Northern Mountains, the Great Plains, the Peninsular Plateaus and the Islands (IIASES 1971). The drainage is mainly divided into the Bay of Bengal and the Arabian Sea systems with a distinct dividing line running along with Sahyadri, Amarkantak, Amaranath and the Sullej-Yamuna divide. Major part of the country drains into the Bay of Bengal (Pofali and Shankaranarayana 1982).

4.3.1 The Northern Mountains

This covers the northern border of India extending from Pakistan in the west to Burma in the east occupying the Himalayan ranges, which started to build the Himalayas in early tertiary times as a result of Indian continent-Eurasian continent collision (Yin and Harrison 2000). This resulted into three major fold axes viz. the Greater Himalayas (Himadri), the Lesser Himalayas (Himanchal) and the Outer Himalayas (Siwaliks). The rivers Ganga, Indua, Kosi and Brahamputra have cut deep gorges to enter the Great Plains. The troughs stretching the ranges are, however, occupied by the longitudinal valleys of streams in respective upper reaches (Sinha et al. 2005).

Kashmir Region (Western Himalayas) It is located between 32° 17′ and 37° 05′ N latitudes and 72° 40′ and 80° 30 E longitudes. It is represented by mountain ranges with snow-covered peaks and large longitudinal valleys of subsequent streams. Deep transverse gorges are formed by antecedent streams entering through the ranges. The region has complex geological history from the Archaean to the Recent. A complete stratigraphic record of sedimentation is exhibited in this region. Structurally, it is a folded mountain belt with anticlinoriums and synclinoriums. The basement complex consists of metamorphosed Archaeans, intrusive granites, gneisses and schists. The rocks of Dharwar, the oldest sedimentary system, are also present. The major marine sedimentation is buried by the volcanic activity. Traps and lime stones are observed in the Karakoram ranges. The rocks of the tertiary period are observed in the outer mountains and in the Indus valley.

The Vale of Kashmir, a structural basin lying between Pir Panjal and Himadri, has flat topped terraces called Karewas. These are the Pleistocene sediments of lacustrine origin in which bands of marl, loess and lenticles of conglomerates of old deltaic fans are found. The rivers Jhelum and Ravi have developed a narrow strip of highly dissected plain located in the south-west part of Kashmir along the Pakistan border (Bose 1961).

The Akasai-Chin region, situated in the north-east portion of Kashmir, represents the denuded peneplained surface of an intermontane plateau with impressions of excessive glacial erosion. A number of salt lakes of various sizes appear, dotting the whole terrain. The Jammu hills with steep escarpment towards the Vale of Kashmir rise up to 600 m from the Punjab plain.

Himanchal Region This region is lying between 32° 22′ and 33° 12′ N latitudes and 75 40′ and 79° 47′ E longitudes presents the most complicated zone of the Himalayas. The elevation increases from west to east and from south to north. The region covers parts of the Greater Himalayas, the Lesser Himalayas and the Outer Himalayas or Siwaliks.

The Greater Himalayas (elevation 5000–6000 m) runs along the eastern boundary and is cut across by the Sutlej. The easternmost Zaskar range separates Spiti and Kinnaur from Tibet. There are many glaciers throughout this section with terminal moraines, hanging and ‘U’ shaped valleys and glacial lakes. The topography of the Lesser Himalayas is marked by ranges and valleys. The Kangra valley is a longitudinal valley whereas the Kulu valley runs transverse to the main alignment. The valleys are carved by the rivers Chenab, Ravi, Beas and Sutlej. Glaciers exist south of Lahul (Rasoul 2010).

The Siwaliks slope to the south with scarps and piedmont plains bordering and Gangetic plains built up by the outgoing streams known as ‘Choas’. The Siwaliks dip into elongated structural valleys to the north known as Duns.

Uttar Pradesh Region This region falls between 29° 05′ and 31° 25′ N latitudes and 77° 45′ and 81° 0′ longitudes also cover parts of the Greater Himalayas, the Lesser Himalayas and the Outer Himalayas in the northern part of Uttar Pradesh (Owen 2014). The Greater Himalayas are separated from the Lesser Himalayas by the main Central thrust, which is well outlined in the Kali gorge and the valleys of the Goriganga and the Pindar rivers. The major rock formations present here are quartzites, migmatities, gneisses, garnet-schists and dioritic amphibolites. Relief averages between 4800 and 6000 m and the important peaks are Nanda Devi and Kamat. There are a series of glaciers.

The main structural features of the Lesser Himalayas are the Krol belt, the Deoban-Tejam belt and the Almora-Dudatoli Crystalline thrust. The ranges are separated by deep valleys. The lake basin of Kumaon is located at the fringe of the Lesser Himalayas. The Outer Himalayas consists of lower, middle and upper Siwaliks. The beds of sandstones and shales are separated from the Eocene bed of the Lesser Himalayas by the main boundary fault. In between the ranges are the 'Duns', the Dehra Dun being the largest of them.

The rivers have formed deep valleys with dendritic drainage pattern. The Ganga, the Yamuna and the Kali constitute the major river systems of the area. Glacial topography is well preserved above the elevation of 3000 m. Flat basins formed by ancient lake bottoms due to uplift of the Himalayan ranges are found in a number of places. Intermittent upheavals witnessed by the valleys have given rise to river terraces, incised meanders and knick-points in the form of waterfalls.

Eastern Himalayas

The Eastern Himalayas, lying between 26° 30′ and 29° 30′ N latitudes and 88° 02′ and 97° 05′ E longitudes, cover the eastern portions of the Greater Himalayas, the Lesser Himalayas and the Outer Himalayas in the lateral form in Bhutan and Arunachal Pradesh. Western part of the Eastern Himalayas is characterised by the great Himalayan peaks, the Central basin and the Darjeeling ridges (Karan and Jenkins 1963).

The greater Himalayas in this part consist of snow-capped ranges about 7000 m high and distinguish itself from the flat or undulating tableland of Tibet on the northern side. Southward spurs form the watersheds of the Sankosh and the Manas rivers dividing Bhutan into two parts. The valleys are at elevations of 3000–4000 m in, extending 60–70 km in the southward direction. Proceeding eastwards, the ridges of Lesser Himalayas more often tend north–south in line with the general change of direction in the axes of the mountains around the Sino-India—Burma Plateau. The Siwaliks rise to a height of 300 m from the plain of Assam and extend to about 10–15 km northward in the form of a series of knife-like ridges (Rasoul 2010).

The region is dissected by numerous rivers and their tributaries. The great Brahmaputra basin forms the natural drainage of the region. The rivers flow between high rocky mountains confining their channels to narrow valleys. Important river systems are the Tista, the Toras, the Lohit, the Wong Chu, the Me Chu, the Manas, the Dihang, the Karla, the Subansiri, the Sankosh and the Siang.

Assam-Burma Ranges

The region between 21° 57′ and 28° 23′ N latitudes and 91° 13′ and 97° 25′ E longitudes is a component of the Assam-Burma geological province which was a part of Tethys Sea in the Archaean Era. A final orogenic phase during early part of Pleistocene raised the upland to its present level. Folding has been so intense that the rock beds are vertical, resulting in bare hill sides. The region is still unstable with several faults and thrusts.

The ranges in Assam running east–west take a hairpin bend crossing the Lohit and further extend southwards across the region beyond which they are known as the Arkan Yoma. The region has north–south parallel ranges with narrow valleys.

The general elevation of this region increases towards north–east, and ranges from 900 to 2100 m above the mean sea level in the large part of Manipur hills and Nagaland. The eastern districts of Arunachal Pradesh are located at higher altitudes. Naga Range marks the eastern frontier of Nagaland and is a water-divide between the rivers of India and Burma. This range has several peaks of more than 3000 m height with the highest peak rising to 4500 m (Rasoul 2010). The Tripura and the Cachar plains which form part of Burma valley built by detrital material are 150 m below the range height, but these are pierced by a series of spurs projecting from Mizo hills. The surface is dotted with a number of lakes and marshes. Another important valley is the Imphal Valley which is roughly oval in shape.

4.3.2 The Great Plains

The Great Plains are often known as Ganga-Satlej plains which are transitional belt between the Himalayas and peninsular India. It is one of the world's greatest stretches of flat alluvium. This is an aggradational plain with the Ganga and the Brahmaputra rivers. The thickness of the alluvium varies but it is more in the Ganga Plain. The cones of the Kosi in the north and the Sone in the South have thicker alluvium while the intracone areas have relatively shallower deposits (Singh and Singh 1971). The plain extends from extremely arid and semi-arid environments of Rajasthan in the west to the humid and per humid delta plain and the Brahmaputra valley in the east. Topographical uniformity is a common feature throughout except in the arid western Rajasthan. The nature of the material brought down by the rivers, however, signify the local geomorphic variations. The elevation varies from about 150 m (Bengal delta) to nearly 300 m (Punjab Plains) with extremely low gradients.

Along the northern margin of the plain lies three narrow but distinct strips, 'the Bhabar', 'the Tarai' and 'the Choa' plains. The Bhabar, a piedmont plain 10–15 km wide, is composed of unassorted debris from the Himalayas. The surface streams disappear in this zone of boulders and sand. Immediately south of this is the 15–30 km wide, relatively low lying Tarai region characterised by medium-textured sediments, natural tall grass cover, emergent and ill-defined water channels and high water-table. The Choa plains are the coalescing alluvial fans built by the torrential streams flowing down the southern slopes of Siwaliks along the Punjab plains.

The southern margin of the plains, in contact with the edge of the southern uplands, is often encroached by the projections of the Peninsular masses sometimes up to the bank of the Ganga.

(i) *Indus Plains*

The Indus plains lie between 24° 30′ and 32° 30′ N latitudes and 69° 45′ and 77° 36′ E longitudes. They are mainly fluvio-aeolian plains and are divided into two parts on the basis of dominant geomorphic processes.

Rajasthan Plain The Rajasthan plain lying between 24° 30′ and 30.12′ N latitudes and 69° 45′ and 76° 45′ E longitudes forms the marusthali eastern portion of the Thar and the adjoining bagar (steppe land) to the west of the Arvallis. It is formed of aeolian deposits of Pleistocene and Recent epoch. The large amount of sand in this area is considered to have originated from long-continued aridity and drifting of sand. Similarly, it is either rock weathering product or had originated in mountain-tinges and transported and deposited by the rivers in the geologic past (Moharana et al. 2013). Aggraded alluvio-fluvial deposits have been found covering wind sorted fine sands and vice versa. Besides, the aeolian and alluvial deposits mentioned earlier, marine deposits of Jurassic and late Tertiary are exposed in Barmer, Jaisalmer and Bikaner; rocks of Vindhyayan systems and Malani volcanics and Jalore-Siwana granites are also encountered (Bakliwala and Wadhawan 2003).

Thus, the relief features of the plain are the products of both fluvial and aeolian processes. The sand covered desert extends from the Great Rann along the Pakistan border to Punjab in the north–west. Transverse and longitudinal dunes and interdunal sandy plains are the characteristic features of this area. To the east are rocky plains with playa lakes having cetripetal drainage (Dhir and Singhvi 2012).

Eastermost part of the plain is the semi-arid bagar land drained by the river Luni. The older rocks here protrude above the surrounding sandy surface. To the north lies the semi-arid Shekhawati tract, a plain characterised by inland drainage and strewn with salt lakes such as the Sambar lake. In the extreme north lies the Ghaggar plain (Babar et al. 2012).

Punjab Plain It is in the north–west between 27° 39′ and 32° 30′ N latitude, and 73° 51′ and 77° 36′ E longitudes in the states of Punjab, Haryana and the Union Territory of Delhi. Eastern border is marked by the river Yamuna, while the Siwaliks forms the northern boundary. There are the rivers Ravi and Sutlej in the West and desert in the south. Major part of the plain is recent and its surface has been built up by the silting action of its wayward streams flowing from the Siwaliks and the Himalayan region. The old high bank of the Yamuna in the north Delhi forms the summit level of the plain (Sinha and Peter 2007). The topographical differences in the north and south have resulted in the formation of a depression near the eastern margin of Rohtak district.

Greater part of the plain lacks marked relief variations. The main physiographic features of the plains are the interfluvial zones which are characterised by higher ground in the middle and the flood plains on either banks of the principal rivers, viz., the Ravi, the Beas and the Sutlej. In the remaining part of the plain also similar pattern follows, i.e. former stream courses at lower level interspersed by the interfluves of higher relief which are locally termed as Khadar, Dharia or Nardak. At places, sand dunes and sand ridges form these interfluves. Southern part of the plain bordering the arid tract is sandy (Srivastava et al. 2014). The southern slopes of the Siwaliks are indented by a large number of hill torrents that flow down the submontane belt and form the sandy piedmont plains of the Siwaliks. In the extreme south of the plain, the continuation of the Aravalli hills extends towards north and northeast as low detached ranges with gaps and a few depressions at places (Bakliwala and Wadhawan 2003). The Khadar of the Yamuna lies between the present watercourse and the old high bank of the river. The old high bank runs parallel to the river throughout its greater length in the Punjab plain. It narrows down rather quickly towards south.

(ii) *Ganga Plains*

The Ganga plains occupy the vast Indo-Ganga trough, located between 21° 25′ and 30° 17′ N latitudes and 73° 03′ and 89° 58′ E longitudes in the States of Uttar Pradesh, Bihar and West Bengal. The plains are generally the products of the alluvium filling the trough in between the foothills of the

Siwaliks in the north, subterranean Delhi ridge in the south–west and the peninsular uplands in the south. The depth of the alluvium is between 4000 and 6000 m. In eastern Uttar Pradesh and Bihar, the average filling is estimated at 1300–1400 m (Shukla and Bora 2003). It is deeper towards Siwaliks and thins out towards peninsular margin in the south. The lower Ganga plain in general and the delta in particular in West Bengal are also products of fluvial filling of the trough. The delta plains are of recent alluvium whereas the extensive northern region is of old alluvium.

Topographically significant and complex part of the region at the foot of the Siwaliks is the submontane belt running from west to east consisting of two parallel strips of the piedmont zone namely, the bhabar and the adjoining relatively gently sloping tarai belt. The bhabar is a zone of alluvial fans of unassorted sediments built up by seasonal torrents traversing the Siwalik scarps. The tarai is an alluvial zone which is a nearly level piedmont plain with high ground water table.

In the south–west, the landscape is related to degradation of the river Chambal with its ephemeral affluents which have stripped the alluvial deposits of the Vindhyayan and the Himalayan origin giving a picture of broken topography with residual hills. In the east, the area is dotted with residual hills of Bhagelkhand plateau, narrow flood-plain strip of the Sone, and a number of hills and elongated ridges, running in south–west north–east directions.

The upper Ganga plain can be distinguished as the Ganga-Ghaghara doab, the Ganga—Yamuna doab and the Yamunapara—a rivinous tract. The doab consists of the bhangar or the interfluves (old flood plain) that are more stable and sloping into the khadar (recent flood plain) on either side. The old flood plain also shows preponderance of local micro variations in slopes that can be distinguished in smaller facets. The old and the recent flood-plains are also shown with levees, marshes, oxbow lakes and meander scars. The old flood plains maybe sometimes 15–60 m above the adjoining flood-plain. Old flood plain is also called the established *Diara* land, whereas recent flood plain is known as active *Diara* land in Bihar (Mishra et al. 2001a, b). The Tal land soils are occurring south of the Ganga almost in strip from Buxar to Bhagalpur in Bihar (Tiwary et al. 1989; Mishra et al. 2001b).

The middle Ganga plain which covers eastern Uttar Pradesh and Bihar can be divided into the Ganga-Ghaghara doab, the Ghaghara interfluve, the Gandak-Kosi interfluve and the Kosi-Mahanada interfluves (Mohindra et al. 1992). The whole plain is less than 100 m above the mean sea level and the heterogeneity within the drainage system is produced due to local features such as natural levees, oxbow lakes, marshes, old river channels and low alluvial terraces.

The physiographic diversities in the lower Ganga plain can be distinguished by the northern basin that adjoins the duars or the tarai and the inactive delta comprising the land of older alluvium on the northern part and the active delta of the Sunderbans and recent deltaic regions comprising Hoogly, Howrah and parts of adjoining districts. The drainage system consists of the tributaries and distributaries of the Ganga river discharging into the Bay of Bengal.

(iii) *Brahmaputra Valley*

The Brahmaputra Valley or the Assam Valley situated between 25° 44′ and 27° 55′ N latitudes and 89° 41′ and 96° 02′ E longitudes is a well-demarcated physical unit within the girdle formed by the Eastern Himalayas, the Patkai Bum and Naga Hills and Garo-Khasi-Jaintia and Mikir hills. It extends from the easternmost tip of Assam near the syntaxial bend of the Eastern Himalayas to the west of Dhubri on the border of Bangladesh (Mukhopadhyay et al. 2012). The Brahmaputra Valley is built mostly by the aggradation of the sag formed during the rise of the Himalayas and brought down by the Brahmaputra and its tributaries. It is almost a level flat plain from its north–east corner at Sadiya (130 m) to Dhubri in the West (30 m). The valley is well defined between the main boundary fault on the north and the Naga thrust in the south.

The valley in its northern margin is characterised by steep slopes, but the southern margin has a gradual fall from the southern hill ranges. It is 80–100 km wide in upper Assam and about 55 km in its middle part where the river encounters the granite gneissic projection of the Mikir hills. Beyond this, it widens westwards, for the plain of the Kopili enjoins the main valley. However, the valley again narrows down to about 65 km on an average when it runs in the gap between the Shillong Plateau and the Bhutan Himalayas. In the valley, a good number of isolated hillocks present on both the banks of the river, right from Tejpur and Mikir hills to as far west as Dhubri, are presumed to be detached from the Meghalaya plateau by the degradational work of the river.

There is a marked difference between the physiography of the north and south banks of the Brahmaputra River. Innumerable tributaries running down from the Arunachal and Bhutan form a series of alluvial fans and flow in almost southerly direction. But, before finding their way into the Brahmaputra, the tributaries run almost parallel to the mainstream as they encounter its levees. This leads to the formation of oxbow lakes and huge marshy tracts. The alluvial fans formed by the coarse alluvial debris in the northern fringe of the valley have given rise to tarai or semi-tarai conditions with dense forest cover.

The southern part of the valley is less wide and the tributaries in the south–east are considerably larger. The headstreams of the tributaries, the Dhansiri and the Kopili, have by their headward erosion almost isolated the Mikir and the Rengma hills from the main mass of the Meghalaya plateau. But meandering in the eastern part of the valley is conspicuous with a number of bits and oxbow lakes.

The Brahmaputra, itself highly braided, has given rise to innumerable river islands including Majli, the biggest (929 km^2) river island in the world. The Brahmaputra with its tributaries carries tremendous volume of water and silt during monsoon period. Active bank—erosion results in filling of the river beds every year. In addition, sometimes earthquakes disturb the natural course of the river creating additional forces for flooding.

4.3.3 The Peninsula

The Peninsula region has remained above the sea level after the forces of erosion acted upon it for hundreds of millions of years. This landscape consists of several cycles of denudation with orogeny, epeirogeny and cymatogeny, effusion, metamorphism of deep-seated rocks, tearing, custatism and widespread resurrection (Singh 1967). Peninsular India, a remnant of the Gondwana land, is the oldest part (pre-Cambrian) of India and is one of the stable land masses of the world except for some marine transgression at a few places. The whole terrain is built up of large and small undulating plateaus with their summits seldom rising to more than 1200 m above the mean sea level. Hard rocks have withstood the forces of denudation. Gentle gradients are due to prolonged weathering and erosion and hence, the low hills are either the remnants of the old mountain system as are the Aravalli hills or the harder part of the plateau itself that has withstood erosion as in the case of Western Ghats. It consists of the oldest shield of the world, occupied by gneisses and granites of the Archaean system of the pre-Cambrian age that covers about half of the Peninsula. The Vindhyayan, the Aravalli, the Cuddapah, the Gondwana, the Decean trap and the Alluvium of Narmada and Tapti basins are the other formations (Radhakrishna 1983). The configuration and drainage of the peninsula have been influenced by fracturing and tilting of the massif. The fault in which Narmada river flows, divides the plateau into two parts, the Central Uplands in the north comprising the Aravalli range, the east Rajasthan upland, the Madhya Bharat plateau and the Bundelkhand uplands and the Deccan plateau in the south comprising Satpuras, Western and Eastern Ghats and a large number of minor plateaus (Balasubramanian 2017).

The northern part is tilted north with its drainage towards the Ganga basin through the rivers Chambal, Sone, Damodar, Ken and Betwa. The southern part is tilted to the east giving eastward trend to the Mahanadi, the Godavari, the Krishna, the Cauvery and the Pennar rivers draining into the Bay of Bengal.

(i) *Central Uplands*

The elevated land to the south of Ganga trough, rising from 300 to 1200 m above the mean sea level, is known as Central Uplands. It comprises the Malwa Plateau, Aravalli ranges, Madhya Bharat Pathar, east Rajasthan uplands, the Vindhyayan, the Bundelkhand uplands and the Narmada valley. The Malwa plateau north of Vindhyatan has wide expanse of lava landscape overlying the pre-cretaceous surface. The general elevation is 300 m but not more than 600 m above the mean sea level. There are rolling plains dotted with sandstone and trap hills covered with forest.

To the west and north of the Malwa plateau are the worn-down Aravalli ranges running over 800 km, terminating on Delhi ridge in the north–east. It acts as a water divide between the Ganga river system and the rivers flowing to the Arabian Sea. In the north, it consists of hard quartzite ridges and strike valley, carved out of soft phyllites. In the middle there are highly eroded narrow ridges while in the south folded hills of Aravalli quartzites with alternating valleys on soft mica schists are found. The highest peaks of Mount Abu hills in the south–west is 1090 m while in the north–east, it is 1722 m above the mean sea level.

The tract east of Aravalli ranges, at an elevation of 250–500 m above the mean sea level, is known as East Rajasthan uplands (Kar and Ghose 1984). This is an extensive alluvial plain underlain by igneous and sedimentary rocks and dotted with low monodnocks. The tract is drained by the Banas, the Kali, Sindh and the Parbati rivers, tributaries of the river Chambal. The Madhya Bharat pathar formed of Vindhyayan sandstone, limestone and shale is a plateau country lying to the east of Aravalli ranges. It has a rolling surface with rocky terrain. The land-form constitutes table-lands, scarps, gorges, ridges, piedmont plains and hills. To the south of the Malwa plateau are the Vindhyayan uplands. The ranges stretch across nearly the whole width of 1050 km of the Peninsular India with an average elevation of 300 m. Shales, slates, sandstones and limestones are the major rocks. They are undisturbed with an almost horizontal stratum (Singh and Singh 1971). Vindhyan uplands constitute the northern boundary of Deccan plateau. They slope gently towards north and are devoid of well-marked spurs. To the south of the Bundelkhand are the Vindhyayan scarp lands with three important plateaus, namely, Bhander, Kaimur and Rewa. This region is an erosional surface and landscape is marked by Plateau summits, valley bottoms, ridges, isolated hills, mesas, pediments, gorges and the like, conditioned by lithology and structure. The area between the Yamuna and northern scarps of Vindhyayan plateau is Bundelkhand plateau. It is an old erosional surface formed by Bundelkhand gneiss. It slopes to the north–east in broad step-like form with occasional tors and reefs. The northern portion is alluviated with deep ravines along the rivers Chambal and Betwa. Narmada flows through an asymmetrical valley enclosed between the Vindhyayan range in the north and Satpura range in the south. Its thalweg has many slope breaks. The river flows alternatively through alluvial plains and rocky gorges.

(ii) *Deccan Plateau*

The Deccan Plateau consists of Satpura range and Maharashtra plateau in the north and Karnataka and Telangana plateaus in the south.

Satpura Range South of the Narmada, rises the Satpura range which essentially is a lava—capped plateau of Archaean gneisses and schists abutting steep-sided hills of the Gondwana sandstones in the west. The trends of the ancient folds are in east, north–east, west and south–west directions which conform to the present topographical features. The highest surface above 1000 m has developed partly on the upper Gondwana sandstones and partly on trap rocks. The hard sanstones, though quite resistant to weathering, have been cut through, to form canyons. Structurally the Satpura range has three parts, (i) Rajpipla hills (width 60 km) in the western part with steep slopes and pointed tops arc covered by the Deccan lava, (ii) The Mahadeo hills are composed of quartzites of the Gondwana system and pink sandstones. The Mahadeo hills slope gently northwards but are cut off in the south by a stupendous line of cliffs overlooking the low hills formed of shales alternating with earthy sandstones and coal seams of the lower Gondwana age. The Mahadeo hills act as a water divide between the rivers Narmada, Tapti and Godavari. (iii) the eastern part consisting of Maikal range (width 200 km) with elevation from 600 to 1000 m and is covered by trap flows of about 150 m thick in the neighbourhood of Amarkantak which is composed of Gondwanas and Archaean gneisses (Balasubramanian 2017). It presents structural diversity and looks like a jumble of hills and valleys. On the western side, the Satpura range consists of two blocks separated by the wide gap of Burhanpur through which Tapti river flows.

The hills of the western block have been considerably affected by orogenic disturbances soon after the outpouring of the lava. This block is also traversed by a large number of dykes forming parallel ridges rising 50 m above the ground level. To the east of Burhanpur gap are the traps forming Gawilgarh hills, Betul plateau and the Kalibhit hills.

Maharashtra Plateau The Maharashtra plateau is built up of nearly horizontal sheets of basaltic lava, the Trap, which slopes very gently eastwards conforming to the dip of the lava flows (Fig. 4.4). The plateau is terraced and scarped. The south–east flowing Godavari, Bhima and Krishna rivers are primarily responsible for the characteristic landscape. The scarps bordering the higher strips owe their origin to compact basalt overlying softer types of the trap. All the three major rivers of the region have meandering courses with fairly steep banks ranging in height from 10 to 15 m (Fig. 4.5). South of the Bhima valley, the Mahadeo range fans out to form another group of lava plateaus with an average elevation of about 700 m. Types of basaltic terrain on varying topography are evidenced from southern part of Bijapur towards Krishna valley. Low mounds of volcanic ash beds and fragmental volcanic products also appear near Bijapur. Dominant drainage pattern is of trellis type, and at places the river courses are structurally controlled. The area has a number of dykes which have affected the course of the tributary streams. The Maharashtra Plateau is further subdivided into 5 micro-units: (i) the Ajanta Hills, (ii) the Godavari valley, (iii) the Ahmednagar-Balaghat plateau, (iv) the Bhima Basin, and (v) the Mahadeo Upland.

Karnataka Plateau The larger a part of the upland may be a plain in numerous stages of development (Fig. 4.5). Generally, it's a rolling surface on mild slopes with occasional monadnocks as found close to Kolar and Mysore. Transverse to this plateau region from south–west to south–east are a minimum of four erosional surfaces (Chatterjee 1961). The highest surface is encountered on the south–west within the Malnad region and therefore the lowest on the north–east in the vale plains of Tungabhadra and Hagari rivers. The current surfaces are sliced out of the traditional folded schistose rocks of Dharwar age and intrusive gneisses, granites and charnockites (Ramarao 1962) (Fig. 4.6).

Landform-Soil relationship Catenary sequence analysed through landform cross-section which could be used for establishing soil landform relationship (Fig. 4.7). The study conducted in Homnabad, Bidar district trending SE–NW direction with the elevation ranging between 450 and 600 m above MSL. Three distinctly different landforms have been demarcated based on contours and image characteristics. They are the upper plateaus, lower plateaus and valley floors. Pedon 1 was positioned on the centre of the upper plateaus and it is shallow, well-drained red ferruginous soil with gravelly sandy clay surface and gravelly clay subsoil. It has a hard and compacted colour as well as textural cambic horizon. The slope ranges between 3 and 8%. The entire profile has 40–70% iron and quartz gravel. Pedon 2 and 3 are located on the lower plateau. Here the slopes are shallower, ranging between 3 and 5% and slope lengths are much more than that of the upper plateaus. Hence, both the pedons are very well developed, very deep, well-drained red ferruginous

Fig. 4.5 Polygonal structure of basalt. *Source* NBSS&LUP Archives 2017

soils with slightly acid gravelly sandy loam to gravelly clay surface horizons and slightly and slightly acid gravelly clay subsurface horizons with hard and compact thick argillic horizons.

The subsoils have 40–60% iron and quartz gravels. Pedons 4 and 5 are positioned on the narrow valley floors just west of Homnabad on either side of the tributary stream. Both the pedons are very deep, moderately well drained to imperfectly drained, calcareous cracking clay black soils with well developed shrink-swell potential. Lithological discontinuity occurs at about 100 cm on pedon 4 and at about 80 cm on Pedon 5. Below the discontinuity, the subsurface horizons have 5Y and 2.5Y hues indicating the presence of thoroughly mixed powdery lime and the availability of permanent through flow water.

Catenary Sequence of Granite/Gneiss Landform

The granite gneiss landform covers an area of about 8.1 M ha in parts of Bangalore, Kolar, Tumkur, Mandya, Mysore, Hassan, Chikmagalur, Shimoga, Chitradurga, Raichur, Bellary and Gulbarga districts (Fig. 4.8).

Steep high hill ranges are represented by Sullya pedon associated with rock outcrop, deep, well-drained, gravelly sandy clay loam to clay in texture, classified as Fine, Kaolinitic, isohyperthermic, Ustic Kandihumults. Steep high hill ranges owing to heavy rainfall and hot humid tropical climate on granite gneissic landform allowed deep to very deep soils with low activity clays and consequent illuviation of clay with low base saturation and low CEC.

Steep low hill ranges, isolated hills and dissected hills and valleys is represented by Beltangadi pedon which has very deep, well-drained, gravelly sandy clay to sandy clay in textures and classified as Fine-loamy, kaolinitic, isohyperthermic, Ustic Kanhaplohumults. Owing to heavy rainfall

Fig. 4.6 Laterite brick cutting at Kumta, Udupi district, Karnataka Catenary sequence of lateritic terrain of Karnataka plateau. *Source* NBSS&LUP Archives 2017

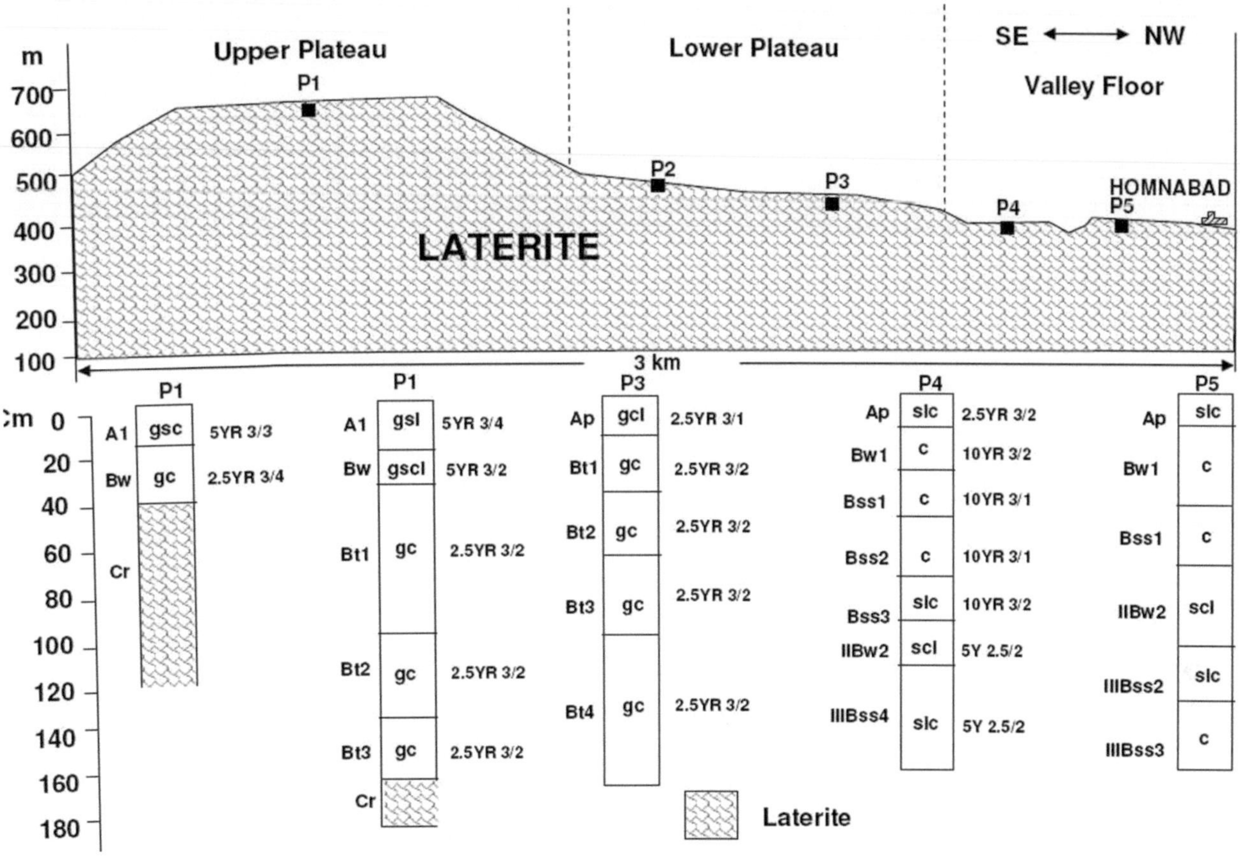

Fig. 4.7 Landform-soils relationship in Lateritic terrain, Homnabad, Bidar. *Source* NBSS&LUP Archives 2017

and hot humid tropical climate over steep low hill ranges, soil development has taken place with display of kandic horizon with low bases and CEC. This land form tends to have multinutrient deficiencies.

Moderately steep sloping elongated ridges and foot hill slope is represented by Kollur pedon where deep, well-drained, gravelly sandy clay loam to gravelly clay textures and classified as Fine, mixed, isohyperthermic, Kandic palehumults. This area is associated with ferrugious schist landform and dominated by soils having high organic carbon and high degree of soil profile development with low CEC and base saturation. Undulating upland is physiography is represented by Molahalli pedon where deep, well-drained, gravelly sandy Loam to gravelly clay loam in textures classified as kaolinitic, isohyperthermic, kanhaplic haplustults.

Telangana Plateau Telangana plateau contains a wide range of geological formations from the oldest Dharwar schists to the recent alluvium (Gade 2016). Among these archean gneissic and granite complex dominate the rock formation in Telangana (Fig. 4.9). The rolling surface and impervious granitic rocks additionally produce favorable conditions for the storage of water. It is drained by two rivers, the Godavari and therefore the Krishna. A north–south running escarpment is faintly visible separating the upper Golconda plateau from the lower Nalgonda plateau. The center Godavari basin could be a faulted syncline of Gondwana rocks. The complete region is split Ghats and peneplains and has valuable coal mines within the districts of Adilabad, Karimnagar, Warangal and Khammam.

(iii) *Eastern Plateau*

Eastern plateau is represented by Baghelkhand plateau, Chotanagpur plateau, Garhjat hills, Mahanadi basin and Dandakaranya upland. It is developed on granite gneisses and massive granite. The landscape exhibits a peneplained surface with rugged topography.

Baghelkhand Plateau Baghelkhand plateau lies within the east of the Maikala range and the south of the river sone. There are two coal seams of Gondwana age separated by hills of Gondwana sandstone and archean metamorphic rocks. It's the oldest dissected plateau comprising of granite and gneiss spreads over five districts of Uttar Pradesh and four districts of Madhya Pradesh. Streams like Betwa, Dhasan and Ken flow through the plateau (Kumar and Rai 1981).

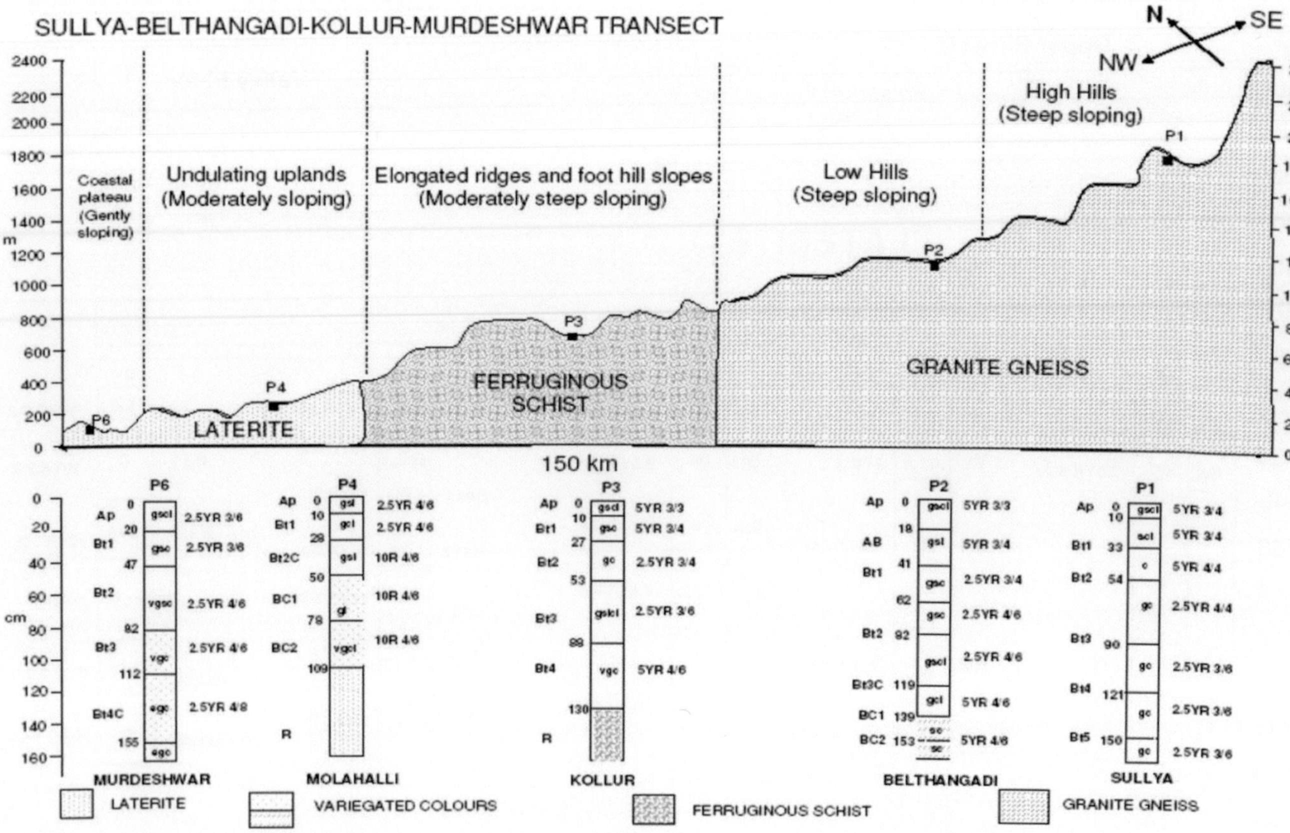

Fig. 4.8 Landform soil relationship in Sullya-Beltangadi-Kollur-Murdeshwar transect. *Source* NBSS&LUP Archives 2017

Chotanagpur Plateau The Chotanagpur region is a plateau in eastern India, covering a lot of the recently created Jharkhand state, yet because the bordering areas of state, West Bengal, Bihar, and Chhattisgarh. The Gondwana substrates attest ancient origin to the plateau. Geohistorically, it absolutely was a part of the Deccan Plate, that bust free from the southern continent throughout the Cretaceous period to embark upon a 50-million-year journey that was violently interrupted by the northern Eurasian continent (Singh 2012). The average elevation on the upper half, i.e. Ranchi, is 700 m (Fig. 4.10). It conjointly includes Basra plateau and Simdega plateau at much lower elevations in south and south–west. Rounded hills of huge granite and slightly elevated terraces of older flood plains diversify the topography of those plateaus. Excluding these low hills, the horizon seems to be quite flat. Eastern a part of the plateau is usually a rolling plain. All the important streams together with the Koel and Subarnarekha have their supply close to Ranchi city and radiate altogether directions while not being influenced by present relief and are characterised by variety of rapids and waterfalls within the thalwegs. The Ranchi plateau is deeply dissected around its edges, giving rise to steep escarpments domestically called ghats plateau (Kumar and Rai 1981).

Garhjat Hills Extending from the southern limit of the Ranchi plateau are the Garhjat hills close to the Mahanadi valley. The elevation is far less than the Ranchi plateau, however, local relief is more pronounced, dividing the hills into hillocks, plateaus and uplands that are remnants of the older peneplains. Scarps, isolated hills of inter-hill valleys, intermediate and lower plateaus, valley bottoms and hill terraces are among the landforms that may be distinguished.

Mahanadi Basin South of Maikala range is that the Mahanadi basin, the upper portion of that is designated as Chattisgarh basin. The floor of the basin, a concave depression, consists of horizontally bedded or low dipping limestones and shales of Cuddapah system (Murthy et al. 2012). The ridges that form the southern rim of the basin are the inward dipping sandstones and quartizites. Coal beds occur in this basin close to Korba. The landforms distinguished are hill ranges, scarps, isolated hillocks, inter-hill valleys, pediments and flood-plains.

Fig. 4.9 Cuddapha stone cutting at Mudhol, Bagalkot District, Karnataka. *Source* NBSS&LUP Archives 2017

Dandakaranya Upland The region as a whole has an undulating topography with well-marked elevations and depressions. It's a well-demarcated unit, representing the dominant plateau character that is highly dissected, giving rise to low hills and intensive plains. The region ranges between 200 and 900 m in elevation. Abundant of the landscape of the region is made of the Dharwar gneisses and Cuddapah limestone and shale (Fig. 4.11). The Bastar plateau that is a part of the region has huge localised ore deposits related to banded haematite and quartizite (Ranganathan and Jayaram 2006). The major landforms related to the region are plateaus, hill ranges, hillocks, pediments and river basins crammed with deposit.

(iv) *Western Hills (The Sahyadri)*

The Western Hills consist of North Sahyadri, Central Sahyadri, Nilgiris and South Sahyadri, running parallel to the west coast.

North Sahyadri The great escarpment of the Deccan called Sahyadri which appears to be a sea cliff, runs north–south to a distance of about 480 km with an average width of 30–45 km (Joshi 2017). Its crest-line runs in broad curves, forming two re-entrants carved by the headwaters of the Godavari and the Bhima rivers and two easterly bulges marked by Harischandragarh and Mahabaleshwar peaks and two important passess

Fig. 4.10 Fuller's earth mining in Pakur district, Jharkhand. Adapted from http://nitishpriyadarshi.blogspot.in/2014/03/fullers-earth-mining-in-pakur-district.html

Fig. 4.11 Precambrian sedimentary beds at Dharwars. *Source* NBSS&LUP Archives 2017

Thalghat and Bhorghat that link the Maharashtra plateau and the Konkan plain (Matmon et al. 2002). Fingerlike spurs jet out from the main crest and give rise to narrow enclosed valleys. In this mountainous range, long and fairly unbroken lines of cliff of younger trap-flows rising almost vertically to about 1000 m from western front, are observed. The eastern flank, however, drops to only 300–450 m giving this mountain an asymmetrical shape. Flat summits, terraced surfaces and uneven slopes characterise the landscape of the Sahyadri. The well-defined rock terraces on the mountain slopes are caused by heterogeneous basaltic lava flows.

Fig. 4.12 Monazite sand in coastal belt of Karunagappally, Kerala, India. Adapted from http://hps.org/hpspublications/journalarchive/118-2009.html

Central Sahyadri It runs very close to the coast and its western scarp appears to have been considerably dissected by the headward erosion of the west-flowing streams, some of them with waterfalls. The Jog falls of the Sharavati river near Gersoppa is an example. Kudremukh is the highest peak rising to 1892 m (Radhakrishna 1983).

Nilgiris Nilgiris, lying between Karnataka plateau and south Sahyadri rises abruptly to more than 2000 m and presents a rolling landscape derived from Charnockites. To the south–west, the hills are more elevated and dissected giving rise to the gigantic escarpment of Kundah hills. On descending the slopes, the streams enter into deep gorges (Kale and Subbarao 2004). The hill exhibits a radial type of drainage. To the south is the Palghat gap through which flows the Ponnani River. The western ghats are joined by the eastern ghats in the Nilgiri plateau (Radhakrishna 1967; Gunnell and Fleitout 1998).

South Sahyadri It lies south of the Nilgiris. It is composed of charnockites and granitoid gneiss. Besides the main longitudinal ranges, are the Anaimalai and the Palni hills with undulating summits and innumerable waterfalls. Farther south, the landscape consists of parallel ranges and valleys carved out of an uplifted peneplain.

(v) *Eastern Hills (Ghats)*

The highly dissected Eastern Hills running in semi-circular fashion border the east coast of India and consist of Eastern Ghats and the Tamil Nadu Upland.

Fig. 4.13 Columnar joints on basalt off Udupi coast. *Source* NBSS&LUP Archives 2017

Eastern Ghats Much of the landscape within the eastern a part of the peninsular India is dominated by belts of Eastern Ghats mountains that are semi-circular in form and significantly dissected (Vaidyanadhan 2002). The meanders of the Krishna are entrenched into Cuddapah sedimentary. Rocky terraces flank the Krishna, particularly wherever quarzites alternate with slates. The Vinukonda range presents a steep scarp towards the Krishna valley. The sixty kilometer long transverse gorge of the Godavari is an proof of the antecedent character of the watercourse. The plateaus of Machkund and Koraput, the scarp of Madugula Kondas, and the gneissic piedmont plateau of Malakanagiri are the other important features of the Eastern Ghats. The Cauvery has cut deep into the Eastern Ghats exposing on its bed a hard structural platform underlain by soft gneisses. The three main tributaries of the Pennar, viz., the Chitravati, the Papagni and the Cheyyeru cut across the ridges.

Fig. 4.14 The exposed Coral reef at Interview Island, west coast of Middle Andaman Island. *Source* Bandopadhyay and Carter 2017

Tamil Nadu Uplands It is a levelled landscape with low relief of 150–450 m above the mean sea level. There is a sharp break in gradient all around its edges. The Kongunad upland has developed on granitoid gneiss, and is drained by three east flowing tributaries of the Cauvery, the Bhavani, the Noyil and the Amaravati. Extreme ruggedness characterises the landscape of the Shevaroy group of hills of granitoid gneiss which are separated from the Eastern Ghats by pediments (Selvam et al. 2011).

(vi) *Coastal Plains*

There are two coastal plains, one along the Arabian Sea (west coast) and the other along the Bay of Bengal (East coast) (Mukhopadhyay and Karisiddaiah 2014).

West Coastal Plains These kind of narrow belt of 10–25 km stretching from the sea to the Western Ghats and lengthening to about 1500 km from Kanniyakumari to Surat, comprising Kerala plain, Karnataka coast, Konkan coast, Kutch peninsula, Gujarat plain and Kathiawar peninsula (Fig. 4.12).

Kerala plain is about 25 km wide and 10–30 m in elevation and is within the form of terraces, most likely of marine origin. there's a cliffed marine terrace. a number of coastal lagoons or backwaters with their connecting canals stretch from north of Calicut to Trivandrum. Typical examples are the Ashthamudi and the Vembanad tidal lakes that have sand bars at their mouths. The Periyar is the largest watercourse of the world. To the north of Kerala plain is the Karnataka coast, the northern a part of that is narrow and is traversed by flat flat-top spurs. To the south, it widens bent on about 70 km and consists of a narrow belt of recent deposits of coastal sand dunes, estuarine mudflats and valleys and high erosion platform (Prabakaran and Anbarasu 2010). The Konkan coast parallel to the Sahyadri is about 50 km wide. It's distinguished by the coastal belt, rocky and broken by small bays and creeks and fringed with islands, longitudinal ranges, valleys, mesas, buttes and the foot hills of the Sahyadri. The important west-flowing rivers are the Vaitarni, the Ulhas and the Mandavi (Mukhopadhyay et al. 2012).

To the south of Rann lies the Kutch peninsula that is finite on all the three sides by water except the Rann area to the north. The region consists of disintegrated arenaceous rock with intrusive and interbedded volcanic rock skirted with fluvial and aeolian deposits. The hill patches within the cutch that are capped with volcanic rock, rise to more than 300 m. The Gujarat plain is principally formed of sediment deposited by the rivers Banas, Sabarmati, Mahi, Narmada and Tapti and lies between 150 and 300 m contour. The plain forms a uniform natural landscape. To the west of Gujarat plain is that the Kathiawar peninsula that is basically a lava plateau dissected into flat-topped hills and step-like terraces (Chandramohan et al. 2001). The current heterogenous pattern of landscape is the results of marine, stream and aeolian cycles of erosion. The ridges and hills within the central half act as water divide between north and south flowing rivers. There are two bill masses interlinked with one another by compound ridges and bills. These hill masses exhibit circular voidance pattern. The West Coast is more indented with rocky headlands, intervening sandy bays and multiple estuaries and therefore the giant saline wetland and lagoon-barrier complexes (kayals) are some of the noteworthy features on the West Coast (Mukhopadhyay and Karisiddaiah 2014).

East Coastal Plains East Coast is widely curving not like the west and therefore the rivers on the east coast are long with wide and intensive delta formations (Vaidyanadhan 1991). The plains comprise the Utkal plain, Andhra plain and state plain (Kumaran et al. 2012). The Bengal plains continue southward merging with the Utkal plain in Orissa which incorporates Mahanadi delta and smaller plains. The Andhra plain is terraced (Fig. 4.13). Mayurbhanj plain is partly aggradational whereas the Bhadrak plain could be a typical alluvial search formed primarily by the Vaitarni and Salandi rivers. The Mahanadi delta differs somewhat from the Ganga delta, being straight on its seaward margin fringed with sand dunes. Tie delta merges into Chilka plain on the west. Two broad deltas engineered up by the Godavari and therefore the Krishna join one another to make a narrow stretch of coastal plain, the Andhra plain, flanked by outlying spurs of the Eastern Ghats. Within the north, stands a fixed sand drift with whale-back shape providing site for the port city Vishakhapatnam however to the south there's a rocky cliff with wavy head lands. Alternative features are coastal marsh with mud flats and salt pans (Nag 2010). The eastern half of Tamil Nadu lying between the coast line and ISO a contour, is thought as Tamil Nadu plain. On the coast, there's a narrow belt of sand dunes that rises to about 10 m on the Tuticorin coast to the south of which are red sand hills. Rest of the plain is especially formed by the east-flowing Cauvery and Tambraparni rivers that have built up deltas at their swath (Murthy et al. 2012).

4.3.3.1 The Islands

There are two groups of Islands particularly the Arabian Sea Islands and also the Bay Of Bengal Islands. Significantly, the Arabian Sea Islands comprise of Lakshadweep, Amindivi, Minicoy and others, numbering 25, are the remnants of the previous land mass and resulting coral formations (Fig. 4.14). The basin separating the islands from the main land is concerning 2000 m deep. Bay of geographic region islands has 222 islands that form the bay cluster extend over a length of 590 km in crescent form. The northern part of Andaman Islands is physically characterised by central range and narrow valleys whereas middle and southern groups have wider flats flanked by the Ghats (Vaidyanadhan 1991). The Nicobars are hilly and form the summit of the submarine mountain range characteristically irregular in form.

4.4 Conclusions

Soil is a thin strip over earth's crust and its formation, development and maturity are all directly influenced by corresponding geology as well as geomorphology. Soil functions, soil health and quality have close relations with type and nature of parent materials, landform and relief. The extent of soil degradation, pollution and erosion are dictated by weathering stage of parent materials and slope gradients. By and large, the conservation as well as management options for improvement of soil productivity could preferably be planned on background of associated geology and geomorphology in order to understand the pedogenesis. It is true that the landscape, which looks so static and constant, is virtually dynamic as influenced by associated geology and geomorphology.

References

Babar M, Radhakrishna C, Madhusudan G, Yadava BG (2012) Quaternary geology and geomorphology of Terna River basin in West Central India. J Quat Sci 61(2):156–167

Bakliwala BC, Wadhawan SK (2003) Geological evolution of Thar desert in India-issue and prospects. Proc Indian Natl Sci Acad 69 (2):151–165

Balasubramanian A (2017) Geological formations of India. Technical report. Centre for Advanced Studies in Earth Science University of Mysore, Mysore-6. https://doi.org/10.13140/rg.2.2.35307.75042

Bandopadhyay PC, Carter A (2017) Introduction to the geography and geomorphology of the Andaman–Nicobar Islands. In: Bandopadhyay PC, Carter A (eds) The Andaman–Nicobar accretionary ridge: geology, tectonics and hazards, vol 47. Geological Society, London, Memoirs, pp 9–18. https://doi.org/10.1144/M47.2

Bloom AL (1983) Sea-level and coastal morphology through the Late Wisconsin glacial maximum. In: Porter SC (ed) Late quaternary environments of the United States—The Late Pleistocene, vol 1. Longman, London, pp 215–229

Bose SC (1961) Some geomorphological features near Kedarnath and Badrinath. Geogr Rev India 22(4):33–36

Chamuah GS, Gangopadhyay SK, Walia CS, Baruah U (1996) Soils of Jorhat district: physiography relationship. Agropedology 6(2):29–36

Chandramohan P, Jena BK, Sanil Kumar V (2001) Littoral drift sources and sinks along the Indian coast. Curr Sci 81(3):292–297

Chatterjee S, Scotese CR (1999) The breakup Gondwana and the evolution of the Indian plate. In: Sahni A, Loyal RS (eds) Gondwana assembly: new issues and perspectives. Indian National Science Academy, New Delhi, India

Chatterjee SP (1961) National Atlas of India, Plate 36 (Madras), Ministry of scientific research and cultural affairs, Government of India

Daniels RB, Gamble EE, Cady JG (1971) The relation between geomorphology and soil morphology and genesis. Adv Agron 23:51–58

Dhir RP, Singhvi AK (2012) The Thar Desert and its antiquity. Curr Sci 102(7):1001–1008

Elias SA (2013) Earth history. Reference module in Earth systems and environmental sciences. https://doi.org/10.1016/b978-0-12-409548-9.05920-0

Encyclopaedia Britannica (2015) Administrative boundary of India. https://media1.britannica.com/eb-media/98/1598-004-94CC5ED.jpg

Gade JN (2016) Water resources and tourism promotion in Telangana State. Zenon Academic Publishing. ISBN 9385886045

Gebrekidan H, Mishra BB, Kibret K (2006) Missing linkage in rainfall-runoff-soil water relationship for sustainable watershed development: a case study around Hirna, Eastern Ethiopia. J Food Agric Environ 4(1):239–245

Gerrad AI (1981) Soil and landforms. George Unwin Ltd., London

Gunnell Y, Fleitout L (1998) Shoulder uplift of the Western Ghats passive margin, India: a denudational model. Earth Surf Proc Land 23:391–404

Hall GF (1983) Pedogenesis and soil taxonomy: concepts and interpretations. Elsevier Scientific Publishing Co., New York

Harindranath CS, Shivaprasad CR (2009) Geology and its importance in soil survey. In: Bhattacharya T, Sarkar D, Pal DK (eds) Soil survey manual. NBSS&LUP Publication No. 146, Nagpur, India, p 400

International Institute for Arial Survey and Earth Sciences (IIASES) (1971) Climatic geomorphology, a selection of recent articles (1968–1971). International Institute for Arial survey and earth sciences, Enschede. ISBN 81-85460-72-8 (11 sheets) (Scale 1:1 M)

Joshi VU (2017) Geomorphological field guide book on Deccan Inland (edited by Amal Kar). Indian Institute of Geomorphologists, Allahabad

Kale VS, Subbarao KV (2004) Some observations on the recession of the Western Ghat escarpment in the Deccan Trap region, India: based on geomorphological evidence. Trans Jpn Geomorphol Union 25:231–245

Kanchan R (2018) Current scenario of research in Indian geomorphology and its future. J Indian Geomorphol 6:1–10. ISSN 2320-0731

Kar A, Ghose B (1984) Drishadvati river system in India: an assessment and new findings. Geog J 150:221–229

Karan PP, Jenkins WM (1963) The Himalayan kingdoms: Bhutan, Sikkim and Nepal. D. Van Nostrand Co., Princeton, New Jersy, pp 56–78

Krishnan MS (1960) Geology of India and Burma. Higginbothams (Madras) Private Ltd., Mount road, Madras

Kumar P, Rai RK (1981) Has Deccan foreland rejuvenated? In: Sharma HS (ed) Perspectives in geomorphology, vol IV, pp 101–114

Kumaran KPN, Ruta B, Padmalal LD (2012) India's fragile coast with special references to late quaternary environmental dynamics. Proc Indian Natl Sci Acad 78(3):343–352

Limaye SD (2017) Importance of percolation tanks for water conservation for sustainable development of ground water in hard-rock aquifers in India. In: Goyal R, Sivanappan RK (eds) Engineering practices for agricultural production and water conservation: an interdisciplinary approach. Apple Academic Press, Canada, pp 19–30

Matmon A, Bierman P, Enzel Y (2002) Pattern and tempo of great escarpment erosion. Geology 30:1135–1138

Medlictt HB (1965) A manual of the geology of India and Burma. Published by Government of India Press, Nasik

Mishra BB, Mall J, Singh ID (2001a) Assessment and management of *Diara* land soils using remotely sensed data. Indian Farming 50:22–26

Mishra BB, Mall J, Choudhary J, Singh RA (2001b) Nutrient mining in different agro-climatic zones of Bihar. Fertilizer News 46:21–43

Mishra DC, Ravi Kumar M (2016) Proterozoic orogenic belt vs rifting of Indian craton. Geosci Front 5(1):25–41. https://doc.org/10.1016/J.gsf.2013.03003

Moharana PC, Gaur MK, Choudhary C, Chauhan JS, Rajpurohit RS (2013) A system of geomorphological mapping for Western Rajasthan with relevance for agricultural land use. Ann Arid Zone 52(3&4):163–180

Mohindra R, Parkash B, Prasad J (1992) Historical geomorphology and pedology of the Gandak Megafan, Middle Gangetic Plains, India. Earth Surf Proc Land. https://doi.org/10.1002/esp.3290170702

Mukhopadhyay R, Karisiddaiah SM, Ghosh AK (2012) Geodynamics of amirante ridge and trench complex, western Indian Ocean. Int Geol Rev 54:81–92

Mukhopadhyay R, Karisiddaiah SM (2014) The indian coastline: processes and landforms. https://doi.org/10.1007/978-94-017-8029-2_8

Murthy KSR, Subrahmanyam AS, Subrahmanyam V (2012) Tectonics of the eastern continental margin of India. The Energy & Resources Institute New Delhi, p 180. ISBN 9788179934081

Nag P (2010) Coastal geomorphic features around Indian ocean. Indian J Geo-Marine Sci 39(4):557–561

Nair RR, Hashimi NH (1980) Holocene climate inferences from the sediments of the western Indian continental shelf. Proc Indian Acad Sci 89:299–315

NBSS&LUP (2002) Soils of India. NBSS & LUP Publication 94, ICAR-NBSS&LUP, Nagpur

NBSS&LUP Archives (2017) Compilation of legacy soil information from different projects of ICAR-NBSSLUP, Nagpur

Nogiya M, Verma TP, Mohrana PC, Singh RS, Tailor BL, Singh R, Meena RL, Kumar S, Singh SK (2017) Influence of landform on the spatial variability of soil fertility in central state farm, Jetsar, Sri Ganganagar district of Rajasthan. Agropedology 27(02):125–130

Owen LA (2014) Himalayan landscapes of India. In: Kale V (ed) Landscapes and landforms of India. World geomorphological landscapes. Springer, Dordrecht

Pandarinath K, Narayana AC, Yadava MG (1998) Radiocarbon dated sedimentation record up to 2 ka BP on the inner continental shelf off Mangalore, south-west coast of India. Curr Sci 75(7):730–732

Pofali RM, Shankaranarayana HS (1982) Physiography, climate, natural vegetation and land use. In: Murthy RS, Hirekerur LR, Deshpande SP, Venkat Rao BV, Shankaranarayana HS (eds) Benchmark soils of India, morphology, characteristics and classification for resource management. NBSS&LUP, Nagpur

Prabakaran K, Anbarasu K (2010) Coastal geomorphology and evolution of Rameswaram Island, Tamil Nadu, India. Res J Earth Sci 2(2):30–35

Radhakrishna BP (1983) Neogene uplift and geomorphic rejuvenation of the Indian Peninsula. Curr Sci 64(11&12):787–793

Radhakrishna BP (1967) The Western Ghats of the Indian Peninsula. In: Proceedings of the seminar on geomorphological studies in India, Sagar, pp 4–14

Ramarao B (1962) A hand book of the geology of Mysore State, Southern India. The Bangalore Printing & Publishing Co., p 264

Ranganathan N, Jayaram S (2006) Geomorphology of Karnataka. Karnataka Geologists' Association, Department of Mines and Geology, Bangalore, p 350

Rasoul S (2010) Geologic formation of the Himalaya. Himal J 66

Sahoo AK, Chattopadhyay T, Singh RS, Shyampura RL (2003) Characterization of soils and vertical distribution of available micronutrients under different landforms of Malwa plateau, Rajasthan. Indian Agric 47:217–223

Sawhney JS, Deka B, Sharma BD, Sidhu PS (1996) Magnitude of soil variability in morphological and other properties across different landscapes in the Siwalik hills of Punjab. J Indian Soc Soil Sci 44 (3):465–469

Sawhney JS, Sharma BD, Uppal HS (2005) Morphostratification and characterization of Siwalik piedmont of semi-arid tract of Punjab. J Indian Soc Soil Sci 53(3):273–280

Sawhney JS, Verma VK, Sharma BD, Sharma PK (1992) Pedogenesis in relation to physiography in semi-arid and arid tracts of Punjab, India. Arid Soil Res Rehabil 6:93–103

Selvam S, Chindam N, Manimaran G, Sivasubramanian P (2011) Geomorphological and textural characteristics of sediments of St. Marys Island Western continental shelf, India. Arch Appl Sci Res 3 (6):480–487

Sen G (2001, December) Generation of Deccan trap magmas. Proc Indian Acad Sci (Earth Planet Sci) 110(4):409–431

Shukla UK, Bora DS (2003) Geomorphology and sedimentology of Piedmont zone, Ganga Plain, India. Curr Sci 84(8):1034–1040

Sharma BD, Mukhopadhyay S, Sidhu PS (1999) Microtopographic control on soil formation in the Punjab region, India. Geoderma 81:357–368

Sidhu PS, Hall GF, Sehgal JL (1976) Studies on some soils at varying stage of pedogenic development in the central Punjab. 1. Morphology and physic chemical characterization. J Res (Punjab Agricultural University) 13:23–24

Singh AK (2012) Probable agricultural biodiversity heritage sites in India: XIV. The Chotanagpur Plateau region. Asian Agrihist 16 (4):371–392

Singh RL, Singh KN (1971) Middle Ganga plain. In: Singh RL (ed) India–a regional geography. National Geographic Society India, Varanasi, p 992

Singh RP (1967) Landscape cycles of Peninsula, India. Proceedings seminar of Geomorphological studies in India. University, Saugar, p 145

Sinha R, Peter F (2007) Quaternary fluvial systems of India. Editorial/Quatern Int 159:1–5

Sinha R, Tandon SK, Gibling MR, Bhattacharjee PS, Dasgupta AS (2005) Late quaternary geology and alluvial stratigraphy of the Ganga basin. Himalayan Geol 26(1):223–240

Spate OHK, Learmonth ATA (1967) India and Pakistan: a general and regional geography. Methuen, London

Srivastava GS, Singh IB, Kulshrestha AK (2014) Geomorphic and tectonic features of Punjab-Haryana Plain as identified from digital elevation model and surface profiles. Himalayan Geol 35(2):97–109

Tiwary JR, Jha PP, Mishra BB (1989) Characteristics and classification of red, black and yellow soils of Rajmahal Trap of Bihar. J Indian Soc Soil Sci 37:518–523

Vaidyanadhan R (ed) (1991) Quaternary deltas of India. Memoir 22, Geological Society of India, Bangalore, p 291

Vaidyanadhan R (2002) Geomorphology of the Indian sub-continent. Monograph published by the Indian Society of Remote Sensing, Dehra Dun

Van Dorsser HJ (1976) Illustrated appendices to geomorphology. International Institute for Arial Survey and Earth Sciences, Enschede

Wadia DN (1961) Geology of India, 3rd edn. Macmillan & co., New York ST. Martin's Press, London

Wadia DN (1976) Geology of India. Tata McGrow Hill Publishing Company, New Delhi

Yadav SS, Prasad J, Gaikwad ST, Thyalan S (1998) Landform-soil relationship and its impact on soil properties. J Indian Soc Soil Sci 46:332–335

Yin A, Harrison TM (1996) The tectonic evolution of Asia. Cambridge University Press, Cambridge, p 666

Yin A, Harrison TM (2000) Geologic evolution of the Himalayan-Tibetan orogen. Ann Rev Earth Planet Sci 28(211):280

Major Soil Types and Classification

5

K. S. Anil Kumar, M. Lalitha, Shivanand, K. Sujatha, K. M. Nair, R. Hegde, S. K. Singh, and Bipin B. Mishra

Abstract

The chapter is an effort to understand soil types of India and further classifying them since time immemorial from the start of civilization itself. Derived from a wide range of rocks and minerals, a large variety of soils exist in the Indian subcontinent. Soil-forming factors like climate, vegetation and topography acting for varying periods on a range of geological formations and parent materials have given rise to different kinds of soil. The National Bureau of Soil Survey and Land Use Planning, Nagpur, as a premier soil survey institute, has been consistently using benchmark soil series to understand the rationale of the soil taxonomy, keeping in view the soil genesis from different rock systems under various physiographic locations in tropical India. The NBSS & LUP has developed a database on soils with field and laboratory studies over the last 50 years. This has generated maps and soil information at different scales, showing area and distribution of various soil groups in different climatic zones or agro-ecological sub-regions. The 1:250,000 scale map shows a threshold soil variation index of 4–5 and 10–25 soil families per m ha for alluvial plains and black soil regions, respectively. Progress in basic and fundamental research in Indian soils has been reviewed in terms of soils and their formation related to various soil-forming processes. More than 50 years ago, the US soil taxonomy was adopted in India. However, India should have its own system of soil classification at least for the purpose of correlation with international and universal systems.

Keywords

Soil survey • Agro-climatic zone • Agro-ecological region • Soil types • Soil classification • Spectral library • Technology transfer

5.1 Introduction

India represents a land of paradoxes due to the large variety of soils. The physiographic features varied from high mountains, glaciers and thick forests in the north to seas and oceans washing lengthy coasts in the south, variety of geological formations, diversified temperature from extreme cold to equatorial hot, rainfall from barely a few centimetres (<10 cm) in the arid parts to per-humid with world's maximum rainfall of several hundred centimetres (1120 cm) per annum in some other parts and the topography of high plateaus, stumpy relic hills, shallow open valleys, rolling uplands, fertile plains, swampy low lands and dreadly barren deserts (Gajbhiye and Mandal 2000). Such varied natural setting has resulted in a great variety of soils compared to any other country of similar size in the world. The soils of India including the highly weathered soils under humid tropical environments are potentially productive in terms of food production as evidenced by the growing self-sufficiency in food production and food stocks since the independence (Bhattacharyya et al. 2013).

K. S. A. Kumar (✉) · M. Lalitha · Shivanand · K. Sujatha · K. M. Nair · R. Hegde
ICAR-National Bureau of Soil Survey and Land Use Planning, Hebbal, Bangalore 560024, India
e-mail: anilsoils@yahoo.co.in

S. K. Singh
ICAR-National Bureau of Soil Survey and Land Use Planning, Amravati Road, University P.O, Nagpur 440033, India

B. B. Mishra
Bihar Agricultural University, Sabour, Bhagalpur 813210, India

Pedology and Land Use Planning, School of Natural Resources Management and Environmental Sciences, Haramaya University, P O Box 138, Dire Dawa, Ethiopia

International Union of Soil Sciences, Vienna, Austria

B. B. Mishra (ed.), *The Soils of India*, World Soils Book Series,
https://doi.org/10.1007/978-3-030-31082-0_5

5.2 Soil Survey in India

The earliest investigations on soils of India by Voelcker date back to 1893 and by Leather to 1898. They categorized the soils of the country into four major groups, namely the Indo-Gangetic alluvium, the black cotton soil or *regur*, red soil and laterite soil. Schokalskaya published a soil map of India in (1932) following the Russian concept and described 16 soil groups based on climate, vegetation, soil-forming materials, salinity, alkalinity, swamps and peats. In 1935, Wadia and his co-workers compiled a soil map of India with emphasis on geological formations and classified the soils as red, black (*Regur*), laterite and lateritic soils of Peninsular India, delta, desert, *bhabar*, *terai* and alkali soils of the Indo-Gangetic plains (IGP). Vishwanath and Ukil (1943) published a soil map of India by placing the soils in different climatic zones. Integrating the effects of climate, vegetation and topography, 16 major and 108 minor soil regions were identified and brought under 27 units by Raychaudhari et al. (1963). Later, a revised soil map of India was generated with 23 major soil groups under FAO/UNESCO's scheme on World Soil Map project. This map was refined with 25 broad soil classes represented on a 1:7 million scale map (Govindarajan 1971).

Based on information available on the soil classes, using USDA 7th approximation, the first US soil classification system for Indian soils was used to develop at the group level, and a soil map of India on 1:6.3 million scale was published (Murthy and Pandey 1983). During the past 30 years, the National Bureau of Soil Survey and Land Use Planning (ICAR-NBSS & LUP), the premier soil research institute of the Indian Council of Agricultural Research (ICAR), mapped the soil resource of the country on 1:7 million scale at great-group association level. One hundred and three soil great groups were recognized. For sustainable resource management, large-scale mapping of soil and climatic resources was initiated in 1986 using a three-tier approach comprising image interpretation, field mapping and laboratory analysis. This was followed by cartography and printing of maps for all the states and union territories on 1:250,000 scale. One hundred and seventy-six false colour composite (FCC) and B/W infrared imageries on 1:250,000 scale were interpreted visually to prepare pre-field physiography-cum-photomorphic maps. These maps were generalized (categorically and cartographically) to generate a state SRMs that have been published on 1:500,000 scale with association of soil families as the basic mapping units. Sikkim, Goa, Lakshadweep and Andaman and Nicobar Islands were mapped on 1:50,000 scale using TM FCC. Based on these maps, a final soil map of India on 1:1 M scale was prepared using association of subgroups as the basic mapping units and presented (Sehgal 1995). The map units are described in a manner that is intelligible to most land-use planners (NBSS & LUP Staff 2002; Bhattacharyya et al. 2013 and 2015).

5.3 Area and Distribution of Soil Groups

The soil resources mapping programme of the entire country in a span of about 10 years (1986–1996) has generated tremendous database (Table 5.1) in terms of soils, their area and extent, characteristics and grouping following the soil taxonomy (Table 5.2). The data indicate that 13.5% Alfisols, 40% Inceptisols, 4.0% Aridisols, 28.0% Entisols, 8.5% Vertisols and 2.5% Ultisols occur in India. Such information has been utilized in preparing the generalized soil maps of the country showing the geographical distribution of the various soils distinguished by major differentiating characteristics.

Although soils of India occur in five bio-climatic systems, but only a few soil orders are spread in more than one bio-climates. Vertisols belong to arid hot, semi-arid, sub-humid and humid to per-humid climatic environments (Pal et al. 2009). Mollisols belong to sub-humid and also humid to per-humid climates (Bhattacharyya et al. 2006). Alfisols and Ultisols belong to humid to per-humid climates (Chandran et al. 2005). Both Entisols and Inceptisols belong to all the five categories of bio-climatic zones of India, and Aridisols belong mainly to arid climatic environments (Bhattacharyya et al. 2008). This baseline information indicates that except for the Ultisols and Aridisols, the rest five soil orders exist in more than one bio- climatic zones of India. The absence of Oxisols and Ultisols, occupying only 2.56% of the total geographical area of the country, suggests that soil diversity in the geographic tropics, in general and in India in particular, is at least as large as in the temperate zone (Eswaran et al. 1992). Some profiles of Ultisols, Mollisols, Vertisols, Histosols, Alfisols, Inceptisols and Aridisols are depicted in Fig. 5.1.

5.4 Soils in Different Climatic Zones

The climatic zones of India were divided into major zones, viz. northern, western, central, southern, eastern, north eastern and islands with different soil orders (Table 5.3). The northern zone covers 20% area (Jammu and Kashmir, Himachal Pradesh, Punjab, Haryana, Delhi and Uttar Pradesh) of the country, and Inceptisols soil order is found more followed by Entisols. Among the six states, Vertisols are reported only in UP, Aridisols in part of Punjab and Haryana and Mollisols in Jammu and Kashmir, Himachal Pradesh and Uttar Pradesh (Fig. 5.2).

The western zone consisting of three states (Rajasthan, Gujarat and Goa) covers 16.5% of the TGA of the country. Aridisols dominate in Rajasthan, and Ultisols reported in Goa. The soils indicate a change of climate from wetter to dry regimes. The central zone consists of three states (Madhya Pradesh, Maharashtra and Chhattisgarh) and covers 23% area

Table 5.1 Information on different categories of soil taxonomy in soil resource mapping of states on 1:250,000 scale (Soil Survey Staff 1999)

Sl. no.	State	TGA (000 ha)	Soil orders	Sub orders	Great group	Sub group	Families	Soil map units
1	Andaman and Nicobar islands	196	4	8	11		37	42
2	Andhra Pradesh	27506.8	7	11	18	46	132	238
3	Arunachal Pradesh	8374.3	5	10	16	30	79	46
4	Assam	7852	4	9	15	26	83	83
5	Bihar	17387.7	5	9	19	38	72	175
6	Delhi	148.3	2	4	4	6	12	31
7	Goa	370.2	4	7	13	16	21	25
8	Gujarat	19602.4	5	11	20	45	124	370
9	Haryana	4421.2	4	6	11	20	27	199
10	Himachal Pradesh	5567.3	4	6	12	17	43	95
11	Jammu and Kashmir	22222.6	4	8	13	28		140
12	Karnataka	19179.1	7	12	18	28	56	141
13	Kerala	3886.3	5	10	19	29	38	38
14	Lakshadweep	32	1	2	4	6	6	
15	Madhya Pradesh	44344.6	5	7	11	26	130	851
16	Maharashtra	30,769	5	8	10	18		356
17	Manipur	2232.7	4	8	13	22	29	19
18	Meghalaya	2242.9	4	8	14	25		24
19	Mizoram	2108.1	4	7	9	11		31
20	Nagaland	1657.9	4	6	8	14	72	34
21	Orissa	15570.7	4	10	17	41	98	159
22	Pondicherry	44	4	6	7	10	10	
23	Punjab	5036.2	4	7	11	14	124	124
24	Rajasthan	34223.9	5	8	16	32	86	375
25	Sikkim	709.6	3	7	12	26	69	69
26	Tamil Nadu	13005.8	6	12	20	44	94	285
27	Tripura	1048.6	5	9	13	23		43
28	Uttar Pradesh	29441.1	5	11	20	35		321
29	West Bengal	8875.2	3	10	19	36		115

Table 5.2 Total soil orders, suborders, great groups and subgroups identified in India (Soil Survey Staff 1999)

Sl. no.	Orders	Sub orders	Great group	Sub group
1	Alfisols	3	15	41
2	Aridisols	3	9	26
3	Entisols	5	23	51
4	Histosols	1	1	1
5	Inceptisols	4	16	72
6	Mollisols	3	6	14
7	Ultisols	3	11	23
8	Vertisols	2	4	18
Total		24	85	246

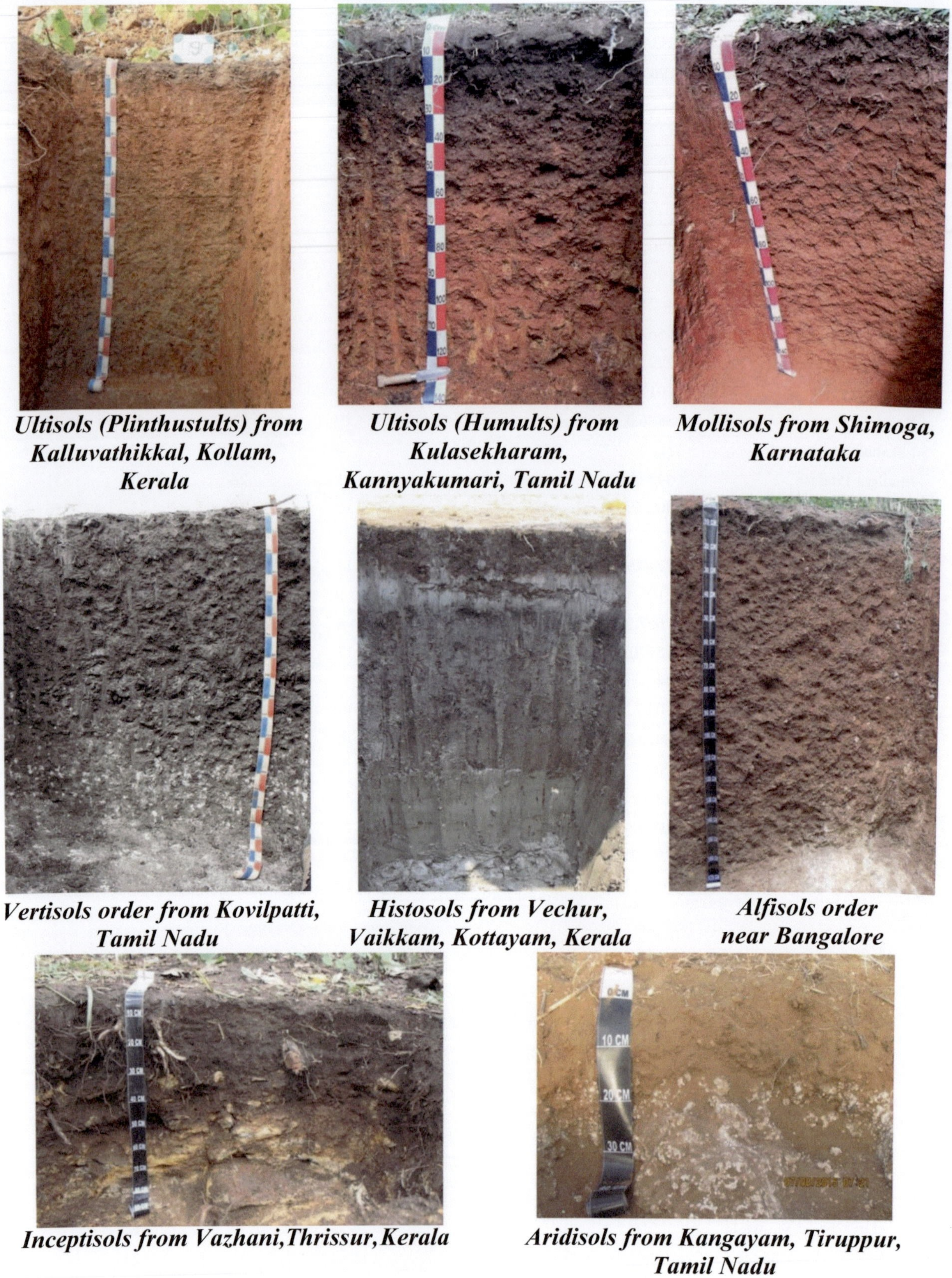

Fig. 5.1 Some soil orders of India. *Source* NBSS & LUP ARCHIVES 2017

Table 5.3 The soil orders in different climatic zones of India

Climatic Zones	Area (in `000 ha) and extent of zone-wise soil orders in India based on 1:2,50,000 scale maps								
	TGA[a]	Vertisols	Ultisols	Aridisols	Mollisols	Alfisols	Inceptisols	Entisols	Others
Northern zone	66,837	415	–	1179	173	2213	29,948	20,008	13,000
Western zone	54,196	2866	42	10,314	–	380	18,708	16,354	5531
Central zone	3,52,058	16,354	–	–	588	7953	29,472	17,948	2791
Southern zone	63,624	5962	3937	1856	762	17,832	21,649	7064	4561
Eastern zone	42,544	1019	–	–	75	12,707	18,526	8688	1527
North-Eastern	25,509	–	4428	–	–	1065	10,904	8372	738
Islands	828	–	–	–	40	49	362	314	62

[a]*TGA* total geographical area
Modified from Bhattacharyya et al. (2013)

of the country. The area under Vertisols is high in MP compared to Maharashtra, and Mollisols are reported both in MP and Maharashtra. The southern zone consists of six states (Andhra Pradesh, Karnataka, Tamil Nadu, Kerala and Puducherry and Karaikal) and covers 19.3% area of the country. The four southern states contain interestingly fertile Mollisols in the humid tropical climate. The eastern zone consists of five states (Bihar, Odisha, West Bengal, Sikkim and Jharkhand) covering 13% area of the country. Vertisols have been reported from Bihar and Odisha, and Mollisols from Sikkim. Jha (1972) established 23 soil associations in Bihar including Jharkhand. *Tal* land soils in south of Ganga river in Bihar are virtually Vertisols (Tiwary and Mishra 1990). Geologically, important Rajmahal traps of Jharkhand is known for red, yellow and black soils in catenary sequence (Tiwary et al. 1987). The north-eastern zone consists of seven states (Arunachal Pradesh, Assam, Nagaland, Manipur, Mizoram, Tripura and Meghalaya) and occupies 7% area of the country. The north-eastern zone is dominantly occupied by weathered soils including Alfisols and Ultisols. Ultisols are found mostly in higher elevation ranges of Arunachal Pradesh, Meghalaya and Manipur. The Indian island consists of two Union Territories (Andaman and Nicobar and Lakshadweep) and covers 0.2% area of the country. Andaman and Nicobar islands have a considerable area under Mollisols other than the dominant Inceptisols and Entisols.

5.5 Soils of Different Agro-ecoregions

The diversified environments across the country are due to macro-relief of high plateau, open valleys, rolling upland, fertile plains, swampy low lands and barren deserts. The systematic appraisal of agro-ecological regions encompassing relatively homogenous regions in terms of climate and physiography resulted in comparatively identical soil in terms of properties and characterization. There are twenty agro-ecological regions identified in India (Fig. 5.2) based on soil, climate, physiography and length of growing period (Table 5.4).

5.6 Types of Soil in India

The grouping of soils could be achieved by using either genetic or soil taxonomy. Genetic system is based on genetic factors and processes, and the soil taxonomy is based on the properties of soils as they exist to date. The major soils of India according to the genetic approach can be classified into few soil groups. They are alluvial, black, red, forest, laterites and lateritic, solonchaks, solonetzs and desert soil (Sehgal 2005). However, the physico-chemical characteristics of soils under different agro-ecological regions are presented in Table 5.5, while extent and distribution of these soils are given in Table 5.6 (soil types) and Table 5.7 (soil classes).

5.6.1 Alluvial Soils

The name Alluvial is given to soils that have developed on alluvium, irrespective of their place of occurrence and degree of profile development. They represent by far the most fertile land and form an important group of soils for agricultural production. The soils are extensively distributed in the Indo-Gangetic plains and Brahmaputra valley and cover an area of about 75 Mha. They also occur in the coastal and deltaic regions. They are predominantly occurring in the states of Punjab, Haryana, Delhi, Uttar Pradesh, Uttarakhand, Bihar, West Bengal, Assam, coastal regions of Orissa, Andhra Pradesh, Tamil Nadu, Kerala and Gujarat and concentrated in the agro-ecological regions, such as 4, 9, 13, 15, 19 and 20 of India. However, for soils derived from flood-plain alluvium, a separate order 'Fluvisols' in USDA soil taxonomy was proposed during 15th WCSS at Acapulco, Mexico (Mishra et al. 1994, Mishra 2000). The floodplain soils are typically the alluvial. The Ganga

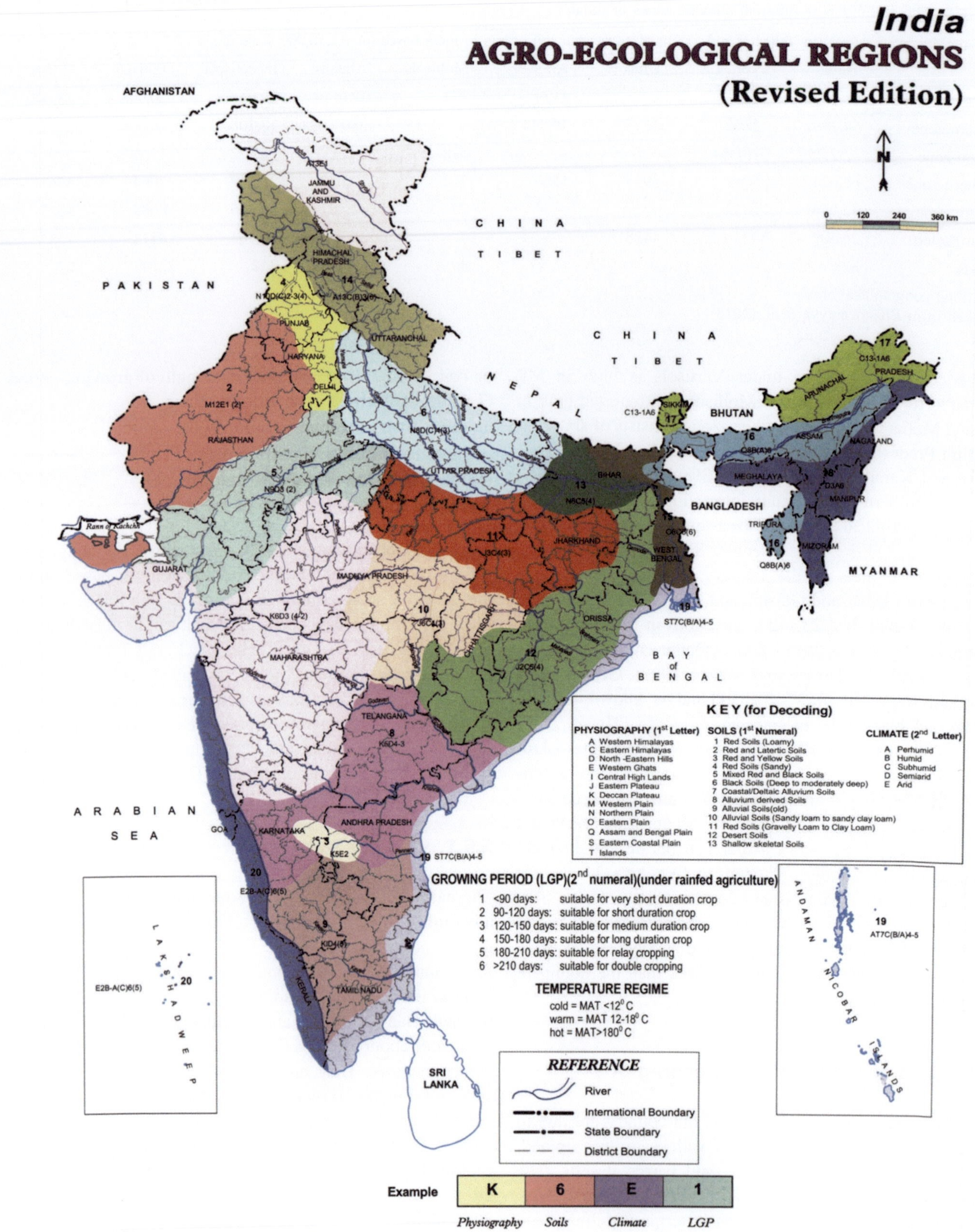

Fig. 5.2 Agro-ecological region map of India. *Source* NBSS & LUP 2016

Table 5.4 Agro-ecological regions of India (NBSS & LUP 2016)

Sl. no.	Agro-ecological region	Soil description	Soil great group
1.	Cold arid agro-ecoregion with shallow skeletal soils	Soils permanently remain under snow cover in northern parts of region, where shallow, sandy to loamy-skeletal calcareous soils occur on gently sloping to nearly level valleys. Soils on the nearly level terraces in the glacio-fluvial valley are very deep, calcareous and sandy clay loam to clay loam	Cryorthents, Cryochrepts
2.	Hot arid agro-ecoregion with desert and saline soils	Soils of the region are deep to very deep, loamy sand and sand in texture, occur in association with saline patches at places on gently to very gently sloping arid plains with intersperse hummocks, sand dunes and barchans	Torrifluvents, Haplocambids, Haplocalcids, Haplosalids, Natrargids
3.	Hot arid agro-ecoregion with red and black soils	The soils on gently to very gently slopes are shallow, sandy loam to sandy clay loam in texture. Deep, clayey Haplustert occurs on level to very gently sloping plains	Rhodustalfs, Haplustalfs, Paleustalfs, Haplusterts
4.	Hot semi-arid agro-ecoregion with coarse loamy alluvial soils	Sandy loam to sandy clay loam soils derived from Indo-Gangetic alluvium occur on moderately to gently sloping plains	Haplustepts, Natrustalfs, Ustipsamments and Ustifluvents
5.	Hot semi-arid agro-ecoregion with old alluvial soils	Deep to very deep sand and sandy loam occur on nearly level and gently sloping lands	Haplocambids, Ustipsamments
6.	Hot semi-arid to dry sub-humid agro-ecoregion with alluvial and *tarai* soils	The soils are derived from the Gangetic alluvium and a part of it locally known as *tarai*, having grey to dark grey colour, sand to clay loam texture and high organic carbon. The water table is high and soils remain saturated or fairly moist during major parts of the year. The soils derived from Gangetic alluvium are neutral to moderately alkaline, calcareous at places, silty loam and silty clay loam in texture. The other deep soils, loam to clay loam developed from mixed alluvium occurring on gentle slopes of alluvial terraces at an elevation of 150 m above MSL, have good available moisture-holding capacity and high productivity potential. The salt-affected soils occupy the low-lying depressions having imperfect drainage with low permeability	Hapludolls, Haplaqualfs, Haplustepts, Ochraqualfs, Haplustalfs, Ustifluvents, Ustorthents
7.	Hot semi-arid agro-ecoregion with moderately deep black soils	Clayey, calcareous, moderately to strongly alkaline, swell–shrink black cotton soils occur on nearly level to gently sloping pediplains, associated with alluvial soils on very gently sloping plains. The black cotton soils have high potential for crop production, however, constrained with management problems	Ustorthents, Haplustepts, Chromusterts, Haplustepts
8.	Hot semi-arid agro-ecoregion with mixed red and black soils	Deep, clay loam to clayey, neutral to slightly acidic soils with moderate permeability, associated with deep, clayey soils on gently sloping plains in the region. These are affected by severe erosion. These soils are agriculturally important due to high water holding capacity	Paleustalfs, Rhodustalfs, Haplusterts
9.	Hot semi-arid agro-ecoregion with red loamy soils	Deep to moderately deep, occasionally shallow, gravelly sandy loam to sandy clay loam soils, having moderate to strong alkalinity constitute the soilscape. These soils are known for low available water holding capacity and their susceptibility to soil erosion and surface crusting	Haplustepts, Haplustalfs, Dystrustepts, Ustifluvents, Haplustalfs, Ochraqualfs

(continued)

Table 5.4 (continued)

Sl. no.	Agro-ecological region	Soil description	Soil great group
10.	Hot sub-humid agro-ecoregion with moderately deep black soils	Medium to deep black soils and red soils constitute the dominant soilscape. Deep black soils are associated with the moderately deep, light brown to dark brown soils, which are characterized with calcareousness, slight alkalinity and high swell–shrink potential. Red soils occurred on eastern Dandakaranya plateaus are shallow to moderately deep, clayey, neutral to slightly acidic	Ustorthents, Haplustepts, Haplustalfs, Haplusterts, Dystrustepts
11.	Hot sub-humid agro-ecoregion with red and yellow soils	The dominant soilscape in the region is moderately to gently sloping uplands. Soils are dark red to reddish-brown having sandy clay loam to sandy clay texture and have moderate to strong acidity. Soils are gravelly, very shallow to shallow and slightly acidic at places	Ustorthents, Haplustepts, Haplustalfs, Rhodustalfs, Plinthustalfs
12.	Hot sub-humid agro-ecoregion with red and lateritic soils	Soils are brown to reddish-brown, fine-textured and developed over very gently sloping pediment surfaces of weathered granite-gneiss complex. These soils are deep, slight to strongly acidic, having low cation exchange capacity. The other soils on gently sloping side slopes have dark reddish-brown colour and strong acidity. Soils of Garjhat hills and eastern plateau are sandy loam to sandy clay loam with reddish to dark reddish colour and moderately acidic reaction. Other soils developed over granitic gneiss are fine-loamy dark reddish-brown to dark yellowish and neutral to slightly acidic reaction on gentle slopes	Haplustepts. Haplustalfs, Plinthustalfs, Paleustalfs, Rhodustalfs, Ustorthents, Endoaquepts
13.	Hot sub-humid agro-ecoregion with alluvial- soils	The soilscapes in the region are level to very gently sloping plain. These occur in association with imperfectly drained. The other soils in this region are greyish brown to yellowish-brown, fine loamy, imperfectly to poorly drained and slightly acidic to neutral in soil reaction. The soils on the old alluvial plains are brown, fine-textured, poorly drained and slightly acidic	Ustifluvents, Haplustepts, Endoaquepts Endoaqualfs
14.	Warm sub-humid to humid (with inclusion of per-humid) agro-ecoregion with sub-montane shallow and skeletal soils	The major soils occurring in the hill and mountain region are shallow to deep, medium to high in organic carbon, classified as brown forest and podzolic soils. The soils are characterized as medium-deep, loamy-skeletal, mixed calcareous with high stoniness (>80%) in the subsoils. The soils of Shiwalik zone are deep fine-loamy with good water holding capacity, nutrient retention capacity and capable to produce good yield for climatically adapted crops. The soils of piedmont *tarai* zone are fine sand, silt and clay occurring on nearly level to gently sloping lands	Eutrudepts, Eutrochrepts, Cryorthents, Udorthents, Hapludalfs, Haplustalfs, Hapludolls, Ustifluvents

(continued)

Table 5.4 (continued)

Sl. no.	Agro-ecological region	Soil description	Soil great group
15.	Hot sub-humid agro-ecoregion with loamy to clayey alluvial soils	Soils are slight to strongly acidic and have low to moderate base saturation occurs on nearly level to gently sloping plains. In the upper part of soilscape, soils are reddish yellow to red, strongly to slightly acidic sandy loam developed over weathered granitic gneiss and are prone to erosion. In the lower part of the soilscape, the soils are greyish brown to grey, deep to very deep in the flood plains. The soils have aquic moisture regimes, slight to strongly acidic and imperfectly to poor drainage in the low-lying plains. These are subjected to seasonal floods during rainy season. At places, deep, sandy loam and dark yellowish-brown soils have low water holding capacity and suffer from drought in summer	Ustifluvents, Endoaquepts, Endoaqualfs, Plinthustalfs, Fluvaquents and Ustipsamments
16.	Warm humid to per-humid ecoregion with loamy to clayey alluvial soils	The dominant soils in the region are rich in organic matter with moderate to low base status. The soilscape of the area is gently sloping *Chaur* lands. These are grey to dark grey in colour, poorly drained with sandy loam to sandy clay loam soils. The other soils on the *chaur* lands are grey to light grey in colour developed over alluvium on nearly level flood plain, subjected to overflow during rainy season	Fluvaquents, Udifluvents, Epiaquepts
17.	Warm per-humid agro-ecoregion with shallow and skeletal red soils	These soils are fine loamy to loamy skeletal, deep, dark brown to yellowish-brown, moderately to strongly acidic and excessively drained with high organic carbon occurs on steeply sloping hills and valleys and are susceptible to severe erosion. The soils are suitable for forest plantations	Dsytrudepts, Hapludolls, Entrudepts, Udorthents
18.	Warm per-humid agro-ecoregion with red and yellow soils	The soils in the area are shallow to deep loamy red and yellow soils. Lithic Hapludalfs and Humic Dystrudepts are some of the extensive soils occurring in the Meghalaya and Mizoram region. These soils are red dark greyish brown to strong brown in colour and are strongly acidic with pH <5.0 and contain fairly good amount of organic carbon at the surface and in the subsurface. The soils are well- drained. Other soils in the valley are Haplaquepts and Humaquepts in the state of Manipur, occurring on nearly level to level inter hilly area. These are dark greyish to yellowish-brown in colour, slight to strongly acidic	Dystrudepts, Entrudepts, Hapludalfs, Udorthents, Palehumults, Paleudults, Kanhapludults
19.	Hot sub-humid (with humid to per-humid inclusion) with coastal and deltaic alluvial soils	The major soils in this region include the alluvium derived from soils of coastal and deltaic plains. The soils developed on deltaic plain (Mahanadi) are deep, susceptible to severe flooding, slight to severely acidic and have wide cracks during non-rainy season. The soils of alluvial plains (Utkal Plains and Bengal Plains) are deep, slightly acidic to neutral, grey to dark grey and are imperfectly to poorly drained. The soils in the semi-arid part of the region are deep, fine, neutral to moderately alkaline and sandy at places	Endoaquepts, Fluvaquents, Haplusterts, Haplustepts, Paleustalfs, Ustipsamments

(continued)

Table 5.4 (continued)

Sl. no.	Agro-ecological region	Soil description	Soil great group
20.	Hot humid/per-humid agro-ecoregion with red and lateritic and alluvial soils	The major soils in this region are red and lateritic soils. Soils of the hills are yellowish red to dark reddish-brown, moderately to strongly acidic, clayey skeletal. The soils of the interhill valleys and low-lying areas in the region are dark brown to dark reddish-brown in colour, moderate to strongly acidic and formed on alluvium with low CEC and low base saturation. The soils of Lakshadweep Island, on the other hand, are highly calcareous and sandy in nature	Dystrudepts, Haplustults, Paleustalfs, Haplustepts, Kanhaplohumults, Kandihumults, Kanhaplustults, Haplustalfs, Ustipsamments

Table 5.5 Soil physico-chemical characteristics under different agro-ecological regions (NBSS & LUP 2016)

Name of AER	pH	EC (dS/m)	OC (%)	$CaCO_3$ (%)	ESP	CEC, cmol(+) kg^{-1}	BS, (%)	CEC/Clay	Clay minerals
Hot Arid	8.0–8.7	0.69.8	0.1–0.83	0–17.15	1.79–31.97	2.2–75.29	76–100	0.34–0.72	Sm, Vm, M, K, Q, F, Ch
Hot Arid	8.4–8.6	0.29–0.46	0.21–0.65	5–11.5	0.75–5.31	15.8–53.2	99.7	0.53–1.20	–
Hot Semi-Arid	6.7–10.3	0.05–120.25	0.1–0.79	0–14.07	0.82–93.9	5.07–37.1	83.5–100	0.27–2.33	Sm, Vm, M, K, F, Ch, Q
Semi-Arid	6.4–8.10	0.15–0.56	0.44–0.96	0–34	0.76–3.72	36–54.8	81.6–100	0.67–2.33	Sm, Vm, ML, M,
Hot Semi-Arid	8.1–9.2	0.18–1.72	0.82–1.05	9.48–13.09	0.91–24.98	44.98–52.56	99	0.75–0.78	Sm, Vm, M, K
Hot Semi-Arid	5–8.4	0–1.05	0.12–0.7	0.94–9.24	0–25	1.6–56.49	45–100	0.31–1.02	Sm, M, K
Hot Sub Humid	6.5–8.3	0.3–1.3	0.17–1.20	0–2.76	0.67–5.63	9.56–22.36	74.7–92.3	0.49–0.89	Sm, Vm, M, K, F, Ch
Hot Sub Humid (D)	6.5–8.7	0.04–0.16	0.32–1.31	0–13.37	0.47–5.85	27.54–65	95	0.64–0.98	Vm, M, K, Ch, Sm
Hot (moist/dry) Sub Humid	5.3–6.9	0.06–0.85	0.31–0.59	0–1.20	0.56–4.83	4.76–39.01	62.3–95	0.23–0.76	–
Hot Sub Humid	4.9–5.70	0.35–0.93	0.41–0.43	0	2.8–3.31	8.90–11.16	51.8–65.4	0.34–0.61	–
Hot Sub Humid (M)	7.7–8.1	8.58–13.53	0.88–1.03	1.4–1.9	3.97–5.79	13.8–15.1	85.6–92	0.42–0.49	–
Warm Sub Humid	4.8–7.9	0.07–0.58	0.4–8.32	0–3.75	0–4.42	7.1–28.3	33.8–100	0.3–1.20	Ch, M, K, Q, F, ML, M(D), Sm, Vm
Hot Sub Humid to Humid	5.0–7.8	0.04–0.97	0.46–3.12	0–2.1	1.25–15.82	5.1–18.17	29–91	0.25–0.54	M, K < Ch, Vm, ML, K, Q, F, Am
Warm Per-Humid	5.6–6.5	0.16–0.17	1.82–2.63	0	1.24–9.17	13.4–17.66	52–58.8	1.07–1.17	Vm, ML, M, K, Q, F, Ch, Am
Warm Per-Humid	4.6–5.13	0.06–0.63	2.2–3.72	0	1.81–3.26	5.2–15.4	12–32	0.42–1.06	Vm, ML, M, K, Ch
Hot Sub Humid to Semi Arid	5.1–8.2	0.57–5.0	0.21–1.11	0–3	0.06–24.43	18.8–62.69	68–97.7	0.83–0.92	Sm, Vm, M, K
Hot Humid to Per-Humid	5–5.9	0.01–0.03	0.29–3.27	0	0.36–2.80	3.8–24.5	16–74.6	0.13–1.26	Vm, ML, M, K, Ch

floodplain soils are typically known as Diara land and Tal land following the distinct genesis and characteristics (Mishra et al. 2001, Mishra 2015c).

5.6.2 Black (Cotton) Soils

The name *black* is given to soils that are very dark in colour and turn extremely hard on drying and sticky and plastic on wetting, and hence, the soils are often very difficult to cultivate and manage. These soils are dark in colour, rich in clay content and have characteristics associated with shrink and swell properties. The high clay content (>30% up to at least 50 cm of the soil surface), preferably in dry regions, develops typical crack that is 1 cm or wider and reaches a depth of 50 cm, or more. Such soils are often called heavy, cracking clay soils. These soils are comparable with the Grumusols of the USA. Their black colour is comparable with the Chernozems of Russia and Prairie soils of USA, but, in their physico-chemical properties, they show

Table 5.6 The extent and distribution of different soil types of India as represented in the soil maps on 1: 250,000 scale along with their USDA/FAO nomenclature system

Soil types	Extent Lakh ha	Distribution in states	Soil orders USDA soil taxonomy	FAO Legend
Alluvial	1000.1	J&K, HP, Punjab, Haryana, Delhi, UP, Gujarat, Goa, MP, MS, AP, Karnataka, TN, Kerala, Puducherry, Bihar, Odisha, WB, ArP, Assam, Nagaland, Manipur, Mizoram, Tripura, Meghalaya, A&N	Inceptisols, Entisols, Alfisols and Aridisols	Cambisols, Fluvisols, Luvisols, Gleysols, Solonchaks, Solonetz
Coastal alluvial	100.5	AP, Karnataka, TN, Kerala, WB, Gujarat, Odisha, Puducherry, Lakshadweep, A&N	Aridisols, Inceptisols, Entisols	Solonchaks, Solonetz, Cambisols, Luvisols, Arenosols
Red	879.9	AP, Karnataka, Kerala, TN, Puducherry, Rajasthan, MP, MS, Gujarat, Goa, ArP, Assam, Manipur, Meghalaya, Nagaland, Mizoram, Tripura, Delhi, UP, HP, A&N	Alfisols, Ultisols, Entisols, Inceptisols, Mollisols, Aridisols	Luvisols, Cambisols, Leptosols, Lixisols, Nitisols
Laterites	180.9	AP, Karnataka, Kerala, TN, Puducherry, MS, Odisha, WB	Alfisols, Ultisols, Inceptisols	Cambisols, Alisols, Leptosols, Plinthosols
Brown forest	5.4	Karnataka, Maharashtra	Mollisols, Inceptisols	Cambisols, Luvisols, Leptosols, Phaeozems
Hill	22.6	Manipur, Odisha, WB, Tripura, Nagaland	Inceptisols, Entisols	
Terai	3.3	UP, Sikkim	Mollisols, Entisols	Chernozoms
Mountain meadow	0.6	J&K	Mollisols	Phaeozems
Sub-montane	1.0	J&K	Alfisols	Luvisols
Black	546.8	MP, MS, Rajasthan, Puducherry, TN, UP, Bihar, Odisha, AP, Gujarat	Vertisols, Mollisols, Inceptisols, Entisols and Aridisols	Vertisols, Cambisols, Leptosols, Regosols
Desert	262.8	Rajasthan, Gujarat, Haryana, Punjab	Aridisols, Inceptisols, Entisols	Cambisols, Leptosols, Calcisols, Arenosols, Fluvisols
Others[a]	283.1	–	–	

[a]Includes glaciers, sand dunes, mangrove swamps, salt waste, water bodies, rock land and rock outcrops. MP, Madhya Pradesh; MS, Maharashtra; UP, Uttar Pradesh; J&K, Jammu and Kashmir; TN, Tamil Nadu; AP, Andhra Pradesh; ArP, Arunachal Pradesh; WB, West Bengal; HP, Himachal Pradesh; A&N, Andaman and Nicobar Islands
Modified from Bhattacharyya et al. (2013)

significant differences. Such black soils are locally termed as *Regur* in central Asia, *Karail* in lower Gangetic Basin in U. P. and *Bhal* in Gujarat. In Bihar, *Tal* lands have comparable soil properties developed on the Ganga alluvium as influenced by back water during flood (Tiwary and Mishra 1990, Mishra 2015c).

Black soils are dominantly distributed in the central, western and southern states of India. According to Sehgal (1996), different members of black cotton soils occupy an estimated area of 72 Mha. Globally, such soils cover an estimated area of 340 Mha, dominantly in the semi-arid tropics of Africa (the Gezira, central Sudan, South Africa, Ethiopia, Kenya and Tanzania), Asia, especially the Deccan Plateau of India and Australia.

5.6.3 Desert (Arid) Soils

The name desert or arid is given to the soils that support almost negligible vegetation, except xerophytic plants, unless irrigated. These are soils of dry places that represent aridic/torric moisture regime. They may, however, occur in

Table 5.7 The extent and distribution of the different soil classes in India as per USDA system of soil classification (Yadav and Singh 2000)

Sl. no.	Indian nomenclature	Area (Km^2)	Distribution in States	USDA system of soil classification
1.	Red loamy soil	2,13,271	Andhra Pradesh, Tamil Nadu, Karnataka, Kerala, Madhya Pradesh, Orissa	Paleustalfs, Rhodustalfs and Haplustalfs
2.	Red sandy soils	3,30,590	Tamil Nadu, Karnataka, Andhra Pradesh	Haplustalfs Rhodustalfs
3.	Laterite soils		Tamil Nadu, Kerala, Karnataka, Andhra Pradesh, Orissa, Maharashtra, Goa, Assam	Plinthaquults
4.	Red and yellow soils	4,03,651	Madhya Pradesh, Orissa	Haplustults Ochraquults
5.	Shallow black soils	31,532	Maharashtra	Ustorthents Ustopepts
6.	Medium black soils	4,30,383	Maharashtra, Madhya Pradesh, Gujarat	Pellusterts Chromusterts
7.	Deep black soils	1,12,060	Maharashtra, Andhra Pradesh, Karnataka, Madhya Pradesh, Gujarat	Pellusterts Chromusterts Pelluderts
8.	Mixed red and black soils	1,62,255	Karnataka, Tamil Nadu, Maharashtra, Madhya Pradesh	Association of Alfisols and Vertisols
9.	Coastal alluvium soils	54,403	Tamil Nadu, Kerala, Andhra Pradesh,	Haplaquents
10.	Coastal sands	4534	Orissa	Ustipsamments
11.	Deltaic alluvium soils	87,045	Tamil Nadu, Andhra Pradesh, Orissa, West Bengal	Quartzipsamments Tropaqualfs
12.	Alluvial soils Khaddar (recent alluvium) Hangar (old Alluvium)	3,56,720	Uttar Pradesh, Punjab, Bihar, West Bengal, Assam	Haplaquents Ustifluvents
13.	Alluvium soils (highly calcareous)	13,611	North-Eastern, Uttar Pradesh, Bihar	Haplustalfs Calciorthents
14.	Calcareous sierozemic Soils	45,080	Punjab	Calciorthids
15.	Grey brown soils	1,01,572	Gujarat, Rajasthan	Calciorthids
16.	Desert soil Rhegosolic	1,54,423	Rajasthan	Calciorthids, Psamments
17.	Desert soils Lithosolic –		–	Lithic Entisols
18.	Terai soils	28,919	Uttar Pradesh, Bihar, West Bengal	Haplaquolls
19.	Brown hill soils, (over sandstones and shales)	81,242	Uttar Pradesh, Bhutan, Sikkim Himachal Pradesh	Palehumults
20.	Sub-mountain soils (Podsolic)	76,695	Uttar Pradesh, Jammu and Kashmir	Hapludalfs
21.	Mountain meadow soils	59,790	Kashmir including Ladakh	Cryoborolls Cryochrepts
22.	Saline and alkali	17,377	Uttar Pradesh, Punjab, Maharashtra, Karnataka, Tamil Nadu	Salorthids, salargids, Natrargids
23.	Peaty and saline peaty soils	2720	Kerala	Histosols
24.	Skeletal soils	79,151	Madhya Pradesh	Lithic Entisols
25.	Glaciers and eternal snow	29,335	Uttar Pradesh, Kashmir	Lithic Entisols

all temperature regimes, including cryic, frigid and/or hyper- and mega-thermic. Because of the limiting rainfall, the desert soils generally show accumulation of salts at or near the surface, forming salic, sodic, gypsic, petrogypsic, calcic or petrocalcic horizons. Large tract of hot and arid region, with a growing period of less than 60 days in a year, representing agro-ecological region (AER) 2 and, lying between the Indus River and the Aravalli Ranges, in North-Western India (Rajasthan, Gujarat, Haryana and Punjab) poses desertic conditions geologically.

5.6.4 Red Soils

Conceptually, red soils are considered to be those that have hues 7.5YR or redder in the series control Sect. (25–150 cm soil depth). They include red loams, red gravelly soils, red earths and latosolic soils. Such soils are moderately (to highly) weathered, enriched in secondary forms of iron and/or aluminium oxides (sesquioxides), poor in humus and generally have a clay enriched B-horizon that develops on stable and higher landforms, experiencing (sub) humid

Fig. 5.3 Red soils of Karnataka and Tamil Nadu. *Source* NBSS & LUP ARCHIVES 2017

Red soils originated from Ironstone, Shimoga district, Karnataka

Red loamy soils from Kangayam, Tiruppur, Tamil Nadu

climatic condition. In India, most typical red soils developed under hot, semi-arid to (sub) humid climatic condition in the subtropical regions are Rhodustalfs (termed *Luvisols* in the FAO Legend). The soils grade from shallow, gravelly and light coloured (in the upland) to fertile, deep and dark reddish-brown (in the plains and valleys). They are relatively high in base status than the lateritic soils.

The red and lateritic soils are the third most important group in the world, covering about 13% of the land area. They occur extensively in the South and SE Asia, western and central Africa, South America especially Brazil, Australia, China, Japan and India. A sizeable portion of these soils occurs in the developing countries where agriculture is the backbone of economy. In India, red and lateritic soils, predominantly occur in areas representing AER's 7, 8, 12, 17 and 19, comprising of north-western hill ranges, Eastern Himalayas, central Highlands, Eastern and Western Ghats, Deccan Plateau and *Konkan* coastal lands (Sehgal 1995). These areas are observed in the states of Andhra Pradesh, Telangana, Tamil Nadu, Karnataka, Maharashtra, Orissa, Goa and NE region. According to a recent survey undertaken by NBSSLUP, the red and lateritic soils cover a total area of about 110 Mha of which 70 Mha is under red soils. In Andhra Pradesh, Telangana, Tamil Nadu, Madhya Pradesh and Karnataka, the red and the black soils occur under similar bioclimatic conditions, but on different parent materials and landforms. The red soils develop on igneous (acidic) rock and occupy higher topographic positions, whereas the black soils develop on basalt (basic) rock or on alluvium derived from basalts and occupy relatively lower topographic positions on the landscape. Typical red soils are presented in Fig. 5.3.

5.6.5 Laterites and Lateritic Soils

The name *Laterite* signifies the soils with laterite formation in the subsoil horizons. The term was originally coined and used by Buchanan (1807) for the highly ferruginous vascular and apparently unstratified deposits observed in the Malabar Hills of Kerala. It is highly weathered material enriched in secondary forms of iron and/or aluminium and devoid of bases and primary minerals. It is either hard or subject to hardening upon exposure to alternate wetting and drying. It includes *Plinthite*. The term lateritic soil is used for soils that do not necessarily have laterites as a subsurface horizon but reflects its occurrence in their morphological features. The term is used to describe red soils that qualify for *Oxic* and/or *Kandic* subgroups and have low base saturation. Most of such soils qualify for *Ultisols* or *Alfisols* with *Kandic* properties in soil taxonomy.

The Lateritic soils are predominantly observed in the southern, SE and NE states and cover an area of about 40 Mha. The well-developed lateritic soils are occasionally observed on hilltops and plateaus in the states of Kerala, Tamil Nadu, Orissa and sparingly in Andhra Pradesh and Karnataka. The lateritic soils are widely distributed in all the above states and cover an area of about 25 Mha (excluding rock-out crops). They are largely observed in AERs 8, 18 and 19 (Fig. 5.4).

Coastal lateritic upland soils from Ullal, Mangalore, Karnataka

Coastal lateritic valley soils from Nenmanikkara, Thrissur, Kerala

Midland lateritic soils under Rubber plantation in Puthupally, Kottayam, Kerala

Fig. 5.4 Laterite and lateritic soils of India. *Source* NBSS & LUP ARCHIVES 2017

5.6.6 Salt Affected Soils

Salt-affected soils are the soils that contain considerable amounts of soluble salts and/or sodium on the exchange complex. They occur where potential evapotranspiration greatly exceeds precipitation, which is in arid and semi-arid regions. They are Intra zonal, as they are interspersed with the other zonal soils dominant of the tract/region (Fig. 5.5).

According to the Central Soil Salinity Research Institute (CSSRI), Karnal (Haryana), salt-affected soils occupy about 10 Mha of land in India. Out of that a major fraction (say two-third) is sodic in nature and occurs in the Indo-Gangetic plains and the Deccan (Peninsula) Plateau of India dominantly supporting black (cotton) soils; the rest (about one-third), occurs in the semi-arid and arid coastal regions, are saline. The saline-sodic soils (recently termed as sodic) of the Indo-Gangetic plains occur dominantly in the AER's 4 and 9 that receives an annual rainfall ranging from 500 to 800 mm. The saline soils, on the other hand, predominate in the AER's 2, 4, 6, 18 and 19 that receive an annual rainfall of less than 500 mm, except in the coastal areas. The worst affected states are: Gujarat, Rajasthan, U.P., Haryana and Bihar (for sodicity),and the coastal region (for salinity). Such soils in the Indo-Gangetic plains have been or are being ameliorated using gypsum technology.

Degraded soils from Gundlupet, Chamrajnagar, Karnataka

Fig. 5.5 Degraded soils. *Source* NBSS & LUP ARCHIVES 2017

5.6.7 Forest and Hill Soils

This name is implied for soils developed under any forest cover, irrespective of the forest species and/or profile development (Fig. 5.6). In India, the total area under different forest species (tropical, deciduous, coniferous, tropical evergreen) is estimated to be 75 Mha and is observed dominantly in the states of Himachal Pradesh, Jammu and

Forest soils from Palakad plains, Kerala

Hilly soils belonging to Inceptisols from Ponmudi, Thiruvananthapuram, Kerala

Hilly soils of Himachal Pradesh

Hilly cold arid region of Lei Ladak, Jammu and Kashmir

Fig. 5.6 Forest and hill soils. *Source* NBSS & LUP ARCHIVES 2017

Kashmir, Uttar Pradesh, Uttaranchal, Bihar, Jharkhand, Madhya Pradesh, Maharashtra, Kerala, Tamil Nadu, North-eastern states and Andaman and Nicobar Islands.

5.6.8 Podzolic Soils

The soils formed under coniferous vegetation, in the presence of acid humus and low base status, show some characteristic association with podzols. Owing to leaching of bases and translocation of sesquioxides, the soils tend to develop a bleached A-horizon. But because of the unfavourable parent material and the breaking down of mineral particles in the unsaturated organic acid, the process of podzolization, in the Himalayas, is restricted up to the mobilization of sesquioxides, and hence, the true podzols (as reported by some earlier workers) are not formed. This distinguishes the podzolic soils of the Himalayas from the true Podzols (observed in Europe, USA, Russia, etc.). Although the soils show evidence of accumulation of organic matter and amorphous Al and Fe in their B-horizons, yet the reported data do not meet the requirement of true Podzols (Spodosols), as laid down in soil taxonomy. Similar reporting has also been made in Kerala state. But, there are no authentic data and/or publication to verify such soils as true Podzols, termed Spodosols in soil taxonomy.

5.6.9 Brown Forest Soils

The other soils, developed on sedimentary and metamorphic rocks and/or alluvium, under sub-humid to humid climatic environment and deciduous or mixed vegetation, are non-calcic Brown or Brown Forest soils (Fig. 5.7). In their pedogenic evolution, climate (cool humid with uniformly distributed rainfall of 1000 mm), topography and vegetation play an important role. The soils have developed a mollic epipedon on stable surfaces, or an ochric epipedon on slopes with cambic or argillic horizon in their subsoils (B-horizon).

5.6.10 Peaty and Marshy Soils

The name peaty and marshy is given to the soils that have developed in the low-lying coastal marshy land or to the soils confined to depressions caused by dried lakes in the alluvial and coastal plain areas, formerly occupied by mangrove swamps. These soils are dominantly observed in

Brownhill soils belonging to Mollisols (Udolls) of Ooty, Nilgris, Tamil Nadu

Soils of Granitic origin from Malanad region ,Narasimharajpura, Chickmangalur, Karnataka

Fig. 5.7 Brown hill soils and soils of Malanad region. *Source* NBSS & LUP ARCHIVES 2017

Fig. 5.8 Peaty soils under waterlogged. *Source* NBSS & LUP ARCHIVES 2017

Kole land soils from below MSL in Mullaseri, Thrissur, Kerala

Gleyed coastal sandy soils from Tiswadi, Goa

the coastal inter-tropical lowlands. Globally, such soils occupy a limited area of about 12 Mha mainly in the coastal tropical lowlands of SE Asia, West Africa and along the north-eastern coast of South America. In India, they occur in localized pockets in the states of Kerala, Orissa, Sunderbans in West Bengal, Goa, South-Eastern Tamil Nadu and NE states where they are mostly associated with mangroves forest vegetation (Fig. 5.8). Acid and acid sulphate soils are also observed in south India (Fig. 5.9).

Fig. 5.9 Acid and acid sulphate soils. *Source* NBSS & LUP ARCHIVES 2017

Acid saline soils (Sulfaquepts) from Vyttila, Eranakulam, Kerala

Acid Sulphate soils from Thiruvarpu, Ettumanur, Kottayam, Kerala

5.7 Soil Classification

The basic requirements of any natural science are to classify the proposed objects studied (Joel 1926). Since soil is a complex and dynamic natural system, grouping of soils is difficult. It was believed earlier that the study of the soils could not advance as a science until a well-accepted classification system was developed (Joel 1926; Marbut 1935). The US soil taxonomy was adopted in India in 1969 (Dhir 2004). In earlier days, soil classification systems were based mainly on geomorphological and geological concepts as reflected in the chemical and mineralogical properties of the parent materials. The United States used systems based on physiography and/or geology. Thereafter, they used zonality-based systems from 1938 until the 1960s. Then, the revised version of Marbut's classification system (Marbut 1928) was published in 1938 (Baldwin et al. 1938) and was adopted by the US soil survey (Brevik 2002). The concept of genetic profiles was used in early and current Russian soil classification schemes (Gerasimov 1975; Gorajichkin et al. 2003). Cline's (1949) basic principles of soil classification were used as the foundation of global soil classification schemes, such as US soil taxonomy (Soil Survey Staff 1975 and 1999) and the World Reference Base (WRB) for soil resources (IUSS Working Group WRB 2006). The WRB was originally intended to be a conversion between national systems rather than being a classification system itself (Krasilnikov et al. 2009).

The US classification schemes permitted classification of soils on the basis of surface and subsurface diagnostic horizons and other characteristics. The concept of US soil taxonomy (Soil Survey Staff 1999) centres on the basic theme of differentiating soils on the basis of the properties of the soils being classified where the factors of soil formation help to describe soils indirectly (Smith 1986; Krasilnikov et al. 2009; Buol et al. 2011; Bockheim et al. 2014). In USDA soil taxonomy, soils are classified into six category levels from broadest to the narrowest: orders, suborders, great groups, subgroups, family and series (Soil Survey Staff 1999). The lowest category (Soil Survey Staff 1999) is the series, which is based on the kind and arrangement of horizons and finer differences in soil properties. The soil series are again divided into phases on the basis of surface stoniness, slope, erosion, and/or other attributes that are not diagnostic in soil taxonomy but are important for land use. In spite of its several advantages, the taxonomy suffers from some drawbacks. The usual criticism signifies that the system is difficult to understand by scientists and end-users working in other branches of science, in general, and soil science (except pedology), in particular.

Until the Seventh Approximation stage of US soil taxonomy, Indian soil scientists classified soils following the best fit. The interesting part of US soil taxonomy has always been its scope for improvement. This, in other words, as felt by many, indicates that this system should not be considered final since appropriate reasoning might always open a new category in the overall framework. Many scientists, actively engaged in soil survey, mapping and classification, have suggested new rationale for US soil taxonomy at various categorical levels. A commentary of such selected novel ideas from the Indian scientists is documented by Bhattacharya et al. (2015). The National Bureau of Soil Survey and Land Use Planning (ICAR) has compiled the soil map of India in 2000 at order (Fig. 5.10) and sub-order (Fig. 5.11) levels.

Mishra (2015a) proposed an Indian soil classification scheme which has 17 soil groups for major soils of India. The subgroup consists of six potentially viable soil features for each soil group, and the soil phase consists of five distinguishing constraints related to soil productivity. The classified soil may enable to identify the associated limitations for correction through the process of improvement measures in order to assign the most remunerative land-use choice preferably in rotation in case of annual crops. The classification scheme has 17 soil groups (inherent soil qualities dominantly controlling the suitability of a soil), viz. Acidisols, Calcisols, Desertisols, Fluvisols, Halosols, Hydrosols, Microbisols, Mixisols, Mollisols, Natrisols, Neutrosols, Plinthosols, Regosols, Vertisols and Uttamisols. The subgroups (soil potentiality in terms of soil productivity for sustainable land use) are effective soil depth, length of growing period, dominant clay minerals, textural arrangement and slope gradient. The soil phase is the limitation and constraint in association with soil features which include soil fertility levels, soil physical constraints, soil chemical/mineralogical constraints, soil sickness/predators and risks, uncertainties and doubts. The soil suitability is by and large associated with specific set of preferred land-use choices (Mishra 2015a).

The team of pedologists at ICAR-NBSS & LUP, Nagpur, established certain rationale and proposed modification for different soil orders of India (Bhattacharyya et al. 2015) as presented in Table 5.8. However, simplified guide to WRB reference soil groups (RSGs) with suggested codes is presented for major Indian soils (Table 5.9).

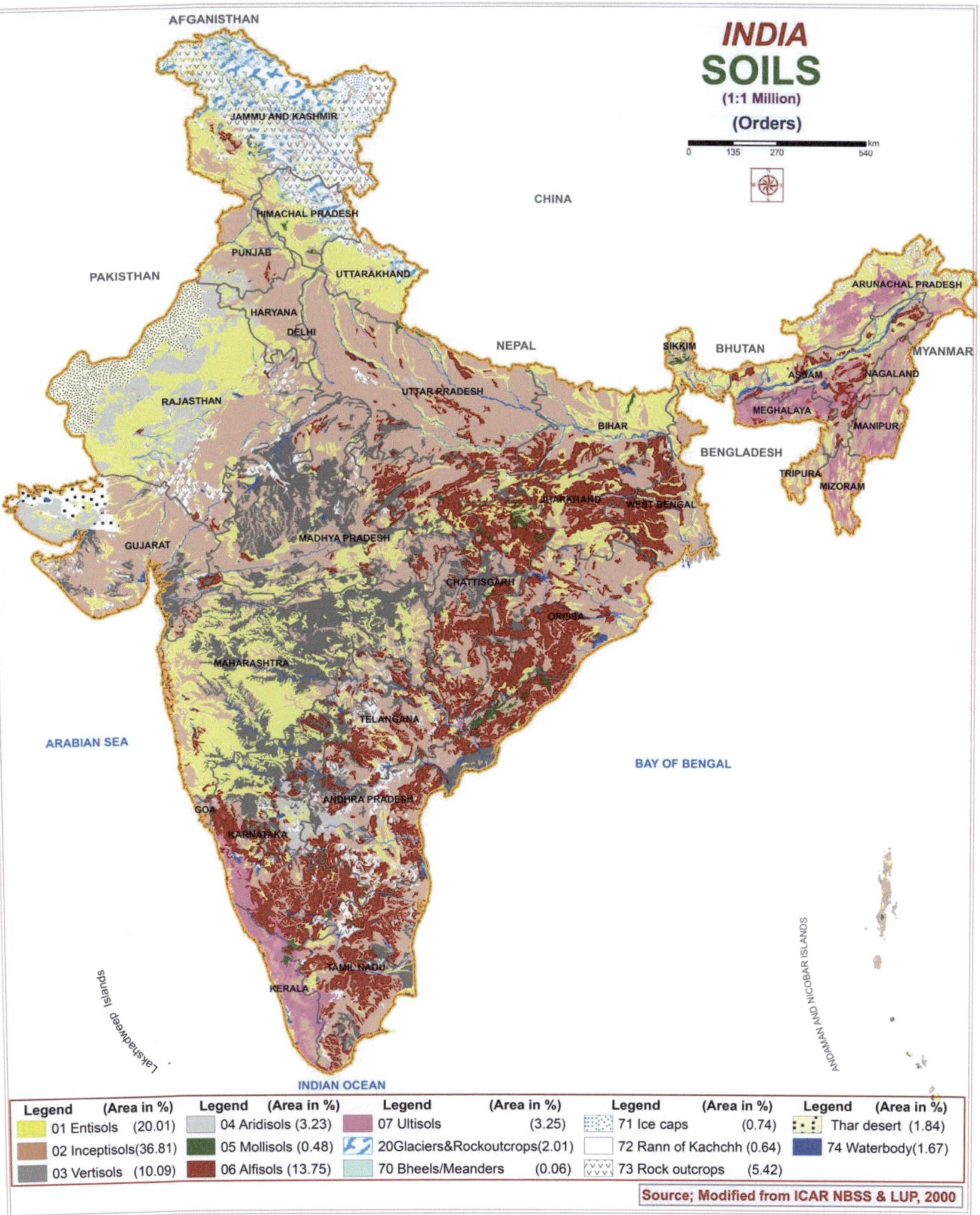

Fig. 5.10 Soil map of India showing Order under US Soil Taxonomy. *Source* NBSS & LUP ARCHIVES 2017

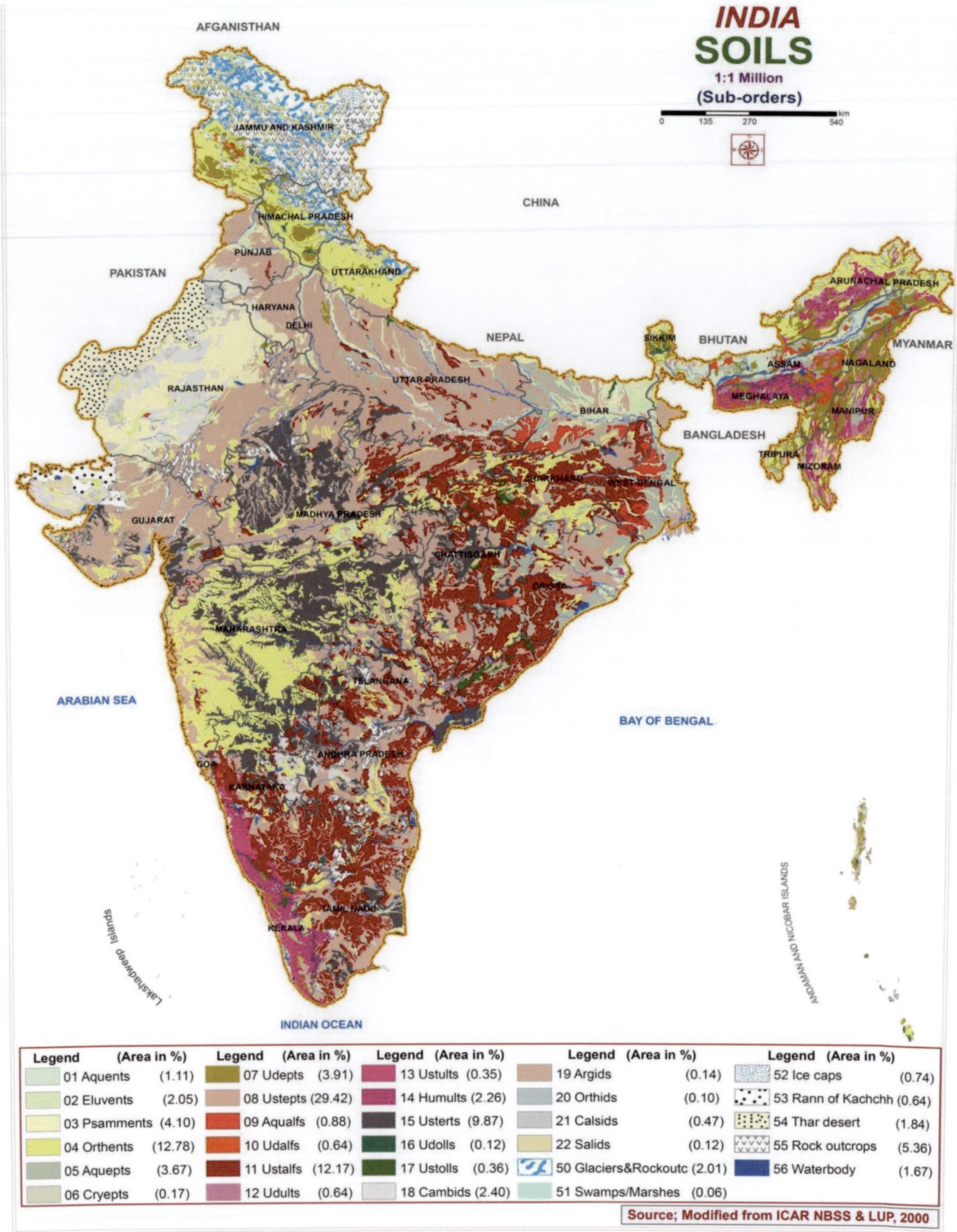

Fig. 5.11 Soil map of India showing Sub-order under US Soil Taxonomy. *Source* NBSS & LUP ARCHIVES 2017

Table 5.8 Rationale and proposed modification for different soil orders of India (Bhattacharyya et al. 2015)

Soil taxonomy category	Rationale	Remarks
Vertisols	Other than thickness ($\geq$25 cm) and depth criteria (within 100 cm), soils should have • Slickensides or • wedge-shaped peds and • $\geq$30% clay and cracks (Soil Survey Staff 2003)	To group a soil in Vertisols, *presence of slickensides not mandatory*
Subgroup	1. For vertic subgroups of any other soil orders should have • $\geq$5-mm-wide cracks within 125 cm or • $\geq$6.0 linear extensibility (LE) within 100 cm (Soil Survey Staff 2003) 2. Indirect evidence of vertic properties in soils if clay smectites $\geq$20% (Shirsath et al. 2000; Bhattacharyya 1997)	1. If cracks not visible in wet field, LE (100 X weighted mean average of COLE) may be determined following standard method (Schafer and Singer 1976) 2. (i) Clay (<2 μm) may be estimated X-ray diffraction (Jackson 1979) or (ii) Clay smectites may be measured indirectly by LE values with the help of equation COLE = 0.263 (smectite %) + 0.771 (Shirsath et al. 2000; Bhuse et al. 2001)
Mollisols	With other criteria of US soil taxonomy (Soil Survey Staff 1999 and 2003), the thickness of mollic epipedon may be $\geq$18 cm (minimum thickness) (Olson et al. 2005)	1. Purpose is to accommodate eroded phase of Mollisols in the soil order 2. The important fact is to adopt soil conservation measures to control soil erosion
Inceptisols, Ultisols, Alfisols	If clay illuviation {ratio of clay (<2 μm)} in B horizons and A horizons ~ 1.2) identified from clay data but clay skins not identified in the field by 10 X lens, then 'Method 1' (under Remarks) or 'Method 2' (under Remarks) will help in deciding soil order either as Alfisols/Ultisols (depending on base saturation criterion, Soil Survey Staff 2003) or as Inceptisols	**Method 1**: Decrease in clay mica (<2 μm) with depth can be considered as incontrovertible evidence of clay illuviation (Pal et al. 1994; Pal 1997; Srivastava et al. 1998) **Method 2:** Determination of total extractable acidity by $BaCl_2$–TEA, 1 N KCl extractable H^+ and Al^{3+}. Estimation of CEC by sum of cations, ECEC, CEC clays and base saturation to confirm the presence of kandic horizon. If yes, soils may be grouped as Alfisols (if BS >35%, BS to be determined by sum of cations) or Ultisols, if otherwise. If not, these soils will be grouped as Inceptisols **Comment 1**: If these two methods are not followed, the soils may be mistakenly grouped as Inceptisols **Comment 2**: If base saturation not determined by sum of cations, these soils (with kandic horizon) may mistakenly be grouped as Alfisols instead of Ultisols (Soil Survey Staff 2003; Bhattacharyya et al. 1994)
Subgroups	1. Calciustepts and Haplustepts may be grouped as (Saxena et al. 2004) • Sodic Petrocalcic Calciustepts or	ESP = (Exchangeable Na/CEC) X 100
	• Sodic Petrocalcic Calciustepts or • Sodic Haplustepts or • Fluventic Sodic Haplustepts if exchangeable sodium percentage (ESP) $\geq$15 2. Haplustepts may be grouped as Sodic Haplustepts if EC (dS m^{-1}) X thickness $\geq$900	
Family level, viz. mineralogy class of kandic Alfisols, and Ultisols and Oxisols	Mineralogy class should be mixed, if • Clay CEC of sum of cations§ in soil control section $\geq$24 cmol(p +) kg^{-1} (Smith 1986) (Method 1), even if • Gibbsite content in <2 mm fraction of soil is >18% in the soil control section (Chandran et al. 2005) (Method 2)	**Method 1**: Clay CEC (sum of cations) = [exchangeable bases (NH_4OAc pH 7.0) + $BaCl_2$ = TEA acidity)/clay %] X 100 **Method 2:** semi-quantitative estimates of gibbsite content through X-ray diffraction analyses of clay samples (<2 μm) (Also see Chandran et al. 2005)
Entisols Subgroup	Ustifluvents and Ustorthents should be grouped as sodic if exchangeable sodium percentage (ESP) $\geq$15 (Verma et al. 2007)	ESP = (exchangeable Na/CEC) X 100

(continued)

Table 5.8 (continued)

Soil taxonomy category	Rationale	Remarks
Family level, viz. Mineralogy class	Mineralogy class should be smectitic for all the shallow soils (Lithic Ustorthents) developed in the weathered basalt in Western and Central India	Mineralogy class may be confirmed by • Detailed investigations on clay minerals through X-ray diffraction technique or • Estimating clay CEC [(soil CEC/clay %) X 100] (Smith 1986; Bhattacharyya et al. 1997) • Determining coefficient of linear extensibility (COLE) (Soil Survey Staff 2003; Schafer and Singer 1976) of soils.

Table 5.9 World reference group, their distinguishing features and distribution

Sl. no.	Distinguishing features	RSG	Code	Distribution in India
1.	**Soils with thick organic layers**	Histosols	HS	Sunderbans (WB), *Kuttanad, Kole, Kaipad, Pokkali* (Kerala)
2.	**Soils with strong human influence**			
	Soils with long and intensive agricultural use	Anthrosols	AT	Not identified based on already available data
	Soils containing significant amounts of artifacts	Technosols	TC	Not identified based on already available data
3.	**Soils with limitations to root growth**			
	Permafrost-affected soils	Cryosols	CR	Not identified based on already available data
	Thin soils or soils with many coarse fragments	Leptosols	LP	Widely distributed on ridges/scrub land
	Soils with a high content of exchangeable Na	Solonetz	SN	Widely distributed on arid/semiarid areas
	Alternating wet-dry conditions, shrink–swell clays	Vertisols	VR	*Regur* or black cotton soils, *tal* land soils. Dominantly in MP, Maharastra, Andhra Pradesh, Gujarat, Tamil Nadu, part of Rajasthan, south Bihar (*Tal* land)
	High concentration of soluble salts	Solonchaks	SC	Widely distributed on arid/semiarid areas
4.	**Soils distinguished by Fe/Al chemistry**			
	Groundwater-affected soils, underwater soils and soils in tidal areas	Gleysols	GL	*Chaur* land
	Allophanes or Al-humus complexes	Andosols	AN	Not identified based on already available data, though localized volcanic activities in Andaman Nicobar
	Subsoil accumulation of humus and/or oxides	Podzols	PZ	Seldom due to lack of specific PM and environment
	Accumulation and redistribution of Fe	Plinthosols	PT	Jharkhand, Kerala, Orissa, Karnataka, TN
	Low-activity clay, P fixation, many Fe oxides, strongly structured	Nitisols	NT	Localized but distributed apart
	Dominance of kaolinite and oxides	Ferralsols	FR	NE Region, Kerala, Jharkhand, Andhra Pradesh, TN, Orissa
	Stagnating water, abrupt textural difference	Planosols	PL	Localized but distributed apart
	Stagnating water, structural difference and/or moderate textural difference	Stagnosols	ST	Not identified based on already available data

(continued)

Table 5.9 (continued)

Sl. no.	Distinguishing features	RSG	Code	Distribution in India
5.	**Pronounced accumulation of organic matter in the mineral topsoil**			
	Blackish topsoil, secondary carbonates	Chernozems	CH	Yet to document
	Dark topsoil, secondary carbonates	Kastanozems	KS	Not identified based on already available data
	Dark topsoil, no secondary carbonates (unless very deep), high base status	Phaeozems	PH	Tarai region (UP and Uttarakhand)
	Dark topsoil, low base status	Umbrisols	UM	Himalayas region
6.	**Accumulation of moderately soluble salts or non-saline substances**			
	Accumulation of and cementation by secondary silica	Durisols	DU	Jharkhand, NE Region, Kerala, TN, Karnataka, Orissa
	Accumulation of secondary gypsum	Gypsisols	GY	Widely distributed on arid/semiarid areas
	Accumulation of secondary carbonates	Calcisols	CL	Widely distributed on arid/semiarid areas
7.	**Soils with a clay-enriched subsoil**			
	Retic properties	Retisols	RT	Not identified based on already available data
	Low-activity clays, low base status	Acrisols	AC	Localized in NE Regions, Jharkhand, Kerala, TN, Karnataka, Orissa
	Low-activity clays, high base status	Lixisols	LX	Not identified based on already available data
	High-activity clays, low base status	Alisols	AL	Not identified based on already available data
	High-activity clays, high base status	Luvisols	LV	Widely distributed. NE Region, Andhra Pradesh, Jharkhand, HP, Maharastra, Karnataka, TN, Orissa, MP.
8.	**Soils with little or no profile differentiation**			
	Moderately developed soils	Cambisols	CM	Widely distributed in udic, ustic, xeric, and aquic except arid region (*Thar* desert in Rajasthan)
	Sandy soils	Arenosols	AR	Most of active flood plains, coastal landforms
	Soils with stratified fluviatile, marine and lacustrine sediments	Fluvisols	FL	Widely distributed in river basins
	Soils with no significant profile development	Regosols	RG	Widely distributed on steep slope, rocky humid and sub-humid mountenous landforms

Modified from Sehgal (2005)

5.8 Correlation of Classification System

The WRB having taken from the framework of the revised FAO Legend as a guide bears many similarities to it. The nomenclature has been adopted and where necessary adapted using the set rules. Many definitions in both the original and revised FAO Legends are adopted from US soil taxonomy (Soil Survey Staff 1999). In most cases, they are summarized and simplified according to the requirement of the legend. The world reference base for soil resources has its stronger bias with the FAO system which in turn has derived some of its definitions from US soil taxonomy (Fig. 5.12). In other words, all these three systems can be interlinked with each other (Table 5.10).

However, soil classification at order level for major benchmark sites (series) is discussed further in Chap. 9. Altogether, there are seven major soil orders found in India, of which Inceptisols covering around 36.8% followed by Entisols (20.01%), Alfisols (13.75), Vertisols (10.09%), Ultisols (3.25%), Aridisols (3.23%) and Mollisols (0.48%). Now, around 18 benchmark sites (series) fall under Alfisols,

Fig. 5.12 Pedons showing taxonomic grouping (US Soil Taxonomy). *Source* NBSS & LUP ARCHIVES 2017

Soils of Typic Fluvaquents in Indo Gangetic Plains of Eastern region (West Bengal)

Soils of Typic Endoaquepts in flood plains of Eastern region (West Bengal)

Patchic Hapludolls in Himalayan region of Sikkim

Patchic Argiudolls in Himalayan region of Sikkim

Typic Paleudalfs in siwalik ranges of northeastern region

Typic Dystrudepts in siwalik ranges of north eastern region

Table 5.10 Correlation of WRB with other systems of soil classification (Sehgal 2005)

US soil taxonomy	FAO/UNESCO legend	WRB for soil resources
Alfisols	Luvisols, Nitisols and high base saturated Acrisols	Luvisols, Nitisols, Acrisols (Phaeozems with argic horizon)
Andisols	Andosols	Andisols, few Nitosols
Aridisols	Yermosols, Xerosols, Solonchaks	Solonchaks, Gypsisols, Calcisols
Entisols	Regosols, Arenosols, Fluvisols, Lithosols, few Rendzims and Rankers	Regosols, Fluvisols, few Cryosols and Leptosols
Gelisols	*Gelic* Podzoluvisols, *Gelic* Planosols, *Gelic* Podsols	Cryosols
Histosols	Histosols	Histosols and few Gleysols
Inceptisols	Cambisols, Andosols, Fluvisols	Cambisols, Andosols, Fluvisols (few Gleysols, Phaeozems, Umbrisols, Anthrosols)
Mollisols	Chernozems, Kastanozems, few Phaeozems, few Gleyzems	Chernozems, Kastanozems, few Phaeozems, few Gleysols
Oxisols	Ferralsols	Ferralsols
Spodosols	Podzols, few Podzoluvisols, Gleyzems and Planosols	Podzols, few Sesquisols
Ultisols	Acrisols, Nitisols and few Luvisols	Lixisols, Acrisols and few Nitisols
Vertisols	Vertisols	Vertisols

followed by 16 benchmark sites in Inceptisols and 13 benchmark series under Vertisols. In addition, Entisols, Aridisols, Ultisols, Mollisols and Histosols have 7, 2, 1, 1 and 1 benchmark sites, respectively (Fig. 5.13). More details are available elsewhere (Chap. 9).

5.9 Spectral Library for Indian Soils

Periodic assessment of soil fertility is required to evaluate the deficiency and excess of nutrients so that the external inputs will be optimized to ensure adequate yield of economic products and plant health. The process involved the surface soil sample collection, analysis in soil testing laboratories and soil management recommendations, but it is time-consuming. Soil spectral measurement is one of the rapid and easy methods since the spectral reflectance of soil is largely influenced by soil colour, texture, organic matter, minerals, soluble salts and moisture. For building of soil spectral library, a systematic database of soils occurring on varied physiographic/climatic zones together with their spectral reflectance properties (measured in field and/or laboratory conditions or acquired through space/air sensors) is a prerequisite. In India, fifty-nine benchmark soils occurring in different physiographic regions were indentified and representative soils were characterized for their morphological, physical, chemical and mineralogical properties. Nearly 600 pedons were studied in the fields, and more than 2500 surface and subsurface samples were collected for laboratory characterization. All the major soil orders Entisols, Inceptisols, Vertisols, Mollisols, Alfisols, Aridisols and Ultisols identified in the country were covered. Wide variation in soil properties, viz. soil colour, texture, organic carbon, soil pH, EC, calcium carbonate content, cation exchange capacity, base saturation, iron content (Fe^{2+}, Fe^{3+} and Free Fe_2O_3) and clay mineralogy was observed in soils occurring on different physiographic regions. The reflectance characteristics of soils were studied both under field and laboratory conditions. Under field conditions, the soil reflectance data showed water band noise around 1400 nm and 1900 nm. Soils with high salt content reflected more energy at different wavebands than the normal soils in the Indo-Gangetic alluvial plains. A characteristic absorption feature around 2200 nm in ferralitic soils near Nelmangala, Karnataka, indicates the dominance of kaolinitic clay mineral, whereas the broad absorption feature around 950 nm indicates the dominance of iron oxides in the soils. The spectral curves (surface and sub-surface) taken in laboratory showed prominent absorption features at 1400, 1900 and 2200 nm. Broad absorption features around 950 nm were observed in red and ferralitic soils. The salt-affected soils rich in Na/Mg chloride salts showed characteristics of U-shaped absorption features at 1900 nm. The soils rich in moisture and/or organic carbon content, in general, has reflectance in all the wavebands. Soil and reflectance data for spectral library at the NBSS & LUP, Nagpur, are presented in Fig. 5.13 (NBSS & LUP 2005).

While applying the spectral data, soil and geomorphology can better be understood using bidirectional reflectance distribution function (BRDF) that takes into consideration

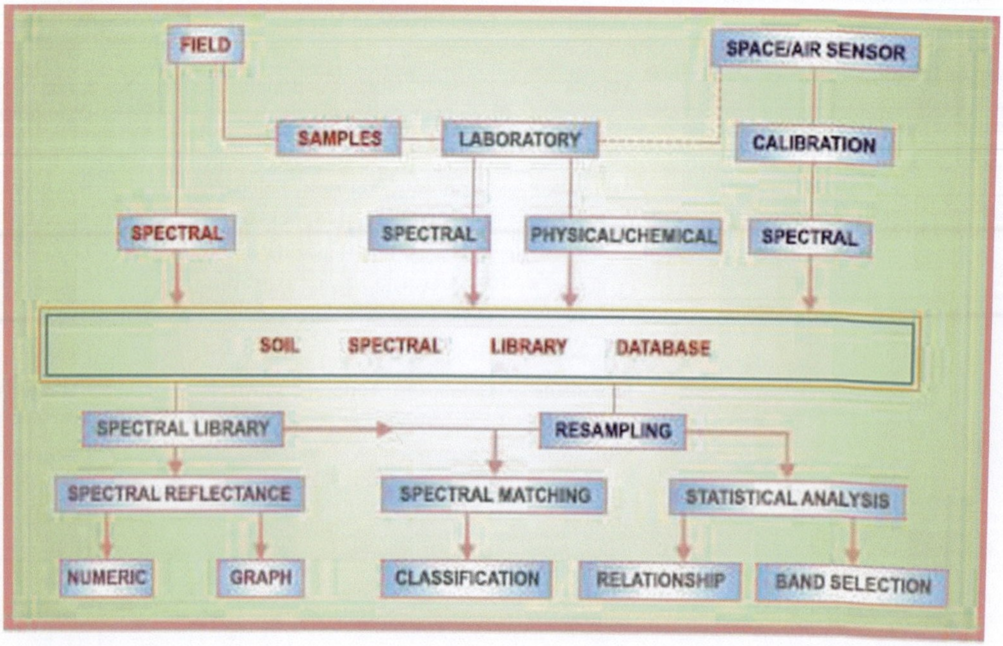

Fig. 5.13 Soil and reflectance data for spectral library. *Source* NBSS & LUP 2005

complete hemisphere of the object. However, there is need of its application in soil and geomorphology in India that can facilitate the scope of BRDF to be used to establish a particular soil type and its geomorphology. Mishra et al. (1990) attempted to fabricate a radiometric device for calculating the BRDF of mica-rich soils using radiometer at IARI, while they were studying the spectral properties of soils (Mishra et al. 1993).

5.10 Soil-Based Agro-technology Transfer

Soil and climate of the particular site decide the genetic potential of any crop cultivar. Management strategy helps in improving the potential of cultivar by overcoming the soil or climatic constraints. Therefore, soil information provides a sound basis for transfer of agro-technology from one site to another site having similar soil and site characteristic which includes climate (Naidu et al. 2015). Soil survey interpretations provide predictions about behaviour of defined kind of soil under stated conditions and expected results of interaction between soil characteristics, crop requirements and management practices (Kellogg 1961). In a review about Benchmark soils Project (BSP), the IBSNT (1983) concluded that the soil information at family level could be used to identify similar soil agro-environments for transfer of agro-technology. The same time, crop performance and management responses were also found similar on soils having similar texture, mineralogy, moisture and temperature regimes (Naidu et al. 1998). Hence, soil classification at family level intent to group the soils within a subgroup having similar physical and chemical properties which affect their responses to management are great enough to meet most of our needs for practical interpretations (Naidu et al. 2015). In such a way, the soil survey may be highly helpful for identifying the sites for agro-technology transfer. So, the soil resource information not only helps in appropriate technology generation but also provides ample scope for technology up-scaling in the target domain (Mishra 2016). For this, the researchable problems need to be systematically discovered and defined to get better production options through smart management strategies (Mishra 2015b). At the same time, lack of large-scale land resources database limits the effort of enhancing the agricultural productivity through introducing novel technologies for optimum utilization of natural resources. Because the efforts made towards both land use/land cover and soil mapping for different regions of the country has been limited up to 1:50,000 scale. To obtain the desired result at village and block level as well as to protect the prime agricultural land conversion to non-agricultural uses, the large-scale data of natural resources on 1:10,000 scale are required. On the other hand, the agricultural productivity of the country as a whole is also lagging behind in terms of global scenario due to large-scale degradation. In order to have a breakthrough in enhancing the agricultural productivity, we need to put the right technology at right time and at right place which requires large-scale database information at block/village level. Moreover, in the process of developing such large-scale maps, areas with salt-affected soils, sodicity development, eroded soils, waterlogged areas, *jhum* lands (shifting cultivation), etc. can be identified and studied in detail so that planning strategies for reclamation and conservation of degraded lands can be developed.

5.11 Conclusions

The soil is highly dynamic in time and space within an open system and so its classification would also be changing in an interval that will vary according to surrounding and inherent conditions. As such, soil classification as an indicator of land utilization type as well as land-use pattern enabling the effective land-use planning would work better for farming communities and justify as a tool towards efforts for ensuring food security, food safety and livelihood in days to come. The road map of the country based on soil classification would now help the policy makers to come up with reliable outcome through land-use planning. The road map of agriculture without proper soil evaluation and land-use suitability classification suffers virtually from a missing gap because ignorance of truth and reality results into disastrous consequences. The classification scheme for Indian soils will integrate soil qualities with soil health, associated limitations and land-use suitability.

References

Baldwin M, Kellogg EC, Thorp J (1938) Soil classification. Soils and men. USDA Agric. Yearb. US Gov. Print. Office, Washington, DC, pp 979–1001

Bhattacharyya T, Pal DK, Deshpande SB (1997) On kaolinitic and mixed mineralogy classes of shrink–swell soils. Aust J Soil Res 35:1245–1252. https://doi.org/10.1071/S96115

Bhattacharyya T, Pal DK, Lal S, Chandran P, Ray SK (2006) Formation and persistence of Mollisols on Zeolitic Deccan basalt of humid tropical India. Geoderma 136:609–620

Bhattacharyya T, Chandran P, Ray SK, Pal DK (2015) Soil classification following the US taxonomy: an Indian commendatory. Soil Horizons:1–6

Bhattacharyya T, Pal DK, Chandran P, Ray SK, Mandal C, Telpande B (2008) Soil carbon storage capacity as a tool to prioritise areas for carbon sequestration. Curr Sci 95:482–494

Bhattacharyya T, Pal DK, Mandal C, Chandran P, Ray SK, Deepak Sarkar, Velmourougane K, Srivastava A, Sidhu GS, Singh RS, Sahoo AK, Dutta D, Nai KM, Srivastava R, Tiwary P, Nagar AP, Nimkhedkar SS (2013) Soils of India: historical perspective, classification and recent advances in knowledge; a review. Curr Sci 104:1308–1323

Bhattacharyya T, Sen TK, Singh RS, Nayak DC, Sehgal JL (1994) Morphology and classification of Ultisols with kandic horizon in north-eastern region. J Indian Soc Soil Sci 42:301–306

Bhuse SR, Vaidya PH, Bhattacharyya T, Pal DK (2001) An improvised method to determine clay smectite in Vertisols. Clay Res 20:65–71

Buol SW, Southard RJ, Graham RC, McDaniel PA (2011) Soil genesis and classification, 6th edn. Wiley-Blackwell, New York

Brevik EC (2002) Problems and suggestions related to soils classification as presented in introduction to physical geology textbooks. J Geosci Educ 50:539–543

Bockheim JG, Genadiyev AN, Hartemink AE, Brevik EC (2014) Soil-forming factors and soil taxonomy. Geoderma 226–227:231–237. https://doi.org/10.1016/j.geoderma.2014.02.016

Buchanan F (1807) A journey from Madras through the countries of Mysore. Canara and Malabar, vol 2. East India Co., London. Geological Magazine, pp 436–460

Chandran P, Ray SK, Bhattacharyya T, Srivastava P, Krishnan P, Pal DK (2005) Lateritic soils of Kerala, India: their mineralogy, genesis and taxonomy. Aust J Soil Res 43:839–852

Cline MG (1949) Basic principles of soil classification. Soil Sci 67:81–91. https://doi.org/10.1097/00010694-194902000-00002

Dhir RP (2004) Whither to pedology and soil survey-Some thoughts. Agropedology 14:1–8

Eswaran H, Kimble J, Cook T, Beinroth FH (1992). Soil diversity in the tropics: Implications for agricultural development. In: Lal R, Sanchez PA (eds) Myths and science of soils of the tropics, pp 1–16. SSSA special publication no. 29. SSSA, In and ACA, Inc., Madison

Gajbhiye KS, Mandal C (2000) Agro-ecological zones, their soil resource and cropping systems, status of farm mechanization in India, National Bureau of soil survey and land use planning, Nagpur

Gerasimov IP (1975) Elementary pedogenic processes as the basis for genetic diagnostics of soils. Sov Soil Sci 5:3–9

Gorajichkin SV, Tonkonogov VD, Gerasimova MI, Lebedeva, II, Shishav LL and Targulian VD (2003) Changing concepts of soil and soil classification. In: Eswaran H, Rice, T Ahrens R, Stewart BA (eds) Russian soil classification: a global desk reference. CRC Press, Boca Raton, FL, pp 187–199

Govindarajan SV (1971) Soil map of India. In: Kanwar JS, Raychaudhuri SP (eds) Review of soil research in India, vol 1, p 7

IBSNT (1983) International benchmark sites network for agro-technology transfer. Department of Agronomy and Soil Science, University of Hawaii

IUSS Working Group WRB (2006) World reference base for soil resources. World Soil Resources Rep. 103. FAO, Rome

Jackson ML (1979) Soil chemical analysis—advanced course, 2nd edn. M.L. Jackson, Madison

Jha PP (1972) Soils of Bihar. Soils of India. Fertiliser Association of India Pub, New Delhi

Joel A (1926) Changing viewpoints and methods in soil classification. Sci Agric 6:225–232

Kellogg CE (1961) Soil interpretations in the-soil survey. United States. Department of Agriculture, Soil Conservation Service. Washington

Krasilnikov P, Ibanez Marti JJ, Arnold R, Shoba S (2009) A handbook of soil terminology, correlation and classification. Earthscan, London

Marbut CF (1928) A scheme of soil classification. In: Proceedings paper, First International Congress Soil Science Commission, 1927, pp 1–31

Marbut CF (1935). Soils of the United States. Part III. In: Atlas of America agriculture. USDA, US Gov. Print. Office, Washington, DC. pp 1–98

Mishra BB, Ghosh, SK, Srivastava SK (1990) Bidirectional reflectance distribution function of mica species and soil derived from mica-rich parent materials. In: Proceedings of Indian society of remote sensing for agric development, IARI, New Delhi, pp 66– 72

Mishra BB, Ghosh SK, Srivastava SK (1993) Spectral reflectance of soil derived from mica-rich parent materials in two different toposequences in relation soil properties. J Indian Soc Soil Sci 41:516–521

Mishra BB, Choudhary CD, Singh NK, Mall J, Singh VN (1994) "Fluvisols" as a proposed order in soil taxonomy for active flood plain soils. In: Transaction 15th world congress of soil sciences, Acapulco, Mexico, vol 6B, pp 55–56

Mishra BB (2000) Fluvisols as the 13th order in USDA soil taxonomy for floodplain soils with *Chemihydropedoturbation* pedongenic process -a proposition. In: Proceeding international conference nature. Resources Sustainable Agric. in 21st Century, Feb 14–18, 2000, New Delhi, part 2, pp 86–88

Mishra BB, Mall J, Singh ID (2001) Assessment and management of *Diara* land soils using remotely sensed data. Indian Farming 50:22–26

Mishra BB (2015a) Soil Classification versus Land Use Planning: Proposed scheme. In national seminar on sustainable management of land resources for livelihood security held during. NBSS & LUP, Nagpur. 28–30 Jan 2015

Mishra BB (2015b). Soil science and land use planning: Myth, reality, evidence and challenge. In: National seminar on sustainable management of land resources for livelihood security held during, NBSS & LUP, Nagpur, 28–30 Jan 2015

Mishra BB (2015c) Soil Based integrated management inputs for flood and flood plain soils of Bihar, India. EC Agriculture 1(2):109–123

Mishra BB (2016). Indian system of soil classification: a way forward. Agri Res & Tech: Open Access J. 2016; 3(2): 555606. DOI:008 https://doi.org/10.19080/artoaj.2016.03.555606

Murthy RS, Pandey S (eds) (1983) Soil map of India (*1: 6.3 m*), National Bureau of soil survey and land use planning, Indian Council of Agricultural Research, Nagpur

Naidu LGK, Ramamurthy V, Niranjana KV, Satyavathi PL, Srinivas S, Dhanorkar BA, Ravindra Chary G, Reddy RS (2015) Evaluation of Regional Benchmark Soils and agro-climate vis-a-vis soils of technology generation sites for agro technology transfer—a case study of Andhra Pradesh. J Indian Soc Soil Sci 63(2):144–158

Naidu LGK, Reddy RS, Sah KD, Bhaskar BP, Niranjana KV, Dhanorkar BA, Srinivas S, Nagaraju MSS, Ray SK, Raghu Mohan NG (1998) Mapping of agro ecological zones of Andhra Pradesh through soil resource data 1998. Indian J Agric Sci 68:661–665

NBSS & LUP Archives (2017). Compilation of legacy soil information from different projects of ICAR-NBSSLUP, Nagpur

NBSS & LUP Staff (2002). Soils of India. NBSS Publications 94. NBSS & LUP, Nagpur, India

NBSS & LUP (2005) Reflectance libraries for development of soil sensor for periodic assessment of state of soil resources. NATP project report, NBSS No. 835

NBSS & LUP (2016). India: agro-ecological region. Revised edition. Publication no. 170

Olson KR, Fenton TE, Smeck NE, Hammer MD, Renson CW, Zanner CW, McLeese R, Sucik MT (2005) Proposed modification of mollic epipedon thickness criteria for eroded condition and potential impacts on existing soil classification. Soil Survey Horizon 46:39–47

Pal DK, Bhattacharyya T, Chandran P, Ray SK (2009) Soils of Indo-Gangetic plains: their historical perpective and management. Curr Sci 9:1193–1201

Pal DK, Kalbande AR, Deshpande SB, Sehgal JL (1994) Evidence of clay illuviation in sodic soils of north-western part of the Indo-Gangetic plain since the Holocene. Soil Sci 158:465–473. https://doi.org/10.1097/00010694-199415860-00008

Pal DK (1997) An improvised method to identify clay illuviation in soils of the Indo- gangetic plains. Clay Res 16:46–50

Raychaudhari SP, Aggarwal RR, Datta Biswas NR, Gupta SP, Thomas PR (1963) Soils of India. Indian Council of Agricultural Research, New Delhi

Saxena RK, Sharma RC, Verma KS, Pal DK, Mandal AK (2004) Salt affected soils, Etah district (UP). NBSS Publications 108. NBSS & LUP, Nagpur, India

Schafer WM, Singer MJ (1976) A new method of necessary shrink-swell potential using soil parts. Soil Sci Soc Am J 40:805–806

Schokalskaya ZJ (1932). In: Spate, OHK, Learmonth, TA (eds) General and regional geography. Methuen and Co. Ltd., New Delhi

Sehgal J (1995) Red and lateritic soils- an overview. In: International seminar on managing red and lateritic soils for sustainable agriculture (ISMARS). Oxford IBH Publication, New Delhi, pp 1–6

Sehgal J (1996) A textbook of pedology: concepts and applications. Kalyani Publishars, Ludhiana, India Revised and Expanded Edition

Sehgal J (2005) a textbook of pedology: concepts and applications. Kalyani Publishars, Ludhiana, India Revised and Expanded Edition, 2005

Shirsath SK, Bhattacharyya T, Pal DK (2000) Minimum threshold value of smectite for vertic properties. Aust J Soil Res 38:189–201. https://doi.org/10.1071/SR99051

Smith GD (1986) The Guy Smith interview: rationale for concepts in *Soil Taxonomy*. SMSS Technical Monogr. 11. SMSS, SCS, USDA. US Gov. Print. Office, Washington, DC

Soil Survey Staff (1975) Soil taxonomy: a basic system of soil classification for making and interpreting soil surveys. Agric. Handb. 436. SCS, USDA. US Gov. Print. Office, Washington, DC

Soil Survey Staff (1999) Soil taxonomy, 2nd edn. USDA-SCS Publication, Washington

Soil Survey Staff (2003) Keys to soil taxonomy, 9th edn. USDA-NRCS, Washington

Srivastava P, Parkash B, Pal DK (1998) Clay minerals in soils as evidence of Holocene climate change, central Indo-Gangetic Plains, northcentral India. Quat Res 50:230–239. https://doi.org/10.1006/qres.1998.1994

Tiwary JR, Mishra BB (1990) Distribution of micronutrients in *tal* land soils (Udic Chromustert) of Bihar. J Indian Soc Soil Sci 38:319–321

Tiwary JR, Jha PP, Mishra BB (1987) Studies on parent rocks of red, black and yellow soils of Rajmahal Traps in Bihar. J Indian Soc Soil Sci 35:539–541

Verma KS, RK Saxena, Bhargava GP (2007) Anomalies in classification of the salt- affected soils under USDA soil taxonomy. J Indian Soc Soil Sci: 1–9

Vishwanath B, Ukil AC (1943) Soil map of India. Indian Council of Agricultural Research, New Delhi, India

Yadav JSP, Singh GB (2000) Natural resource management for agricultural production in India. Indian Society of Soil Science, New Delhi

Soil Mineralogy and Clay Minerals

6

S. C. Datta, Sudhansu K. Ghosh, and Debarup Das

Abstract

Soil mineralogy indicates primarily the parent materials from which the soil has developed. It also translates valuable information regarding the past climatic conditions and how the transition from past to present day climate occurred. This chapter highlights the physical phenomena like swelling and shrinkage, aggregation and thixotropy as strongly related to the mineralogical composition of soils. Similarly, chemical phenomena like adsorption and desorption of heavy metals (e.g. Zn, Cd, Pd etc.), and fixation of P and K depend on a great extent on soil mineralogy, especially clay mineralogy. Clay minerals form bonds of varying strengths with soil humus resulting in clay-humus complexes of varying stability. Distribution of non-labile and labile C in soil is also closely related to the different clay minerals present in soil. Such relations of clay minerals with soil organic matter have great implications on soil C-sequestration. The clays and clay minerals are not only the integral part of soils but also the driving factors in the dynamics of edaphology and pedology. The chapter covers different aspects of mineralogy of Indian soils based on the studies conducted since the 1950s to till date. Information on the mineralogical compositions of all the major soil groups of India has been presented. Efforts are made to link soil mineralogy with pedogenesis and climate. Some site-specific soil features related to soil mineralogy, e.g. effect of zeolites on fertility of calcareous Vertisols, presence of dioctahedral smectites in ferruginous Alfisols, spatial association of red and black soils, interstratification of kaolinite with smectite at some sites have been documented. Thereafter, the impact of soil minerals on different soil physical and chemical properties followed by the interaction of clay minerals with humus and different pesticides and other soil pollutants have also been established. The chapter ends with the findings of some advanced studies on soil mineralogy as carried out by Indian researchers.

Keywords

Soil mineralogy • Clays and clay minerals • Amorphous minerals • Pedogenesis • Clay-humus interaction

6.1 Introduction

In India, the study on soil mineralogy started with the use of X-ray diffraction technique for identification of clay minerals in soil in 1951 by Bagchi under the guidance of Prof. J. N. Mukherjee in the laboratory of Prof. S. N. Bose at Dacca (now in Bangladesh). The X-ray instruments at that time were much less sophisticated than they are today. A large number of Indian soils were analysed in H-saturated condition by Bagchi (1951). He showed that laterite soil contained kaolinite and quartz; black cotton soil from Satara and Mumbai contained montmorillonite; red earth from Coimbatore taluk contained kaolinite and quartz. After that, Das (1956) made an X-ray identification of soil clays isolated from black cotton soils of Indore, brown Matasi soil of Labhandi and alluvial soils of Delhi and Karnal. He found montmorillonite with small amounts of kaolinite and traces of illite in the black cotton soil clays; mainly illite and kaolinite in the brown Matasi soil clays; and illite with small amounts of kaolinite in the Delhi and Karnal soil clays. One year later, Adhikari (1957) reported the presence of illite and small amounts of kaolinite in some soils of West Bengal. Hence, mineralogical research in India initiated in the 1950s. For first few decades after initiation, researchers in India

S. C. Datta (✉) · S. K. Ghosh · D. Das
Division of Soil Science and Agricultural Chemistry, ICAR-Indian Agricultural Research Institute, Pusa,
New Delhi 110012, India
e-mail: samar1953@yahoo.com

S. C. Datta · S. K. Ghosh · D. Das
The Clay Minerals Society of India,
New Delhi, 110012, India

B. B. Mishra (ed.), *The Soils of India*, World Soils Book Series,
https://doi.org/10.1007/978-3-030-31082-0_6

mainly tried to find out the distribution of different minerals in various Indian soils in general, and in the clay fractions of the major soil groups of India in particular. Presently, the applied aspects of mineralogy are gaining more importance.

6.2 Distribution of Clay Minerals in Major Soil Groups of India

Percentage distribution of different clay minerals in different soil groups of India based mainly on X-ray diffraction studies has been discussed in this section.

6.2.1 Alluvial Soils

Alluvial soils cover around 75 Mha area across India and mainly occur in the Indo-Gangetic Plains, Brahmaputra valley and the coastal and deltaic regions (Seghal 1996). States of Delhi, Punjab, Uttar Pradesh, Haryana, West Bengal, Bihar, Assam, Orissa, Tamil Nadu, Kerala, Andhra Pradesh and Gujarat show the presence of alluvial soils. Studies by Pundeer et al. (1978) in some soil profiles of central Punjab revealed the dominance of illite (37–66%) in the clay fraction. The same also contained kaolinite (10–26%), chlorite (15–18%), vermiculite (4–18%), smectite (2–13%), hydrobiotite and metahalloysite. Besides this, alluvial soils of Lakhimpur in Assam; of Cooch Behar, Mursihid-abad, Hoogly, Howrah, Nadia and Burdwan districts of West Bengal; of Patna, Katihar, Sabour, Saharsa, Araria and Rexaul of Bihar have illite as the dominant clay mineral (Adhikari 1957; Prasad et al. 1967; Datta and Das 1972; Ghosh and Datta 1972). Soils of Delhi, Haryana and Punjab showed illite as the dominant mineral, which was associated with kaolinite, chlorite, montmorillonite and some mixed layer minerals (Das 1956; Ghosh 1964; Singh et al. 1972). Some researchers (Sehgal and Coninck 1971) reported that the chlorite found in Punjab, Haryana and Himachal Pradesh was actually chloritized vermiculite or montmorillonite; only the Hapludolls and Hapludalfs of Punjab contained true chlorite. Alluvial soils of Jammu and Kashmir also showed the dominance of illite in the clay fraction (Gupta 1967). Udaipur alluvial soils were also illitic but contained appreciable quantities of chlorite and quartz (Khangarot et al. 1972). Alluvial soils from Rupnagar of Punjab (Verma et al. 1994), and Karnal, Faridabad and Rohtak of Haryana (Ravi Kumar et al. 1991) were found to be predominantly illitic with contents ranging from 36 to 61% at different sites. Some calcareous alluvial soils from Bihar showed the dominance of illite in combination with kaolinite and chlorite (Ghosh 1964; Das and Datta 1969). Clay minerals in Alfisols occurring in Siwan district of north Bihar were almost identical to the clay minerals of most of the alluvial soils of Bihar (Mall and Mishra 2000). Mishra and Ghosh (2000) presented a detailed identification of clay types in mica-rich soils of Bihar and Jharkhand. Mishra et al. (1996) presented a map of Bihar (including Jharkhand) showing clay mineral associations, which needs refinement. The clay minerals in soils of Rajmahal Trap of Jharkhand were also identified in toposequences with red, yellow and red soils (Tiwary and Mishra 1992a, b, 1993). Soils of deltaic alluvium of Canning in West Bengal recorded illite to the tune of 40% of clay fraction, followed by smectite (30%), chlorite (15%), mixed-layer minerals (10%) and kaolinite (5%) (Ghosh and Datta 1972). The *Tal* land soils of Bihar are typical showing vertic features (*Udic Chromustert*) in south of the Ganga alluvium (Tiwary and Mishra 1990).

Illite is not the most dominant mineral in all the alluvial soils. Alluvial soils in many parts of West Bengal showed the dominance of smectite (30–60%) in the clay fraction, followed by mica (10–40%), chlorite (5–40%), mica-smectite or mica-vermiculite mixed-layer minerals (5–20%), kaolinite (6–10%) and a little amount of quartz (Ghosh and Datta 1974; Ghosh et al. 1974, 1976). Soils from different parts of Nadia, Hoogly and Burdwan districts of West Bengal have been found to be dominated by montmorillonite, with the presence of mixed-layer minerals and chlorite (Ghosh and Datta 1972; Ghosh et al. 1972; Ghosh 1973). Sand, silt and clay fractions of four benchmark soils (Ballartop, Chandipur, Narayanpur and Patibunia) occurring on the coastal plains in West Bengal were studied to identify the minerals present in these fractions and also to understand their transformation in the ecosystem (Nayak and Sarkar 2013). Clay fraction of some heavy-textured soils from Patna and Bhagalpur showed the presence of montmorillonite, vermiculite, illite and a few unrecognized minerals. Among them, some contained more montmorillonite than vermiculite, and some had more vermiculite than montmorillonite. Most of the soil clays from Bharatpur and Jaipur showed the dominance of montmorillonite (Khangarot et al. 1972). Some of the alluvial soils also show the dominance of kaolinite in the clay fraction. Soils from Sibsagar of Assam were dominantly kaolinitic with small quantities of illite and quartz (Das and Datta 1969). Sen and Chatterjee (1960, 1963) observed the dominance of kaolinite in pre-monsoon Gangetic silt.

6.2.2 Red Soils

Red soils, sometimes along with lateritic soils, mainly occur in Andhra Pradesh, Tamil Nadu, Maharashtra, Karnataka, Kerala, Goa, Orissa, West Bengal and north-eastern states of India. Red soils are usually of kaolinitic dominance, but sometimes illite and rarely montmorillonite can also become the dominant mineral in the clay fraction. In West Bengal, red soils of Midnapur and Bankura were found to be dominantly kaolinitic (50–55%), with the presence of illite (25–35%), smectite (10–15%), chlorite (0–5%) and mixed-layer minerals (0–5%) (Ghosh et al. 1974). Red soils in Burdwan district of West Bengal also dominantly contained kaolinite (42%), with as high as 30% illite and 17% smectite (Ghosh and Das 1976). Kaolinite contents in red soils of Kerala and Karnataka were as high as 90%. In a study conducted by Bhattacharyya and Ghosh (1990), on red soils derived from granite-gneiss in Bangalore revealed that kaolinite content (48–77%) decreased, while mica content (23–32%) increased down the profile. In the same profiles, smectite was absent in A and B horizons but was present in the C horizon. Red sandy soils of Ganjam, Balasore, Kalahandi, Puri and Koraput districts of Orissa were mainly dominated by kaolinite with some illite content (Das 1972; Datta and Adhikari 1972; Sahu and Nanda 1972). Clay fractions of the red soils from different parts of Bihar were also dominated by kaolinite with considerable amounts of illite (Ghosh and Das 1963; Sinha and Mandal 1963). However, genesis of red soils in Rajmahal Trap and other parts especially in Jharkhand were earlier studied in detail (Tiwary et al. 1987, 1996, 1997; Singh and Mishra 1995).

The loamy textured red soils of Maldah district of West Bengal showed preponderance of illite (60%) along with mixed-layer minerals (20%), chlorite (15%) and kaolinite (5%) (Ghosh and Datta 1972). Similar soils from Orissa also showed dominance of illite in association with kaolinite and montmorillonite (Sahu and Nanda 1972). In Mysore, the red sandy soils are illite dominant with the presence of some kaolinite (Ramkrishnaya 1971). Red and yellow soils in some parts of Tripura, Bihar and Orissa have been reported to contain illite as the most important mineral followed by kaolinite and sometimes chlorite (Singh and Sinha 1972; Ghosh 1973).

6.2.3 Black Soils

Black soils are mostly dominated by smectitic minerals; however, illite may also be present sometimes as the dominant mineral. Black soils present in the basaltic regions of Madhya Pradesh were recorded with 60–78% smectite along with 7–12% kaolinite and 5–12% illite (Chatterjee and Rathore 1976). Apart from montmorillonite and illite, some kaolinite was also detected in soil clays of Guntur and Lam in Andhra Pradesh (Ramkrishnaya 1971). Rao et al. (1983), after studying eight pedons from semiarid regions of Andhra Pradesh reported that clay fraction of soils developed from limestones and buff shales was smectite dominant. They also observed that black soil of Vertisol order contained at least 50% smectite apart from 21 to 30% mica and 6–10% chlorite and kaolinite. Shallow black soils developed on basaltic parent material in Chindwara of Madhya Pradesh revealed the dominance of illite in the clay fraction. Apart from illite, an appreciable quantity of montmorillonite, illite-montmorillonite mixed layer and 2:1–2:2 intergrade minerals were also present (Ghosh 1964). Dubey et al. (1985) studied the mineralogy of different mechanical separates in sodic Vertisols of western Madhya Pradesh. They observed that the fine sand fractions contained 95–97% of light minerals, mainly in the form of feldspars and quartz. In the silt fraction, albite was the major mineral followed by anorthite. In the clay fraction, smectite was the dominant mineral constituting 58–64%.

6.2.4 Laterite Soils

Laterite soils are mainly observed in the southern, south-eastern and north-eastern states covering an area of about 40 Mha (Seghal 1996). The most dominant clay mineral in laterite soils is kaolinite. Apart from kaolinite, presence of illite, chlorite, montmorillonite, vermiculite, quartz and feldspars has also been reported. Laterite soils of Midnapur and Bankura districts in West Bengal were found to be dominant in kaolinite with contents in clay ranging from 49 to 60%. Along with kaolinite, 20–29% mica, 7–20% mixed layer, 4–7% smectite, and 4–7% chlorite were also present (Ghosh and Datta 1974; Ghosh et al. 1974). In some parts of Midnapur, laterite soils showed the dominance of illite (60–70%) with kaolinite as the second dominant mineral (25–35%) and a little amount of quartz (5%). Laterite soils of Belgaum were dominant in kaolinite (50–88%) in association with chlorite (5–10%), illite (0–10%), smectite (3–5%), and tectosilicates like quartz (3–15%) and feldspars (4–10%) (Datta et al. 1973). Apart from kaolinite as the main clay mineral, Ghosh and Tomar (1973) reported the presence of diatom skeletons in laterite soils of Kadoli (Belgaum). Clays of laterite soils from Kerala were reported to contain kaolinite and halloysite as the principal minerals (Datta and Das 1972). Gibbsite as a component of soil clays was present in areas with high and excessive rainfall but not in low rainfall areas (Gowaikar 1972).

6.2.5 Hilly Soils

Hilly soils have a number of different minerals in the clay separates. In soils of Palam valley, the clay fractions contained illite (18–44%), chlorite (17–28%), vermiculite (6–19%), mixed-layer minerals (18–42%) and smectite (7–38%) (Gupta and Tripathi 1988). Clays in hilly soils of Kangra district showed the presence of illite (25–45%), smectite (0–35%), kaolinite (0–25%), chlorite (10–35%), vermiculite (0–25%) and mixed-layer minerals (15–35%) (Gupta et al. 1991). The same workers found 20–50% illite, 10–50% chlorite and 10–50% smectite in some soils from Poonch and Rajouri of Kashmir.

6.2.6 Tarai Soils

Tarai soils found in the northern parts of West Bengal are dominantly illitic (45–55%), with substantial quantities of chlorite (25–40%). Other minerals like mixed layer (5–20%), kaolinite (5%) and quartz (5%) are also present (Ghosh and Datta 1972). Tarai soils of Nainital district of Uttar Pradesh (UP) are similar in mineralogy to those found in West Bengal (Ghosh 1964). Some tarai soils have regular interstratification of mica-vermiculite in the soil clay fraction (Sahu et al. 1977; 1981).

6.2.7 Saline and Alkali Soils

Ghosh and Datta (1974) studied the coastal soils of West Bengal and reported the presence of illite (45%), smectite (25%), vermiculite (10%), chlorite (8%) and kaolinite (5%). Predominance of illite and kaolinite was observed in saline-alkali soils of Lucknow, UP (Raychaudhuri 1952–1953). Some saline soils have been reported to contain sepiolite (Kanwar 1959; 1961). Pandey and Pathak (1972) observed the presence of mica-montmorillonite and vermiculite-chlorite interstratified minerals in the salt-affected soils of Kanpur, UP.

6.2.8 Peaty Soils

Studies in peaty soils from Varanasi, UP revealed the presence of montmorillonite and kaolinite in the clay fraction (Rao 1963). Investigation on acid sulphate soils of Kerala by Ghosh et al. (1973) and Ghosh and Tomar (1973) revealed that kaolinite along with halloysite formed 34–37% of the clay. Apart from these two, smectite (18–32%), illite (6–12%), chlorite (4–11%), vermiculite (0–5%), interstratified minerals (4–12%), gibbsite (6–17%), amphibole (0–4%), quartz (0–2%) and feldspars (0–2%) were also present.

6.3 Amorphous Clay Minerals in Soil

Amorphous minerals mainly occur as coatings on crystalline minerals and are very difficult to separate. However, these can be selectively dissolved. First systematic study of amorphous minerals in Indian soils has been carried out by Krishna Murti et al. (1976). The amorphous minerals selectively dissolved from clay fraction of twenty-six ferruginous soils contained considerable amounts of iron (Fe) in addition to silicon (Si) and aluminium (Al). They termed these as amorphous ferri-aluminosilicates (AFAS) having silica (SiO_2): alumina (Al_2O_3) and SiO_2: sesquioxides (R_2O_3) molar ratio varying from 2.03 to 3.52 and 1.72 to 2.95, respectively. The model of AFAS postulated by them consisted mainly of negatively charged tetrahedrally coordinated silica-alumina phase, $Si_3AlO_6(OH)_4$ containing domains of neutral FeOOH with an outer positively charged hydroxyl-aluminium polymeric component $[Al(OH)_{2.5}]_n$. The calculated mean hydroxyl water content of the AFAS was 17.8%, and CEC ranged from 48.6 to 112 cmol(+) kg^{-1}. The CEC showed a high positive correlation with the ratio of the tetrahedral Si–Al component to the octahedral Al and a high negative correlation with the outer hydroxyl-aluminium octahedral component, but the same had no relationship with the Fe content. The amount of AFAS ranged from 19 to 32% in the clay fraction. They postulated that these AFAS formed preceding the formation of kaolinite in these soils. Initially, the soil solution was rich in bases and high in pH which favoured the tetrahedral coordination. The Fe was the first to precipitate and form colloidal solution on which chemisorption of Si took place which stabilized the amorphous state of Fe-particles. Then solution tetrahedral Al got incorporated in Si-network so that a tetrahedrally coordinated Si–Al gel structure was obtained. The Fe formed the domains in this three-dimensional mass. With time as leaching continued, formation of octahedrally coordinated polymeric hydroxyl-aluminium ions took place, which got attached electrostatically to the negatively charged Si–Al phase and neutralized the negative charges. Another consequence of lower pH is the tendency of Al to change its coordination from four to six which would favour the crystallization of kaolinite. The crystallization of kaolinite from Si–Al gel is very slow, but the presence of Fe domains in this gel mass caused loss of stability of tetrahedrally coordinated Al. In fact, the kaolinite minerals in these soils contained lots of Fe in octahedral layer (Rengasamy et al. 1975a, b).

Considerable amounts of amorphous materials were reported in the clay fraction of laterite soils of Madhya Pradesh (Gaikwad and Govindarajan 1971); in arid soils of Rajasthan (Choudhari and Dhir 1983); in Vertisols (Seshagiri Rao et al. 1992); in soils derived from mica-rich parent materials of Bihar (Mishra and Ghosh 1994); and alluvium-derived soils of Assam (Karmakar and Borah 1996). Karmakar (2012) studied the AFAS in five soils from Assam's lower Brahmaputra valley, namely Ruptic-Ultic Dystrudepts (P1) developed on lower piedmont plain, Aquic Udipsamments (P2) developed on alluvial fan plain, Umbric Dystrudepts (P3) developed on alluvial plain, Oxyaquic Udifluvents (P4) developed on flood plain and Typic Hapludults (P5) developed on monadnock. The clay fraction of these five soils contained 10.14–38.16% AFAS with SiO_2: R_2O_3 ratio of 0.99–2.69 and SiO_2: Al_2O_3 ratio of 1.05–3.18 (Table 6.1). The P5 soils showed the largest amount of AFAS (30.67–38.16%), and the P1 soils had the lowest of AFAS (10.14–12.39%). In the P1 and P5 soils, the Bt

Table 6.1 Amorphous ferri-aluminosilicate minerals in the clay fractions of five different soils from Assam's lower Brahmaputra valley (Karmakar 2012)

Horizon	Neutral FeOOH (*a*)	Tetrahedral $Si_3AlO_6(OH)_4$ (*b*)	Octahedral $Al(OH)_2$ (*c*)	Hydroxyl water	Amorphous material (*a* + *b* + *c*)
	% of soil clay				
P1: Lower piedmont plain (Kochugaon): Ruptic-Ultic Dystrudept					
A	1.29	6.33	2.52	1.81	10.14
BA	1.23	6.67	2.88	1.96	10.78
Bt	1.47	7.08	3.84	2.30	12.39
BC	1.26	6.40	4.22	2.27	11.88
P2: Alluvial fan plain (Gossaigaon): Aquic Udipsamment					
Ap	1.33	10.09	1.52	2.14	12.94
AC	1.57	8.20	2.27	2.08	12.04
C1	3.03	7.82	2.14	2.13	12.99
P3: Alluvial plain (Bhaoraguri): Umbric Dystrudept					
A1	1.39	9.94	4.48	2.93	15.81
A2	1.57	9.40	4.56	2.87	15.53
Bw	1.08	6.90	6.70	2.99	14.68
2C1	1.27	10.09	11.95	4.91	23.31
2C2	1.71	8.69	8.42	3.80	18.82
P4: Flood plain (Khoraghat): Oxyaquic Udifluvent					
A1	0.93	9.77	2.29	2.25	12.99
A2	1.01	9.11	2.03	2.09	12.15
2C1	1.51	5.98	5.41	2.53	12.90
2C2	1.08	8.07	5.56	2.88	14.71
3C3	1.77	10.32	6.05	3.43	18.14
P5: Monadnock (Alamganj): Typic Hapludult					
Ap	1.54	22.15	6.98	5.56	30.67
BA	1.65	23.92	8.71	6.32	34.28
Bt1	2.19	26.39	9.58	6.97	38.16
Bt2	2.29	26.07	9.74	7.00	38.10

horizons had the highest content of AFAS within the entire profile. Tetrahedral $Si_3AlO_6(OH)_4$ component followed by octahedral $Al(OH)_{2.5}$, and FeOOH components mainly constituted the AFAS in the studied soils. The AFAS of the five soils contained 16.4–21.1% (mean, 18.5%) of hydroxyl water, and its variation was very narrow (18.1–18.4%) in Typic Hapludults (P5), which were more developed.

Mineralogy of soil samples collected from different depths of two A-C pedons from two different altitudes of Darjeeling Himalayan region was investigated by CEC determinations, thermogravimetric analysis and selective dissolution analysis (Sahu and Patra 1985). Content of AFAS minerals in the soil clay fractions, determined from Si, Al, and Fe extracted by 0.5 N NaOH, ranged from 12.4 to 21.6% with SiO_2: R_2O_3 molar ratios of 4.54–7.16 and 2.77–5.40 at two sites. Clay fractions were mainly rich in mica and chlorite with good amounts of kaolinite and AFAS. Mica and chlorite in the soil clays inherited from the schists, gneisses and slates present in the parent rocks. Presence of AFAS was due to adsorption of silicic acid monomers on precipitated Fe-hydroxide colloids along with the formation of silica-alumina gel structure in earlier alkaline environment by cross linking. Kaolinite was proposed to be formed from AFAS by crystallization upon ageing. Weathering mean was also calculated for soil clays of different horizons of both the profiles. Frequency distribution of clay minerals of different depths in both the pedons was more or less similar and gave three peaks at 7th, 9th and 11th stages of weathering.

Chatterjee et al. (2015), from an investigation on some Indian soils, concluded that formation or dissolution of short-range order minerals by low molecular weight organic acid depends on the concentration of the acid. They observed that citric acid concentration of 20 mg L^{-1} led to the formation of maximum amount of oxalate extractable allophane

+imogolite and total oxides of colloidal clay in Inceptisols. In Alfisols, 10 mg L^{-1} and in Vertisol, 40 mg L^{-1} citric acid concentration resulted in the highest amount of AFAS. Averaged over the three soil types, 10 mg L^{-1} of citric acid concentration led to the highest amount of AFAS. Increase in citric acid concentration resulted in decrease in crystalline iron content in colloidal clay.

In another study, Chatterjee et al. (2013) investigated the distribution and involvement of short range order minerals or AFAS in rhizosphere and non-rhizosphere zones under three different nutrient status, namely 'depleted' (soil pre-depleted with sorghum-sudan grass hybrid), 'original' (untreated) and 'fertilized' (original soil fertilized with N, P and K) in soils of three different orders (Alfisols, Inceptisols and Vertisols). In fine clay (<0.2 μm), the AFAS content (averaged across nutrient status and order) was significantly higher in rhizosphere than non-rhizosphere (Fig. 6.1a); while in coarse clay, average AFAS was more in non-rhizosphere (Fig. 6.1 b). In fine clay, Inceptisols had the highest mean AFAS content followed by Vertisols, and least content was in Alfisols. In coarse clay, Vertisol showed the highest mean AFAS content followed by Alfisols and Inceptisols. Mean content of AFAS in fine clay was more under 'original' and 'depleted' fertility status than the 'fertilized' one. Mean AFAS content in coarse clay under 'depleted' fertility was significantly higher than that under 'original' and 'fertilized' fertility status. Among the short range order minerals, mean allophane + imogolite content was highest in the fine clay of Inceptisol and coarse clay of Vertisol, while poorly ordered Fe, crystalline Fe and ferrihydrite were maximum in Alfisol. In the same study, they also investigated the non-labile carbon: labile carbon (C_{NL}: C_L) ratio, which is an index of carbon (C)-sequestration (Fig. 6.2). Higher value of this ratio indicates greater C-sequestration. It

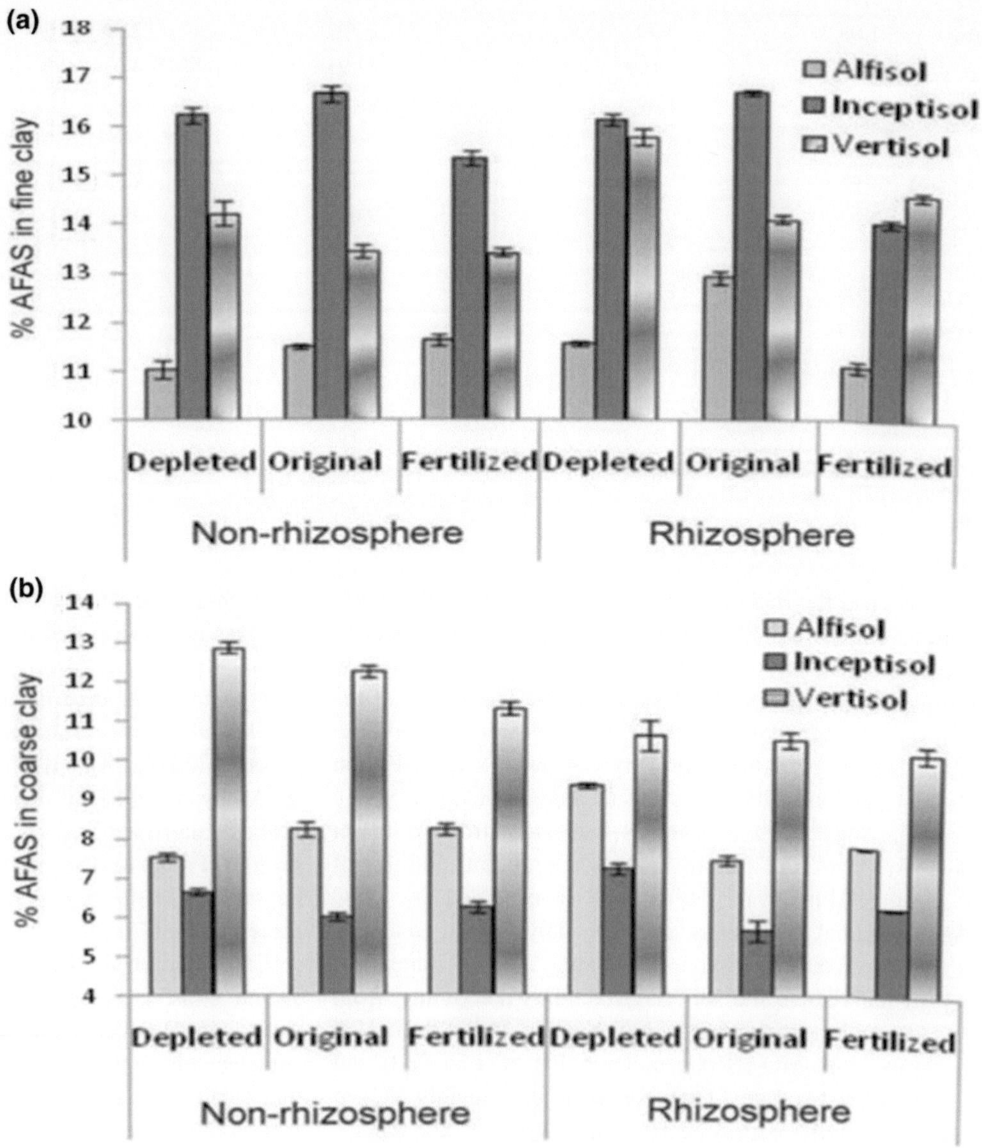

Fig. 6.1 Amorphus ferri-alumino silicate (AFAS) content (%) in **a** fine (<0.2 μm) and **b** coarse (2–0.2 μm) clay-humus complex (Chatterjee et al. 2013)

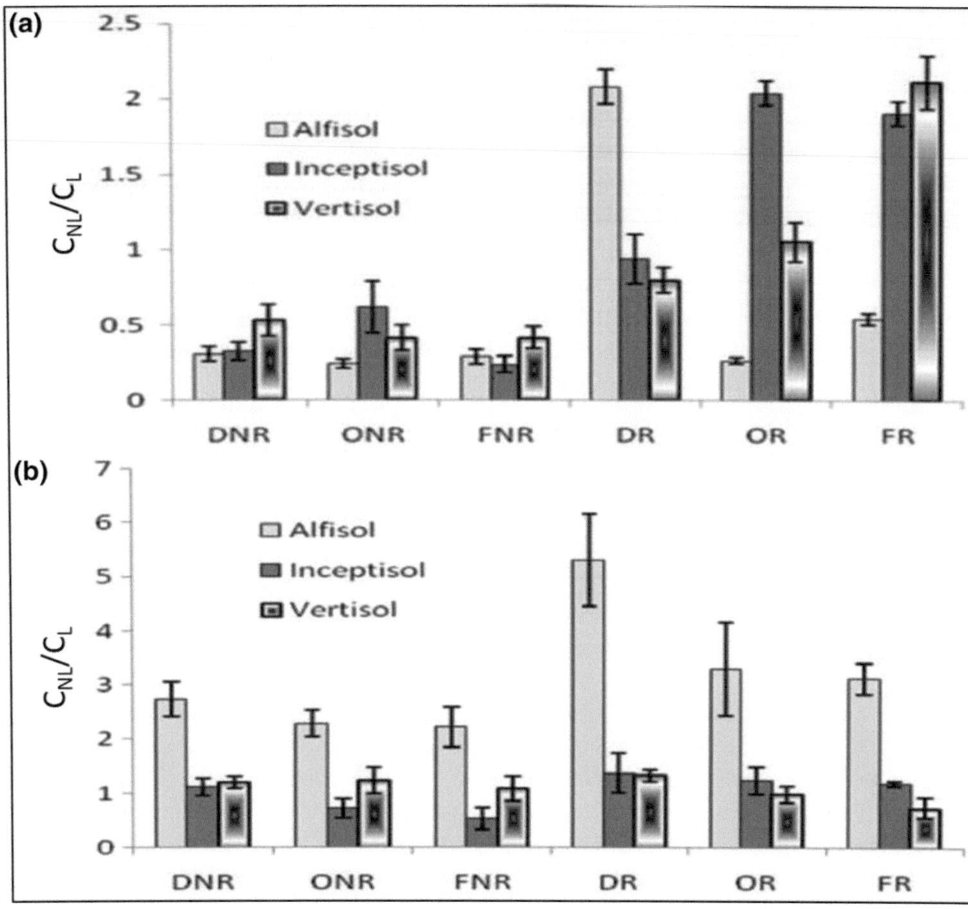

Fig. 6.2 Non-labile carbon: labile carbon (C_{NL}: C_L) ratio of **a** fine (<0.2 μm) and **b** coarse (2–0.2 μm) clay-humus complex. R-Rhizosphere; NR-Non-rhizosphere; D- Depleted; O-Original; F- Fertilized (Chatterjee et al. 2013)

was found that C_{NL}: C_L ratio for both non-rhizosphere and rhizosphere soil was more in Alfisols than Vertisols and Inceptisols. Within Alfisols, the same was significantly higher in rhizosphere soil than non-rhizosphere soil (Fig. 6.2). The content of short-range order minerals was less in Alfisols than the other two orders, but Alfisols had the highest contents of poorly crystalline Fe, crystalline Fe and organically bound Fe. Hence, organo-metallic complexes played more important role in C-sequestration than organo-mineral complexes.

6.4 Mineralogical Impacts on Pedogenesis and Soil Development

6.4.1 Linking Clay Minerals to Parent Material and Climate Change

Ram et al. (2013) studied the rate of weathering in soils evolved over sandstone, quartzite and shale landscapes from the mineralogical composition of mechanical separates of soils collected from five different physiographic divisions of Prakasam district of Andhra Pradesh. The soils belonged to Entisols, Inceptisols, Alfisols and Vertisols. They observed that mica and kaolinitic minerals mainly dominated the soils developed over sandstone and quartzite, while smectite was the dominant mineral in soils evolved from shale. Quartz, K-, Na- and Ca-feldspars and mica were the important minerals found in the sand and silt fractions of these soils. From the weathering index, they observed higher weathering rate in Vertisols than Inceptisols, Entisols and Alfisols. Alfisols showed the least rate of weathering among the four soil orders.

Several Indian researchers have considered minerals of intermediate weathering stage as potential indicators of paleoclimatic changes in parts of central India and Gangetic Plains (Pal et al. 1989; Srivastava et al. 1998; Pal et al. 2009, 2011). They have demonstrated how secondary minerals like di- and trioctahedral smectites, smectite-kaolin interstratified minerals (Sm/Kao), hydroxy-interlayered smectite (HIS), hydroxy-interlayered vermiculite (HIV), pseudo-chlorite (PCh) of intermediate weathering stage and $CaCO_3$ of pedogenic (PC) and non-pedogenic (NPC) origin can be regarded as potential indicators of paleo-climatic changes in major soil types of India. Pal et al. (2012) observed slight spatial variations in type and depth distribution of clay minerals in cores (~50 m deep) from the northern (IITK) and southern (Bhognipur) parts of the Ganga–Yamuna interfluve in the foreland basin of Himalaya, India, which they attributed to the combined effects of climatic transitions and sediment source. The core sediments from the northern part were rich in mica. Their fine clay fractions showed the dominance of dioctahedral low-charge smectites (LCS) with hydroxyl interlayering, while coarse clay and silt fractions revealed the dominance of trioctahedral high-charge smectites (HCS) (Fig. 6.3a). Low mica content was recorded in the core sediments from the southern part. In these sediments, the upper 28 m showed the presence of both LCS and HCS, while the lower part was predominantly LCS (Fig. 6.3b). The paleosols in these cores developed under semi-arid to sub-humid climates revealed the presence of 10–14 Å minerals in the clay fraction, e.g., HCS, LCS, PCh, vermiculite, hydroxyl-interlayered vermiculite (HIV) and kaolinite (formed under earlier humid condition). Large scale conversion of biotite to LCS from cratonic or Himalayan sources and interlayering by hydroxyl groups of smectite are very less likely in the alkaline soil reactions of the contemporary semi-arid climate. Hence, the presence of PCh, LCS, kaolinite and HIV points out that there has been a shift in climate in the Ganga Plains from humid to semi-arid.

The ferruginous Alfisols overlying the saprolites dominated either by dioctahedral smectite or kaolin are relict paleosols (Pal et al. 1989; Chandran et al. 2000), which have been affected by the climatic changes from humid to drier conditions during the Pliocene-Pleistocene transition period. Well-crystallized dioctahedral smectites as the first weathering product of peninsular gneiss partly transformed to kaolin in ferruginous soils (Alfisols) formed in a pre-Pliocene tropical humid climate (Pal et al. 1989). Such kaolin (Kao) is not a discrete kaolinite as XRD diagrams of its Ca-saturated and glycolated sample indicates the broad base of 0.72 nm peak and tails towards the low angle. On heating the K-saturated sample at 550 °C, the 0.72 nm peak disappears, confirming the presence of kaolin and simultaneously reinforces the 1.0 nm region at much higher degree even in presence of 1.4 nm minerals, indicating the presence of Kao-HIV/HIS (kaolin interstratified with either hydroxy-interlayered vermiculite, HIV or smectite, HIS). With the termination of humid climate, both these clay minerals were preserved to the present.

6.4.2 Spatially Associated Red Ferruginous (Alfisols) and Black (Vertisols) Soils

Occurrence of spatially associated red ferruginous (Alfisols) and black (Vertisols) soils on gneiss under similar topographical conditions is very common in semi-arid region of southern peninsular India (Pal and Deshpande 1987). Ferruginous soil clays consist mostly of kaolin and smectite, whereas black soil clays are dominated by low charge dioctahedral smectite. The inverse relation between kaolin and smectite with pedon depth of ferruginous soil clays (Pal et al. 1989) indicated the transformation of smectite to kaolin

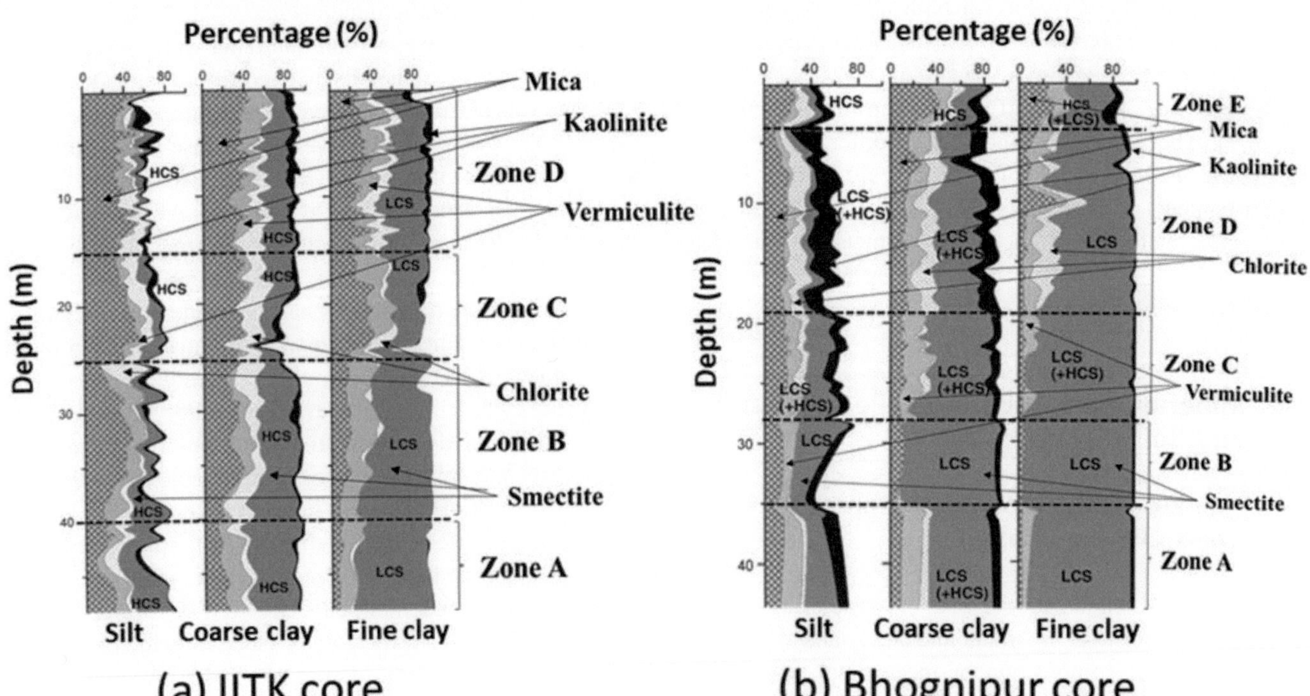

Fig. 6.3 Depth distribution of minerals in the silt, coarse clay and fine clay fraction of the cores collected from the **a** northern part of Ganga–Yamuna interfluve (IITK core), and **b** southern part of Ganga–Yamuna interfluve (Bhognipur core) (Adapted from Pal et al. 2012)

even though prevailing semi-arid climate cannot favour the formation of kaolin at the expense of smectite in slightly acid to moderately alkaline reaction. Similarly, the arid climate cannot yield the huge amount of smectite required for the formation of Vertisols. Earlier studies in southern peninsular India (Murali et al. 1978; Rengasamy et al. 1978) suggested that kaolinite was formed in an earlier geological period with more rainfall and great fluctuations in temperature, as evidenced by the presence of granitic tors all around such area (Pal and Deshpande 1987).

Two associated shrink-swell soils (red and grey coloured) occurring in a catena in the Hingoli district of Maharashtra were studied by Kolhe et al. (2011). The low hydraulic conductivity of red Vertisols was due to high Mg^{2+} and Na^{+} in the exchange complex. Mineralogical properties indicated that smectite was the dominant mineral in both the soils. The presence of kaolin in the grey Vertisols indicated the transformation of smectite to kaolin in the past humid climate. The absence of kaolin in red Vertisols indicated that the parent material (red bole) was not exposed to the earlier humid climate. They have been formed from the red boles mineral. The presence of palygorskite mineral with smectite in red shrink-swell soils and its absence in other associated soils in a catena indicated that this mineral cannot be considered as an index mineral for arid climate. The release of Mg^{2+} from palygorskite mineral disperses the clay and clogs the pores resulting in impeded drainage which adversely affect the crop production.

6.4.3 Smectite-Kaolinite Interstratification

Balbudhhe and Bhattacharyya (2009) studied the mineralogy and genesis of some representative rice-growing soils collected from four districts of eastern Vidarbha region of Maharashtra. They observed the dominance of smectites in the soil clays along with interstratified minerals of smectite-kaolinite (Sm/Kao) (Table 6.2). The interstratified Sm/Kao minerals developed from weathering of smectite minerals. They concluded that smectite present in the soils

Table 6.2 Semi-quantitative estimates of minerals present in total clay fraction (<2μm) of paddy-growing soils of Maharasthra, India (Balbuddhe and Bhattacharyya 2009)

Horizon	Depth (cm)	Minerals (%)[a]		
		Sm	Mica	Sm/Kao
Dighori soil (Bhandara)				
Ap	0–13	66 (72)[b]	8	26 (28)
Bw1	13–27	76 (82)	8	16 (18)
Bw2	27–53	77 (83)	8	15 (17)
Bss2	53–76	79 (85)	8	13 (15)
Bss3	76–105	81 (87)	7	12 (13)
Rajegaon soil (Gondia)				
Ap	0–10	49 (58)	15	36 (42)
Bt1	10–25	62 (72)	13	25 (28)
Bt2	25–55	56 (65)	14	30 (35)
Bt3	55–74	69 (76)	9	22 (24)
Bt4	74–107	74 (82)	10	16 (18)
Haldi soil (Chandrapur)				
Ap	0–14	80 (85)	6	14 (15)
Bw	14–46	83 (88)	6	11 (12)
Bsst	46–82	90 (94)	4	6 (6)
Bsst	82–114	82 (88)	7	11 (12)
Bsst	114–135	85 (90)	6	9 (10)
Kurul soil (Gadchiroli)				
Ap	0–12	88 (91)	4	8 (9)
Bw1	12–28	85 (90)	6	9 (10)
Bw2	28–50	81 (88)	8	11 (12)
Bw3	50–76	92 (95)	3	5 (5)
Bss1	76–101	89 (94)	6	5 (6)

[a]Sm: Smectite; Sm/Kao: Smectite-kaolinite interstratified minerals
[b]Parentheses indicate mineral content on mica-free basis

was developed as the first weathering product from plagioclase feldspars in earlier times under humid climate. The interstratification of Sm/Kao minerals was thought of as the result of post-depositional transformation. Smectite minerals prevailed due to poor drainage conditions, and only transformation was into Sm/Kao interstratified minerals.

Transformation of clay minerals in a soil chrono-association comprising five fluvial surfaces (QGH1 to QGH5) of the Indo-Gangetic Planis (IGP) between Ramganga and Rapti rivers demonstrated that the pedogenic Sm/Kao can be considered as a potential indicator for Holocene climatic changes from arid to humid conditions (Srivastava et al. 1998). The ages of QGH1 to QGH5 were <500 year before present (BP), >500 year BP, >2500 year BP, 8000 year BP and 13,500 year BP, respectively. During soil formation, two major regional climatic cycles were recorded. Relatively arid to semi-arid cycles between 10,000 and 6500 year BP and 4000 year BP till present was punctuated by a warm and humid climate. Biotite weathered to trioctahedral vermiculite and smectite in the soils during arid conditions that were unstable and transformed to Sm/Kao during the following warm and humid climate (7400–4150 year BP). When the humid climate terminated, vermiculite, smectite and Sm/Kao were preserved to the present. During the hot semi-arid climate that followed the humid climate, transformation of biotite into its weathering products like trioctahedral vermiculite and smectite did continue. Initiated by the formation of pedogenic carbonate, fine clay vermiculite and smectite translocated downward in the profile as Na-clay, to make soils calcareous and sodic (Pal et al. 1994, 2003). This pedogenetic process with time became an example of self-terminating process exhibiting their polygenetic features (Yaalon 1971).

6.4.4 Paleosols

Paleosols can act as important terrestrial archives containing information of the paleo-environment. But unscrambling the primeval pedogenic processes from paleosols might become difficult due to diagenetic alterations. Srivastava et al. (2013a) reported some important findings regarding the diagenetic alterations of the clay minerals of the oldest paleosols from the Himalayan foreland. Clay mineralogy of the paleosols revealed the presence of minerals like quartz, feldspar, vermiculite, kaolinite and mica in order of increasing abundance. Clay minerals in these paleosols underwent significant changes in terms of increased crystallinity, transformation and illitization due to burial diagenesis (around 7.5 km depth, and 140 °C temperature). Paleosols found in the fluvial sequences of Dagshai Formation near Koshaliya River, North West Himalaya were 0.5–1.5 m thick $B_k/B_{tk}/B_t/B_w/B_{ss}$ horizons, and showed prominence of pedogenic carbonates, rhizoliths and Fe-rich clay pedo features that belong to contemporary Inceptisols, Entisols, Vertisols and Alfisols (Srivastava et al. 2013b).

6.4.5 Layer Charge Characteristics of Smectitic Minerals in the Vertisols

The black soils of the Deccan Traps are rich in plagioclase feldspars which yield dioctahedral smectites as the first weathering product (Pal and Deshpande 1987; Pal et al. 1989; Bhattacharyya et al. 1993). Earlier review on the Vertisols of India indicates that black soils of India are dominated by beidillite-nontronite type of minerals (Ghosh and Kapoor 1982). Fine clay smectite, when subjected to the Greene-Kelly test (1953) involving heating the clay with lithium (Li) and subjecting to glycerol solvation, expands to about 1.8 nm and contracts to ~0.95 nm indicating this to be a mixture of beidillite/nontronite and montmorillonite in which the amount of the former is more than the latter (Pal and Deshpande 1987; Ray et al. 2003). Information on charge characteristics of Vertisols may be helpful to identify the smectite species in the mineralogical class instead of the group name, i.e., smectite. Another attempt (Bhople et al. 2011) has been made to locate the seat of charge of some Vertisol clays by determining the reduced CEC (CEC_R) of smectitic clays. Six benchmark Vertisols from Maharashtra namely, Linga, Loni, Asra, Paral, Kalwan and Nimone from Nagpur, Yavatmal, Amravati, Akola, Nashik and Ahmednagar districts, respectively, were chosen for the study. Fine clays were separated from the soils, and their original CECs (CEC_{UT}) were determined by exchange with neutral normal $MgCl_2$ followed by $CaCl_2$ and $BaCl_2$. Greene-Kelly test was performed by treating these clays with 3 *N* LiCl. The Li-clays were also treated with the above metal chlorides in the same order to get the reduced CECs. It was observed that the tetrahedral CEC decreased with increasing depth of soil for Linga, Asra, Kalwan and Nimone, whereas opposite trend was observed for Loni and Paral. The octahedral charge increased with depth for all except for Asra and Paral. The study indicated that the CEC determination appeared to be an accurate and less time-consuming approach to measure charge reduction. It can be used to locate the seat of charge and also to study the changes in the proportion of tetrahedral and octahedral charge during the pedogenic processes of soil formation.

6.4.6 Effect of Zeolite in Sand Fraction of Calcareous Vertisol

Formation of calcium carbonate ($CaCO_3$) has been reported to be a signature of dry (arid and semi-arid) climate (Pal et al. 2000; Raja et al. 2009). The increase in $CaCO_3$ content in soils is related to the development of sodicity as reflected by high exchangeable sodium percentage (ESP) (Balpande et al. 1996). The depth function of $CaCO_3$ and ESP (Pal et al. 1994, 2000) suggests that due to the formation of $CaCO_3$, sodicity develops initially in the subsoils. This subsoil sodicity impairs the hydraulic properties of soils. The initial impairment of the percolative moisture regime in the subsoils results in a soil where gains exceed losses. This self-terminating process (Yaalon 1983) subsequently leads to the development of sodic soils where ESP decreases with depth. Thus, the formation of pedogenic $CaCO_3$ is a basic process of initiation of chemical soil degradation (Pal et al. 2000, 2006). Against this background, it is experienced that some soils such as Nimone soils (Bhattacharyya et al. 2009), despite being calcareous and qualifying for Sodic Haplusterts, have not lost their productivity as evident from successful growing of agricultural crops. The reason for this has been attributed to zeolite in sand fraction of the soils. The sand fractions showed the presence of zeolites along with quartz and feldspar. Changes in phases of these zeolites on thermal treatments as indicated by shifting of 0.90 nm peak to 0.86 and 0.83 nm at 300 °C and its disappearance at 450 °C suggested that these zeolites belonged to Si-poor heulandites type (Bhattacharyya et al. 1993, 1997, 2006a; Pal et al. 2006). These base-rich zeolites have been reported to act as saviours for the Vertisols in the black soil region (Bhattacharyya et al. 1993; Pal et al. 2006). The presence of these zeolites helps in maintaining the productivity of these soils by two ways, viz. (i) supplying nutrient elements (Ca, Mg) in appropriate proportions and (ii) improving drainage for bringing right pedo-edaphic environment for crop growth. The presence of these zeolites is also responsible for the persistence of Alfisols (Bhattacharyya et al. 1999) and Mollisols in the Western Ghats and the Satpura (Bhattacharyya et al. 2006b). According to Bhattacharyya et al. (2009), mineralogical class of Soil Taxonomy should be based on pedo-edaphic datasets supporting resilience rather than any arbitrary nomenclature in view of contemporary natural chemical degradation process in terms of sequestration of inorganic C (as $CaCO_3$) and geogenic presence of natural modifiers (e.g. zeolites and gypsum).

6.5 Properties of Soil as Affected by Clay Minerals

6.5.1 Soil Physical Properties as Affected by Clay Minerals

Clay fraction controls many physical properties of soil which are agriculturally important. Pore size distribution of soils in relation to their clay mineralogy was studied for two Vertisols and two Alfisols of Andhra Pradesh by Nagarajarao and Prasadini (1991). Pore size distribution for the Vertisols was bimodal with peaks near aeration porosity and in the pore-size range of <0.01 μm. Montmorillonite with COLE (coefficient of linear expansion) value of >0.085 was the dominant mineral in Vertisols. The Alfisols showed a unimodal pore-size distribution with a peak in the aeration porosity range. The latter soils had lower clay contents with COLE values ranging from 0.008 to 0.027. Free-swell behaviour of bentonite clay was studied by Sridharan et al. (1992) after homoionizing it with different mono-, di- and tri-valent cations. Increase in cation valency led to decrease in free-swell index. For same valency, increase in hydrated size increased the free-swell index, and such effect was most prominent when bentonite was saturated with monovalent cations. The negative effect of cation valency on free-swell index could be due to its influence on the thickness of diffuse double layer, formation of quasi-crystals by divalent cation-saturated bentonites through the creation of outer-sphere complex between a pair of divalent cations and a pair of opposing siloxane ditrigonal cavities and aggregate formation by particles through hydrolysed Al or Fe ions in case of trivalent cation-saturated bentonites. Rengasamy (1999) studied the rheological properties of clays from Gondwana clay deposits of India. The clay dispersions in water showed non-Newtonian flow when subjected to different shearing stresses (0.0061–60 Pa) with different thixotropic critical stresses (1.5–13.5 Pa) and Bingham yield points (0.005–5.8 Pa) depending on the sample and its dispersion concentration. Creep responses of the dispersions were measured following the four element Burger's mechanical model.

6.5.2 Soil Chemical Properties as Affected by Clay Minerals

Adsorption of Zn at 25 °C and pH 5.6 on Na-kaolinite and Mg-kaolinite and kaolinitic soil clays from Alfisols was studied by Pal and Sastry (1985). They observed preferential Zn-adsorption throughout the range of surface coverage on both Na- and Mg-saturated clays. Very small amount of Zn (mole fraction 0.005) in solution was capable of replacing about 80% of Na from the surface. But at similar Zn: Mg ratio in solution, much less amount of Mg was exchanged. Muralidharudu and Raman (1985) studied the adsorption behaviour of Zn on homoionic soil clays saturated with Cu, Ca, Mg, Na and K at three temperature levels (30, 40 and 50 °C). The adsorption isotherms were well fitted to Freundlich model. Zinc adsorption per unit amount of clay increased in the order Na < K < Ca < Mg < Cu, and the amount adsorbed in each case was dependent on temperature and Zn concentration. Desorption studies in the same investigation showed that not entire amount of adsorbed Zn could be extracted into solution as some of the adsorbed Zn was fixed irreversibly in the system. Such fixation indicates 'ion trapping effect', which is likely to be more in clays where accessibility to interlayer spaces is more. Desorption decreased in the order Na-clay > K-clay > Ca-clay > Mg-clay > Cu-clay. Due to greater randomness and lower bond strength in Na-system, Zn adsorption was less, and desorption of adsorbed Zn was more. Zinc adsorption pattern on soil clays obtained from nine different soils (belonging Vertisols and Inceptisols) of Tamil Nadu was studied by Krishnaswamy et al. (1985). Their adsorption data conformed to both Langmuir and Freundlich adsorption equations. However, Langmuir gave better predictions of Zn adsorption capacities and adsorption energies. Among the nine soils, montmorillonite dominated soil clays showed higher values of adsorption maxima. Adsorptive capacities for Zn were positively correlated with clay CEC. Studies were carried out by Jain et al. (1985) on dielectric dispersion of adsorbed water on clay surfaces. Different clays showed different frequencies at which highest dielectric loss occurred. Kaolinite gave the maxima at around 500 Hz at low hydrations. Behaviour of illite in this regard was similar to that of kaolinite. Halloysite gave a peak at around 800 Hz and a second peak at around 5 kHz. Montmorillonite also gave the first peak at 800 Hz but the second peak at around 2 kHz. With increase in water content, the first peak initially became less prominent, then disappeared, while there was a shift in the second peak towards higher frequency. The authors concluded that the water molecules remain very closely attached to the clay surface at very low wetness and undergo Debye polar relaxation around 500 Hz frequency. Exchange equilibria on illite surfaces involving Na–Pb, Ca–Pb and Mg–Pb pairs at 30 and 60 °C were investigated at pH 5.5 using thermodynamic models (Khan and Khan 1985). Thermodynamic parameters (i.e. changes in enthalpy, entropy and standard free energy) showed that the preference of Pb for illite surface increased in the order Mg-illite < Ca-illite < Na-illite. The Mg–Pb exchange was found to be least ordered followed by Ca–Pb and Na–Pb exchanges.

Chatterjee et al. (2014) studied the fixation of P and K by non-colloidal (2–0.2 μm) clay-humus complex (NCH) and colloidal (<0.2 μm) clay-humus complex (CH) isolated from three contrasting soil orders, viz. Alfisol, Inceptisol and Vertisol. Alfisol showed the highest P fixation capacity (67.26% for NCH and 58.72% for CH), while Vertisol showed the highest K fixation capacity (37.94% for NCH and 32.57% for CH) for both the clay-humus complexes. They also observed that P fixation capacity of NCH was significantly and positively correlated with the contents of kaolinite, allophane, ferrihydrite, crystalline Al and Fe and poorly ordered Fe. For CH, P fixation capacity was positively related to the contents of kaolinite, ferrihydrite, crystalline Fe and poorly ordered Fe, but negatively related to organically bound Al. Potassium fixation capacity of NCH as well as CH showed significantly positive correlation with amorphous ferri-alumino silicate (AFAS) minerals. Potassium fixation in NCH was negatively correlated with the contents of kaolinite, ferrihydrite, crystalline Al and Fe and poorly ordered Fe.

6.6 Interaction of Clay Minerals with Humus

A range of amorphous aluminosilicates (AA) with different Si/Al molar ratios and their organic complexes with fulvic and humic acids were prepared and characterized by Basak and Ghosh (1996). The synthesized complexes were amorphous in nature mainly involving carboxylic groups in bonding between AA and organics. Allophane-humic acid complexes showed higher C contents than allophane-fulvic acid complexes. Organic C content was positively correlated with moisture in allophane-fulvic acid complexes and negatively correlated in allophane-humic acid complexes. Chattopadhyay et al. (2010) studied the interactions of soil fulvic acid (FA) and humic acid (HA) fractions with a siliceous allophane at different pH levels. Fixation of FA and HA by allophane decreased with increasing pH values due to the simultaneous development of negative charges on all of them. At any pH value, fixation of HA was much greater than that of FA due to less negative charges and hydrophilic nature of the former than the latter. Fixation of HA by allophane did not bring any considerable change in the X-ray diffraction pattern as the quantity of HA in the allophane-HA complex was not sufficient enough to bring such change. Response to infrared light of allophane-HA also remained almost same to that of allophane except in the 3500–3450 cm^{-1} zone, suggesting the interaction between HA and allophane to be weak in nature.

Ghosh and Mishra (1989) reviewed the research findings on clay-organic interactions. Das et al. (2013) studied the effect of humic acids separated from Alfisols, Entisols and Mollisols on the varying release of cations from olivine and tourmaline, rate of solubilisation, stability and characteristics of the residual products. Kinetics of solubilisation of the two minerals indicated a sequence of alternating dissolution and re-precipitation. They concluded that position of the cation in the crystal lattice and the stability of the cation-humic acid complex govern the release of the cation from the minerals. After the reactions with HAs, it was found that both the minerals, i.e., easily weatherable olivine and fairly resistant tourmaline have undergone significant changes. Infrared spectra indicated HA deposition on mineral surface which might have occurred after complexation. Das et al. (2014) also studied the interaction of HAs extracted from Alfisols,

Entisols and Mollisols with biotite (phyllosilicate) and hornblende (inosilicate) and release of nutrients from them. Drastic changes in terms of reduction of Ca^{2+} and $Fe^{2+/3+}$ and augmentation of Si^{4+}, Al^{3+} and Mg^{2+} were observed in hornblende, while biotite showed minimal alteration. Prominent XRD bands of hornblende either attenuated or disappeared in the reacted residues with the formation of several new crystalline phases. Biotite, on the contrary, showed slight alteration in the crystalline phases. Infrared spectra revealed that HA was deposited on the surface of both the minerals, with more prominent deposition being observed on hornblende.

6.7 Interaction of Clay Minerals with Pollutants

Pesticides used in intensive agricultural practices to control insect pests, weeds, diseases, etc., may sometimes persist in the soil too long to cause damage to soil health. Soil components like clay minerals can alter their bioactivity by specific interactions. The metal ions present in clay colloids have the ability to influence adsorption and degradation of such compounds. However, the extent of such activity depends on the clay minerals present in the soil where the pesticides have been applied. In this regard, Agnihotri and Raman (1982) studied the adsorption of metribuzin, a herbicide, on homoionic chlorite dominant clay at 5 and 35 °C temperatures. Freundlich equation well described the adsorption isotherms. Extent of adsorption followed the order: Ca-clay < Zn-clay < Na-clay < Cu-clay < Al-clay. In all the cases, adsorption was more at 5 °C temperature than that at 35 °C temperature.

Carbofuran and bendiocarb are broad spectrum systemic insecticides. Garg and Agnihotri (1984) carried out a study to observe the adsorption of carbofuran and bendiocarb on soils of varying mineralogy (alluvial, red, black and forest soil) and reference clay minerals, namely kaolinite, vermiculite and bentonite at three different pH (5.0, 7.0, 9.0) levels. Freundlich equation best described the adsorption data. Amount of adsorption increased in the order: kaolinite < vermiculite < bentonite and also increased with decrease in pH. Shifts in bands at 2920, 1745 and 1225 cm^{-1} of IR spectra due to adsorption of insecticides indicated the involvement of $-CH_2$, $-C{=}O$ and $-C{-}O{-}C-$ groups. Desorption study showed that not more than 43% of carbofuran and 23% of bendiocarb got desorbed indicating the irreversible nature of the sorption reactions of insecticides on soil solid surfaces.

Raman and Rao (1984) studied the adsorption behaviour of metoxuron, a herbicide, on homoionic montmorillonitic clays. The charge and the polarizing power of the saturating cations mainly governed the extent of metoxuron adsorption on the clay surface. Adsorption was maximum on H-clay, Fe-clay and Al-clay followed by Ni-clay, Zn-clay, Cu-clay, Mg-clay, Ca-clay and Ba-clay. Clays saturated with Na, K or NH_4 showed the least adsorption. Adsorption maxima showed a positive correlation with crystalline ionic potential indicating ionic potential to be the main driver in binding of metoxuron to metal clays. Adsorption maxima values also indicated that mostly the edges and less likely the interlayers were involved in adsorbing metoxuron.

Sharma et al. (1993a) studied the adsorption of Topsin-M, a systemic fungicide belonging to thiophanate group on Na- and H/Al-saturated illites and montmorillonites at 20 and 40 °C temperatures. Adsorption of the fungicide, at 20 and 40 °C temperatures, followed the order: Na-illite < H/Al-illite < Na-montmorillonite < H/Al-montmorillonite. Increase in temperature led to decrease in adsorption of Topsin-M. They concluded the adsorption to be more of physical than chemical in nature. Thermodynamic parameters studied by Sharma et al. (1993b) for Topsin-M adsorption on Na- and H/Al- saturated illite and montmorillonite showed that in all the cases thermodynamic equilibrium constants (ln K°) were more than one, which indicate high affinity of Topsin-M (Thiophanate-methyl) for the clay surfaces. This affinity for adsorption decreased with increase in temperature. The $\Delta G°$ values were negative confirming that the reactions were spontaneous, and the spontaneity decreased with rising temperature. Narayanan et al. (2014) observed the effect of clay on leaching of three pesticides, viz. chlorpyriphos, chlorothalonil and pendimethalin in an Inceptisol. Their study showed that in normal soil, leachates contained around 2% pesticides, but in the same soil after partial removal of clay fraction, the leachates contained as high as 5–25% pesticides.

Chloramine-T (sodium N-chloro paratoluene sulphonamide) is a very common oxidizing agent used in analytical chemistry. Jai Prakash and Jaya Kumar (1984) studied the adsorption behaviour of chloramine-T on Zn-montmorillonite at two different temperatures (5 and 40 °C). Freundlich isotherm showed better fit to adsorption data than Langmuir isotherm. The heat of adsorption (334 cal mol^{-1}) indicated that forces like van der Waals or H-bonding between the oxygen on the mineral surface and the amino group of the adsorbate might have caused the adsorption to occur.

Heavy metals in high concentrations have toxic effects on agriculture, ecosystem and human health and pose very serious environmental problems. Kumararaja et al. (2014) evaluated the effectiveness of a bentonite clay in removing two heavy metals (Zn and Cu) from aqueous systems and their subsequent immobilization. They found maximum adsorption of Zn at pH 7 and that of Cu at pH 6 (Fig. 6.4).

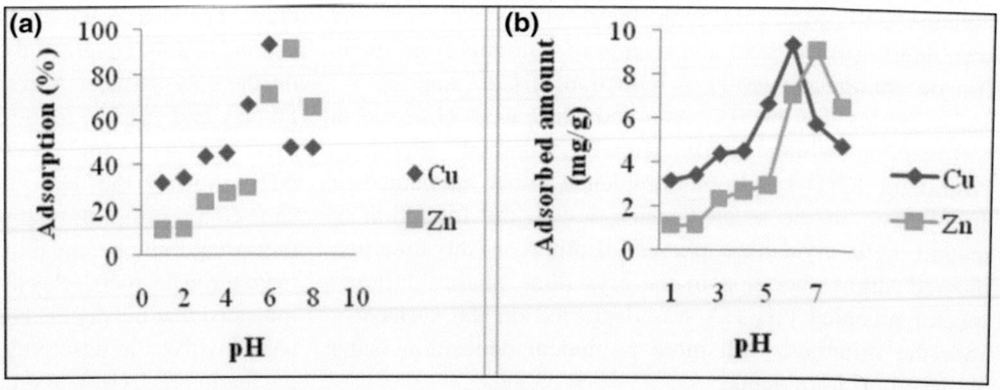

Fig. 6.4 Effect of pH on **a** removal (%) and **b** adsorption capacity (mg/g) of bentonite clay for Zn and Cu (Kumararaja et al. 2014)

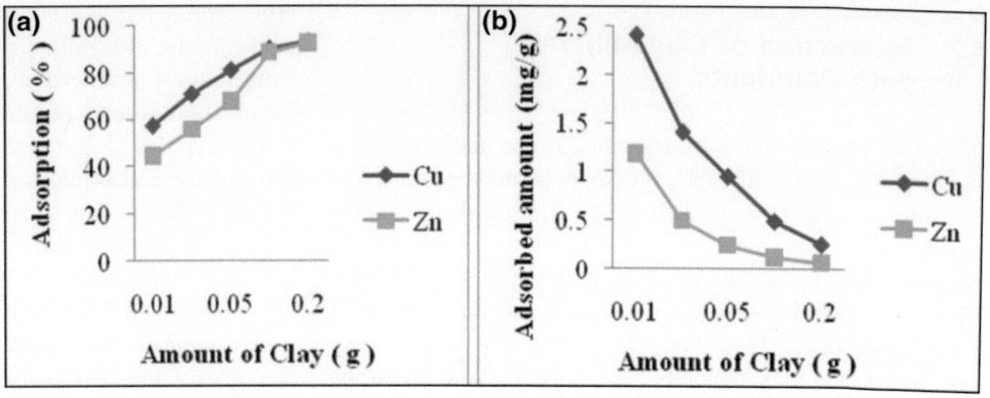

Fig. 6.5 Effect of amount of adsorbent on **a** removal (%) and **b** amount of metal (Zn and Cu) adsorbed per unit amount of adsorbent (Kumararaja et al. 2014)

Proportion of heavy metals removed from aqueous system increased with increase in the amount of adsorbent and decreased with increase in the initial concentration of the heavy metals in the aqueous system (Fig. 6.5). They concluded that bentonite could be useful in removing heavy metals from water through adsorption.

Radionuclides entering into soils and water sources may put at risk the ecosystem stability and become a serious threat to animal and human health. Radio-caesium, the most important artificial radionuclide, has found its entry point to the terrestrial environment for many years through the testing of nuclear weapons, accidental leakage from nuclear facilities, and authorized expulsion of nuclear waste. Agricultural product grown in such contaminated soils acts as the primary route for this radionuclide to enter into the food chain. Uptake of such radionuclide by plants depends on its retention in soil by interactions with the solid phase. In this context, Sandeep and Manjaiah (2009) studied the sorption-desorption behaviour of radio-caesium in soils and clay minerals (kaolinite, halloysite, attapulgite, nontronite, vermiculite, mica and illite) under the influence of oxalate and sodium tetraphenylboron. They found that more than 90% of radio-caesium (^{134}Cs) got adsorbed by soils and clays even in the presence of competing solutions (e.g. sodium oxalate). Halloysite and attapulgite showed the minimum adsorption (96.7%) followed by kaolinite (97.1%), vermiculite (97.2%), illite (98.0%), mica (98.7%), and highest adsorption was shown by nontronite (99.4%). Red and laterite soils showed maximum desorption; while among the clays, maximum desorption was observed for attapulgite. Sorption-desorption studies of ^{134}Cs on clay fractions and clay-sized aggregates separated from seven important soil orders found in India showed that sorption of ^{134}Cs was lesser in clay-sized aggregates with the lowest values being observed in Ultisols and Aridisols (Fig. 6.6) (Chari and Manjaiah 2010). Sorption significantly increased when organic matter was removed from the aggregates. On the contrary, ^{134}Cs sorption drastically reduced when sesquioxides were removed from clay samples. Sorption was more in smectite and mica than kaolinite. In waste mica, sorption increased by around 1.2 times with decrease in size of the particles. Desorption study showed that 1.2–7.3% of the sorbed ^{134}Cs could be extracted and the effectiveness of the extractants in this regard was in the order water < $CaCl_2$ < KCl < NH_4Cl (Fig. 6.7).

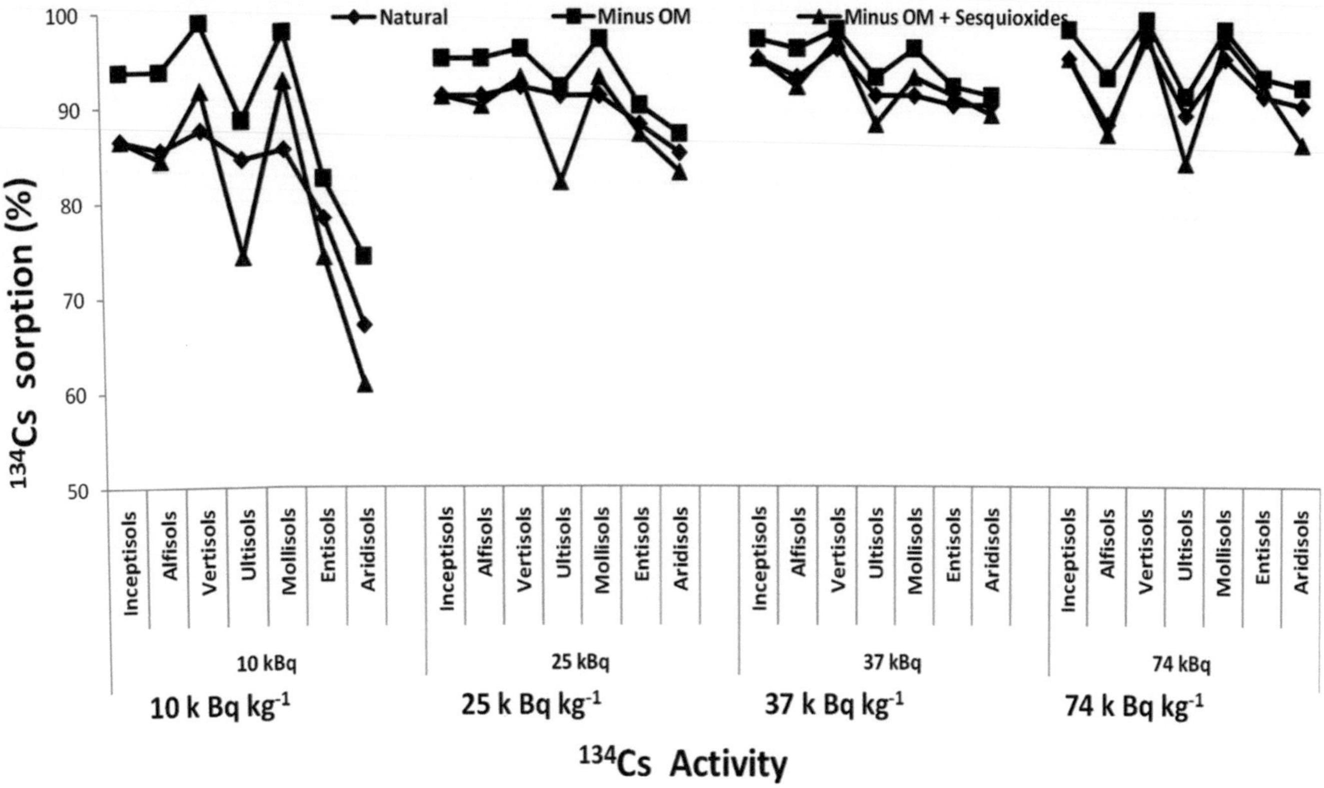

Fig. 6.6 ^{134}Cs sorption (%) on clay aggregates as influenced by its activity levels (Chari and Manjaiah 2010)

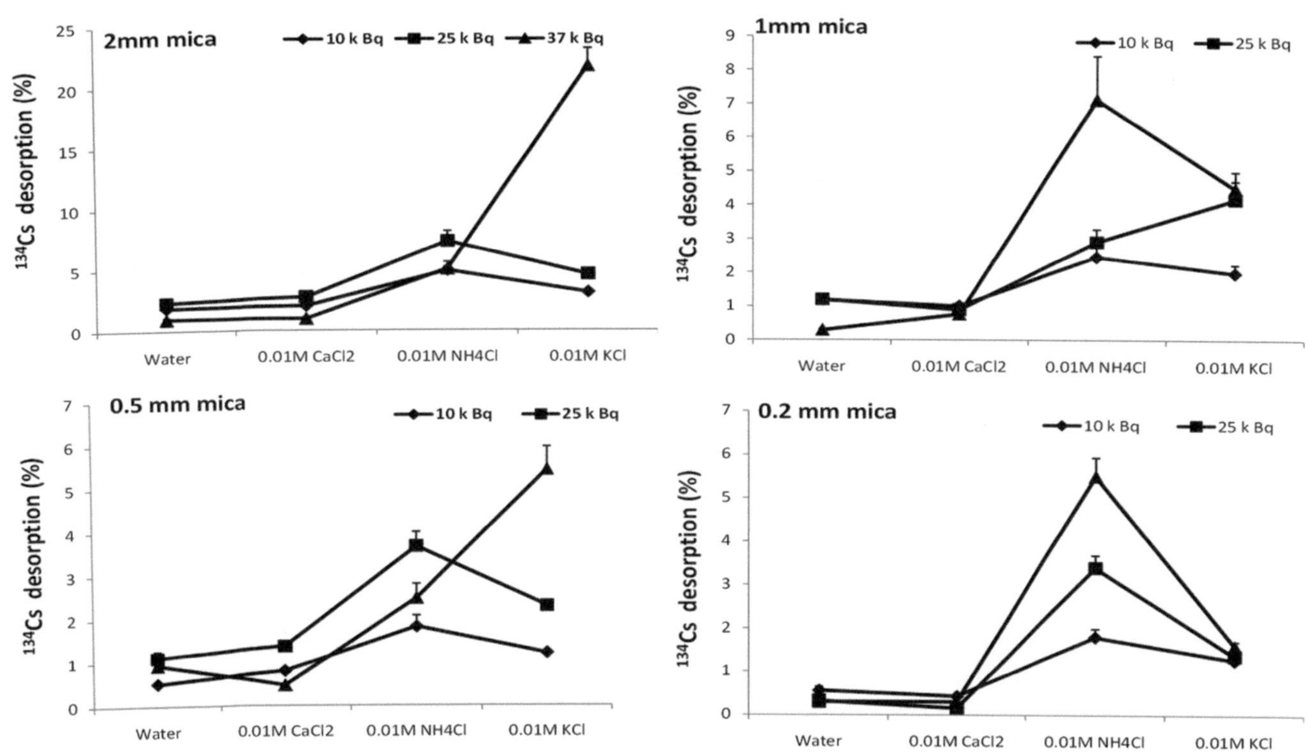

Fig. 6.7 ^{134}Cs desorption (%) from waste mica as influenced by extractants (Chari and Manjaiah 2010)

Table 6.3 Characteristics of resolved peaks obtained from deconvolution and curve fitting of XRD profile of soil clay (<0.02 μm) in the range of 4–11° 2θ (Datta 1996)

Peak no.	d-spacing (Å)	Mineral	Width		Mean thickness of crystallites (no. of units)	Strain (%)	Peak area (% of total)
			Gaussian	Cauchy			
			2θ				
Udic Ustochrept							
1	9.96	Mica	0.12	0.34	260	1.3	12
2	10.3	Mica-chlorite	0.66	1.54	–	–	53
3	13.9	Chlorite	0.76	1.68	53	11.9	35
Vertic Ustochrept							
1	15.1	Smectite	0.53	1.39	63	9.0	78
2	12.1	Mica-smectite	1.1	2.68	–	–	17
3	10.6	Mica-smectite	0.69	0.87	–	–	5
Typic Halpustalf							
1	14.6	Chlorite	0.487	1.156	77	8.0	22
2	11.66	Chlorite-mica	0.649	2.12	–	–	47
3	9.93	Mica	0.362	0.866	102	4.0	31

6.8 Advanced Studies on Clay Mineral Structure, Size and Crystallinity

Effect of heating on absorption of infrared light by layer silicates was studied in clay fractions separated from one kaolinitic and four illitic soils (Banerjee 1992). It was observed that the IR absorption bands specific to OH groups of layer silicate minerals were present when the samples were kept at room temperature. The same bands eliminated when the clay samples were heated to 1050 °C. However, the heated samples showed broad bands at 3420 and 1620 cm^{-1}. Origin of such bands could be attributed to the development of structural defects at high temperature and residual water after rehydration.

By curve-fitting and deconvolution of XRD profile, Datta (1996) characterised some micaceous minerals present in soils for size and strain of crystallites. For this purpose, oriented samples of fine clay fraction (<0.2 μm), isolated from Udic Ustochrepts (57% illite), Vertic Ustochrepts (49% smectite) and Typic Haplustalfs (81% kaolinite), were scanned by X-ray diffractometer (Table 6.3). The effect of instrumental broadening was eliminated from the observed X-ray peak by an iterative deconvolution programme followed by Lorentz-Polarization (LP) factor correction. Then the cluster of peaks was resolved into individual peaks by fitting the LP corrected digital data to a pseudo-Voigt function by a non-linear regression programme. This programme optimized the parameters, e.g., true position, intensity, Cauchy and Gaussian components of each of the peaks present in the cluster. Strain and particle size were then determined from Gaussian and Cauchy components, respectively. The study showed that micaceous mineral found in the Udic Ustochrepts was bigger and more crystalline than those present in Vertic Ustochrepts and Typic Haplustalfs.

6.9 Conclusions

Soil mineralogy with special emphasis of clays and clay minerals plays vital role in the study of soil resource. Some of the clay minerals like kaolinite, montmorillonite, vermiculite, etc., have been found to have the ability to adsorb different pesticide molecules and reduce their leaching into the groundwater. Sometimes, the bioactivity of such pesticide molecules is altered upon interaction with clay minerals. Such properties of clay minerals are very important for the maintenance of soil health and groundwater quality.

References

Adhikari M (1957) Physico-chemical properties of some West Bengal Clays. J Indian Soc Soil Sci 5:199–204

Agnihotri AK, Raman KV (1982) Adsorption of metribuzin on chlorite dominant clays: role of different metal ions. Clay Res 1:41–48

Bagchi SN (1951) Minerals present in hydrogen clays from Indian soils, kaolins and bentonites. Part I. X-ray studies. Indian Soc Soil Sci Bull 6:19–41

Balbudhhe DV, Bhattacharyya T (2009) Mineralogy and genesis of some paddy-growing soils of eastern Vidarbha, Maharashtra. Clay Res 28:111–142

Balpande SS, Deshpande SB, Pal DK (1996) Factors and processes of soil degradation in Vertisols of the Purna Valley, Maharashira, India. Land Degrad Dev 7:313–324

Banerjee BK (1992) Some aspects of IR absorption band at 3420 cm^{-1} in heated layer silicates. Clay Res 11:19–21

Basak UK, Ghosh SK (1996) Preparation and characterization of synthetic allophane-organic complexes. Clay Res 5:28–37

Bhattacharyya T, Ghosh SK (1990) Genesis and transformation of minerals in an Alfisol derived from granite gneiss. J Indian Soc Soil Sci 38:279–285

Bhattacharyya T, Chandran P, Ray SK, Mandal C, Pal DK, Venugopalan MV, Durge SL, Srivastava P, Dubey PN, Kamble GK, Sharma RP (2006a). Characterization of benchmark spots of selected red and black soils in semi-arid tropics, India. Working report of "Identifying systems for carbon sequestration and increased productivity in semi-arid tropical environments (RNPS-25) (NATP, ICAR), p 370

Bhattacharyya T, Pal DK, Deshpande SB (1993) Genesis and transformation of minerals in the formation of red (Alfisols) and black (Inceptisols and Vertisols) soils on Deccan basalt. J Soil Sci 44:159–171

Bhattacharyya T, Pal DK, Deshpande SB (1997) On kaolinitic and mixed mineralogy classes of shrink-swell soils. Aust J Soil Res 35:1245–1252

Bhattacharyya T, Pal DK, Srivastava P (1999) Role of zeolites in persistence of high altitude ferruginous Alfisols of the Western Ghats, India. Geoderma 90:263–276

Bhattacharyya T, Pal DK, Lal S, Chandran P, Ray SK (2006b) Formation and persistence of Mollisols on zeolitic Deccan basalt of humid tropical India. Geoderma 136:609–620

Bhattacharyya T, Ray SK, Pal DK, Chandran P (2009) Mineralogy class of calcareous zeolitised Vertisols. Clay Res 28:95–109

Bhople BS, Pal DK, Ray SK, Bhattacharyya T, Chandran P (2011) Seat of charge in clay smectites of some Vertisol Maharashtra. Clay Res 30:15–27

Chandran P, Ray SK, Bhattacharyya T, Krishnan P, Pal DK (2000) Clay minerals in two ferruginous soils of southern India. Clay Res 19:77–85

Chari MS, Manjaiah KM (2010) Radiocesium sorption-desorption on soil clays and clay-organic complexes. Clay Res 29:23–45

Chatterjee D, Dutta SC, Manjaiah KM (2013) Clay carbon pools and their relationship with short-range order minerals: avenues to mitigate climate change? Curr Sci 105:1404–1410

Chatterjee D, Dutta SC, Manjaiah KM (2014) Fractions, uptake and fixation capacity of phosphorus and potassium in three contrasting soil orders. J Soil Sci Plant Nutr 14:640–656

Chatterjee D, Dutta SC, Manjaiah KM (2015) Characterization of citric acid induced transformation of short range order minerals in Alfisol, Inceptisol and Vertisol of India. Eur J Mineral 27:551–557

Chatterjee RK, Rathore KS (1976) Clay mineral composition, genesis and classification of some soils developed from basalts in Madhya Pradesh. J Indian Soc Soil Sci 24:144–157

Chattopadhyay T, Varadachari C, Ghosh K (2010) Complexation of humic substances by allophane. Clay Res 29:15–22

Choudhari JS, Dhir RP (1983) Nature and distribution of amorphous aluminosilicates in arid soils of western Rajasthan. J Indian Soc Soil Sci 31:94–98

Das DK (1972) Proceedings of the symposium on clay minerals in Indian soils, INSA

Das DK, Datta NP (1969) Mineralogy of soil clays and its reaction with phosphates. Indian J Appl Chem 32:143–148

Das S, Ghosh K, Varadachari C (2013) Weathering of silicate minerals by humic acids: I. Nature of cation solubilisation from olivine and tourmaline and characteristics of the residual products. Clay Res 32:58–75

Das S, Ghosh K, Varadachari C (2014) Weathering of silicate minerals by humic acids: II. Nature of cation solubilisation from hornblende and biotite and characteristics of the residual products. Clay Res 33:46–63

Das SC (1956) An X-ray investigation of the mineralogical composition of some soil clays. J Indian Soc Soil Sci 4:135–140

Datta B, Adhikari M (1972) In: Proceedings of the symposium on clay minerals in Indian soils, INSA

Datta B, Das SC (1972) Proceedings of the symposium on clay minerals in Indian soils, INSA

Datta NP, Lotse EG, Tomar KP, Motsara MR (1973) Proceedings of the international symposium on use of isotopes and radiation in agriculture and animal husbandry research (Nov 30–Dec 2, 1971) p 370

Datta SC (1996) Characterization of micaceous minerals in soils for strain and size of crystallites through deconvolution and curve fitting of XRD profile. Clay Res 15:20–27

Dubey DD, Sharma OP, Banger KS (1985) Mineralogical studies on sodic Vertisols. Clay Res 4:96–99

Gaikwad ST, Govindarajan SV (1971) Nature and distribution of silicon, aluminium and iron oxides in the lateritic soils from Durg district, Madhya Pradesh. Indian J Agric Sci 41:1079–1084

Garg PK, Agnihotri AK (1984) Adsorption and desorption of carbofuran and bendicarb on soils and clay minerals. Clay Res 3:16–22

Ghosh SK (1964) Ph.D. Thesis, Post Graduate School, IARI, New Delhi

Ghosh SK (1973). Proceedings of the international symposium on use of isotopes and radiation in agriculture and animal husbandry research (Nov 30–Dec 2, 1971) p 337

Ghosh SK, Das SC (1963) Nature of minerals in the soil clays of some red soils of. Ranchi. J Indian Soc Soil Sci 11:69–72

Ghosh SK, Das SC (1976) Mineralogy of clay, silt and fine sand fractions of two acid soil profiles. Indian Soc Soil Sci Bull 11:145–155

Ghosh SK, Datta NP (1972) Proceedings of the symposium on clay minerals in Indian soils, INSA

Ghosh SK, Datta NP (1974) X-ray investigations of clay minerals in the soils of West Bengal. In: Proceedings of the indian national science academy, 40(B), pp 138–150

Ghosh SK, Kapoor BS (1982) Clay minerals in Indian soils. In: Review of soil research in india. Trans12th International congress of soil science, New Delhi 2, pp 703–710

Ghosh SK, Mishra BB (1989) Clay-organic interactions. Clay Res 8:84–90

Ghosh SK, Tomar KP (1973) Proceedings of the symposium on use of isotope in agriculture, biology and animal science, Chandigarh Jan 1–2, 1973, pp 73–78

Ghosh SK, Das DK, Deb DL (1973) Abstract of papers, symposium on acid sulphate and other acid soils of India, Trivandrum

Ghosh SK, Ghosh G, Das SC (1976) Smectite in some Gangetic alluvial soils of West Bengal. J Indian Soc Soil Sci 24:263–269

Ghosh SK, Sahu SS, Das SC (1972) Proceedings of the symposium on clay minerals in Indian soils, INSA

Ghosh SK, Sahu SS, Das SC (1974) Proceedings of the Indian national science academy, 40(B) p 200

Gowaikar AS (1972) Influence of moisture regime on the genesis of laterite soils in south India II. Clay composition and mineralogy. J Indian Soc Soil Sci 20:59–66

Greene-Kelley R (1953) The identification of montmorillonoids in clays. J Soil Sci 4:233–237

Gupta RD (1967). M.Sc. Thesis, Ranchi University

Gupta RD, Tripathi BR (1988) Mineralogy of a Haplohumult and a Paleudalf of Himachal Pradesh developed under west temperate climate. Clay Res 7:24–34

Gupta RD, Raman KV, Tripathi BR (1991) Himalayas: effect of inheritance and pedogenesis. Clay Res 10:57–63

Jai Prakash BS, Jaya Kumar H (1984) Adsorption of N-chloro para toluene sulphonamide on Zn-montmorillonite. Clay Res 3:10–12

Jain MC, Lal K, Krishna Murti GSR (1985) Low frequency dielectric dispersion of water adsorbed on clays. Clay Res 4:61–67

Kanwar JS (1959) Two dominant clay minerals in Punjab soils. J Indian Soc Soil Sci 7:249–254

Kanwar JS (1961) Clay minerals in saline alkali soils of Punjab. J Indian Soc Soil Sci 9:35–40

Karmakar RM (2012) Amorphous ferri-aluminosilicates in soils developed on different landforms in lower Brahmaputra valley of Assam. Clay Res 31:27–37

Karmakar RM, Borah G (1996) Amorphous material in some Inceptisols of Assam. Agropedology 6:43–48

Khan S, Khan NN (1985) Thermodynamics of Pb exchange with Na, Ca and Mg on illite surfaces. Clay Res 4:75–80

Khangarot AS, Tank DR, Veniwal RK, Rathore DS (1972) Proceedings of the symposium on clay minerals in Indian soils, INSA

Kolhe AH, Chandran P, Ray SK, Bhattacharyya T, Pal DK, Sarkar D (2011) Genesis of associated red and black shrink-swell soils of Maharashtra. Clay Res 30:1–11

Krishna Murti GSR, Sarma VAK, Rengasamy P (1976) Amorphous ferri-aluminosilicates in some tropical ferruginous soils. Clay Miner 11:137–145

Krishnaswamy R, Krishnamoorthy KK, Manickam TS (1985) Zinc adsorption isotherms for soil clays. Clay Res 4:92–95

Kumararaja P, Manjaiah KM, Datta SC, Shabeer TPA (2014) Potential of bentonite clay for heavy metal immobilization in soil. Clay Res 33:19–32

Mall J, Mishra BB (2000) Identification and genesis of clay minerals of some Alfisols of northern Bihar. J Indian Soc Soil Sci 48:586–593

Mishra BB, Ghosh SK (1994) Amorphous constituents of clays in soils derived from mica rich parent materials. J Indian Soc Soil Sci 42:668–670

Mishra BB, Ghosh SK (2000) Identification of Clay minerals in soils derived from muscovite dominant parent material. J Nucl Agric Biol 29:136–141

Mishra BB, Mall J, Singh RA, Choudhary J (1996) Clay mineral associations occurring in major soil groups of Bihar. Clay Res 15:15–24

Murali V, Krishna Murthi GSR, Sarma VAK (1978) Clay mineral distribution in two toposequences of tropical soils of India. Geoderma 20:257–269

Muralidharudu Y, Raman KV (1985) Zinc adsorption equilibria on soil clays as influenced by some saturating cations. Clay Res 4:81–86

Nagarajarao Y, Prasadini RP (1991) Mercury intrusion porosimetry of some Vertisols and Alfisols of Andhra Pradesh in relation to clay mineralogy. Clay Res 10:11–17

Narayanan N, Gajbhiye VT, Gupta S, Manjaiah KM (2014) Leaching behaviour of chlorothalonil, chlorpyrifos and pendimethalin in soil: effect of soil organic matter and clay. Clay Res 33:15–25

Nayak DC, Sarkar D (2013) Mineralogy of some benchmark soils of the coastal plain, West Bengal. Clay Res 32:1–16

Pal D, Sastry TG (1985) Sorption of zinc on kaolinitic soil clays. Clay Res 4:55–60

Pal DK, Deshpande SB (1987) Genesis of clay minerals in a red and black complex soils of southern India. Clay Res 6:6–13

Pal DK, Bhattacharyya T, Wani SP (2011). Formation and management of cracking clay soils (Vertisols) to enhance crop productivity: Indian experience. In: Lal R, Stewart BA (eds) World soil resources, Francis and Taylor, pp 317–343

Pal DK, Bhattacharyya T, Chandran P, Ray SK, Satyavathi PLA, Durge SL, Raja P, Maurya UK (2009) Vertisols (cracking clay soils) in a climosequence of Peninsular India: evidence for Holocene climate changes. Quatern Int 209:6–21

Pal DK, Bhattacharyya T, Ray SK, Chandran P, Srivastava P, Durge SL, Bhuse SR (2006) Significance of soil modifiers (Ca-zeolites and gypsum) in naturally degraded Vertisols of Peninsular India in redefining the sodic soils. Geoderma 136:210–228

Pal DK, Bhattacharyya T, Sinha R, Srivastava P, Dasgupta AS, Chandran P, Ray SK, Nimje A (2012) Clay minerals record from Late Quaternary drill cores of the Ganga Plains and their implications for provenance and climate change in the Himalayan foreland. Palaeogeogr Palaeoclimatol Palaeoecol 356–357:27–37

Pal DK, Dasog GS, Vadivelu S, Ahuja RL, Bhattacharyya T (2000) In: Lal R, Kimble JM, Eswaran H, Stewart BA (eds) Global climate change and pedogenic carbonates. Lewis Publishers, Boca Raton, pp 149–185

Pal DK, Deshpande SB, Venugopal KR, Kalbande AR (1989) Formation of di and trioctahedral smectite as an evidence for paleoclimatic changes in southern and central Peninsular India. Geoderma 45:175–184

Pal DK, Kalbande AR, Deshpande SB, Sehgal JL (1994) Evidence of clay illuviation in sodic soils of north-western part of the Indo-Gangetic plains since the Holocene. Soil Sci 158:465–473

Pal DK, Srivastava P, Bhattacharyya T (2003) Clay illuviation in calcareous soils of the semi-arid part of the Indo-Gangetic Plains, India. Geoderma 115:177–192

Pandey RN, Pathak AN (1972) Proceedings of the symposium on clay minerals in Indian soils, INSA

Prasad RN, Sinha H, Mandal SC (1967) Fractions of potassium in Bihar soils. J Indian Soc Soil Sci 15:173–179

Pundeer GS, Sidhu PS, Hall KF (1978) Mineralogy of soils developed on two geomorphic surfaces of the Sutlej Alluvium in the central Punjab, N.W. India. J Indian Soc Soil Sci 26:151–159

Raja P, Maurya UK, Ray SK, Chandran P, Bhattacharyya T, Pal DK (2009) Clay pedofeatures as indicator of neotectonic stress in the soils of the Indo-Gangetic plains. In: ISSS platinum jubilee celebrations and 74th annual convention, Dec 22–25, New Delhi (Abstracted)

Ram RL, Sharma PK, Ahmed N, Datta SC, Sharma RK (2013) Mineralogical transformation in benchmark soils of Prakasam district, Andhra Pradesh as influenced by weathering and pedogenic processes. Clay Res 32:34–57

Raman S, Rao PC (1984) Adsorption of metoxuron by some soils and their constituents. Clay Res 3:6–9

Ramkrishnaya BV (1971) Ph.D. Thesis, IARI, New Delhi

Rao MVR (1963) Identification of clay minerals in some Indian soil clays by X-ray diffraction. J Indian Soc Soil Sci 11:321–323

Rao YS, Ghosh SK, Vasudeva Rao AE (1983) Studies on sand and clay mineralogy of soils in Cuddhapah basin. Clay Res 2:1–5

Ravi Kumar DS, Singh M, Ghabru SK (1991) Soil mineralogy and potassium distribution in some alluvial soils of Haryana. Clay Res 10:64–70

Ray SK, Chandran P, Bhattacharyya T, Durge SL, Pal DK (2003) Layer Charge of two Benchmark Vertisol clays by alkylammonium method. Clay Res 22:13–27

Raychaudhuri SP (1952–53) Scientific report, IARI, New Delhi

Rengasamy P, Krishna Murti GSR, Sarma VAK (1975a) Isomorphous substitution of iron for aluminium in some soil kaolinites. Clays Clay Miner 23:211–214

Rengasamy P, Sarma VAK, Krishna Murti GSR (1975b) Quantitative mineralogical analysis of soil clays containing amorphous materials; a modification of the Alexiades and Jackson procedure. Clays Clay Miner 23:78–80

Rengasamy P, Sarma VAK, Murthy RS, Krishna Murti GSR (1978) Mineralogy, genesis and classification of ferruginous soils of eastern Mysore plateau, India. J Soil Sci 29:431–445

Rengasamy S (1999) Rheological properties of clays from Gondwana clay deposits of India. Clay Res 18:47–57
Sahu GS, Nanda SSK (1972) Proceedings of the symposium on clay minerals in Indian soils, INSA
Sahu SS, Patra SK (1985) The nature, properties and distribution of amorphous ferri-aluminosilicates in two pedons from Darjeeling Himalayan region, West Bengal. Clay Res 4:24–30
Sahu SS, Roy S, Ghosh SK (1977) Clay mineralogy of a terai soil profile from Mohitnagar, Jalpaiguri district, West Bengal. Indian Agric 21:187–194
Sahu SS, Roy S, Ghosh SK (1981) Clay mineralogy of a terai soil profile. Indian Agric 25:231–239
Sandeep S, Manjaiah KM (2009) Radiocesium sorption on soils and clay minerals: effect of oxalates and sodium tetraphenylboron. Clay Res 28:41–67
Seghal J (1996) A text book of pedology: concepts and applications. Kalyani Publishers, New Delhi
Seghal JL, Coninck FD (1971) Identification of 14Å and 7Å clay minerals in Punjab Soils. J Indian Soc Soil Sci 19:159–166
Sen RK, Chatterjee B (1960) Properties of Ganges silt. J Indian Soc Soil Sci 8:139–142
Sen RK, Chatterjee B (1963) Clay minerals in Ganges silt. J Indian Soc Soil Sci 11:325–328
Seshagiri Rao M, Chatterjee RK, Raghu Mohan NG (1992) Amorphous constituents of clays in soils derived from mica rich parent materials. J Indian Soc Soil Sci 40:603–605
Sharma DK, Singh GR, Lata S (1993a) Adsorption of topsin-M (thiophanate-methyl) on clays. Clay Res 12:52–58
Sharma DK, Singh GR, Lata S (1993b) Thermodynamics of adsorption of topsin-M (thiophanate-methyl) on clays. Clay Res 12:59–64
Singh RN, Sinha H (1972) Proceedings of the symposium on clay minerals in Indian soils, INSA
Singh S, Garalapuri GN, Khanna SS (1972) Proceedings of the symposium on clay minerals in Indian soils, INSA
Singh VN, Mishra BB (1995) Changes in morphological and other features of some red soils developed on granite gneiss in toposequence. J Indian Soc Soil Sci 43:139–142
Sinha MK, Mandal SC (1963) A mineralogical study of some acidic red loam soils of Chotanagpur. J Indian Soc Soil Sci 11:329–332
Sridharan A, Rao SM, Murthy NS (1992) Free-swell behaviour of homoionized bentonites. Clay Res 11:13–18
Srivastava P, Banerjee B, Aruche M, Ahmed N (2013a) Clay mineralogy of the oldest paleosols from the Himalayan foreland: implications of diagenetic overprinting and paleo-environments. Clay Res 32:17–24
Srivastava P, Parkash B, Pal DK (1998) Clay minerals in soils as evidence of Holocene climatic change, central Indo-Gangetic Plains, northcentral India. Quatern Res 50:230–239
Srivastava P, Patel S, Singh N, Jamir T, Kumar N, Aruche M, Patel RC (2013b) Early Oligoscene paleosols of the Dagshai Formation, India: a record of the oldest tropical weathering in the Himalayan foreland. Sed Geol 294:142–156
Tiwary JR, Jha PP, Mishra BB (1987) Studies on parent rocks of red, black and yellow soils of Rajmahal Traps in Bihar. J Indian Soc Soil Sci 35:539–541
Tiwary JR, Mishra BB (1992a) Clay mineralogy of some yellow soils developed in Rajmahal trap of Bihar. J Indian Soc Soil Sci 40:406–408
Tiwary JR, Mishra BB (1992b) Clay mineralogy of some red soils developed on Rajmahal trap of Bihar. J Indian Soc Soil Sci 40:833–836
Tiwary JR, Mishra BB (1993) Clay mineralogy of black soils of Rajmahal trap. J Indian Soc Soil Sci 41:535–540
Tiwary JR, Mishra BB (1990) Distribution of micronutrients in *tal* land soils (Udic Chromustert) of Bihar. J Indian Soc Soil Sci 38:319–321
Tiwary JR, Singh ID, Mishra BB, Mall J (1996) Characterization of red soils derived from five different parent rocks in Bihar. Agropedology 6:13–20
Tiwary JR, Singh ID, Prasad SN, Mishra BB (1997) Genesis of red soils developed in Rajmahal Trap of Bihar. Indian J Agric Chem 30:7–13
Verma VK, Sawhney JS, Jassal HS (1994) Clay mineral composition in relation to physiography in soils of Rupnagar district (Punjab). Clay Res 13:38–43
Yaalon DH (1971) Soil forming processes in time and space. In: Yaalon DH (ed) Paleopedology. Israel University Press, Jerusalem
Yaalon DH (1983) Climate, time and soil development. In: Wilding LP, Smeck NE, Hall GF (eds) Pedogenesis and soil taxonomy I. Concepts and Interactions Elsevier, Amsterdam, pp 233–251

7 Soil Micromorphology

Sayantani Neogi and Sean Taylor

Abstract

Soil micromorphology is a branch of Soil Science and its application is very relevant in pedology, soil classification and other fields including earth surface processes and archaeology. The thin section technique is commonly used in direct evaluation of soil structure in situ under the microscope. Soils observed under a microscope reveal many pedological features which are not visible with the naked eye. This chapter explores the kinds of information that soil micromorphology can advance for the better understanding of soils. The chapter highlights the micromorphological evidences with relevant examples from different parts of India for important thematic contemporary issues on advancement in Indian agriculture. It describes its applied usage in agriculture including Quaternary research (buried soils, palaeosols, palaeoclimate and palaeoenvironmetal issues, human-environmental relationship) and archaeology. This is because the soils can retain a whole range of information which provides the opportunity to investigate the complex interplay between natural and human agency. It is often very difficult to perceive many aspects of soil development when field studies are conducted. It is only through micromorphology that we can understand whether soil is formed in situ or it represents reworked alluvial or colluvial sedimentation that has been transported and then undergone further pedogenesis. The chapter highlights the micromorphological traits and features in soils of the Indo-Gangetic alluvium, *Thar* area of Rajasthan, Haryana, Eastern and western Uttar Pradesh, Kashmir valleys, Coimbatore, Tamil Nadu and other parts with reference to formation of soils in situ, horizontal orientation due to sedimentation, breakdown of organic matter, clay-rich argillic subsoil horizon, pedofeatures like clay and iron coatings, translocation of iron and calcium carbonate nodules, diagenesis of organic matter, passage features as a result of movement of soil fauna, polygenesis, pedogenic and non-pedogenic calcium carbonates including b-fabric in Vertisols.

Keywords

Soil micromorphology • Thin section studies • Pedology and soil classification • Quaternary research • Palaeoclimate • Palaeoenvironment • Archaeology • Indian agriculture

7.1 Introduction

Soil micromorphology is the study of soils and sediments in thin section. It is developed out of geological sciences and in particular mineralogy (Fitzpatrick 1984). It uses the principles that soils have optical properties, which can be viewed under the petrographic microscope. Thin sections of a soil are observed under increasing magnification with polarized and reflected light in order to understand, quantify and characterize the soil properties (Stoops 2003). These properties include the structure, texture, porosity, minerals, parent material and their weathering products, organic matter, soil animals, pedogenic features and many more besides. Soil thin sections are made from intact soil block samples that have been collected from soil pits and other exposures. It is very important to preserve the integrity of the sample as it is in the field (Kubiëna 1938). Sometimes special equipment is needed such as 'Kubiena tins' which can be inserted into sandy soils thus making the extraction and transport of the soil-less liable to failure. Once sampled and wrapped carefully, the soil blocks are taken to a laboratory where they are dried and impregnated with resin. They are then allowed to harden and then are cut with a diamond saw. Specialized equipments are needed such as a vacuum to draw out the trapped air and make the impregnation of resin successful.

S. Neogi (✉) · S. Taylor
McDonald Institute for Archaeological Research, University of Cambridge, Downing Street, Cambridge, CB2 3ER, UK
e-mail: sayantani.neogi@cantab.net

B. B. Mishra (ed.), *The Soils of India*, World Soils Book Series,
https://doi.org/10.1007/978-3-030-31082-0_7

Some of the chemicals are hazardous, so it is advisable to use a fume cupboard and gloves to protect the technician. One of the most important items of equipment is the machine for grinding down the impregnated blocks on a glass slide (Dalrymple 1957). It is necessary to make the thin sections of correct thickness (30 μ), otherwise it is impossible to discern the optical properties. This is a very skillful and is typically carried out in geological departments of universities where they have the expertise and machinery for making petrographic thin sections of rocks. The most important thing about soil micromorphology is that the relationships between all the soil components are maintained as they are in the landscape. In other words, the sample preserves not only the majority of soil properties that were present before being sampled but also retains the spatial relationships between all components. Therefore, a soil block will preserve the architecture of the soil. Biswas and Mukherjee (2001) stressed on uses of this technique for characterization and evaluation of soil structure directly in the field.

Historically, the application of soil micromorphology was confined only to pedology and soil classification. However, more fields of its application have emerged with new developments in light microscopy and submicroscopy, microchemical analysis and quantitative approaches (Kooistra 1990). Now we are looking for micromorphological applications in soils to establish the relations between biotic and abiotic processes, the effects of land uses and agricultural practices, the reconstruction of past environments and of human-induced events in sedimentology and archaeology. Kooistra (1990) further opined that if the weak points of micromorphology are improved and the level to which micromorphologists are trained is promoted, this branch of soil science would expand its scope further. Some emerging issues like micromorphology of slope deposits in relation to geomorphology, geology and palaeopedology need to be promoted since slope deposits depend on lithology as well as pedofeatures. Saprolite, for example, is weathered bedrock retaining original lithic fabric that is influenced by rock species involved. Minerals like olivines, amphiboles, pyroxenes, feldspars and biotite may also indicate alteration. Kaolinite, for example, is a common weathering product commonly of silicate minerals (Stoops et al. 2018).

Srivastava et al. (2009a, b) recorded that the current trend in pedological research indicates that soil micromorphology has become an indispensable tool providing finer details of the pedogenic processes and genesis of the soils that are difficult to understand without thin section studies in soil surveys. Soil-micromorphological studies at NBSS&LUP, Nagpur have not only added more concise and descriptive information about the soil resource inventory but also provide evidence of climate change, polygenesis and tectonics. Such experience of working with soil micromorphology has helped to resolve several issues of pedogenesis and determination of various key micromorphic features such of clay illuviation, b-fabric, pedogenic carbonate, polygenetic pedofeatures and weathering of minerals that cannot be recognized with traditional field investigations of soil survey (Kalbande et al. 1992; Pal et al. 1994, 2003; Srivastava and Parkash 2002).

7.2 Soil Complexity and Micromorphology

The soil is a very complex system (Phillips 1998) on the earth's surface. In order to understand its complexity, we must preserve and observe the soil properties and work out how they relate to each other. The famous Austrian soil scientist, Kubiena used the analogy of a watch to explain the principles of micromorphology (Kubiëna 1970). His insight was that it was much more useful to understand soils from the perspective of the soil components working as part of a system (much like the cogs of a watch) than to separate the elements into discrete categories. The latter include particle size, organic matter, nitrogen and potassium and a myriad of other soil properties. What is important is to understand how the different components work together. Of course, laboratory analysis of soils is important, but in many cases, micromorphology provides control for understanding the geochemical and physical datasets of soils. Further its application helps to resolve issues of pedogenesis with special reference to micromorphic features viz. pedogenic carbonate, clay illuviation, b-fabric, polygenetic pedofeatures including weathering of minerals (Kalbande et al. 1992; Srivastava et al. 2009a). Soil micromorphology has thus a unique role to play in this regard revealing the nature and complexity of soils in detail and explaining its formation.

When examining the soils, it is important to observe them at different scales of resolution and to make links between them (Bullock et al. 1985). It is often very difficult to perceive many aspects of soil development when field studies are conducted. It is only through micromorphology that we can really understand whether the soil is formed in situ or whether it represents colluvial material that has been brought in and then undergone further pedogenesis. A thin section from organic-rich horizon of an alluvial soil formed in situ at Meerat in Uttar Pradesh (Fig. 7.1) indicates a porous but humified organic soil that retains horizontal orientation due to sedimentation, subangular quartz grains with mica and opal phytoliths derived from the breakdown of organic matter (Neogi 2014).

In India, we have soils formed in diverse environments (Lal et al. 1994). Soils can have formed relatively recently as in the case of young alluvial soils of the Indo-Gangetic

Fig. 7.1 Thin section from organic-rich horizon of an alluvial soil formed in situ, Meerut (UP) showing a porous but humified organic soil that retains horizontal orientation due to sedimentation, subangular quartz grains with mica and opal phytoliths derived from the breakdown of organic matter. *Source* Neogi (2014)

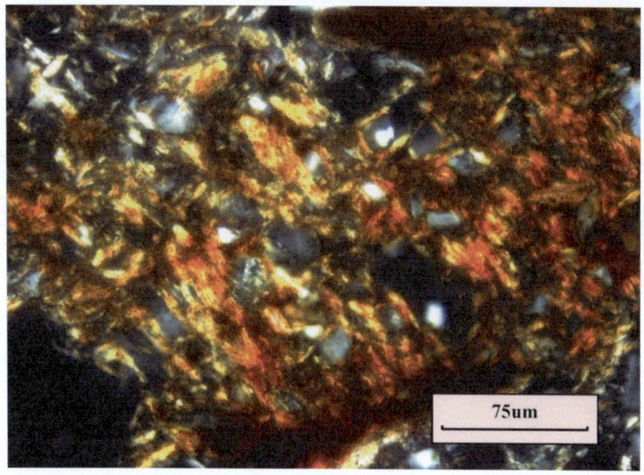

Fig. 7.2 Photomicrograph from a clay-rich argillic subsoil horizon as a result of weathering showing highly birefringent limpid clay throughout the groundmass in Jalpaiguri, West Bengal. *Source* Neogi (2014)

plains or the Himalayan soils which are in a perpetual stage of youthfulness due to processes of erosion. On the other hand, they can be much older, such as those formed on the basaltic topography of the Deccan Trap. Micromorphology allows us to understand how different soil properties are developed through the operation of soil-forming processes under a set of particular and often evolving environmental factors. Therefore, soil micromorphology provides the necessary data to understand the genesis of soils. For example, the very beginning of soil formation in a region such as the Sunderbans started as a result of the extension of the Gangetic delta outwards into the Bay of Bengal (Banerjee 1998). Initially silts were still being deposited and these soils had very few pedological properties. With time, as hydrological conditions changed along with other environmental factors, soil structure developed and horizons were formed. This process of haploidization occurs because the soil-forming processes such as weathering, leaching, translocation of soil components particularly iron compounds, clay and organics down the profile and the action of soil fauna and biological processes were dominant. Photomicrograph of a clay-rich argillic subsoil horizon as a result of weathering indicates highly birefringent limpid clay throughout the groundmass in Jalpaiguri district, West Bengal (Fig. 7.2).

In another photomicrograph, a number of pedofeatures consisting of birefringent clay coatings (c) and iron coating (ic) around a void (v) are indicative of the leaching of fine particles down the profile with quartz grain (q) in eastern Uttar Pradesh (Fig. 7.3).

Similarly, the translocation of soil components particularly iron compounds is displayed in photomicrograph,

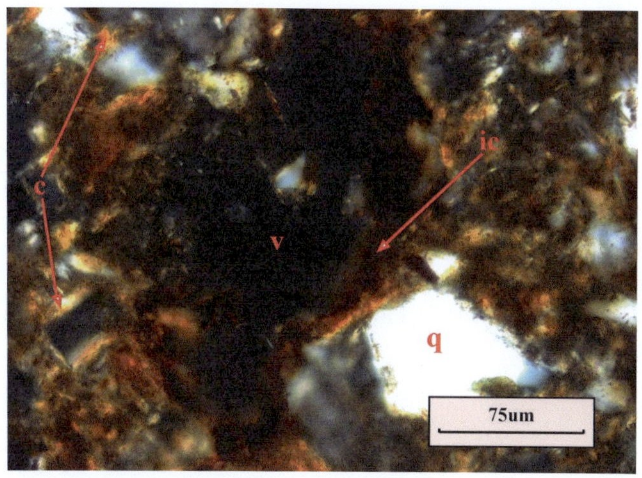

Fig. 7.3 Photomicrograph showing a number of pedofeatures consisting of birefringent clay coatings (c) and iron coating (ic) around a void (v), indicating the leaching of fine particles down the profile with quartz grain (q) in eastern Uttar Pradesh. *Source* Neogi (2014)

wherein a calcium carbonate nodule is superimposed with impregnative iron (cp) indicating that the movement of iron post-dated the precipitation of calcium carbonate and coating (cc) of impure clay (argillan) around the channel (v) indicating that impure clay has moved down the profile through agricultural practices such as ploughing, which might have shown detrimental effect on soil structure within the subsoil (Fig. 7.4).

In another photomicrograph (Fig. 7.5), the clay and organics within the profile being horizontally stacked as alluvial crusts (cr) with coarser and finer strata, upwardly indicate a platy-microstructure impregnated with iron compounds due to the diagenesis of organic matter, as evidenced

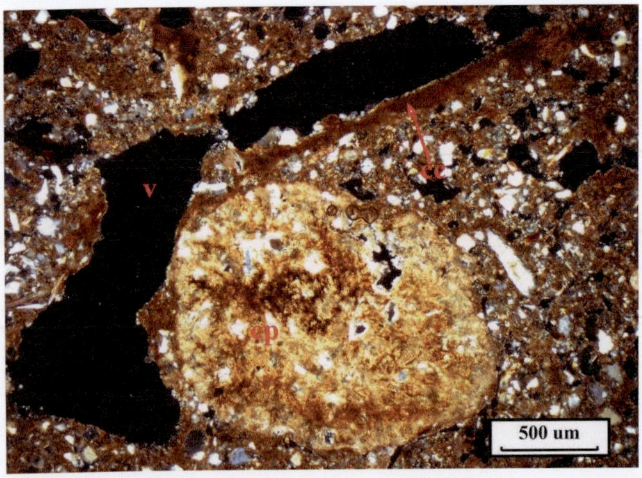

Fig. 7.4 Photomicrograph showing translocation of soil components particularly iron compounds in which a calcium carbonate nodule is superimposed with impregnative iron (cp) indicating the movement of iron post-dated the precipitation of calcium carbonate besides coating (cc) of impure clay (argillan) around the channel (v) showing downward movement of impure clay in profile of Western Uttar Pradesh. *Source* Neogi (2014)

within the matrix with voids (v) being either related to biological processes or plant pseudomorphs.

In similar observation, photomicrograph showing a typical passage feature is formed as a result of the movement of the soil fauna within a sandy topsoil indicating highly melanised fine groundmass, a sign of the presence of organic matter within the topsoil (Fig. 7.6).

The major soils in India are placed in different Orders in USDA Soil Taxonomy (Soil Survey Staff 1975, 2014).

Fig. 7.5 Photomicrograph showing clay and organics within the profile. Horizontally stacked alluvial crusts (cr) with coarser and finer strata, upwardly fining crust with a platy-microstructure impregnated with iron compounds due to diagenesis of organic matter. Within the matrix, voids (v) are either related to biological processes or plant pseudomorphs in Meerut, Uttar Pradesh. *Source* Neogi (2014)

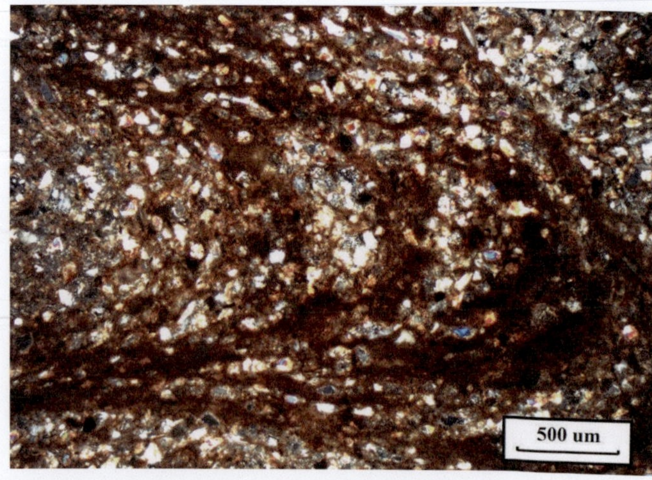

Fig. 7.6 Photomicrograph showing a typical passage feature formed as a result of the movement of the soil fauna within a sandy topsoil indicating highly melanised fine groundmass, a sign of the presence of organic matter within the topsoil of Haryana. *Source* Neogi (2014)

However, these soils and their spatial distribution across the landscape are highly generalized. In reality, soils of a particular group have a range of distinguishing properties, even over which, in many cases, relate to catenary relationships and particular landscape contexts. In India, this has been expressed using Soil Taxonomy, but those definitions as provided in The World Reference Base for Soil Resources (WRB 2014) are being increasingly popularized. The latter system is a simple means of expressing the properties of a particular soil to a great level of detail. The World Reference Base is the international system which is easily understood by scientists from around the world and can make successful comparisons among the soils in India and those around the world. The Indian system of soil classification proposed recently (Mishra 2015) deserves an extensive follow-up based on the framework developed. The approach seems to be not only simple and straight forward, but acceptable to the common farmers and policymakers also as it aims at linking soil resource to land use planning and avoids many taxonomic terms that might be suitable for academic purposes. The system is flexible with a very simple structural framework. It has a strong academic base and it will help to establish a strong background of land use planning in order to ensure productivity, safety, profitability and sustainability in Indian agriculture in the coming years (Mishra 2016). Micromorphology is a powerful tool for distinguishing processes and properties that allows us to define a particular soil with great accuracy. This is particularly useful when considering soils as a continuum across landscapes (Gerrard 1992). It is only when examining soils in thin section that we can appreciate the subtle differences in them, which typically reflects a particular soils position within the catena. We can better appreciate the variation in soils in this way. For example, when considering the Vertisols of India (Black cotton soils, locally called *Regur*, we can distinguish higher categories of differentiation by identifying additional properties. In this way, we may classify two Vertisols, one as a gleyic calcic Vertisols and another as a colluvic gypsic Vertisols. Micromorphology allows us to distinguish, and interpret the incremental differences between soils within a landscape and allows the generation of vocabulary to explain those differences. It is a technique which reveals the huge variation in the great soil groups. This also allows for higher resolution mapping of soils using GIS applications for a variety of purposes (Lagacherie and McBratney 2006). However, three basic terms are frequently used while micromorphology is applied to study the soils.

(i) ***Clay Illuviation***

Clay illuviation, in the process during pedogenesis, represents the mechanical translocation of clay with water from the surface horizons towards lower horizons. The water gets filtered out as it moves down in the horizon but the clay particles remain stick to walls of the voids. Such transported clays on the walls of the voids are called either clay skins, or clay coatings or clay cutans or argillans or clay pedofeatures (Soil Survey Staff 1975; Srivastava et al. 2009b) and used for establishing the presence of argillic horizon or illuvial clay horizon or Bt horizon in Soil Taxonomy (Soil Survey Staff 1975, 2014). To identify such illuvial features and differentiate them to their types and pedogenesis, thin section study of the soil under microscope is the best way (Pal et al. 1994, 2003; Srivastava et al. 1994; Srivastava and Parkash 2002). Scientists at NBSS&LUP recorded that the illuvial clay in the microscope can be observed as void/grain coatings that may be colourless or yellowish or reddish due to iron whereas platy nature of the clay platelets get oriented parallel to the walls over each other indicating the extinction process related to the degree of orientation and nature of clay platelets (Pal. 1994, 2003; Srivastava et al. 1994, 2009b; Srivastava and Parkash 2002). Pure clay pedofeatures, on the other hand, indicate strong orientation with prominent extinction bands, whereas impure clay pedofeatures are marked by poor orientation and lack of distinct extinction bands as recorded by Srivastava et al. (2009a, b) who further reported that the polygenetic nature of soils can also be examined. Pal et al. 2003) while using micromorphological study in characterization of 28 Alfisols from semiarid part of the Indo-Gangetic Plains (IGP); it was reported that identified clay pedofeatures are typical of the type "impure clay pedofeatures" that resulted from the impairment of the parallel orientation of the clay platelets induced by dispersion of both clay and silt size layer silicates in slightly to highly sodic environment. Srivastava et al. (2009a, b), however, observed that the impure clay pedofeatures are commonly associated with pedogenic calcium carbonate. It is believed that the formation of impure clay pedofeatures and pedogenic $CaCO_3$ are two pedogenic processes occurring simultaneously in soils of the Indo-Gangetic plain under the semiarid climate (Pal et al. 2003; Srivastava et al. 2009b). The pedologists in India further demonstrated with application of micromorphology that the soils formed after 13500 BP are polygenetic by showing different phases of illuvial activity (Srivastava et al. 2009b), as recorded degraded and fragmented illuvial clay pedofeatures, pedogenic calcrete, unaltered thick illuvial clay pedofeatures, and weathered minerals as a set of polygenetic features (Srivastava and Parkash 2002). The studies show that micromorphology is not only useful in identification and characterization of Bt horizon for Alfisols but it may also be used in detailing polygenesis induced by climate changes and tectonics (Srivastava and Parkash 2002; Pal et al. 2003).

(ii) ***$CaCO_3$ Pedofeatures***

Calcium carbonate is by and large commonly occurring in soils with different names and in variety of forms viz powdery, nodular or even indurated. The mineralogical and chemical nature, size, shape, distribution, and orientation of carbonate glaebules in two Chromusterts and two Pellusterts of India were studied. Micromorphology analysis of these features indicated their pedogenic origin (Mermut and Dasog 1986). In the Indo-Gangetic plains, Vertisols and ferruginous soils from arid, semiarid, humid and per-humid regions of the country, calcium carbonates are widely distributed with diverse origin (Srivastava 2001; Pal et al. 2003). However, only the micromorphological observation could make a reliable distinction between pedogenic calcium carbonate and non-pedogenic calcium carbonate (Srivastava 2001; Srivastava et al. 2002; Pal et al. 2003). In thin sections, they can be observed as coatings and nodules that have diffused boundaries with the groundmass. The pedogenic carbonate in these soils is fine textured and impure in nature with assimilation of soil constituents showing skelton mineral grains and weathering features similar to those in basic soil mass (Srivastava 2001; Srivastava et al. 2002). On other hand, non-pedogenic carbonate in soils is not related to pedogenic processes and it presence is considered to be inherited in soils developed either in strongly calcareous parent material or in young geomorphic surfaces (Pal et al. 2000; Srivastava et al. 2002). In thin sections non-pedogenic carbonate can be demarcated as calcite features showing sharp boundaries with the groundmass in form of nodules, fragments, concretions and skeleton grains that are relatively pure with little or no assimilation of soil constituents and often with Fe–Mn coatings. Micromorphological investigations of calcareous soils of the Gangetic plains have helped in resolving the issue of different origin of illuviation and formation of calcium carbonate in soils (Pal et al. 2003). Here impure clay pedofeatures are intimately associated with pedogenic carbonate suggesting that formation of impure clay pedofeatures and pedogenic $CaCO_3$ are two pedogenic processes occurring simultaneously in soils of the Gangetic Plains as contemporary pedogenic events in the semiarid climate (Pal et al. 2003). Thin section studies of the 23 soil series of Vertisols representing sub-humid, semiarid and arid climatic regions of India indicate that Vertisols contain both pedogenic calcium carbonate and non-pedogenic calcium carbonate irrespective of the ecosystems to which they belong. The non-pedogenic calcium carbonates are part of the parent material of Vertisols. The dissolution of the non-pedogenic calcium carbonate to release Ca^{2+} ions and their recrystallization as $CaCO_3$ (pedogenic calcium carbonate) is the prime chemical reaction of the pedogenesis of these Vertisols. Formation of pedogenic calcium carbonate is the prime chemical reaction responsible for the increase in pH, the decrease in the Ca/Mg ratio of exchange site with depth and in the development of subsoil sodicity (Srivastava and Parkash 2002).

(iii) ***b-Fabric***

Examination of the soil fabric is crucial part of micromorphological study. Srivastava et al. (2009a, b) at NBSS&LUP, Nagpur undertook initiatives with parallel arrangement of the clay domains in the groundmass is related to stress and especially the elongated patterns of clay domains corresponding to slickensides in Vertisols known as porostriated or parallel striated b-fabrics. Where less developed and randomly oriented domains of 20–50 μm size are called speckled b-fabrics found in weakly developed soils. In soils of the Indo-Gangetic plains, microfabric analysis shows systematic changes from speckled b-fabric in most of the soils that were grouped in Inceptisol order to strongly striated b-fabric in soils showing strong pedogenesis classified as Alfisols (Srivastava et al. 1994; Srivastava and Parkash 2002). The b-fabric in Vertisols is dominantly porostriated or parallel striated due to shrink swell phenomena. Plasma separation in Purna valley Vertisols has been found to be varying in response to increasing aridity and polygenesis (Pal et al. 2001). The b-fabric analysis of the soils form IGP and Vertisol has been proved to be useful not only in characterization of the soils and the pedogenic processes but also provide evidence of Holocene climate change and polygenesis (Srivastava et al. 2007; Pal et al. 2009). Micromorphological investigation of endopedons in six pedons of Bundelkhand region (Devagan-Ultic Haplustalf, Bahilpurwa-Udic Ustochrept, Mariahu-Udic Ustochrept, Khraund-Fluventic Ustochrept, Anuwan-Vertic Epiaquept and Badausa-Udic Haplustert), Uttar Pradesh highlighted that the related distribution was porphyric in red-soil and porphyric-chitonic in alluvial soils. Skelsepic-mosepic fabric orientation was dominantly observed in red soils and mosepic in alluvium-derived soils. Porosity mainly consisted of vughs, round pores, frequent channels and packing pores in red soils reflecting greater root and faunal activity (Walia et al. 2000).

7.2.1 Soil Micromorphology in Soil Forming Factors and Processes

When examining the properties of soils in the field, we are recording very important characteristics such as the number of horizons and their thickness, colour and texture (Sahai 2011). Some properties such as soil structure may or may not be possible to determine in the field. However, there are many soil properties that are difficult or impossible to observe with naked eyes. This is where soil micromorphology really comes into its own. For example, properties such as porosity can be studied closely (Sander 2002). Imaging software allow us to precisely calculate the shape and volume of pores and make comparisons between soils more accurate. The complexity of subsoils is revealed with micromorphology. Illuvial horizons result with the translocation and accumulation of soil components down the profile. Photomicrograph showing illuvial horizons as a result of the translocation and accumulation of soil components down the profile in a sandy soil of well-sorted angular quartz grains with a fine clay matrix indicates humified organics and a shell fragment (sh) with channel infilling (ca) of calcium carbonate showing carbonation as the dominating process with micaceous groundmass (gm) in alluvial soil of Haryana (Fig. 7.7).

Iron compounds, responsible for brunified B horizons can be recognized in the field but clay is usually not visible. Photomicrograph (Fig. 7.8) showing a series of three stacked sedimentary crusts (cr) associated with alluvial soils in Meerut, UP with a platy microstructure indicates sedimentation processes, where calcium carbonate gets precipitated within the voids. Iron compounds, responsible for brunified B horizons can be recognized in the field but clay is usually not visible. Crusts comprise of fine-grained material and vescicles (v) due to trapped air within the matrix of the crusts (Fig. 7.8).

However, micromorphological observation allows us to observe illuvial horizons as silicate clays have optical properties very similar to rock-forming minerals. We can understand the amount and type of clay that allows us interpret the genetic history of a particular soil and the length

Fig. 7.7 Photomicrograph showing illuvial horizons as a result of the translocation and accumulation of soil components down the profile in a sandy soil of well-sorted angular quartz grains with a fine clay matrix, having humified organics and a shell fragment (sh) with channel infilling (ca) of calcium carbonate indicating carbonation as the dominating process with micaceous groundmass (gm) in alluvial soil, Hisar district, Haryana. *Source* Neogi (2014)

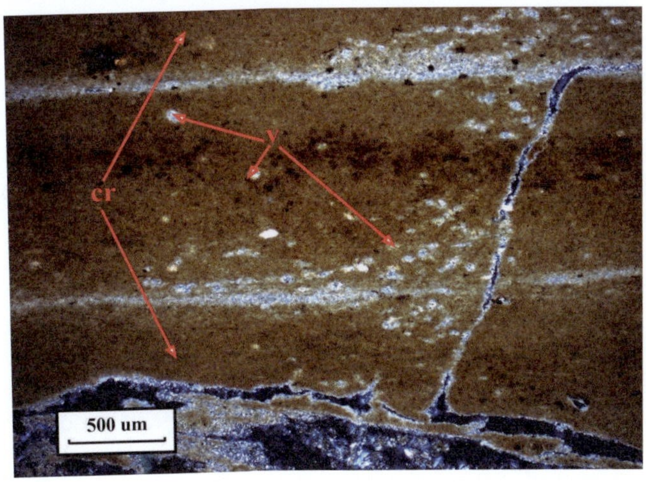

Fig. 7.8 Photomicrograph showing a series of three stacked sedimentary crusts (cr) associated with alluvial soils in Meerut, UP with a platy microstructure indicating sedimentation processes, where calcium carbonate gets precipitated within the voids and iron compounds, responsible for brunified B horizons can be recognized in the field but clay is usually not visible but crusts comprise of fine-grained material and vesicles (v) due to trapped air within the matrix of the crusts. *Source* Neogi (2014)

of the time it has been taken to form. Micromorphology, by providing a fuller range of soil properties, allows us to attempt to solve Jenny's equation (Jenny 1941). The challenge is to determine the environmental factors that are in operation during the life of the soil to initiate soil-forming processes, thus imprinting soils with a unique set of interrelating soil properties (Tsatskin 2012). Mishra (2015, 2016) included two additional soil-forming factors viz. incoming solar radiation (interacting with soil and soil-forming other factors) and sustainable management (input, process and frequency) besides five known factors (parent material, climate, relief, biosphere and time or maturity or stage of development) that dictate the mode of soil-forming processes on way to develop a soil as a pedon in the landscape. Micromorphology may thus strengthen the scope of pedology in soil science.

7.2.2 Soil Micromorphology and Polygenetic Soils

It is true that most soils are polygenetic which means that they have developed during a number of different soil-forming episodes (Federoff et al. 2010; Federoff and Courty 1999). The soils of *Diara* and *Tal* land developed under fluvial processes are typically the polygenetic in nature with a central concept of Chemihydropedoturbation (Mishra et al. 1994; Mishra 2000, 2015). Most stable soils are the product of changing processes and soil properties. It is only when soils are examined in great detail through a microscope that these polygenetic features can be appreciated. Photomicrograph from Fatehabad district, Haryana shows the polygenetic features, in which an iron or manganese-rich hypocoating (c) has formed on a channel void through the diagenesis of organic tissue originally in the pore space (Fig. 7.9). Thus, a number of soil properties are simultaneously imprinted on the soil and then followed by other properties.

Micromorphology allows one to understand the evolution of the soil through the hierarchy and relationships among properties and their interactions. There is however evidence

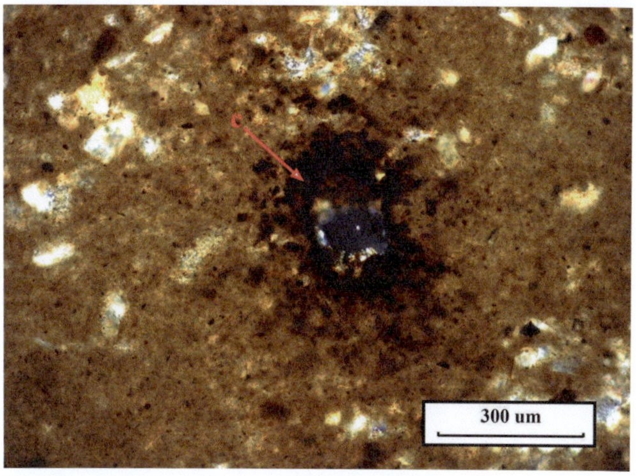

Fig. 7.9 Photomicrograph from Fatehabad district, Haryana showing polygenetic features wherein an iron or manganese-rich hypocoating (c) has formed on a channel void through the diagenesis of organic tissue originally in the pore space, but showing a very little porosity due to fine groundmass. *Source* Neogi (2014)

with alluvial soils wherein sedimentation of a well sorted mineral component is operative. Photomicrograph showing the quasi-coating of iron (qc) within a deep alluvial, well-sorted subsoil in western Uttar Pradesh is surrounding a soil pore, wherein pedofeatures relate to the diagenesis of organic matter under particular redox conditions associated with a fluctuating water table (Fig. 7.10).

With time, well-defined B horizons such as clay or iron-rich accumulations develop. Through micromorphological analysis, we are in a position to understand the sequence of their evolution. This might be perceived through a platy microstructure and horizontally orientated silt and sand particles. It facilitates an understanding of the changing processes resulting in a variety of soil properties. Soil forming processes change over time and are mostly controlled by external environmental factors. Therefore, climate changes including variation in the monsoon have had an effect on soil processes. There are internal processes that are in operation which also lead to changes in soil properties. For example, when pores become blocked and percolating water is inhibited from free movement within the soil profile, hydromorphic soil properties will develop (Hallaire and Curmi 1993). We often see this in very well developed illuvial clay-rich B horizons where gleyic properties develop as a consequence of a perched water table. With time these changes might lead to the complete destruction of the stable system, not because of an external change but because particular internal intrinsic thresholds have been reached (Cady and Daniels 1968). It is only through micromorphology that the tripartite relationship between environmental factors, soil properties and soil processes can be fully evaluated.

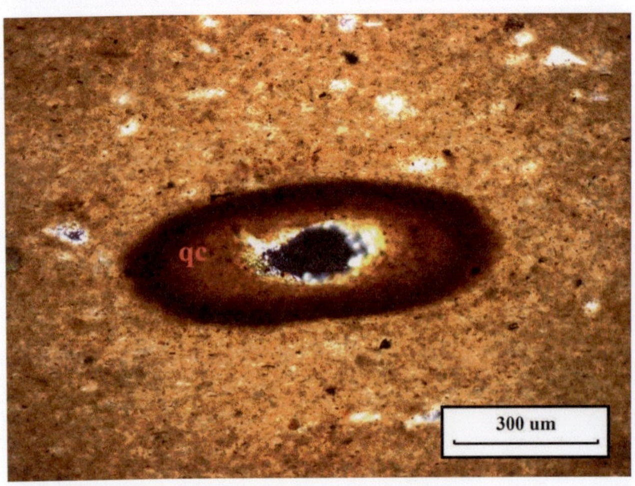

Fig. 7.10 Photomicrograph showing the quasi-coating of iron (qc) within a deep alluvial, well-sorted subsoil in western Uttar Pradesh surrounding a soil pore, wherein pedofeatures relate to the diagenesis of organic matter under particular redox conditions associated with a fluctuating water table. *Source* Neogi (2014)

7.2.3 Soil Micromorphology in Quaternary Research

Micromorphology has been successfully applied to the study of buried soils in India. There are several examples of this kind of study. Some of the best have focused on the 25 m thick loess deposits of Kashmir which contain palaeosols dating to the middle and late Pleistocene. They have been dated using thermoluminescence technique. Bronger et al. (1987) investigated some of these buried soils and found that most were polygenetic in their properties indicating that the soils had formed in changing climatic contexts. Of importance was the last interglacial soil which was thought to have formed under a deciduous forest in a 'xeric' soil moisture regime. Three warm and mostly humid climatic episodes between ca. 80,000 and 50,000 yr B.P were expressed as humus-rich Aht horizon indicating climatic conditions conducive for the illuviation of clay.

Dar et al. (2015a, b) carried out isotopic and micromorphological studies of the Karewa Group in Kashmir. They examined the isotopic signature of carbonate nodules contained within a sequence of Late Quaternary palaeosols. The stable carbon isotopic analysis ($\delta 13C$) reveals a gradual increase of the C4 vegetation towards the top of the sections. The development of C4 plant biomass towards the top of sections is reflective of water stress through increasing aridity. The micromorphological analysis of palaeosol horizons corroborated the isotopic data and suggested that cold arid to semiarid climatic conditions prevailed during the Late Quaternary in the valley. Dar et al. (2015a) carried out further micromorphological investigations of soils at three representative loess-palaeosol sections at Shankerpora, Khan Sahib and Pattan localities of the Karewa Basin of Kashmir. These paleosols have well developed Ah and Bk horizons representing prolonged stability when pedogenic processes outpace loess deposition. On the other hand, comparatively thin to thick palaeosol profiles represent weak to moderate pedogenic maturity indicating short stratigraphic breaks with rapid loess deposition. Micromorphological observations of thin sections suggested that clay illuviation and $CaCO_3$ accumulation have operated within the palaeosol profiles. $CaCO_3$ features are often associated with clay coatings, suggesting decalcification of carbonates followed by clay illuviation. The pedogenic $CaCO_3$, illuvial clay, mottles, iron manganese features, pedal microstructure and blocky aggregates reveal variation in the loess-palaeosol sections. The micromorphological attributes of the palaeosols thus clearly indicate variation of climatic conditions during the Late Quaternary period in the Karewa Basin of Kashmir Valley, India.

Sinha and Sarkar (2009) carried out a review of climate-induced variability in the Late Pleistocene-Holocene sedimentary successions within the Ganga plains of India. The study also included the conduction of micromorphological analysis. They found that the geomorphic diversity that has existed over most of the Late Quaternary is manifested in significant spatial variability in the alluvial architecture developed below the plains. They concluded that the sedimentary successions in the Ganga basin were influenced by climatic variability.

Pal et al. (2003) carried out a micromorphological study on 28 Alfisols of the semiarid part of the Indo-Gangetic plains. The study indicates that clay pedofeatures were characterized by impure clay suggesting that they had developed in a sodic environment. The subsequent accumulation in the Bt horizons suggests that the formation of impure clay pedofeatures and pedogenic $CaCO_3$ are two pedogenic processes occurring simultaneously in soils in the semiarid climate since the last 4000 years B.P.

Khadkikar et al. (2000) examined a sequence of palaeosols in the *Thar* Desert. A section preserved 12 significant moist intervals separated by 11 drier intervals during the past 200 thousand years. They analyzed the morphology of various varieties of calcrete include pedogenic calcrete, groundwater calcrete, calcrete conglomerate (transported calcrete) and rhizogenic calcrete. Pedogenic calcrete nodules associated with Vertisols and the red-soil show marked difference in morphology, dimensions and the distribution of microscopic features. These differences were interpreted to be controlled by the changing monsoonal activity.

Srivastava (2001) investigated the paleoclimatic implications of pedogenic carbonates in Holocene soils of the Gangetic Plains. The study attempted to reconstruct the Holocene climatic history of the Indo-Gangetic plains based on micromorphological characteristics and stable isotope composition of calcretes. Formation of this calcrete took place during the pedogenesis in an arid to semiarid climate. A change to warm and humid phase at 6500 year BP led to dissolution and reprecipitation of the calcretes. This led to the formation of blocky crystals and needles of calcite in voids and coarsening of the fabric in the lower horizons. The younger soils (2500 year BP) show secondary carbonate accumulation associated mainly along the voids in lower parts of the profiles. These are relatively coarse-textured resulting from capillary rise from a fluctuating water table in a sub-humid climate similar to present.

Thomas et al. (2002) investigated palaeosols of the Middle and Upper Siwalik Groups in Himachal Pradesh. This is a 2800 m thick succession and has been subdivided into Unit M1 of the Middle Siwalik and four units of the Upper Siwalik Group, on the basis of facies associations, and type and degree of development of palaeosols. A warm and humid climate between 5.3 and 2.6 Ma led to the formation of red Alfisols with calcrete nodules at places. Slightly cooler and drier climate starting at about 2.6 Ma and approximately coinciding with the onset of global-scale glaciation, produced poorly developed yellow soil with common development of nodular calcretic horizon and calcite material disseminated in the groundmass.

Srivastava et al. (2013) investigated Early Oligocene palaeosols of the Dagshai Formation. Micromorphological analysis reveals that this formation contains four pedofacies (named Pedofacies A-D) of ferruginous palaeosol sequences contained within overbank sediments. The Dagshai Formation unconformably overlies the marine Subathu Formation. Pedofacies A consists of 3–4 well-developed ferruginous palaeosols overlain by gray sandstone beds. Pedofacies B-D is marked by a progressive decrease in pedogenesis. These palaeosols occur as 0.5–1.5 m thick Bw/Bt/Btk/Bk/Bss horizons that are marked by extensive development of rhizoliths, pedogenic carbonate, and iron-rich clay pedofeatures that correspond to modern Entisols, Inceptisols, Alfisols and Vertisols. The information that was gathered from the palaeosols was used to locate India's early Oligocene paleogeographic position. Furthermore, it indicated a tropical climate (paleoprecipitation of 947–1256 mm and paleotemperature of ~ 25 °C) with an initial phase of monsoonal conditions during pedogenesis.

Srivastava and Parkash (2002) investigated the micromorphology of polygenetic soils of the Gangetic Plains. Distinct palaeosols dated to 13500, 8000, 2500, >500 and <500 BP show degraded illuvial clay pedofeatures of an early humid phase (13500–11000 BP) and thick (150–200 mm) microlaminated clay pedofeatures of a later humid phase (6500–4000 BP). The earlier clay pedofeatures were degraded by bleaching, loss of preferred orientation, development of a coarse speckled appearance and fragmentation, whereas those of the later phase are thick, smooth and strongly birefringent microlaminated clay pedofeatures. The illuviation was more extensive during the later phase, as indicated by enrichment of groundmass as discrete pedofeatures of clay intercalations. Pedogenic carbonate was formed during the intervening dry phase from the early Holocene to 6500 BP. It forms the irregularly shaped nodules of micrite and diffuse needles with inclusions of soil constituents. The subsequent change to wetter conditions caused dissolution-reprecipitation, which resulted in partial to complete removal of carbonate from soils over large areas.

Srivastava et al. (2007) investigated the micromorphology of Quaternary palaeosols of the Himalayas and Gangetic plains. They found out that climatic changes during late Quaternary time had been a significant control on the formation of soils. The palaeosols formed within rapidly aggrading sediments of the alluvial fans of the Dehradun valley resulted in response to the reduced rate of sedimentation and climatic changes and correspond to the MIS3 (Marine Isotope Stage 3) and MIS2 stages. Distinctive micromorphological features of these palaeosols provided the details of the prevalent pedogenesis in response to the palaeoclimatic changes during 50 ka. Microfabrics of these palaeosols show reorganization of the pedality from massive and/or subangular blocky to platy and prismatic structures, strong to very strong mobilization of the plasma, different types of textural pedofeatures along with faunal activities. These pedofeatures are indicative of cold-humid climate with subsequent change to even colder but drier conditions during the last Glaciation. Comparison of the micromorphological characters of the paleosols of the NW Himalayas and the polygenetic soils of the Gangetic plains show the same degree of soil development indicating 5–10 ka pedogenic intervals in alluvial fans of the Dehradun Valley. However, the difference in their pedofeatures is attributed to different pedogenic environments. The paleosols of the Dehradun Valley show predominance of the illuvial features with superposed impure silty clay on earlier clay pedofeatures and banded clay fabric features without any pedogenic calcium carbonate. The bordering Gangetic Plains are covered with polygenetic soils developed on stable surfaces. Srivastava and Sauer (2014) studied the lithified palaeosols from Dagshai formation of the Himalayan Forestland revealing that it contains four types of well-developed palaeosols. Despite diagenetic alteration, evidence of palaeopedogenic processes is still well-preserved in these fossil soils in the form of microstructures, b-fabrics, pedogenic calcite, bioturbation, and textural pedofeatures. Thin-section analysis helped to distinguish pedogenic and diagenetic features of lithified paleosols and to infer the paleoenvironment of the Dagshai palaeosols. The paleopedological characteristics of the fossil soils suggest humid to sub-humid conditions during their formation in early Oligocene. The paleoclimate inferred is consistent with prevalence of tropical paleovegetation reported from Dagshai sediments.

Achyuthan (2003) applied micromorphological analysis on the Late Neogene-Early Quaternary hardpan calcretes of Western Rajasthan. They form a distinct element of the Quaternary landscape and the study was to determine the processes governing their development of hardpan calcretes and to evaluate the local and regional controls on their formation. The micromorphology included pedogenetic and groundwater features within the hardpan calcretes. Thickening of calcite laminae downward and tapering at the sideward edges around the unweathered minerals of quartz and feldspars indicated cumulative and compound pedogenesis, which probably occurred locally following the downward movement of carbonate solution and pore water. The source of most of the calcite is groundwater. Dust was found to be a major source for carbonate precipitation. The study showed with the presence of palaeolithic stone tools that the assumption that the powdery calcretes are younger in age compared to the more complex forms, it was found that the morphology of calcretes is not a reliable indicator of age.

Shankar and Achyuthan (2007) carried out an investigation of calcic and petrocalcic horizons in the area around Coimbatore, Tamil Nadu. The calcic horizons represent the Bk horizons that occur as thick complex profiles in the foothill regions while the laminar petrocalcic horizons representing the K horizon formed on hard rock in the topographic low lying area. Micromorphological study of the calcic horizons show the occurrence of alveolar septal structures, calcified filaments, coated grains, spherulites, calcified root cells and calcispheres that indicate biogenic origins, mainly induced by plant root related microbial activity. The calcic nodules within the calcic horizons consist of quartz sand grains cemented by finely crystalline, grain-coating, often glaebular and pore-filling micrite. This development has taken place in phases of soil formation, erosion and reworking. The inter-relationships between these processes have caused variations in the phases of accretion of soil profiles developed in the foothill region.

Pal et al. (2009) investigated colluvial Vertisols associated with transported smectitic parent material from the weathering of Deccan basalt. The soils show a change in their morphological, physical, chemical, and micromorphological properties in the climosequence. Soils of the lower horizons indicate a sub-humid moist to aridic climatic environments with the formation of pedogenic $CaCO_3$. Higher in the succession, the illuviation of clay indicates a climate change of the Holocene period. These processes impaired the hydraulic properties of soils in general, and in soils of drier climates in particular. As a result, cracking pattern, chemical composition and plasmic fabric were more modified in soils of the drier climates. The study demonstrates how the intrinsic soil properties of the cracking clay soils in a clinosequence may help in inferring the change in climate in a geologic period.

7.2.4 Soil Micromorphology in Archaeology

Apart from providing insight into buried soils and landscape, soil micromorphology is also useful in studying archaeological sediments. This helps in understanding human behaviour and activity within a particular culture. Anthropogenic sediments are very distinct from natural soils, though both of these share common formation processes and thus can provide some information about soil development (Catt and Weir 1976). Micromorphological investigations of anthropogenic sediments are the fundamental sources for indications of human activity. One of the first works in investigating human activity from the micromorphology of sediments in India has been undertaken by the author on the Harappan archaeological sites of Haryana, Uttar Pradesh and Rajasthan (Neogi 2014). Since then the authors have been employing the techniques of micromorphological thin section analysis to trace a diverse range of past domestic activities architectural materials, site formation processes and use of space in several other archaeological sites in India.

Major changes in the environment have often been postulated as a critical factor for the decline of the Bronze Age Harappan civilization of the Indian subcontinent (Staubwasser et al. 2003). Speculations as to the causes have included shifting of river beds, drying up of rivers or increasing aridity (Berkelhammer et al. 2012). In order to understand these issues better, some pioneering sedimentological studies have been carried out along the abandoned channel depressions of the Saraswati and Chautang rivers (Courty 1990). Their main aim had been to relate the structural relations to the changing geography of the Ghaggar-Hakra plain, for which they had undertaken the soil micromorphological work from the archaeological site of Mitathal (Courty and Fedoroff 1985). Through the employment of the techniques of micromorphology, Courty was able to identify reduced seasonal flooding and dune encroachments prior to the Harappan times. She concluded that a shift to drier conditions and somewhat similar rainfall distribution patterns and environment to the present had been in action from during the period of Harappan occupation. Though Courty's work had received criticism, it certainly did pave the way for the application of micromorphological techniques in Indian archaeology.

Micromorphological technique is one of the key methods for ethnoarchaeological studies. It can be applied to understand some specific questions relating to the hunter–gather societies. Such societies have distinct social perceptions and practices which are expressed in unique use of space and material deposition patterns. Ethnographic data, in combination with micromorphological analysis of thin sections from abandoned and the contemporary sites can yield information about archaeological site formation processes related to hunter-gatherer's ways of living. Micromorphological analysis from his samples, accompanied by geochemical analysis, has shown that post-depositional processes in tropical forests result in poor preservation of archaeological materials due to acidic conditions and intensive biological activity within the sediments.

7.3 Soil Micromorphology and Indian Agriculture

Micromorphology has been a technique used in India for agricultural research (Srivastava et al. 2009a, b), but it has a lot of potential applications in the future. There are aspects of soil health which are of particular concern at the moment and will increase in attention in the future. These concerns include food security, climate change, the impact of agricultural practice on soil health and desertification (Randhawa 1982). It provides a technique where soil properties can be monitored in response to environmental changes. For example, changes in agricultural practice may have detrimental effects on aspects of soil properties including compaction, loss of organic matter and the composition of soil flora and fauna. It has the potential to monitor consequences of climate change on soil properties. The relationship between soil properties and agricultural output is a topic of importance. It can be integrated into sustainability studies, monitoring soil degradation due to agricultural practice. It can be utilized to assess the progress of land restoration in terms of characterizing the qualities of health and productivity of soils. The maintenance of optimal level of a soil is very important. It has a unique capability to integrate all the other aspects of soil science and to understand the relationships between soil components. For example, micromorphology may help to characterize soil porosity, structure, texture and other components such as the amount and status of organic matter and soil hydrology, and compare soils across the landscapes and their development through time.

Evaluation of soil structure in the field is, for example, unreliable without examining it under a petrographic microscope that clearly shows the nature, shape and size of the secondary particles and voids. Permanent records may be made through photographs in which peds and voids are shown, (Biswas and Mukherjee 2001). Micromorphological studies of undisturbed soils in soil survey and mapping provide wealth of information regarding genesis and weathering processes for its application in soil classification, soil-landscape relationship that forms the basis for soil-geomorphic evolution over time (Srivastava et al. 2009a, b).

Climate change is of direct relevance to agriculture and is topical in India at present. For example, the likely scenario of sea-level rise and its impact on alluvial soils of the Gangetic delta of the Bengal region would likely see the salinization or sodiumization of soil. This might perhaps lead to the loss of agricultural productivity. Of course, soil chemistry and other techniques are important for monitoring soil changes, but micromorphology would be important in showing how soil properties change and we can use this information to mitigate against their negative consequences. It would be important to know at what point within the soil profile, salts are being precipitated. Likewise, desertification is more accurate and descriptive know-how about the soil resource inventory besides providing soil based evidences of climate change as well as polygenesis. This unique branch of soil science also plays vital roles to understand weathering processes, field variability, spatial soil diversity, and landscape dynamics s another important problem affecting agriculture adversely. It is important to be able to monitor the effect of changes in the monsoonal climate on soils in order to formulate guidance of best practice for farmers to maintain optimum soil health (Srivastava et al. 2009a, b).

7.4 Conclusions

Soil micromorphology is well proven to answer many research problems between two extremes, i.e. pedology and edaphology as well as in soil classification and thus it helps in generating a reliable database for better understanding of diversified soil resource. Application of soil micromorphology in soil survey and classification is of high relevance wherein the technique is useful in recognizing the diagnostic micromorphic features in order to differentiate and classify the soils at all categorical levels of Soil Taxonomy. However, the techniques need to be more popularized among students and researchers at graduate levels.

References

Achyuthan H (2003) Petrologic analysis and geochemistry of the Late Neogene-Early Quaternary hardpan calcretes of Western Rajasthan, India. Quatern Int 106:3–10

Banerjee A (1998) Environment, population, and human settlements of Sundarban Delta. Concept Publishing Company

Berkelhammer M, Sinha A, Stott L, Cheng H, Pausata FSR, Yoshimura K (2012) An abrupt shift in the Indian Monsoon 4000 years ago. In Giosan L, Fuller D, NNichol K, Flad R, Clift P (eds) Climates, Landscapes and Civilizations, vol 198. The American Geophysical Union, Washinfton, pp 75–88

Biswas TD, Mukherjee SK (2001) Text book of soil science. Tata McGrow-Hill Education, New Delhi

Bronger A, Pant RK, Singhvi AK (1987) Pleistocene climatic changes and landscape evolution in the Kashmir Basin, India: paleopedologic and chronostratigraphic studies. Quatern Res 27(2):167–181

Bullock P, Fedoroff N, Jongerius A, Stoops G, Tursina T (1985) Handbook for soil thin section description. Waine Research

Cady JG, Daniels RB (1968) Genesis of some very old soils-the Paleudults. In Holmes J (ed) Transactions of the 9th international congress soil science, vol 4, International Society of Soil Science and Angus and Robertson, Adelaide, pp 103–112

Catt JA, Weir AH (1976) The study of archaeologically important sediments by petro-graphic techniques. In: Davidson DA, Shackley ML (eds) Geoarchaeology: earth science and the past. Duckworth, London, pp 65–91

Courty MA (1990) Pedogenesis on Holocene calcareous parent materials under semi-arid conditions (Ghaggar Plain, NW India). Dev Soil Sci 19:361–366

Courty MA, Fedoroff N (1985) Micromorphlogy of recent and buried soils in a semi-arid region of northwestern India. Geoderma 35 (4):287–332

Dalrymple J (1957) Preparation of thin sections of soils. J Soil Sci 8 (1):161–165

Dar RA, Chandra R, Romshoo SA, Kowser N (2015a) Micromorphological investigations of the Late Quaternary loess–paleosol sequences of the Kashmir Valley, India. J Asian Earth Sci 111:328–338

Dar RA, Chandra R, Romshoo SA, Lone MA, Ahmad SM (2015b) Isotopic and micromorphological studies of Late Quaternary loess paleosol sequences of the Karewa Group: inferences for palaeoclimate of Kashmir Valley. Quat Int 371(12):2e134

Fedoroff N, Courty M (1999) Soil and soil forming processes under increasing aridity. In: Singhvi A, Derbyshire E (eds) Palaeoenvironmental reconstruction in arid lands, Oxford and IBH Publishing, New Delhi, pp 73–108

Fedoroff N, Courty M, Guo Z (2010) Palaeosoils and relict soils. In: Stoops G, Marcelino V, Mees F (eds) Interpretation of micromorphological features of soils and regoliths, Elsevier, Amsterdam, pp 623–662

Fitzpatrick EA (1984) The micromorphology of soils. Chapman and Hall, London and New York

Gerrard J (1992) Soil geomorphology: an integration of pedology and geomorphology. Springer, London

Hallaire V, Curmi P (1993) Image analysis of pore space morphology in soil sections, in relation to water movement. In: Ringrose-Voase A, Humphreys G (eds) Soil micromorphology: studies in management and genesis, proceedings of the {IX} international working meeting on soil micromorphology, Townsville, Australia, July 1992, vol 22 of Developments in soil science, pp 559–567

Jenny H (1941) Factors of soil formation: a system of quantitative pedology. Dover Publications, New York

Kalbande AR, Pal DK, Deshpande SB (1992) b-fabric of some benchmark Vertisols of India in relation to their mineralogy. J Soil Sci 43:375–385

Khadkikar AS, Chamyal LS, Ramesh R (2000) The character and genesis of calcrete in Late Quaternary alluvial deposits, Gujarat, western India, and its bearing on the interpretation of ancient climates. Palaeogeogr Palaeoclimatol Palaeoecol 162(3):239–261

Kooistra MJ (1990) The future of soil micromorphology. Dev Soil Sci 19:1–7. https://doi.org/10.1016/S0166-2481(08)70310-6

Kubiëna WL (1938) Micropedology. Collegiate Press, Ames, Iowa

Kubiëna WL (1970) Micromorphological features of soil geography. Rutgers University Press, New Brunswick New Jersey

Lagacherie P, McBratney A (eds) (2006) Digital soil mapping: an introductory perspective, vol 31. Elsevier, Amsterdam

Lal S, Deshpande S, Sehgal J (1994) Soil series of India. Number 40. National Bureau of Soil Survey & Land Use Planning (ICAR)

Mermut AR, Dasog GS (1986) Nature and micromorphology of carbonate glaebules in some Vertisols of India. Soil Sci Soc Am J 50:382–391

Mishra BB, Choudhary CD, Singh NK, Mall J, and Singh VN (1994) Fluvisol as a proposed Order in soil Taxonomy for active flood plain soils. Trans. XVth. world congress soil science, Acapulco, Mexico, 6B, pp 55–56

Mishra BB (2000) Fluvisol as the 13th order in USDA soil taxonomy for floodplain soils with Chemihydropedoturbation-Pedogenic process-A proposition. In: Proceedings of international conference on natural resource & sustainable agriculture in 21st century, vol 2, 14–18 Feb 2000, New Delhi, pp 86–88

Mishra BB (2015) Soil Classification: Issues and Opportunities for Indian System, The 26th Dr. SP Raychaudhuri Memorial Lecture, 19th October, 2015, NBSS & LUP, Nagpur, Journal of the Indian Society of Soil Science, Vol. 63 (Supplement) pp S41-S52

Mishra BB (2016) Indian System of Soil Classification: A way Forward. Agri Res & Tech: Open Access J. 2016; 3(2): 555606. https://doi.org/10.19080/artoaj.2016.03.555606002

Neogi S. (2014). Geoarchaeological investigations of Indus settlements in the plains of northwest India. PhD thesis, University of Cambridge, UK

Pal DK, Bhattacharyya T, Chandran P, Ray S, Satyavathi P, Durge S, Raja P, Maurya U (2009) Vertisols (cracking clay soils) in a climosequence of Peninsular India: evidence for Holocene climate changes. Quatern Int 209(1):6–21

Pal DK, Srivastava P, Bhattacharyya T (2003) Clay illuviation in calcareous soils of the semiarid part of the Indo-Gangetic Plains. India. Geoderma 115(3):177–192

Pal DK, Kalbande AR, Deshpande SB, Sehgal JL (1994) Evidence of clay illuviation in sodic soils of the Indo-Gangetic plain since the Holocene. Soil Sci 158:465–473

Pal DK, Dasog GS, Vadivelu S, Ahuja RL, Bhattacharyya T (2000) Secondary carbonate in soils of arid and semi-arid regions of India. In: Lal R, Kimble JM, Eswaran H, Stewart BA (eds) Global climate change and pedogenic carbonates. Lewis Publishers, Boca Raton, FL, pp 149–185

Pal DK, Balpande SS, Srivastava P (2001) Polygenetic vertisols of the purna valley of central India. Catena 43:231–249

Phillips JD (1998) On the relations between complex systems and the factorial model of soil formation (with discussion). Geoderma 86 (1):1–21

Randhawa N (ed) (1982). Review of soil research in India: managing soil resources to meet the challenges to mankind. In: 12th international congress of soil science, New Delhi, India 8–16 February 1982. Indian Society of Soil Science

Sahai V (2011) Fundamentals of Soil. Kalyani Publishers, India

Sander H (2002) The porosity of tropical soils and implications for geomorphological and pedo- genetic processes and the movement of solutions within the weathering cover. CATENA 49(1):129–137

Shankar N, Achyuthan H (2007) Genesis of calcic and petrocalcic horizons from Coimbatore, Tamil Nadu: Micromorphology and geochemical studies. Quatern Int 175(1):140–154

Sinha R and Sarkar S (2009) Climate-induced variability in the Late Pleistocene–Holocene fluvial and fluvio-deltaic successions in the Ganga plains, India: a synthesis. Geomorphology 113(3):173–188

Soil Survey Staff (1975) Soil Taxonomy. USDA-NRCS, US Government Printing Office, Washington, DC

Soil Survey Staff (2014) Keys to Soil Taxonomy, 12th edition, USDA, NRCS, Natural Resources Conservation Service, United States Department of Agriculture, Government Printing Office, Washington, DC

Srivastava P (2001) Paleoclimatic implications of pedogenic carbonates in Holocene soils of the Gangetic Plains, India. Palaeogeogr Palaeoclimatol Palaeoecol 172(3):207–222

Srivastava P, Parkash B (2002) Polygenetic soils of the north-central part of the Gangetic Plains: a micromorphological approach. CATENA 46(4):243–259

Srivastava P, Patel S, Singh N, Jamir T, Kumar N, Aruche M, Patel RC (2013) Early Oligocene paleosols of the Dagshai Formation, India: a record of the oldest tropical weathering in the Himalayan foreland. Sed Geol 294:142–156

Srivastava P, Sauer D (2014) Thin-section analysis of lithified paleosols from Dagshai formation of the Himalayan Foreland: identification of paleopedogenic features and diagenetic overprinting and implications for paleoenvironmental reconstruction. CATENA 112:86–98

Srivastava P, Singh AK, Parkash B, Singh A, Rajak M (2007) Paleoclimatic implications of micromorphic features of Quaternary paleosols of NW Himalayas and polygenetic soils of the Gangetic PlainsâĂŤa comparative study. CATENA 70(2):169–184

Srivastava P, Parkash B, Sehgal JL, Kumar S (1994) Role of neotectonics and climate in development of the Holocene geomorphology and soils of the Gangetic Plains between Ramganga and Rapti rivers. Sed Geol 94:119–151

Srivastava P, Bhattacharyya T, Pal DK (2002) Significance of calcium carbonate minerals in the pedogenesis and management of cracking clay soils (Vertisols) of India. Clays Clay Miner 50:110–125

Srivastava P, Rajak M, Singh LP (2009a) Late Quaternary alluvial fans and paleosols of the Kangra Basin, NW Himalaya: tectonic and paleoclimatic implications. Catena 76:135–154

Srivastava P, Pal DK, Kalbande AR (2009b) Soil micromorphology and its usefulness in soil surveys. In: Bhattacharyya T, Sarkar D, Pal DK (eds) Soil survey manual, NBSS&LUP Publication No. 146, Nagpur, India, pp 400

Staubwasser M, Sirocko F, Grootes PM, Segl M (2003) Climate change at the 4.2 ka BP termination of the Indus valley civilization and Holocene south Asian monsoon variability. Geophys Res Lett 30 (8):1–4

Stoops G (2003) Guidelines for analysis and description of soil and regolith thin sections. Soil Science Society of America, Madison, Wisconsin

Stoops G, Marcelino V, Mees F (eds) (2018) Interpretation of micromorphological features of soils and regoliths, 2nd Edn. Elsevier Science, The Netherlands, p 752, ISBN 10-0444638482. https://doi.org/10.1016/B978-0-444-63522-8.00009-7

Thomas J, Parkash B, Mohindra R (2002) Lithofacies and palaeosol analysis of the middle and upper Siwalik Groups (Plio–Pleistocene), Haripur-Kolar section, Himachal Pradesh, India. Sediment Geol 150 (3):343–366

Tsatskin A (2012) On Russian concepts of soil memory-expansion of Dokuchaev's pedological paradigm. In: EGU general assembly 2012, vol 14, held 22–27 Apr 2012 in Vienna, Austria, p 38

Walia CS, Rao YN, Rao YS (2000) Micromorphology of endopedons of some soils of Bundelkhand region. J Indian Soc Soil Sci 48 (4):813–818

WRB (2014) World reference base for soil resources. World Soil Resources Report No. 106. FAO, Rome

Soil Biogeochemistry

8

Debjani Sihi and Biswanath Dari

Abstract

Depending upon type and pedogenic stage, soils are subject to biotic and abiotic interactions of complex nature depending on type, nature and specific properties. Soil biogeochemistry involves the study of elemental cycling as mediated by complex and inseparable interactions between the biotic (living) and abiotic (non-living) components of soils. Human activities have substantially altered biogeochemical cycling of several key elements including carbon, nitrogen, phosphorus, potassium, and other secondary and minor nutrients over the past few decades, which, in turn, had serious environmental consequences. The present chapter outlines the soil biogeochemical investigations conducted within Indian subcontinent in both natural ecosystems and managed agricultural systems and addresses the state-of-the-art in order to understand the undergoing biogeochemical reactions. Here, we sought to clarify complex interactions generally occurred during biogeochemical transformations of an element (or compound) of interest within the type-specific soil. Further, we emphasized the importance of advancing our understanding of feedback loops in soil biogeochemical processes as altered by anthropogenic perturbations in tropical and sub-tropical soils of India. Overall, this chapter is broadly focused on the nutrient cycling, which is followed by more specific topics like "soil microbiology," "soil biodiversity," and "soil biotechnology."

Keywords

Biogeochemical cycles • Soil organic matter • Nutrient cycling • Microbial and enzymatic activity • Xenobiotics

8.1 Introduction

Today's soil science without the understanding of soil biogeochemistry is incomplete, since it forms the sandwich between pedology and edaphology. Soil biogeochemistry lies at the interface between soil microbiology, soil biodiversity, and soil biotechnology. Soils may differ in geology, micromorphology, and mineralogy, but their inherent potential productivity lies with the nature, extent, and degree of the biogeochemical cycles of soil nutrients. It is the most relevant topic after geology and geomorphology, soil groups, type and classification, soil mineralogy and clay minerals.

Biogeochemistry is an emerging interdisciplinary field that deals with the fluxes of chemical elements through living and non-living components of an ecosystem by exploring physical, chemical, biological, and geological processes (Fig. 8.1). Although the field of biogeochemistry was formally recognized only a few decades ago (Gorham 1991), the term was coined quite a long time ago by Vernadsky (1924) as a sub-discipline of geochemistry. Soil serves as a key nexus in the exchange of elements between boundaries of four spheres (lithosphere, hydrosphere, biosphere, and atmosphere). Thus, broader definition of ecosystem biogeochemistry also applied to soil biogeochemistry, which integrates knowledge and techniques from a wide array of research fields within the domain of soil sciences.

Soil biogeochemistry investigates the factors that drive complex biogeochemical processing of crucial elements as it moves through the soil (Fig. 8.2). Earlier insights of Jenny (1941) on the formation of soil organic matter because of

D. Sihi (✉)
Oak Ridge National Laboratory, Environmental Sciences Division, Climate Change Science Institute, Oak Ridge, TN, USA
e-mail: sihid@ornl.gov

B. Dari
Aberdeen Research and Extension Center, University of Idaho, Aberdeen, ID, USA

B. B. Mishra (ed.), *The Soils of India*, World Soils Book Series,
https://doi.org/10.1007/978-3-030-31082-0_8

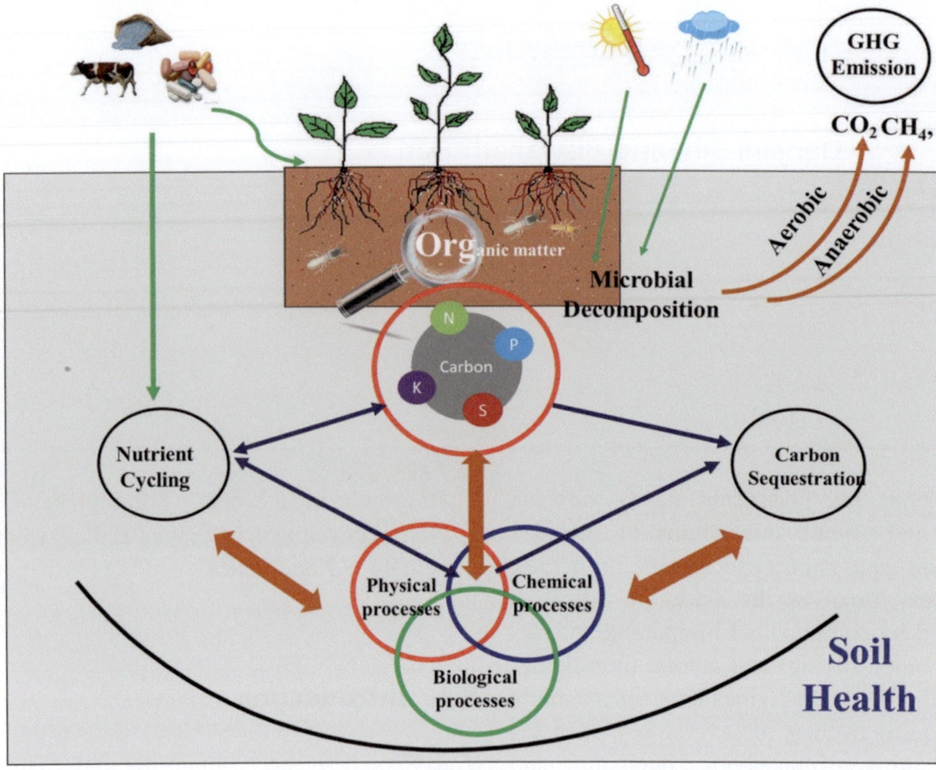

Fig. 8.1 Linkage between physical, chemical, and biological processes in soil and ecosystem functioning. Modified from Reddy and DeLaune (2008)

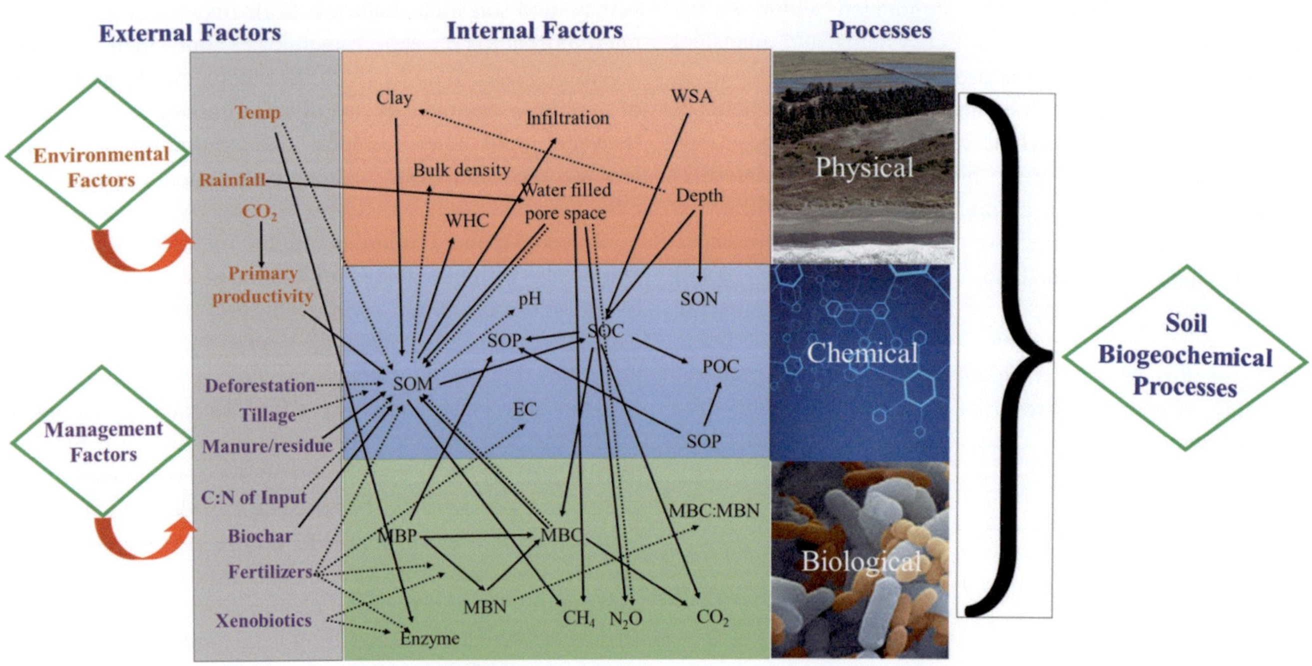

Fig. 8.2 Factors affecting soil biogeochemical processes and associated complex feedback loop. Solid line represents positive feedback and dotted line represent negative effect. WSA: Water stable aggregates, WHC: Water holding capacity, SOM: Soil organic matter, POC: Particulate organic C, SOC: Soil organic C, SON: Soil organic N, SOP: Soil organic P, MBC: Microbial biomass C, MBN: Microbial biomass N, MBP: Microbial biomass P

interactions between climate, geology, and biology serve as a foundation work for soil biogeochemistry. Advancement of analytical tools paved the pathways for soil biogeochemistry in modern times and revealed novel mechanisms related to exchange of elements between inorganic and organic forms within soil matrix. Another aspect of soil biogeochemical studies targets in determining a quantitative relationship among different system components for an element of interest and extrapolate those relationships to advise management decisions related to restoration and preservation of soil resources.

8.2 Research Thrust in Soil Biogeochemistry in India

Major tropical and subtropical soils of the Indian subcontinent are inherently low in organic C and nutrient content due to high weathering rates under prevalent climates with high temperature. Large-scale conversion of these virgin lands to agriculture since the Green Revolution initiated the process of rapid decline of soil organic carbon (SOC) and nutrient stocks. The disparity between food security and soil nutrient stocks further challenged the need to provide nutrition to the ever-growing population of the country, which was contingent upon the efficiency of functioning of national agricultural systems (Jones et al. 2013). For instance, cultivation of cereals and other food crops further exacerbated the situation due to their high nutrient demand which resulted into mining of soil nutrients. This created an imbalance between nutrient stripping by crop off-take and subsequent replenishment by mineral weathering of soils, which, in turn, was mediated by the external addition of NPK (nitrogen-phosphorus-potassium) fertilizers. India remains the third-largest producer and consumer of fertilizers (Singh et al. 2012). But extensive use of NPK fertilizers failed to effectively replenish mined soil due to a wider range of nutrients (both macro- and micro-nutrients) stripping by agricultural crops and exerted more pressure on soil resources than ever before. Excessive application of fertilizers along with other chemicals (xenobiotics like pesticides to kill unwanted insects, plants, and rodents) were credited with higher yield, but these yield gains were far below the potential capacity of these soils and the productivity stagnated after a while due to reduction of overall soil fertility, more so in P-poor tropical soils in India (Kumar 2011). For example, K stripping in Indian soils limited productivity directly or indirectly (by reducing N use efficiency) in many regions (Brar et al. 2011; Tandon and Tiwari 2011). The story continued with more reports for deficiency in zinc (Zn), sulfur (S), boron (B), iron (Fe), molybdenum (Mo), and copper (Cu) in 49, 40, 33, 12, 11, and $\sim$5%, respectively, for soils in India on an area basis (Singh 2008).

Potential mitigation strategies advocated to solve this, and were to improve tight cycling of nutrients by diversification of land-use types (e.g., agro-forestry), crop diversification, use of leguminous cover crops, zero or no-tillage practices, restoration of riparian forests (e.g., mangroves along coast), and wetlands (Nair 1993; Shukla et al. 2010). Holistically redress this nutrient deficit issue requires attempt to improve the soil organic matter (SOM) status and associated soil structure, biological activity, and water retention (Lal 2004). However, the lack of available resources (e.g., organic manures; Samra and Sharma 2009) seemed to offer limited potential, more so with increasing rate of urbanization (with the consequence of additional deforestation) and called for more attention to future food security in the face of climate change.

Based on the above background, research thrust areas in soil biogeochemistry in India were aimed to identify and minimize negative outcomes of land-management practices in both natural and agricultural systems on biogeochemical cycles of essential elements to maintain sustainability of human–environment interactions. Soil biogeochemistry studies with a focus on SOM mostly relates to SOM turnover, soil C sequestration, and greenhouse gas (GHG) emission under different land-use managements as a means of climate change abatement (Tables 8.1 and 8.2). Studies on nutrient transformations and cycling in agroecosystems and the associated increase of soil productivity have been a major focus for a long time to feed the ever-increasing population (Table 8.2).

In recent decades, a handful of studies on soil health (or quality) management aim to identify sustainable agricultural production systems for present and future generations (Table 8.2). Concurrently, there is a trend to monitor soil microbial and biochemical parameters as a function of xenobiotics application in agricultural fields in the form of chemical pesticides and herbicides (Table 8.3).

8.3 Overlapping Domains of Soil Biogeochemistry

8.3.1 Soil Organic Matter: The Hub of Soil Biogeochemistry

Soil organic matter (SOM) is a central facet of soil biogeochemistry as it serves as a storehouse for carbon (C) and other elements as depicted in Fig. 8.1 (Reddy and DeLaune 2008). Thus, N, P, K, and S cycling in soil are often coupled with the cycling of soil organic carbon (SOC). In addition to these major elements, cycling of trace elements such as Zn, Fe, Mn, and Cu are also influenced by SOM content. Thus, dynamics of SOM are intimately related to the rate and extent of many of these chemical transformations that either

Table 8.1 Soil biogeochemical cycling as a function of land-use practices

Study type	Study area/soil type	Main findings	Reference
Land-use impacts on SOC fractions and aggregate stability	Typicustochrepts of Northwest India	Surface layer SOC: grassland > forest > agricultural > eroded land Subsurface layer SOC: forest > grassland > agricultural > eroded land WSA > 2 mm: highest in grasslands and lowest in eroded lands WSA < 0.25 mm: highest in eroded lands and lowest in forest lands	Saha et al. (2011)
SOC stock and fractions in relation to land-use and depth	Shiwaliks hills of lower Himalayas	SOC stock (1.05 m profile): forest > grassland > cultivated > eroded lands SOC in micro-aggregates (WSA < 0.25 mm) and POC fraction accumulated under ecosystems with sufficient C inputs	Saha et al. (2014)
SOC stocks and dynamics under different land-use	Arid and semi-arid regions of Rajasthan	Intensive agriculture without proper management caused rapid SOC depletion in cropland than untilled soils under scrub vegetation and agro-forestry	Singh et al. (2007)
SOC stocks after deforestation and land cover changes	Western Ghats biodiversity hotspot	Overall SOC stock of the area was maintained after deforestation as SOC loss from deforested area was compensated by sequestration elsewhere (where degraded forests, wastelands or non-forest lands were transformed into tree plantations and irrigated agriculture were practiced)	Lo Seen et al. (2010)
Soil nutrients and fertility in three traditional land-use systems	Khonoma, Nagaland	Soil nutrient index: natural forest > Jhum fallow > paddy cultivation Soil nutrient index was based on soil fertility indicators like pH, Av. N, Av. P, Av. K, SOC, and SOM content	Chase and Singh (2014)
Effects of deforestation and cultivation on soil biochemical and microbial indices	South Andaman	Deforestation and cultivation markedly reduced microbial and enzyme activities (qCO2, PHO, BPHO, BG, urease, BAA-protease, casein-protease, arylsulfatase, invertase, and carboxymethylcellulase) due to significant depletion of organic matter and nutrients at the plantation sites than the forest sites	Dinesh et al. (2004a)
Biochemical properties of soils of undisturbed and disturbed mangrove forests		Significant reductions in microbial and biochemical indicators of soil fertility at disturbed sites. Number and activity of soil microorganisms were contingent on the quantity of mineralizable substrate and the availability of nutrients in these mangrove soils	Dinesh et al. (2004b)
Effects of the mono-cropping of rice, forestry, and agro-forestry on MBC	ICAR-Central Soil Salinity Research Institute, Karnal	Greater MBC(42%) and MBN (13%) were observed in agro-forestry systems as compared to mono-cropping	Kaur et al. (2000)
Status of microbial diversity in agro-forestry systems	Tamil Nadu	Microbial diversity pattern was bacteria (64%) > actinomycetes (23%) > fungi (13%). SOM, vegetation, and soil nutrients altered microbial community (structure and composition) under agro-forestry systems. Bacterial and actinomycetes count positively correlated with physicochemical properties (SOC, moisture, pH, N, P, K, Fe, Cu, Zn, Mg, and Ca)	Radhakrishnan and Varadharajan (2015)
Soil microbial responses as a function of *Jatropha* plantation	ICAR-Indian Agricultural Research Institute, New Delhi	*Jatropha* plantation improved soil health (greater SOC, MBC, MBN, soil enzymes: urease, DH and lower C:N ratio)	Mahmoud et al. (2016)
Impact of herbivore grazing on soil microbes and carbon sequestration	Grazing ecosystem Trans-Himalaya, Spiti, Ladak	Grazer influence on microbial abundance and community composition collectively played a crucial role in net soil C dynamics, but the effect was constrained by environmental factors (e.g., moisture availability)	Bagchi et al. (2017)

(continued)

Table 8.1 (continued)

Study type	Study area/soil type	Main findings	Reference
Comparative evaluation of three contrasting land-use systems for soil quality indicators	Mollisol of Uttarakhand	Long-term fertilizer experiment (continuous pulse and organic system), mango (horticulture) and Dalbergia (agro-forestry) system appeared as more sustainable than guava (horticulture) based on soil carbon, biochemical (DH, glomalin related soil protein), and microbiological (MBC and BSR)	Bhattacharjya et al. (2017)
Seasonal changes and influence of soil moisture on MBC	Sal forest and mixed forest of Madhya Pradesh	MBC inversely related with soil moisture. Higher protozoan grazing increased MBC turnover during rainy season when plant productivity was at its max. Clay increased survival rate of MBC at low humidity	Srivastava (1992)

SOM: Soil organic matter, SOC: Soil organic C, POC: Particulate organic C, BSR: Basal respiration, WSA: Water stable aggregates, Av. N: Available N, Av. P: Available P, Av. K: Available K, qCO2: metabolic quotient, PHO: phosphomonoesterase, BPHO: phosphodiesterase, BG: glucosidase, MBC: Microbial biomass C

Table 8.2 Soil biogeochemical cycling as a function of management practices in different soil types

Study type		Study area/soil type	Main findings	Reference
Effects of management practices on soil C sequestration	Long-term effect	Punjab Agricultural University, Ludhiana, Punjab	Integrated nutrient management (100% NPK + FYM) should be used to sustain maize–wheat cropping system	Kaur et al. (2008)
		ICAR-Indian Agricultural Research Institute, New Delhi	Balanced fertilization (with NPK) improved the SOC level, with pronounced effect under double rice crop	Mandal et al. (2007)
		Alfisols and Inceptisols of North-west India	Long-term rice–wheat system stored more SOC than maize–wheat system in both surface (Ap horizon) and subsurface (B-horizon) layers	Kukal et al. (2016)
		TypicUstochreptsof Punjab	Long-term intensive cultivation and fertilization significantly increased surface soil SOC content; however, stability of SOC varies with nature of added organic manures	Chaudhary et al. (2017)
		Indo-Gangetic plains and Vindhyan highlands	Strong correlation was observed between NPP and soil C density	Kumar et al. (2014)
		ICAR-Indian Agricultural Research Institute, New Delhi	100% NPK + FYM efficiently increased different SOC fractions and decreased qCO_2	Rudrappa et al. (2006)
		SardarVallabhbhai Patel University of Agriculture & Technology, Meerut	Zero-tillage increased POC and macro-aggregates and thus improved soil C in rice–wheat rotation	Kumari et al. (2011)
	Short-term effect	ICAR-Indian Agricultural Research Institute, New Delhi	Addition of K salts at lower rate stimulated C mineralization but a decline was noticed at higher concentrations	Chandra et al. (2002)
Effects of management practices on soil health (or fertility) indicators	Long-term effect	Indo-Gangetic plains	TC,TN, Av. P and K was either maintained or increased with time for FYM treatment but not in chemical treatments, but the system was limited by K	Bhandari et al. (2002)
		Sandy loam soil of Hawalbagh, Almora	Balanced fertilization-maintained productivity and increased SOC of soybean–wheat, but performance was contingent on the K input	Kundu et al. (2007)

(continued)

Table 8.2 (continued)

Study type		Study area/soil type	Main findings	Reference
		Cambisol of North-west India	SOC, MBC, and enzymatic activity improved by balanced application of nutrients with FYM	Kanchikerimath and Singh (2001)
		Inceptisol of West Bengal	SOC (HWC, POC, MBC, hydrolysable carbohydrates) and SON (PON, MBN) fractions in surface soil trend was: NPK + FYM > NPK > NP > N > control	Manna et al. (2005, 2006)
		Indo-Gangetic plains	Physical (BD, WHC), chemical (SOC, macro and micro-nutrients), and biochemical (DH, PHO, and FDHA) soil health indicators and system productivity were greater in organic versus conventional farming system	Sihi et al. (2012, 2017a)
		ICAR-Indian Agricultural Research Institute, New Delhi	Puddled rice with irrigation after 3 days of drainage and 25% N substitution by FYM and no-tillage, two irrigations, and domestic sewage sludge in wheat-maintained soil quality	Bhaduri et al. (2014)
			Non-puddling significantly enhanced DH, MBC, Av. N over puddling in rice and no-tillage increased soil biological indicators in wheat	Bhaduri et al. (2017)
	Short-term effect	ICAR-Indian Agricultural Research Institute, New Delhi	FYM application increased SOC in rice–wheat system and SOC was suggested to be used as a proxy for MBC	Banerjee et al. (2006)
			Organic manure improved DH, PHO, and FDHA, but their biannual application was not beneficial beyond three rice–wheat crop cycles	Gaind and Singh (2016)
			Higher thermal sensitivity (Q_{10}) of micro-aggregates than macro-aggregates indicated better preservation of WSA fractions in maize–wheat systems	Sandeep and Manjaiah (2014), Sandeep et al. (2016)
			Bed planting was better than conventional tillage for stability (measured by $\delta^{13}C$ signatures) of SOC pools in maize–wheat system	
			Maize biochar enhanced soil C, N and P and while wheat biochar improved soil K status. In contrast, rice biochar increased CO_2 loss	Purakayastha et al. (2015b)
		Vertisol of, Bhopal, Madhya Pradesh	Sustainability and productivity of the system could be maintained by returning residues in direct seeding of rice with conventional tillage	Mohanty et al. (2007)
		TypicUstochrept of CSSRI	Enzyme (protease, amidase, and deaminase) activities enhanced by organic amendment in salt-affected soil	Pathak and Rao (1998)
		Bahour, Puducherry	All measured soil (physical, chemical, microbial, and fauna) properties were improved in organic vs conventional farming fields	Padmavathy and Poyyamoli (2011)
		Experimental farm, Karnal	Soil physical properties, nutrient dynamics, and microbial activity were greater after bio-manure application, but pronounced effect was observed with vermicompost and FYM than pressmud application	Singh et al. (2007)

(continued)

Table 8.2 (continued)

Study type		Study area/soil type	Main findings	Reference
Greenhouse gas emission from soils	CH_4	Rice–wheat cropping system of Indo-Gangetic plains	CH4 production and emission showed positive correlation with organic manure addition as well as soil physicochemical properties (CEC, total N, Av. N, Av. K, clay, WSC, AnMC) Trade-off between improved yield and soil health vs GHG emissions (higher GWP) were observed as a result of organic amendment (FYM, GM). Intermittent wetting and drying of soil in rice was reported to have reduced CH4 emission potential	Bhatia et al. (2005), Mitra et al. (2002), Pathak et al. (2003)
	N_2O		N_2O emission interacted with temperature and moisture content (or WFPS) Lower N2O emissions were observed during submergence and substantially higher emission was observed during intermittent wetting and drying of soil in rice and after drainage of standing water. No-tillage increased N_2O emission in wheat crop Nitrification inhibitor application (DCD, nimin, thiosulphate, neem coated urea, SBT butanoate, and SBT-furoate) and LCC-based urea application were effective in reducing N_2O emission	Bhatia et al. (2010, 2012a), Kumar et al. (2000a, b), Oyeogbe et al. (2017), Majumdar (2002), Majumdar et al. (2002), Pathak (2009)
	Both CH_4 and N_2O		Replacement of existing conventional system CTW-TPR by ZTW-DSR (or NOCU-ZTW) + Rice straw residue-DSR (or SRI) along with GS-based N management were suggested to significantly reduce GWP, increase SOC and SON content without any significant yield reduction	Gupta et al. (2016), Jain et al. (2014), Nath et al. (2017)

SOC: Soil organic C, WSC: Water soluble C, HWC: Hot water soluble C, MBC: Microbial biomass C, MBN: Microbial biomass N, SON: Soil organic N, PON: Particulate organic N, qCO_2: Microbial metabolic quotient, TC: Total C, TN: Total N, Av. N: Available N, Av. P: Available P, Av. K: Available K, BD: Bulk density, WHC: Water holding capacity, DH: dehydrogenase, PHO: Phosphatase, FDHA: Fluorescein di-acetate hydrolysis, WSA: Water stable aggregates, CTW: Conventional Tillage Wheat, TPR: Transplanted puddled rice, ZTW: Zero tillage wheat, DSR: Direct seeded rice, NOCU: Neem oil coated urea, SRI: system of rice root intensification, GS: $\text{GreenSeeker}_{\text{TM}}$, FYM: Farmyard manure, GM: Green manure, GWP: Global warming potential, GHG: Greenhouse gas, CEC: Cation exchange capacity, WSC: Water soluble C, AnMC: C mineralized under anaerobic condition, DCD: Dicyandiamide, SBTbutanoate: S-benzylisothiouroniumbutanoate, and SBT-furoate: S-benzylisothiouroniumfuroate, LCC: Leaf color chart, WFPC: Water filled pore space

store or release energy, which, in turn, sustain the structure and function of largest diversity of soil organisms and associated physicochemical processes.

8.3.2 Soil Microorganisms: Primary Agent of Soil Biogeochemical Cycles

Soil microorganisms are a crucial component in the biogeochemical cycling of elements. It plays the dual role of being a major C and nutrient sink during immobilization and a source during mineralization. Thus, turnover of SOM pool and associated elements is contingent upon the rate at which soil microorganisms consume it. Enzymes produced by soil microorganisms (e.g., bacteria and fungi) act as a catalyst that process C and nutrients into a biologically useful form (Sinsabaugh et al. 1991) and facilitate nutrient cycling in natural and managed ecosystems (Dick 1997). Thus, understanding complex interactions between soil microorganisms and its physical environment need careful attention due to their essential role in ecosystem functioning.

8.3.3 Soil Macro-Organisms: Secondary Agent of Soil Biogeochemical Cycles

Soil invertebrates such as arthropods, coleoptra, heteroptera, and araneae also indirectly influence SOM decomposition by altering the composition and activity of primary decomposers (e.g., bacteria and fungi) by either feeding on or from

Table 8.3 Soil biogeochemical cycling in response to xenobiotic application in different soils

Study type	Study area/soil type	Main findings	Reference
Impact of pesticide residues on microbial and biochemical parameters	Tea garden soils of Darjeeling-Dooars, WB	Organochlorine and organophosphate pesticide residues have strong negative impact on the microbiological (MBC, BSR, SIR) and biochemical (FDHA, BG) parameters of soil quality	Bishnu et al. (2008, 2009)
	ICAR-Directorate of groundnut research, Junagadh, Gujarat	Application of post-emergence herbicides (imazethapyr and quizalofop-p-ethyl) affected activities of different soil enzymes (FDHA, DH, Acid, and alkaline PHO) and microbial biomass (MBC) differently. However, both herbicides exerted transient toxic effects on soil microbiological parameters, especially at recommended and double recommended dose	Saha et al. (2016)
	Silt loam soil of the Kashmir, Himalaya	Different soil enzyme activities respond differently to mancozeb fungicide application: activities of urease and asparaginase were inhibited but DH, protease and amidase activities were stimulated. PHO showed highly variable response to different rates of pesticide application	Rasool and Rashi (2010)
	Tropical soils of WB and Karnataka	MBC and soil clay content inversely related to half-lives of metalaxyl, but SOC did not show any correlation with metalaxyl persistence	Sukul and Spiteller (2001)
	ICAR-National Rice Research Institute, Cuttack	Pretilachlor at recommended dose did not significantly alter microbial properties (number of bacteria, actinomycetes, fungi, nitrogen fixers, and MBC) at recommended dose, rather stimulated certain microbial characteristics. However, application at higher rate significantly reduced these microbial parameters	Sahoo et al. (2016b)
		Pretilachlor caused short-term transitory changes in soil enzyme activities (BG, cellulase, invertase, and FDHA); however, negative impact on soil enzymes were observed at very high dose of pretilachlor application	Sahoo et al. (2016a)
		2,4-D was inhibitory to microorganisms at higher rate. Commercial formulation of HCH increased MBC distinctly in treated vs untreated soil under both flooded and non-flooded conditions	Rath et al. (1998)
Persistence/dissipation of pesticides and its interaction with soil properties	Hiriyur, Mandya and Mudigere, Karnataka	Higher disappearance of butachlor herbicide in non-sterile soils than in sterile soils under both field capacity and submerged conditions indicated potentiality of microbial degradation. Disappearance was more rapid under field capacity than under submergence even with the addition of farmyard manure and paddy straw as organic amendments	Prakash et al. (2000)
	ICAR-Indian Agricultural Research Institute, New Delhi	Rapid degradation of pendimethalin was observed in mineral salt as mediated by the supply of sole C source in mineral solution. However, degradation of sterilized culture and uninoculated control was negligible	Rai et al. (1999), Barua et al. (1990)
		Inhibition of microbial activity at high rate resulted into slow dissipation from soil. Soil type (i.e., magnitude of adsorption and desorption of flufenacet in different soils) affected rate of dissipation. Negligible difference in dissipation rate was observed between soils at field capacity and submerged condition moisture regimes	Gupta and Gajbhiye (2002)

(continued)

Table 8.3 (continued)

Study type	Study area/soil type	Main findings	Reference
	ICAR-National Rice Research Institute, Cuttack	Flooded conditions favored degradation of organophosphorus insecticides than under non-flooded conditions. Nitro group reduction was the major pathway of degradation for all the three insecticides after the first addition and the rate of hydrolysis increased after each successive addition	Adhya et al. (1987)
		Pronounced accumulation of parathion was observed in nonsaline soil vs saline soil. Inhibition of nitro group reduction in saline soils was related to low microbial activities as evidences by decreased DH activity and slow Fe reduction	Reddy and Sethunathan (1985)
	Alluvial soil of WB	Main metabolic pathways of microbial degradation of oxadiazon by soil fungus *Fusariumsolani* involved oxadiazoline ring cleavage without dechlorination reaction	Chakraborty et al. (1995)
	Different Agro-Climatic regions of India	SOC dominantly control adsorption–desorption behavior of different triazoles fungicides. Soil clay and silt content were also affected the Freundlich adsorption constants of fungicides	Singh (2002)
	Alluvial soil (TypicUdifluvent) and coastal saline soil (Typic endoaquept)	In general, pencycuron degraded faster in coastal saline soil and in soil amended with decomposed cow manure at 60% of maximum WHC. Half-life values depended on the initial concentrations of pencycuron	Pal et al. (2005)
	Alluvial soil of WB and Acid sulfate saline (Kari and Pokkali) soil	Degradation of the vinclozolin fungicide was more rapid in all the soils irrespective of flooding conditions. However, under non-flooded conditions, soil acidity and salinity significantly affected the persistence of the fungicide	Banerjee et al. (1999)

DH: Dehydrogenase, FDHA: Fluorescein di-acetate hydrolysis, BSR: Basal respiration, MBC: Microbial biomass C, HCH: hexachlorocyclohexane, WHC: Water holding capacity

their communities of plant detritus and fecal deposition, thereby increasing surface area for microbial attack. Additionally, soil macro-fauna (especially earthworm and termite) are also known to influence soil nutrient status. For instance, burrowing activities of earthworms generally create stable and continuous soil macrospores that act as a conduit for water and improve sol physical structure (by improving infiltration) as well as facilitate gas exchange and favor aerobic microbial activities. Earthworm cast (i.e., mucus secreted by earthworms) is considered as nutrient-rich (e.g., N, P, K, and Ca), which further increase the biologically available nutrients. Likewise, termites are also known to influence soil nutrient status by their foraging activities, which tend to concentrate nutrients and organic matter inside termite-worked mounds, and thereby affect nutrient availability of soil microbes and plants. Forest degradation and agricultural intensification (mostly due to mechanical tillage and application of agricultural chemicals) also negatively impacted the diversity of these beneficial soil fauna and further affected soil fertility in tropical soils of India (reviewed by Beare et al. 1997; Bhadauria and Saxena 2010; Black and Okwakol 1997; Kale and Karmegam 2010).

8.3.4 Major Biogeochemical Cycles in Soil

(i) ***Carbon cycle***

Carbon, the basic building block of life, can be readily obtained from the atmosphere in the form of carbon dioxide (CO_2), but it needs to be transformed into a biologically usable form before its incorporation into living organisms. This carbon fixation mechanism, also known as photosynthesis is mediated by the activity of microorganisms like cyanobacteria. Fate of these organic compounds represents a major, but dynamic reservoir of C in soil. Depolymerization of polymeric soil organic carbon (SOC) to dissolved organic carbon (DOC) is mediated by hydrolytic (or extracellular) enzymes secreted by soil microorganisms. DOC is then assimilated by microorganisms, part of which is used to build their tissue (i.e., microbial biomass C, MBC) and the rest is return to the atmosphere as a by-product of microbial metabolism, also known as respiration (CO_2) (Fig. 8.3). Changes in the rate of these processes as a result of climate or land-use change potentially impact the size of soil C pool as well as atmospheric concentrations of greenhouse gases

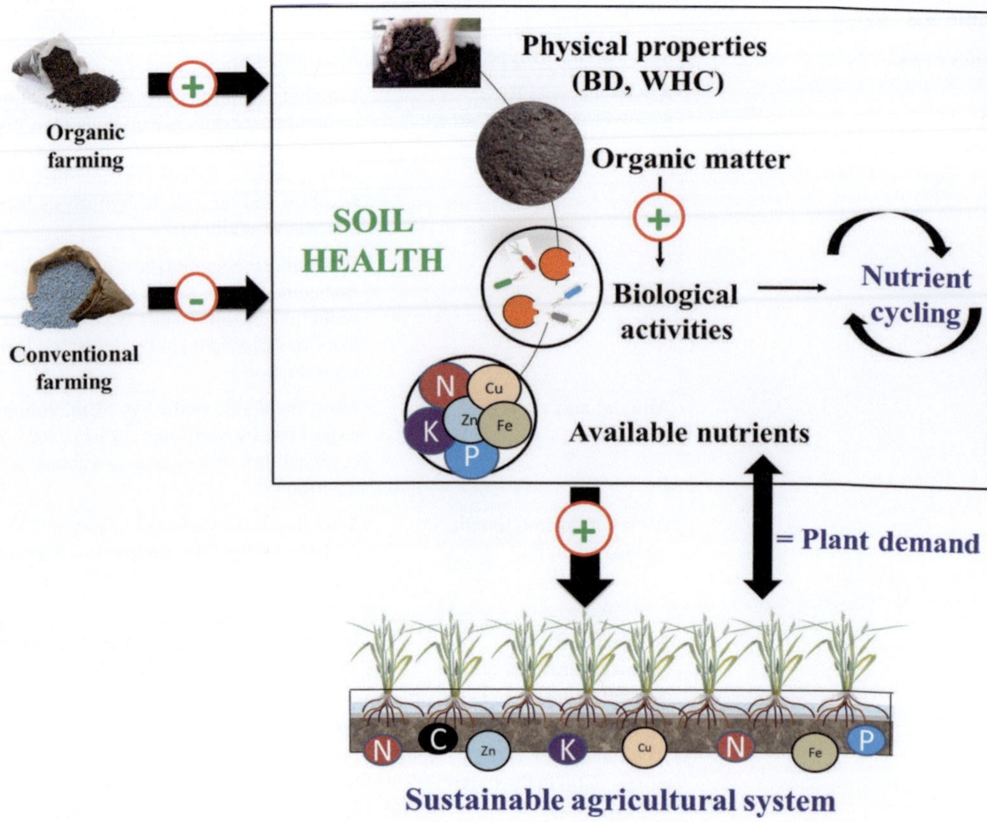

Fig. 8.3 Interdependencies of soil health indicators and environmental sustainability. *Source* Sihi et al. (2017a)

(GHGs) such as CO_2 (under aerobic decomposition) and methane (CH_4, under anaerobic decomposition).

(ii) ***Nitrogen Cycle***

Nitrogen is abundant in atmosphere, in the form of dinitrogen gas (N_2). Processing of this biologically unusable form by either chemical or biological (e.g., biological nitrogen fixation) converts gaseous N_2 into a biologically usable form. Mineralization of organic nitrogen compounds in soil (SON) produce ammonia by so bacteria ammonification. This step is followed by nitrifying microorganisms that oxidize ammonia to nitrite and then to nitrate (nitrous oxide can be a by-product in this step). Denitrifying bacteria reduce nitrate to nitrous oxide (N_2O), a potent GHGs under incomplete reduction process and N_2 after complete reduction. Both N_2O and N_2, along with a series of intermediate gaseous nitrogen oxide (NOx) products escape to atmosphere. Alteration of nitrogen cycle (e.g., application of artificial nitrogen fertilizers in agricultural lands) can greatly influence nitrogen availability for primary production as well as its loss to downstream water bodies.

(iii) ***Phosphorus Cycle***

Unlike C and N, movements of P are not mediated much through atmosphere, rather sediment chemistry plays a significant role in biogeochemical cycling of P. Only a small fraction of total soil P is generally available to plant uptake as either most of the P are bound to mineral matrix (Fe-P, Al-P, and Ca-P) or excess P slowly lost with runoff water. Thus, P management is important to maintain productivity, more so in P-poor soil of India, where inorganic mineral-bound P contributes to 54–84% of total P (Sanyal et al. 2015). P solubilizing microorganisms are effective in dissolving insoluble inorganic P by releasing organic acids (Gyaneshwar et al. 2002; Munda et al. 2016). Phosphatase enzymes also play a major role in cycling of P by converting organophosphorus compounds to plant-available form (orthophosphate, PO_4^{-3}).

8.4 Soil Biogeochemical Cycles: Alteration by Anthropogenic Perturbations

8.4.1 Soil Biogeochemical Cycles as a Function of Land-Use Types

Case studies that evaluated the influence of land-use category on different components of soil biochemical cycles are listed in Table 8.1. An understanding of the dynamics of soil C stock (or C density), as influenced by different land-use categories, has been a major research focus for a long time to lock C in soils at a level critical for mitigating climate change effects. Studies that investigated the processes of aggregations and C sequestration indicated that C stored in micro-aggregates was generally found to be less vulnerable to change in response to land-use changes (Saha et al. 2014). Soil profile stocks for native forest and grass lands were always greater than deforested, cultivated, and degraded (eroded) lands (Saha et al. 2011, 2014). Intensive cultivation in agricultural lands without proper management in tropical and subtropical systems was the major cause for rapid depletion of SOC as compared to the native vegetation, and the effect was influenced by other physicochemical properties like clay content, parent materials, quality of available substrates, and nutrient status, etc. (Table 8.1).

Other than C, nutrient index and microbial properties in terms of their pool size (Kaur et al. 2000), activities (Dinesh et al. 2004a), abundance (Dinesh et al. 2004b), and diversity (Radhakrishnan and Varadharajan 2016) were also greater in undisturbed sites as compared to disturbed and cultivated systems. Importantly, factors that affect microbial population (e.g., grazing, Bagchi et al. 2017; Srivastava 1992) influenced net-C balance of the system, which were again dependent on environmental conditions (e.g., rainfall and soil moisture).Plantation of perennial trees along with agricultural crops (i.e., agro-forestry system) was effective in restoring fertility status of soil (Kaur et al. 2000; Mahmoud et al. 2016).

Interestingly, marginal (or barren) lands with lower fertility status have been reported to evolve with their own survival mechanism to resist from perturbations (e.g., wind erosion) either by forming cyanobacterial crusts in arid lands (Tirkey and Adhikary 2005) and mycorrhizal mats in salt-affected soils (Yadav et al. 2017), which, in turn, help to sustain soil fertility by either fixing C and N (for cyanobacteria by photosynthetic activity and nitrogen fixation) or acquiring P in P-deficient tropical soils of India (for mycorrhizal fungi). Within this context, mangrove ecosystems along the coastal areas of India acted as a significant sink of P by immobilizing P in sediments, the capacity of that was also dependent on the amount of organic matter, sediment C:P ratio, and the type of clay minerals present (Resmi et al. 2016; Singh et al. 2015).

8.4.2 Soil Biogeochemical Cycles as a Function of Agri-Management

Case studies that evaluated the influence of agricultural management practices on soil biochemical cycles are listed in Table 8.2. Monitoring of soil fertility indicators in both long-term (15–30 yrs) and short-term (2–3 cropping sequence) field trials was in practice for decades to evaluate the impact of management practices on the fertility of agricultural soils. Overall, organic amendments and balanced fertilization (i.e., integrated nutrient management, NPK along with organic manures) were found to be effective in either maintaining or increasing the soil C sequestration potential in longer term (Kaur et al. 2008; Rudrappa et al. 2006), but the stability of SOC varies based on the choice of organic manure (e.g., FYM vs. vermicomposting, Chaudhary et al. 2017) as well as the cropping sequence in question, with more SOC storage in rice–wheat versus maize–wheat cropping system (Kukal et al. 2016), and more storage under double rice crop (Mandal et al. 2007) due to prolonged anaerobic condition (which slows down aerobic decomposition) and enhanced subsurface accumulation of C (resulting from interaction with iron-oxides) leached from top Ap (or puddled) layer. Long-term practices of zero-tillage were also reported to significantly increase POC and water stable soil aggregates in rice–wheat cropping systems of Indo-Gangetic Plains (Kumari et al. 2011). Additionally, bed planting system in maize–wheat systems were also showed to increase thermal stability of SOC than conventional tillage practices in a short-term study (Sandeep et al. 2016).

Along with SOC dynamics, other physicochemical (Av. N, P, and K, particulate organic nitrogen PON) soil fertility indicators also indicated better performance of manured soil and sols receiving minimum or no-tillage, and system productivity were constrained by other limited factors (e.g., K availability, Bhandari et al. 2002; Kundu et al. 2007). Biological (e.g., microbial biomass) and biochemical (e.g., soil enzymes) indicators of soil health further proved the importance of agricultural management practices with respect to residue application, crop diversification, inclusion of legumes as cover crop, (Kanchikerimath and Singh 2001; Manna et al. 2005, 2006; Sihi et al. 2012) and reduced tillage (Bhaduri et al. 2014, 2017) for sustaining long-term productivity of agricultural soil. To evaluate the effect of management practices on a short-term basis, use of microbial and biochemical indicators were enhanced which were generally effective in showing any early warming of perturbations (Gaind and Singh 2016; Mohanty et al. 2007; Banerjee et al. 2006). Recently, applications of biochar in agricultural soils were encouraged to increase stabilization of SOC and other soil health indicators, which seemed to be a promising strategy, but again the effect was dependent on the feedstock used for biochar, pyrolysis temperature employed, initial SOC status, and other soil properties (Purakayastha et al. 2015b; Dari et al. 2016).

Bhatia et al. (2005) reported that improved soil health in puddled rice–wheat system of Indo-Gangetic plains may also come at the expense of increased greenhouse gas emission, particularly CH_4. Organic amendments could stimulate activity of methanogens by providing readily usable organic C substrates under prolonged anaerobic condition which is the ideal environment for CH_4 production. Thus, restoring SOM may counteract with soil C sequestration potential by increasing overall global warming potential (GWP). Adoption of direct-seeded rice (DSR) and system of rice root intensification (SRI) proved as an alternative cropping practice to reduce CH_4 emission with additional benefit of optimizing productivity by altering irrigation management and thereby reducing usage of limited water resource (Dari et al. 2017). However, intermittent wetting and drying seemed to enhance N_2O emission rates by favoring N_2O production as a by-product of the nitrification process (Pathak et al. 2001, 2003). Complete submergence on the other hand; again decrease N_2O flux due to its reduction of N_2O to N_2 under very reduced conditions (Kumar et al. 2000a, b). Dynamics of N_2O further gets complicated under different temperature and moisture regimes of soils with peak N_2O emission were obtained at greater temperature at the field capacity than at the submerged condition (Pathak 2009). Application of nitrification inhibitors (Bhatia et al. 2010) and demand driven nitrogen fertilizer addition have been claimed to be effective in reducing N_2O emission from agricultural soils (Oyeogbe et al. 2017).

Understanding these complex interactions between different components of soil biogeochemistry and environmental factors are prerequisites in developing regional and national budgets of nutrients and GHGs emissions. This could be used as a tool to identify hotspots and hot moments of excess nutrient load, mining, and loss, which would increase our ability to predict environmental and socio-economic consequences of current management practices (Gupta et al. 2009; Pathak et al. 2010). This way, environmental fate of applied chemicals in agricultural soils can be assessed such that sustainability of agricultural production is maintained.

Xenobiotic compounds (e.g., insecticides and herbicides) are another class of chemicals intensively applied in conventional agricultural practices to kill unwanted pests (insects, weeds) and control diseases and maintain crop productivity. Indiscriminate use of these xenobiotics raised concerns as most of these pesticides contain organochlorine and organophosphorus compounds which could escape microbial degradation and residues may accumulate in soil and hamper biogeochemical cycling of nutrients. Thus, identifying the environmental fate of these compounds and their potential toxic effect on soil microbial properties has drawn attention of many environmentalists.

Microbial degradation of pesticide residues is the primary removal pathways from soil (see Chowdhury et al. 2008 for more information on different degradation pathways). Persistence or dissipation of pesticides in soil could also be dependent on soil properties such as adsorption or desorption mechanisms of soils (Gupta and Gajbhiye 2002), flooding conditions (Adhya et al. 1987), WHC (Pal et al. 2005), soil salinity (Banerjee et al. 1999; Reddy and Sethunathan 1985), and SOC (Singh 2002) and clay content (Sukul and Spiteller 2001) (Table 8.3). Effect of xenobiotics on microbial and biochemical parameters showed mixed results, with some reported strong negative effects (Bishnu et al. 2008, 2009; Chowdhury et al. Rath et al. 1998), while others reported either a stimulatory effect (Sahoo et al. 2016b) or a short transitory toxic effect (Saha et al. 2016; Sahoo et al. 2016a) when applied at low rates. However, applications at very high rates (several magnitudes higher than recommended dose) were detrimental to microbial and biochemical properties (Sahoo et al. 2016a, b).

8.5 Effective Microorganisms for Sustainable Soil Management in India

Higa and Parr (1994) first discovered the usefulness of effective microorganisms for sustainable soil and environmental management. Effective microorganisms are a group of beneficial microorganisms generally coexist in the natural environment and play an instrumental role in improving the soil and environmental quality. The consortium could be consisting of both aerobic and anaerobic microorganisms including lactic acid bacteria, yeasts, actinomycetes, fermenting fungi, as well as photosynthetic bacteria (Diver 2001).

Recently, the application of effective microorganisms to increase the productivity of the unfertile soils is a common practice in Indian agricultural systems. For example, Packialakshmi and Yasotha (2014) reported that effective microorganism treatment increased the soil microbial population, which in turn, increased the productivity of agricultural soil and the yield of agricultural crops under cultivation. The application of efficient microorganism-based compost in Indian horticultural farms under calendula and marigold cultivation was also known to improve soil health (Sharma et al. 2017). Additionally, Pushpa et al (2016) demonstrated an eco-friendly solution of solid waste disposal by composting solid organic waste with activated effective microorganisms, where the effective microorganisms-based composts were proved to be of better quality than the conventional composts.

In addition to agricultural soils, application of the efficient microorganisms-based formulations was also found to be fruitful for the restoration of degraded soils in the forest soils of the Eastern Ghats in India (Ramachandran and Radhapriya 2016). Thus, use of effective microorganisms for the sustainable soil and environmental management has been proved to be a viable option in India (Kumar and Gopal 2015). Hence, effective microorganisms can trigger the formation of the so-called Microbisols, a soil group proposed for the Indian system, which is supposed to be rich in soil microbial biomass and promote stabilization of soil organic matter through enhanced organo–mineral interaction (Mishra 2016).

8.6 Modeling Studies in Soil Biogeochemistry

Soil biogeochemistry modeling studies in India have been mostly centered on agroecosystems using either simple excel-based models that calculate the input–output budget of a system or complicated crop growth simulation models. On a regional scale, rice–wheat cropping systems of Indo-Gangetic Plains were major focus to simulate overall performance of different resource conservation technologies (RCT) with respect to productivity, econometric analysis, biocide residues, nitrogen balance, and GHGs emission. To that end, InfoRCT (Information on Use of Resource-Conserving Technologies, Pathak et al. 2011, 2013) and Info Nitro (Information on Nitrogen Management Technologies in Rice, Pathak et al. 2010) models have been developed by integrating biophysical, agronomic, and socio-economic information that were aimed to establish input–output budget related to water, fertilizer, labor, and biocide use, and their associated environmental and socio-economic impacts under different cropping systems, soil conditions, and climate. InfoRCT model performance under farmers' participatory mode was generally able to reproduce expected dynamics of GHGs emission under different tillage and water management practices (Pathak et al. 2011, 2013). Further, simulated yield trend and plateauing of rice–wheat system productivity by InfoRCT (Saharawat et al. 2012) were in agreement with what have been observed in this region since last few decades. Thus, InfoRCT served as an excellent decision support system (DSS) that allowed farmers to adopt a certain RCT that would be beneficial in terms of its environmental impact as well as net income.

Complex computer simulation models like DNDC (DeNitrification-DeComposition, Pathak et al. 2005, 2006) and infoCrop (Bhatia et al. 2012b) were also applied at both regional and national scales to determine C and N biogeochemistry in agroecosystems. Both of these models were able to reproduce the trade-off mechanisms for CO_2, CH_4, and N_2O emission (see Sect. 8.4.2, Pathak et al. 2005; Bhatia et al. 2012b). Additionally, Pathak et al. (2006) were able to identify hot spots of N loss and uptake in the rice–wheat systems of the Indo-Gangetic Plains, with greater yields and N uptake in north-western part (e.g., Punjab and Haryana) than eastern transects (e.g., West Bengal, Uttar Pradesh, and Bihar) due to higher amount of N use and favorable climatic conditions in the former. Efforts to model soil P biogeochemistry in Indian subcontinent focused on its adsorption–desorption dynamics as a function of soil pH, time (De and Datta 2007), soil solution P concentration (Silberbush and Barber 1983), and P-solubilization by organic acid (Bhattacharyya and Datta 2005).

Although output of all of the above-mentioned models was in general agreement with observation for current climatic conditions, these models do not explicitly represent microbial and enzymatic dynamics (see Sihi et al. 2016, 2017b) which are intimately related to soil organic matter decomposition and proved to better represent soil organic matter dynamics worldwide.This should be the future research thrust for modeling soil biochemistry in Indian subcontinent under changing climate. Higher the nutrient content of soil, higher will be the rate of reaction and more so in nutrient-limited warmer tropical and subtropical soils (Sihi et al. 2016, 2017b).

8.7 Conclusions

Soil biogeochemistry study involves complex processes occurring at the nexus between the physical, chemical, and biological component of type-specific soil. A better understanding of soil biogeochemical cycling is needed as it controls ecosystem functioning of both natural and managed (i.e., agricultural) ecosystems, productivity, and sustainability. Soil organic matter and soil microorganisms are key components of soil biogeochemistry that determine the overall flow of C and nutrients release from soil. Maintenance of soil health (or quality) is essential to reduce negative environmental consequences of anthropogenic perturbations.

References

Adhya TK, Wahid PA, Sethunathan N (1987) Persistence and biodegradation of selected organophosphorus insecticides in flooded versus non-flooded soils. Biol Fertil Soils 5(1):36–40

Bagchi S, Roy S, Maitra A, Sran RS (2017) Herbivores suppress soil microbes to influence carbon sequestration in the grazing ecosystem of the Trans-Himalaya. Agr Ecosyst Environ 239:199–206

Banerjee B, Aggarwal PK, Pathak H, Singh AK, Chaudhary A (2006) Dynamics of organic carbon and microbial biomass in alluvial soil with tillage and amendments in rice-wheat systems. Environ Monit Assess 119(1):173–189

Banerjee A, Padhi S, Adhya TK (1999) Persistence and biodegradation of vinclozolin in tropical rice soils. Pestic Sci 55(12):1177–1181

Barua AS, Saha J, Chaudhuri S, Chowdhury A, Adityachaudhury N (1990) Degradation of pendimethalin by soil fungi. Pestic Sci 29 (4):419–425

Beare MH, Reddy MV, Tian G, Srivastava SC (1997) Agricultural intensification, soil biodiversity and agroecosystem function in the tropics: the role of decomposer biota. Appl Soil Ecol 6(1):87–108

Bhaduri D, Purakayastha TJ, Patra AK, Chakraborty D (2014) Evaluating soil quality under a long-term integrated tillage–water–nutrient experiment with intensive rice–wheat rotation in a semi-arid Inceptisol India. Environ Monitor Assess 186(4):2535–2547

Bhaduri D, Purakayastha TJ, Patra AK, Singh M, Wilson BR (2017) Biological indicators of soil quality in a long-term rice–wheat system on the Indo-Gangetic plain: combined effect of tillage–water–nutrient management. Environ Earth Sci 76(5):202

Bhadauria T, Saxena KG (2010) Role of earthworms in soil fertility maintenance through the production of biogenic structures. Appl Environ Soil Sci 2010:7

Bhandari AL, Ladha JK, Pathak H, Padre AT, Dawe D, Gupta RK (2002) Yield and soil nutrient changes in a long-term rice-wheat rotation in India. Soil Sci Soc Am J 66(1):162–170

Bhatia A, Pathak H, Jain N, Singh PK, Singh AK (2005) Global warming potential of manure amended soils under rice–wheat system in the Indo-Gangetic plains. Atmos Environ 39(37):6976–6984

Bhatia A, Pathak H, Jain N, Singh PK, Tomer R (2012a) Greenhouse gas mitigation in rice–wheat system with leaf color chart-based urea application. Environ Monit Assess 184(5):3095–3107

Bhatia A, Aggarwal PK, Jain N, Pathak H (2012b) Greenhouse gas emission from rice- and wheat-growing areas in India: spatial analysis and upscaling. Greenhouse Gases: Sci Technol 2(2):115–125

Bhatia A, Sasmal S, Jain N, Pathak H, Kumar R, Singh A (2010) Mitigating nitrous oxide emission from soil under conventional and no-tillage in wheat using nitrification inhibitors. Agr Ecosyst Environ 136(3–4):247–253

Bhattacharjya S, Bhaduri D, Chauhan S, Chandra R, Raverkar KP, Pareek N (2017) Comparative evaluation of three contrasting land use systems for soil carbon, microbial and biochemical indicators in North-Western Himalaya. Ecol Eng 103(Part A):21–30

Bhattacharyya P, Datta SC (2005) Modification of phosphate uptake model considering changing buffering capacity of soil and release threshold limit of phosphorus in soil due to rhizosphere effect. Commun Soil Sci Plant Anal 35(13–14):1773–1792

Bishnu A, Saha T, Mazumdar D, Chakrabarti K, Chakraborty A (2008) Assessment of the impact of pesticide residues on microbiological and biochemical parameters of tea garden soils in India. J Environ Sci Health, Part B 43(8):723–731

Bishnu A, Saha T, Ghosh PB, Mazumdar D, Chakraborty A, Chakrabarti K (2009) Effect of pesticide residues on microbiological and biochemical soil indicators in tea gardens of Darjeeling Hills India. World J Agric Sci 5(6):690–697

Black HIJ and Okwakol MJN (1997) Agricultural intensification, soil biodiversity and agroecosystem function in the tropics: the role of termites. Appl Soil Ecol 6(1):37–53

Brar MS, Bijay-Singh Bansal SK, Srinivasarao C (2011) Role of potassium nutrition in nitrogen use efficiency in cereals. Electr Int Fertil Corresp 29:20–27

Chakraborty SK, Chowdhury A, Bhattacharyya A, Ghosh S, Pan S, Waters R, Adityachaudhury N (1995) Microbial degradation of oxadiazon by soil fungus Fusariumsolani. J Agric Food Chem 43 (11):2964–2969

Chandra S, Joshi HC, Pathak H, Jain MC, Kalra N (2002) Effect of potassium salts and distillery effluent on carbon mineralization in soil. Biores Technol 83(3):255–257

Chase P, Singh OP (2014) Soil nutrients and fertility in three traditional land use systems of Khonoma, Nagaland, India. Resour Environ 4 (4):181–189

Chaudhary S, Dheri GS, Brar BS (2017) Long-term effects of NPK fertilizers and organic manures on carbon stabilization and management index under rice-wheat cropping system. Soil Tillage Res 166:59–66

Chowdhury A, Pradhan S, Saha M, Sanyal N (2008) Impact of pesticides on soil microbiological parameters and possible bioremediation strategies. Indian J Microbiol 48(1):114–127

Dari B, Sihi D, Bal SK, Kunwar S (2017) Performance of direct-seeded rice under various dates of sowing and irrigation regimes in semi-arid region of India. Paddy Water Environ 15(2):395–401

Dari B, Nair VD, Harris WG, Nair PKR, SollenbergerL Mylavarapu R (2016) Relative influence of soil- vs. biocharproperties on soil phosphorus retention. Geoderma 280:82–87

De N, Datta SC (2007) Relationship between phosphorus sorption and soil acidity as affected by bicarbonate and silicate ions. Commun Soil Sci Plant Anal 38(5–6):679–694

Dick RP (1997) Soil enzyme activities as integrative indicators of soil health. In: Pankhurst CE, Doube BM, Gupta VVSR (eds) Biological indicators of soil health. CABI, Wellingford, pp 121–156

Dinesh R, Chaudhuri SG, GaneshamurthyANandPramanik SC (2004a) Biochemical properties of soils of undisturbed and disturbed mangrove forests of South Andaman (India). Wetlands Ecol Manage 12(5):309–320

Dinesh R, Ghoshal CS, Sheeja TE (2004b) Soil biochemical and microbial indices in wet tropical forests: Effects of deforestation and cultivation. J Plant Nutr Soil Sci 167(1):24–32

Diver S (2001). Nature farming and effective microorganisms. Retrieved from Rhizosphere II: publications, resource lists and web links from Steve Diver. http://ncatark.uark.edu/~steved/Nature-Farm-EM.html

Gaind S, Singh YV (2016) Short-term impact of organic fertilization and seasonal variations on enzymes and microbial indices under rice–wheat rotation. CLEAN—Soil, Air, Water 44(10):1396–1404

Gyaneshwar P, Naresh Kumar G, Parekh LJ, Poole PS (2002) Role of soil microorganisms in improving P nutrition of plants. Plant Soil 245(1):83–93

Gorham E (1991) Biogeochemistry: its origins and development. Biogeochemistry 13(3):199–239

Gupta DK, Bhatia A, Kumar A, Das TK, Jain N, Tomer R, Sandeep KM, Fagodiya RK, Dubey R, Pathak H (2016) Mitigation of greenhouse gas emission from rice–wheat system of the Indo-Gangetic plains: through tillage, irrigation and fertilizer management. Agr Ecosyst Environ 230:1–9

Gupta PK, Gupta V, Sharma C, Das SN, Purkait N, Adhya TK, Pathak H, Ramesh R, Baruah KK, Venkatratnam L, Singh G, Lyer CSP (2009) Development of methane emission factors for Indian paddy fields and estimation of national methane budget. Chemosphere 74(4):590–598

Gupta S, Gajbhiye VT (2002) Effect of concentration, moisture and soil type on the dissipation of flufenacet from soil. Chemosphere 47 (9):901–906

Higa T, Parr JF (1994) Beneficial and effective microorganisms for a sustainable agriculture and environment (vol 1). Atami, International Nature Farming Research Center

Jain N, Dubey R, Dubey DS, Singh J, Khanna M, Pathak H, Bhatia A (2014) Mitigation of greenhouse gas emission with system of rice intensification in the Indo-Gangetic Plains. Paddy Water Environ 12 (3):355–363

Jenny H (1941) Factors of soil formation. McGraw-Hill, New York

Jones DL, Cross P, Withers PJ, DeLuca TH, Robinson DA, Quilliam RS, Harris IM, Chadwick DR, Edwards-Jones G (2013) Review: nutrient stripping: the global disparity between food security and soil nutrient stocks. J Appl Ecol 50(4):851–862

Kale RD, Karmegam N (2010) The role of earthworms in tropics with emphasis on Indian ecosystems. Appl Environ Soil Sci 2010:16

Kanchikerimath M, Singh D (2001) Soil organic matter and biological properties after 26 years of maize–wheat–cowpea cropping as affected by manure and fertilization in a Cambisol in semiarid region of India. Agr Ecosyst Environ 86(2):155–162

Kaur T, Brar BS, Dhillon NS (2008) Soil organic matter dynamics as affected by long-term use of organic and inorganic fertilizers under maize–wheat cropping system. Nutr Cycl Agroecosyst 81(1):59–69

Kaur B, Gupta SR, Singh G (2000) Soil carbon, microbial activity and nitrogen availability in agroforestry systems on moderately alkaline soils in northern India. Appl Soil Ecol 15(3):283–294

Kukal SS, Saha D, Sharma P, Sharma BD (2016) Profile distribution of carbon fractions under long-term rice-wheat and maize-wheat production in Alfisols and Inceptisols of Northwest India. Land Degrad Dev 27(4):1205–1214

Kumar BL, Gopal DS (2015) Effective role of indigenous microorganisms for sustainable environment. 3 Biotech 5(6):867–876

Kumar M (2011) North East India: soil and water management imperatives for food security in a changing climate. Curr Sci 101:1119

Kumar M, Singh RP, Panigrahy S, Raghubanshi AS (2014) Carbon density and accumulation in agroecosystem of Indo-Gangetic Plains and Vindhyan highlands, India. Environ Monitor Assess 186 (8):4971–4985

Kumar U, Jain MC, Kumar S, Pathak H, Majumdar D (2000a) Role of nitrification inhibitors on nitrous oxide emissions in a fertilized alluvial clay loam under different moisture regimes. Curr Sci 79 (2):224–227

Kumar U, Jain MC, Pathak H, Kumar S, Majumdar D (2000b) Nitrous oxide emission from different fertilizers and its mitigation by nitrification inhibitors in irrigated rice. Biol Fertil Soils 32(6): 474–478

Kumari, MD, Chakraborty D, Gathala MK, Pathak H, Dwivedi BS, Tomar RK, Garg RN, Singh R, Ladha JK (2011) Soil aggregation and associated organic carbon fractions as affected by tillage in a rice–wheat rotation in North India all rights reserved. No part of this periodical may be reproduced or transmitted in any form or by any means, electronic or mechanical, including photocopying, recording, or any information storage and retrieval system, without permission in writing from the publisher. Permission for printing and for reprinting the material contained herein has been obtained by the publisher. Soil Sci Soc Am J 75(2):560–567

Kundu S, Bhattacharyya R, Prakash V, Gupta HS, Pathak H, Ladha JK (2007) Long-term yield trend and sustainability of rainfed soybean–wheat system through farmyard manure application in a sandy loam soil of the Indian Himalayas. Biol Fertil Soils 43(3):271–280

Lal R (2004) Soil carbon sequestration impacts on global climate change and food security. Science 304:1623–1627

Lo Seen D, Ramesh BR, Nair KM, Martin M, Arrouays D, Bourgeon G (2010) Soil carbon stocks, deforestation and land-cover changes in the Western Ghats biodiversity hotspot (India). Glob Change Biol 16(6):1777–1792

Majumdar D (2002) Suppression of nitrification and N_2O emission by karanjin—a nitrification inhibitor prepared from karanja (Pongamiaglabra Vent.). Chemosphere 47(8):845–850

Majumdar D, Pathak H, Kumar S, Jain MC (2002) Nitrous oxide emission from a sandy loam Inceptisol under irrigated wheat in India as influenced by different nitrification inhibitors. Agr Ecosyst Environ 91(1–3):283–293

Mahmoud A, Singh SD, Muralikrishna KS, Patha H, Saha ND (2016) Soil microbial responses as influenced by Jatropha plantation under rainfed condition in north-west India. Agrofor Syst, pp 1–12

Mandal B, Majumder B, Bandyopadhyay PK, Hazra GC, Gangopadhyay A, Samantaray RN, Mishra AK, Chaudhury J, Saha MN, Kundu S (2007) The potential of cropping systems and soil amendments for carbon sequestration in soils under long-term experiments in subtropical India. Glob Change Biol 13(2):357–369

Manna MC, Swarup A, Wanjari RH, Ravankar HN, Mishra B, Saha MN, Singh YV, Sahi DK, Sarap PA (2005) Long-term effect of fertilizer and manure application on soil organic carbon storage, soil quality and yield sustainability under sub-humid and semi-arid tropical India. Field Crops Res 93(2–3):264–280

Manna MC, Swarup A, Wanjari RH, Singh YV, Ghosh PK, Singh KN, Tripathi AK, Saha MN (2006) Soil organic matter in a West Bengal Inceptisol after 30 Years of multiple cropping and fertilization. Soil Sci Soc Am J 70(1):121–129

Mishra BB (2016) Indian system of soil classification: a way forward. Agric Res Technol Open Access J 3(2):555606

Mitra S, Wassmann R, Jain MC, Pathak H (2002) Properties of rice soils affecting methane production potentials: 2. Differences in topsoil and subsoil. Nutr Cycl Agroecosyst 64(1):183–191

Mohanty M, Painuli DK, Misra AK, Ghosh PK (2007) Soil quality effects of tillage and residue under rice–wheat cropping on a Vertisol in India. Soil Tillage Res 92(1–2):243–250

Munda S, Shivakumar BG, Rana DS, Gangaiah B, Manjaiah KM, Dass A, Layek J, Lakshman K (2016) Inorganic phosphorus along with biofertilizers improves profitability and sustainability in soybean (Glycine max)–potato (Solanum tuberosum) cropping system. Journal of the Saudi Society of Agricultural Sciences (in press)

Nair PKR (1993) An introduction to agroforestry. Kluwer Academic Publishers/ ICRAF, Dordrecht

Nath CP, Das TK, Rana KS, Bhattacharyaa R, Pathak H, Paul S (2017) Greenhouse gases emission, soil organic carbon and wheat yield as affected by tillage systems and nitrogen management practices. Archives Agron Soil Sci, pp 1–17

Oyeogbe AI, Das TK, Bhatia A, Singh SB (2017) Adaptive nitrogen and integrated weed management in conservation agriculture: impacts on agronomic productivity, greenhouse gas emissions, and herbicide residues. Environ Monitor Assess 189(4):198

Packialakshmi N, Yasotha C (2014) Role of effective microorganism in unfertilesoil. Int J Phytopharm 4(1):25–27

Padmavathy A, Poyyamoli G (2011) Effects of conventional and organic management strategies on soil quality and biodiversity in agricultural fields of Bahour, Puducherry, India. Am-Eur J Agric Environ Sci 10(4):644–652

Pal R, Chakrabarti K, Chakraborty A, Chowdhury A (2005) Pencycuron dissipation in soil: effect of application rate and soil conditions. Pest Manag Sci 61(12):1220–1223

Pathak H (2009) Emission of nitrous oxide from soil in relation to temperature, fertilizer application and moisture regime. Curr Adv Agric Sci (Int J) 1:1

Pathak H, Li C, Wassmann R (2005) Greenhouse gas emissions from Indian rice fields: calibration and upscaling using the DNDC model. Biogeosciences 2(2):113–123

Pathak H, Li C, Wassmann R, Ladha JK (2006) Simulation of Nitrogen Balance in Rice-Wheat Systems of the Indo-Gangetic Plains. Soil Sci Soc Am J 70(5):1612–1622

Pathak H, Mohanty S, Jain N, Bhatia A (2010) Nitrogen, phosphorus, and potassium budgets in Indian agriculture. Nutr Cycl Agroecosyst 86(3):287–299

Pathak H, Prasad S, Bhatia A, Singh S, Kumar S, Singh J, Jain MC (2003) Methane emission from rice–wheat cropping system in the Indo-Gangetic plain in relation to irrigation, farmyard manure and dicyandiamide application. Agr Ecosyst Environ 97(1–3):309–316

Pathak H, Rao DLN (1998) Carbon and nitrogen mineralization from added organic matter in saline and alkali soils. Soil Biol Biochem 30(6):695–702

Pathak H, Saharawat YS, Gathala M, Ladha JK (2011) Impact of resource-conserving technologies on productivity and greenhouse gas emissions in the rice-wheat system. Greenhouse Gases: Sci Technol 1(3):261–277

Pathak H, Sankhyan S, Dubey DS, Bhatia A, Jain N (2013) Dry direct-seeding of rice for mitigating greenhouse gas emission: field experimentation and simulation. Paddy Water Environ 11(1): 593–601

Prakash NB, Suseela DL (2000) Persistence of butachlor in soils under different moisture regime. J Indian Soc Soil Sci 48:249–256

Purakayastha TJ, Kumari S, Pathak H (2015a) Characterisation, stability, and microbial effects of four biochars produced from crop residues. Geoderma 239–240:293–303

Purakayastha TJ, Kumari S, Sasmal S, Pathak H (2015b) Biochar carbon sequestration in soil-A myth or reality? Int J Bio-Resour Stress Manag 6(5)

Pushpa TB, Sekaran V, Basha SS, andJegan J (2016). Investigation on preparation, characterization and application of effective microorganisms (EM) based composts-an ecofriendly solution. Nat Environ Poll Technol 15(1):153

Radhakrishnan S, Varadharajan M (2016) Status of microbial diversity in agroforestry systems in Tamil Nadu, India. J Basic Microbiol 56 (6):662–669

Rai A, Chhonkar PK, Agnihotri NP (1999) Degradation of pendimethalin and anilofos by enrichment cultures from different soils. Pesticide Res J 11(2):127–131

Ramachandran A, Radhapriya P (2016) Restoration of degraded soil in the Nanmangalam reserve forest with native tree species: effect of indigenous plant growth-promoting bacteria. Scientif World J, p 5465841

Rasool N, Reshi ZA (2010) Effect of the fungicide Mancozeb at different application rates on enzyme activities in a silt loam soil of the Kashmir Himalaya, India. Tropical Ecol 51(2):199

Rath AK, Ramakrishnan B, Rath AK, Kumaraswamy S, Bharati K, Singla BP, Sethunathan N (1998) Effect of pesticides on microbial biomass of flooded soil. Chemosphere 37(4):661–671

Reddy KR, DeLaune RD (2008) Biogeochemistry of wetlands: science and applications. CRC press

Reddy BR, Sethunathan N (1985) Salinity and the persistence of parathion in flooded soils. Soil Biol Biochem 17(2):235–239

Resmi P, Manju MN, Gireeshkumar TR, Ramseh Kumar CS, Movitha M, Shameem K, Chandramohanakumar N (2016) Phosphorous fractionation in mangrove sediments of Kerala, south west coast of India: the relative importance of inorganic and organic phosphorous fractions. Environ Monit Assess 188(6):366

Rudrappa L, Purakayastha TJ, Singh D, Bhadraray S (2006) Long-term manuring and fertilization effects on soil organic carbon pools in a TypicHaplustept of semi-arid sub-tropical India. Soil Tillage Res 88 (1–2):180–192

Saha A, Bhaduri D, Pipariya A, Jain NK (2016) Influence of imazethapyr and quizalofop-p-ethyl application on microbial biomass and enzymatic activity in peanut grown soil. Environ Sci Pollut Res 23(23):23758–23771

Saha D, Kukal SS, Bawa SS (2014) Soil organic carbon stock and fractions in relation to land use and soil depth in the degraded shiwaliks hills of lower Himalayas. Land Degrad Dev 25(5):407–416

Saha D, Kukal SS, Sharma S (2011) Landuse impacts on SOC fractions and aggregate stability in typicustochrepts of Northwest India. Plant Soil 339(1):457–470

Saharawat YS, Ladha JK, Pathak H, Gathala M, Chaudhary N, Jat ML (2012) Simulation of resource-conserving technologies on productivity, income and greenhouse gas GHG emission in rice-wheat system. J Soil Sci Environ Manag 3(1):9–22

Sahoo S, Adak T, Bagchi TB, Kumar U, Munda S, Saha S, Berliner J, Jena M, Mishra BB (2016a) Effect of pretilachlor on soil enzyme activities in tropical rice soil. Bull Environ Contam Toxicol 98 (3):439–445

Sahoo S, Adak T, Bagchi TB, Kumar U, Munda S, Saha S, Berliner J, Jena M, Mishra BB (2016b) Non-target effects of pretilachlor on microbial properties in tropical rice soil. Environ Sci Pollut Res 23 (8):7595–7602

Samra JS and Sharma PD (2009) Food security—Indian scenario. In: Proceedings IPI-OUAT-IPNI international symposium. International Potash Institute, Horgen, Switzerland and International Plant Nutrition Institute, Norcross, GA, USA, pp 15–43

Sandeep S, Manjaiah KM (2014) Thermal stability of organic carbon in soil aggregates of maize-wheat system in semi arid India. J Soil Sci Plant Nutr 14:111–124

Sandeep S, Manjaiah KM, Mayadevi MR, Singh AK (2016) Monitoring temperature sensitivity of soil organic carbon decomposition under maize–wheat cropping systems in semi-arid India. Environ Monit Assess 188(8):451

Sanyal SK, Dwivedi BS, Singh VK, Majumdar K, Datta SC, Pattanayak SK, Annapurna K (2015) Phosphorus in relation to dominant cropping sequences in India: chemistry, fertility relations and management options. Curr Sci 108:7–10

Sharma A, Saha TN, Arora A, Shah R, Nain L (2017) Efficient microorganism compost benefits plant growth and improves soil health in Calendula and Marigold. Hortic Plant J 3(2):67–72

Shukla AK, Singh VK, Dwivedi BS, Vinod K (2010) Sustainability of rice-wheat system: a means of food security in India. Studium Press (India) Pvt. Ltd., New Delhi, India, pp 499–542

Sihi D, Dari B, Sharma DK, Pathak H, Nain L, Sharma OP (2017a) Evaluation of soil health in organic vs. conventional farming of basmati rice in North India. J Plant Nutr Soil Sci: n/a-n/a

Sihi D, Gerber S, Inglett PW, Inglett KS (2016) Comparing models of microbial–substrate interactions and their response to warming. Biogeosciences 13(6):1733–1752

Sihi D, Inglett PW, Gerber S, Inglett KS (2017b) Rate of warming affects temperature sensitivity of anaerobic peat decomposition and greenhouse gas production. Glob Change Biol 24:e259–e274

Sihi D, Sharma DK, Pathak H, Singh YV, Sharma OP, LataNain Chaudhary A, Dari B (2012) Effect of organic farming on productivity and quality of basmati rice. ORYZA-Int J Rice 49:24–29

Silberbush M, Barber SA (1983) Prediction of Phosphorus and Potassium uptake by soybeans with a mechanistic mathematical model1. Soil Sci Soc Am J 47(2):262–265

Singh MV (2008) Micronutrient deficiencies in crops and soils in India. In: Micronutrient deficiencies in global crop production, pp 93–125

Singh G, Chauhan R, Ranjan RK, Prasad MB, Ramanathan AL (2015) Phosphorus dynamics in mangroves of India. Curr Sci 108(10):1874

Singh N (2002) Sorption behavior of triazole fungicides in Indian soils and its correlation with soil properties. J Agric Food Chem 50 (22):6434–6439

Singh SK, Singh AK, Sharma BK, Tarafdar JC (2007) Carbon stock and organic carbon dynamics in soils of Rajasthan, India. J Arid Environ 68(3):408–421

Singh SJ, Krausmann F, Gingrich S, Haberl H, Erb KH, Lanz P, Martinez-Alier J, Temper L (2012) India's biophysical economy, 1961–2008. Sustainability in a national and global context. Ecol Econ 76:60–69

Sinsabaugh RL, Antibus RK, Linkins AE (1991) Anenzymic approach to the analysis of microbial activity during plant litter decomposition. Agric Ecosyst Environ 34:43–54

Srivastava SC (1992) Microbial c, n and p in dry tropical soils: seasonal changes and influence of soil moisture. Soil Biol Biochem 24 (7):711–714

Sukul P, Spiteller M (2001) Persistence, fate, and metabolism of [14C] Metalaxyl in typical Indian soils. J Agric Food Chem 49(5):2352–2358

Tandon HLS, Tiwari KN (2011) Fertiliser use in Indian agriculture—an eventful half century. Better Crops India 1:3–5

Tirkey J, Adhikary SP (2005) Cyanobacteria in biological soil crusts of India. Curr Sci 89(3):515

Vernadsky W (1924) La géochimie. Alcan, Paris

Yadav RS, Mahatma MK, Thirumalaisamy PP, Meena HN, Bhaduri D, Arora S, Panwar J (2017) Arbuscular Mycorrhizal Fungi (AMF) for sustainable soil and plant health in salt-affected soils. In: Bioremediation of salt affected soils: an Indian perspective. Springer, pp 133–156

Benchmark Soils in Agro-ecological Regions

9

K. S. Anil Kumar, K. S. Karthika, M. Lalitha, R. Srinivasan, Shivanand, K. Sujatha, K. M. Nair, R. Hegde, S. K. Singh, and Bipin B. Mishra

Abstract

The benchmark soils in twenty agro-ecological regions of the country are briefly explained in this chapter using the established soil series to interpret soils, their physical, chemical characteristics, problems and potentials. This would be helpful in identifying the use of soil resource inventory and classification in optimal land use and production system. In fact, a benchmark soil is widely extensive, holds a key position in the soil classification system and is of special significance to farming, engineering or other uses and focuses on its agronomic concepts for wider acceptability of interpretations and for extrapolation of research data. It is representative of the most extensive soils in major land resource area or agro-ecological zone. This chapter highlights the benchmark soils in conducting soil correlation, standardization of legends, prediction of soil behaviour, agro-technology transfer and planning for further research in soil science and allied disciplines. However, further refinement in its applications using GIS tools is of priority.

Keywords

Benchmark soils • Agro-ecological regions • Soil series • Soil classification • Resource inventory • Agro-technology transfer

K. S. Anil Kumar (✉) · K. S. Karthika · M. Lalitha · R. Srinivasan · Shivanand · K. Sujatha · K. M. Nair · R. Hegde
ICAR-National Bureau of Soil Survey and Land Use Planning, Regional Centre, Hebbal, Bangalore 560024, India
e-mail: anilsoils@yahoo.co.in

S. K. Singh
ICAR-National Bureau of Soil Survey and Land Use Planning, Amravati Road, University P.O., Nagpur 440033, India

B. B. Mishra
Bihar Agricultural University, Sabour, Bhagalpur 813210, India

Pedology and Land Use Planning, School of Natural Resources Management and Environmental Sciences, Haramaya University, P O Box 138 Dire Dawa, Ethiopia

International Union of Soil Science, Vienna, Austria

B. B. Mishra (ed.), *The Soils of India*, World Soils Book Series,
https://doi.org/10.1007/978-3-030-31082-0_9

9.1 Introduction

India is gifted with a variety of soils due to varying geological formations, diversified climate, topography and relief. Such diversifications have given rise to varied physiographic features. This extends from the snow-clad mountainous peaks, thick forests in the north, to the lengthy coasts in the Peninsula. The temperature varies from arctic cold to equatorial hot, and rainfall varies from few centimetres in the arid regions to per-humid regions with maximum rainfall of several hundred centimetres per annum. These variations in temperature and rainfall have provided for high plateaus, hills, valleys, rolling uplands, swampy lowlands, fertile plains and barren deserts. These varied natural environments have resulted in a large variety of soils distinctly different from one another creating soil as a valuable resource of India. Eighty-two benchmark soils have been identified in the 27 different major land resource areas of Kerala for which soil monoliths were collected (GOI 2019).

According to USDA (2017), a benchmark soil is one of large extent within one or more major land resource areas (MLRAs), one that holds a key position in the soil classification system, one for which there is a large amount of data, one that has special importance to one or more significant land uses or one that is of significant ecological importance. Benchmark soils, while being important soils in their own right, are also intended to serve as proxies for other similar soils. Their purpose is to focus on data collection and the investigative effort on soils that have the greatest potential for extending collected data and resultant interpretations to other soils. In fact, this purpose is relevant both in making soil surveys and to soil survey customers in their goal to extend findings of their research. The cost of investigation and the large number of combinations of soil uses and management practices preclude laboratory and field studies of all soils; therefore, studies of benchmark soils are an essential component of developing nationwide soil databases and soil interpretations. Data obtained by studying benchmark soils can be used to help estimate important properties in similar soils. Benchmark soils can also be used to test new interpretations or to provide input to regional-scale models. Knowledge of the properties and behaviour of benchmark soils contribute to the understanding and interpretation of other soils with similar properties. This knowledge is important to soil technology and the use of soil surveys (USDA 2017).

The soil series receives a benchmark status when the established soil series is widely extensive, possesses a key position in the soil classification system and owns special importance to agricultural uses. Benchmark soil status is given after thorough identification and checking of soils, their characterization and classification in order to highlight their extensive occurrence, associated problems and potential for sustainable agriculture (Sehgal 1994).

Importantly, a benchmark soil is a reference point which is adequately characterized in terms of its properties and environmental conditions. The information about these soils can be extended to other soils closely related in classification and geography. In USDA, these soils are selected from among the existing and established soils that represent typical range of characters (USDA 2017) based on major land resource areas (MLRAs) or agro-ecological zones (AEZs). A list of the benchmark soils helps the research community to focus their investigative efforts on key soils that have the greatest potential for applying new technology across a large area and also for transferring new technologies to similar soils, thereby optimizing cost–benefit ratios. Establishment of the benchmark soils has been a hard task. The designation of "benchmark" facilitates the selection of soils that meet research and other study objectives while also allowing for maximum extension of study data. Benchmark soils can be selected for their representativeness and included in studies of single soils or a suite of soils, such as those representing a gradient in temperature or moisture across a region.

9.2 Benchmark Soils in Agro-Ecological Regions (AERs)

Agro-ecological region is characterized based on soils, fauna, aquatic systems, vegetation, etc., indicating AER as that land unit carved out of agro-climatic region (ACR). The approach of agro-ecological region is based on climatic conditions, length of growing period (LGP), landforms and soils (Sehgal et al. 1992). The formation and development of soils are based on several differing factors. Such groupings would be of great assistance in understanding the problems and prospects of these soils in relation to the crop production and productivity. This would be further useful to attain optimum crop production potential and cropping sequence.

As already documented, the benchmark soils are occurring in the 20 agro-ecological regions in India and the soil series are accordingly identified in these regions. Soil series refers to the lowest category in soil taxonomy which is the largest landscape unit and aids in distinguishing the features and characteristics relevant to soil formation (Sehgal 1994). In this chapter, one series from each category of soils is explained with their selected physico-chemical characteristics and their taxonomic classification at the soil family level.

The benchmark soils in different AER are presented in Table 9.1. The soils 9.2.1–9.2.3 represent arid ecosystem, 9.2.4–9.2.9 represent semi-arid ecosystem, 9.2.10–9.2.15 represent sub-humid ecosystem, 9.2.16–9.2.18 define humid, per-humid agro-ecosystem, and 9.2.19 and 9.2.20 are coastal ecosystems (east and west coast, respectively) as reported by Mandal et al. (2016). These soils are briefly described hereunder as benchmark soils associated with agro-ecological regions. The soil series qualifying for benchmark meaning "a soil occupying a key interpretative position in soil classification framework and/or covering a large area" in different agro-ecological regions are chosen in this chapter to explain these soils (Murthy et al. 1982). The benchmark soils in different agro-ecological regions are represented in Fig. 9.1, and their extent of occurrence in Mha is given in Fig. 9.2.

9.2.1 Cold Arid Shallow Skeletal Soils

Area: 18.50 Mha (5.6% of total geographical area of India).

Location and AER

Agro-ecological Region 1

These soils occur in cold arid eco-region encompassing areas of Ladakh, Chilas Wazarat, parts of Muzaffarabad, North Kashmir and Gilgit districts of Northwest Himalayas.

Description of Soils

Soils are shallow, sandy to loamy-skeletal calcareous on gently sloping to nearly level valleys in the northern parts of the AER and remain under snow cover permanently (*sandy-skeletal Cryorthents*). These soils are excessively drained with moderate erosion and moderate stoniness. Very deep, calcareous, sandy clay loam to clay loam soils occur on nearly level terraces in the glacio-fluvial valley (fine-loamy *Cryochrepts*) (Fig. 9.3).

9.2.2 Hot Arid Desert and Saline Soils

Area: 24.7 Mha (7.5% of total geographical area of India).

Table 9.1 Benchmark soils in agro-ecological regions

Sl. No	AER	Benchmark Soils	Soil series
9.2.1	1	Cold arid shallow skeletal soils	–
9.2.2	2	Hot arid desert and saline soils	Masitawali and Thar series
9.2.3	3	Hot arid red and black soils	Teligi series
9.2.4	4	Hot semi-arid coarse-loamy alluvial soils	Fatehpur, Kanjili, Nabha, Zarifa Viran, Sadhu and Khoh series
9.2.5	5	Hot semi-arid old alluvial soils	Chirai, Pali, Chomu and Singpura series.
9.2.6	6	Hot semi-arid to dry sub-humid alluvial and *Tarai* soils	Hirapur, Sakit, Itwa, Bijaipur and Basiaram series
9.2.7	7	Hot semi-arid moderately deep black soils	Sarol, Kamliakheri, Jambha, Linga, Barsi, Jhamkhandi and Dandi series
9.2.8	8	Hot semi-arid with mixed red and black soils	Adesar, Achmatti, Hungund, Kagalgomb, Raichur, Kadirabad, Chinnaloni, Kasireddipalli, Patancheru and Lakhpat series
9.2.9	9	Hot semi-arid with red loamy soils	Tyamagondalu, Palathurai, Coimbatore, Channasandra and Vijayapura series
9.2.10	10	Hot sub-humid with moderately deep black soils	Kheri series
9.2.11	11	Hot sub-humid red and yellow soils	Marha series
9.2.12	12	Hot sub-humid red and lateritic soils	Chougel, Mrigindihi and Bhubaneswar series
9.2.13	13	Hot sub-humid alluvial soils	Hathiapathar series
9.2.14	14	Warm sub- humid to humid sub-montane shallow and skeletal soils	Gogji Pather, Wahthora, Shinwali and Haldi series
9.2.15	15	Hot sub-humid loamy to clayey alluvial soils	Kanagarh, Madhpur, Hanrgram and Jagdishpur series
9.2.16	16	Warm sub-humid to per-humid loamy to clayey alluvial soils	
9.2.17	17	Warm per-humid shallow and skeletal soils	Jaihing and Gemotali series
9.2.18	18	Warm per-humid red and yellow soils	Phullen, Dialong and Selsekgiri series
9.2.19	19	Hot sub-humid coastal and deltaic alluvial soils	Motto and Kalathur series
9.2.20	20	Hot humid/per-humid red and lateritic and alluvial soils	Thekkady, Trivandrum, Kunnamangalam and Ambalapuzha series

Location and AER

Agro-ecological Region 2

These soils occur in hot arid regions of south-western parts of Punjab, Haryana, western parts of Rajasthan and Kutch peninsula of Gujarat. These soils are distributed in Bathinda, Mansa, Ferozepur and Muktsar districts of Punjab, Sirsa, Hisar, Bhiwani and Fatehabad districts of Haryana, Bikaner, Jaisalmer, Sri Ganganagar, Hanumangarh, Churu, Jhunjhunu, Sikar, Nagaur, Barmer, Jalor, Jodhpur districts of Rajasthan and Kutch District of Gujarat.

Description of Soils

Soils of the region are deep to very deep, loamy sand, and sand in texture occurs in association with saline patches at places on gently to very gently sloping arid plains with intersperse hummocks, sand dunes and barchans. Soils are grouped with *Torrifluvents, Haplocambids, Haplocalcids* and *Natrargids* great groups of *Entisols* and *Aridisols* soil orders. The major soil series identified in this are Thar series and Masitawali series (Fig. 9.4).

Masitawali Series: *Torrifluventic Haplustepts*

These soils are formed in Ghaggar alluvium of Shivalik origin in the old flood plain of Northwest Rajasthan. These soils are deep, coarse-loamy, well-drained, calcareous and moderately alkaline. These soils have pale brown to brown, moderately alkaline, sandy loam Ap horizons, and brown to dark yellowish brown, strongly alkaline, sandy loam B horizons. The CEC is 7.0–9.7 cmol (+) kg^{-1} of soil, with low to medium available moisture capacity and moderately rapid permeability. These soils are mainly cultivated to cotton, wheat, mustard and cluster bean. These soils are associated with Shakhi series which is a *Typic Torripsamments* (Ray et al. 2014a).

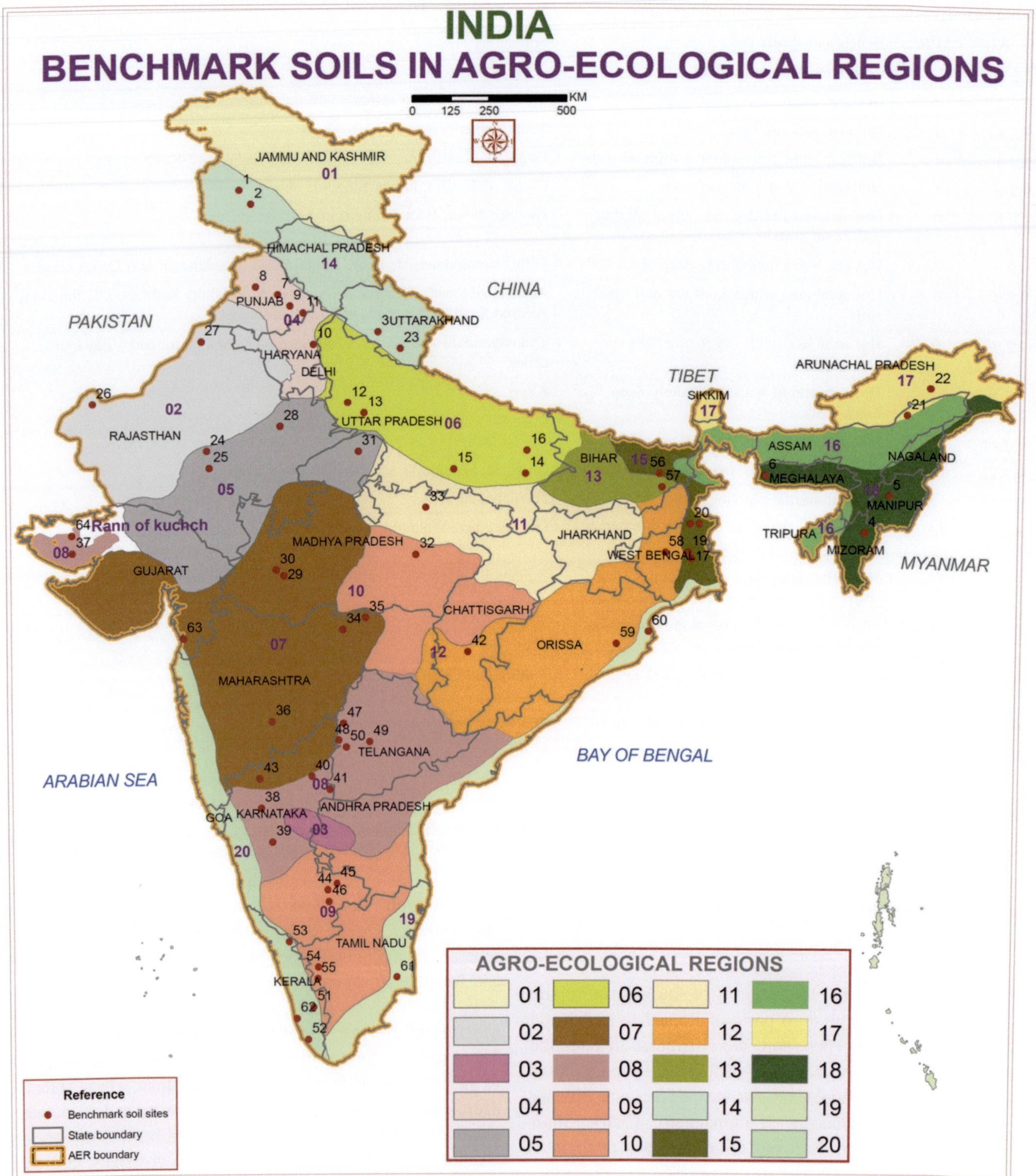

Fig. 9.1 Benchmark soils in agro-ecological regions *Source* NBSS and LUP (2015)

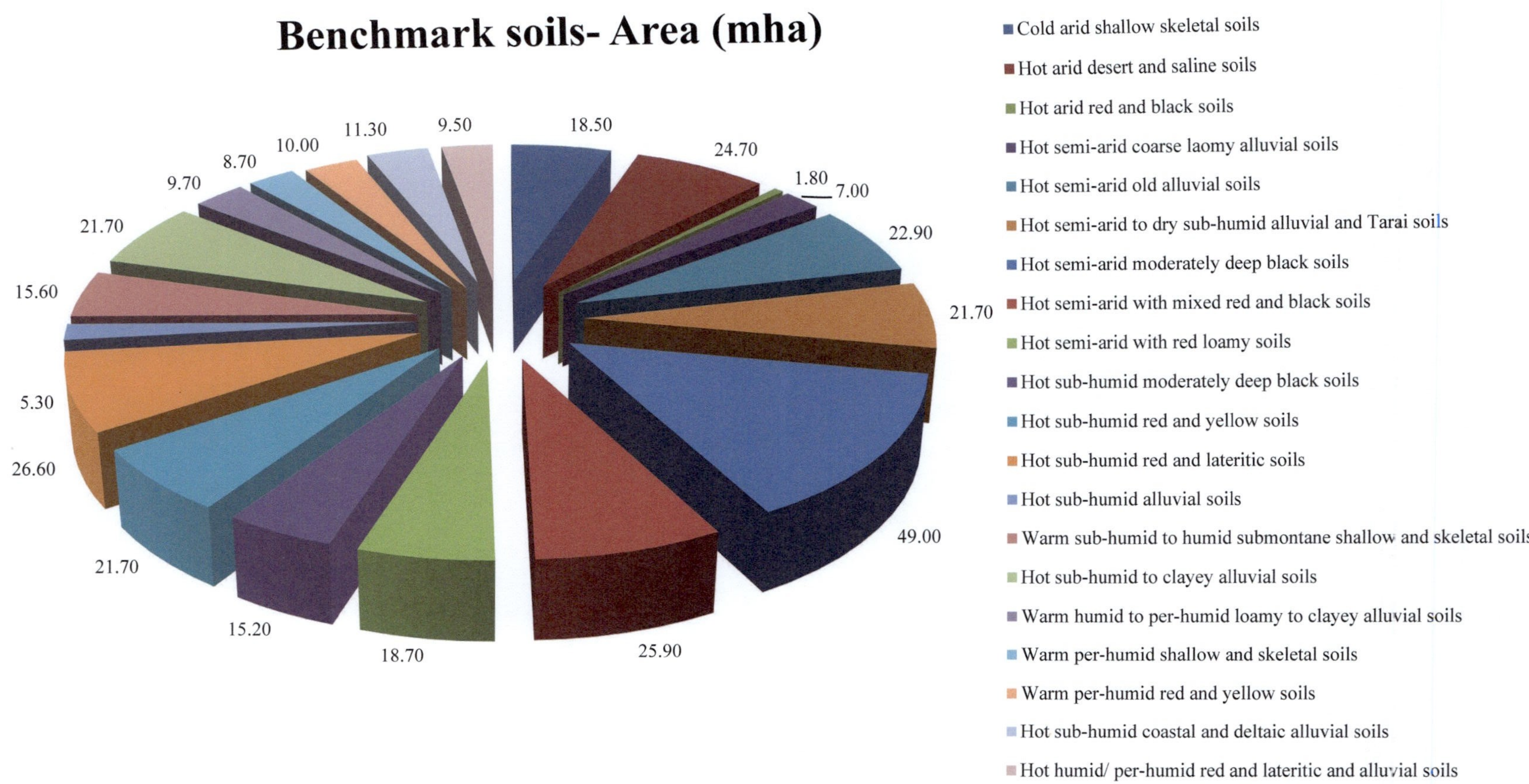

Fig. 9.2 Extent of benchmark soils in different agro-ecological regions (Area in mha). *Source* NBSS & LUP ARCHIVES 2017

Thar Series: *Typic Torripsamments*

Thar soils are deep, sandy, excessively drained, calcareous and moderately alkaline with low available moisture capacity. Aridity is the major limitation in these soils. The productivity is also low (Murthy 1982; Murthy et al. 1982).

9.2.3 Hot Arid Red and Black Soils

Area: 1.80 Mha (0.6% of total geographical area of India).

Location and AER

Agro-ecological Region 3

These soils occur in hot arid regions in parts of Deccan Plateau distributed in Anantapur District of Andhra Pradesh and in Bellary District of Karnataka.

Description of Soils

The soils on gently to very gently slopes are shallow, sandy loam to sandy clay loam in texture (*Rhodustalfs, Haplustalfs, Paleustalfs*). Deep, clayey *Haplusterts* occur on level to very gently sloping plains.

Teligi Series: *Sodic Haplusterts*

These soils have very dark greyish brown, moderately alkaline clay A horizons and strongly to very strongly alkaline clay B horizons (Fig. 9.5). These occur on very gently sloping plain and are moderately well-drained with slow permeability (Bhattacharyya et al. 2014).

Selected physico-chemical properties of series representing arid ecosystem (soils 9.2.2 and 9.2.3) are represented in Table 9.2.

9.2.4 Hot Semi-arid Coarse-Loamy Alluvial Soils

Area: 7 Mha (2.1% of total geographical area of India).

Location and AER

Agro-ecological Region 4

These soils occur in hot semi-arid regions in districts of Punjab, Haryana, south and north parts of Delhi. The districts in Punjab include Amritsar, Firozpur, Faridkot, Sangrur, Patiala, Fatehgarh Sahib, Rupnagar, Nawanshahr, Kapurthala, Ludhiana and Hoshiarpur and in Haryana include Panchkula, Ambala, Yamunanagar, Kurukshetra, Kaithal, Karnal, Jind, Sonipat, Rohtak, Jhajjar, Mahendragarh, Rewari, Gurgaon and Faridabad.

Description of Soils

Sandy loam to sandy clay loam soils derived from Indo-Gangetic alluvium occur on moderately to gently sloping plains. These are classified under the great groups of *Haplustepts* and *Natrustalfs* of *Inceptisols* and *Alfisols* soil orders, respectively. The associated soils are *Ustipsamments* and *Ustifluvents* of Entisols soil order. The major soils in this group are Fatehpur series, Kanjili series, Nabha series, Zarifa Viran series, Sadhu series and Khoh series (Fig. 9.6).

Fatehpur Series: *Inceptic Haplustalfs*

These soils are formed in old alluvium overlain by Aeolian sandy material. Fatehpur soils are deep, coarse-loamy, well-drained with rapid permeability. These soils have pale brown and brown, neutral, loamy sand A horizons, brown to strong brown, slightly to moderately alkaline, loamy sand to sandy loam B horizons. The soils have low moisture retentivity and are also susceptible to wind erosion. These soils are cultivated to paddy, sunflower, ground nut, pearl millet and maize (Ray et al. 2014a).

Zarifa Viran: *Typic Natrustalfs*

These soils are formed in old alluvium and occur in micro depressions in Karnal and adjoining districts of Haryana. They are deep, fine-loamy, moderately to strongly alkaline, imperfectly drained with moderately slow permeability. The CEC ranges between 9 and 15 cmol (+) kg^{-1} of soil. The soils are saline. These soils have light brownish grey to brown, strongly alkaline, silty loam A horizons; and dark yellowish brown to greyish brown, strongly to very strongly alkaline, silt loam Bt horizons, over dark yellowish brown to yellowish brown, very strongly alkaline, silt loam BC horizons (Ray et al. 2014a).

Khoh Series: *Typic Ustipsamments*

These soils are formed in the Aeolian deposits of mixed origin on gently sloping plains in adjoining districts of Gurgaon, Haryana. They are very deep, brownish yellow to yellowish brown in colour, moderately alkaline in reaction, sand to loamy sand in texture, well-drained or somewhat excessively drained with very rapid permeability and low available moisture capacity. The CEC ranges from 3 to 6 cmol (+) kg^{-1} of soil (Ray et al. 2014a).

9.2.5 Hot Semi-arid Old Alluvial Soils

Area: 22.9 Mha (7% of total geographical area of India).

Fig. 9.3 Cold arid shallow skeletal soils of Ladakh region of Jammu and Kashmir (AER-1). *Source* NBSS & LUP ARCHIVES 2017

Location and AER

Agro-ecological Region 5

These soils occur in hot semi-arid regions in districts coming under Central Highlands of Malwa region covering Gujarat plains, western parts of Madhya Pradesh and southern parts of Rajasthan.

Description of Soils

Deep to very deep sand and sandy loam occur on nearly level and gently sloping lands. *Haplocambids* and *Haplustepts* occur in association together with *Ustipsamments.*

The major soil series found in this are Chirai series (Fig. 9.7), Pali series, Chomu series and Singpura series.

Singpura Series: *Vertic Haplustalfs*

These soils are deep, fine-loamy, well-drained, moderately to strongly alkaline. The CEC ranges between 10 and 19 cmol (+) kg^{-1} of soil. The soils are medium to high in its available moisture capacity. These soils have yellowish brown, loamy, moderately alkaline A horizon and dark yellowish brown to light yellowish brown, clay loamy, strongly to moderately alkaline, B horizons (Ray et al. 2014a).

9.2.6 Hot Semi-arid to Dry Sub-humid Alluvial and Tarai Soils

Area: 21.7 Mha (6.6% of total geographical area of India).

Location and AER

Agro-ecological Region 6

These soils occur in hot semi-arid to sub-humid regions in many districts of Uttar Pradesh and western Champaran, Gopalganj and Siwan in Bihar.

Fig. 9.4 Hot arid desert and saline soils of Chattargarh tehsil, Bikaner District, Rajasthan Thar series; *Typic Torripsamments* (AER-3). *Source* NBSS & LUP ARCHIVES 2017

Description of Soils

The soils are derived from the Gangetic alluvium and a part of it locally known as *tarai*, having grey to dark grey colour, sand to clay loam texture and high organic carbon. The water table is high, and soils remain saturated or fairly moist during major parts of the year. The tarai soils belong to *Hapludolls*, *Haplaqualfs* and *Haplustepts*. The soils derived from Gangetic alluvium are neutral to moderately alkaline, calcareous at places, silty loam and silty clay loam in texture. These soils belong to *Haplustepts*, *Ochraqualfs*, *Haplustalfs*, *Ustifluvents* and *Ustorthents*. The other deep soils, loam to clay loam developed from mixed alluvium occurring on gentle slopes of alluvial terraces at an elevation of 150 m above MSL, have good available moisture holding capacity and high productivity potential. The salt-affected soils occupy the low-lying depressions having imperfect drainage with low permeability (Fig. 9.8).

The major soil series include Hirapur, Sakit, Itwa, Bijaipur and Basiaram series.

Hirapur Series: *Vertic Natrustalfs*

These soils occur on level to gently sloping alluvial lands in Aligarh and adjoining districts of Uttar Pradesh. These soils are deep, fine-loamy, very strongly alkaline and imperfectly drained with moderately slow permeability. Hirapur soils have light brownish grey to light olive brown, very strongly alkaline, loam A horizons, silt loam Bw horizons, silty clay loam Btn horizons over light olive brown, very strongly alkaline, loam to silty loam BCk horizons. The CEC is 7–12 cmol (+) kg^{-1} of soil. These soils are highly saline and highly sodic with ESP greater than 80 (Ray et al. 2014a).

Sakit Series: *Vertic Haplustalfs*

These soils occur on flat old alluvial terraces in Etah and adjoining districts of UP. These soils are very deep, fine-silty, imperfectly drained with slow permeability, calcareous and very strongly alkaline. The CEC ranges from 12 to 15 cmol (+) kg^{-1} of soil. These soils are highly saline and sodic with ESP greater than 80 (Ray et al. 2014a).

Fig. 9.5 Hot arid red and black soils of Siruguppa, Bellary, Teligi series, *Sodic Haplusterts* (AER-3). *Source* NBSS & LUP ARCHIVES 2017

Itwa Series: *Typic Endoaqualfs*
These soils are formed in the Gangetic alluvium. These are deep, fine-loamy, imperfectly drained, moderately to strongly alkaline. The CEC ranges from 10 to 15 cmol (+) kg^{-1} of soil. The AWC is medium to high, with medium productivity potential (Ray et al. 2014b).

Bijaipur Series: *Udic Ustochrepts*
These soils are formed in mixed alluvium and occur on terrace and uplands of the Ganga plains in Fatehpur District of Uttar Pradesh. They are deep, fine-loamy, well-drained and neutral with CEC ranging from 6 to 14 cmol (+) kg^{-1} of soil. The available water capacity is medium to high (Ray et al. 2014a).

Basiaram Series: *Udic Haplustepts*
These soils occur on nearly level terraces and levees in Azamgarh District of Uttar Pradesh. They are deep, fine-loamy, neutral and somewhat poorly drained with moderately slow permeability. The CEC ranged between 6 and 11 cmol (+) kg^{-1} of soil with medium to high AWC. Basiaram soils have light brownish grey to light yellowish brown, neutral to mildly alkaline, silt loam A horizons and yellowish brown to brown, mildly alkaline to neutral silt loam to silty clay loam B horizons (Ray et al. 2014a).

9.2.7 Hot Semi-arid Moderately Deep Black Soils

Area: 49.0 Mha (14.9% of total geographical area of India).

Location and AER

Agro-ecological Region 7

These soils occur in parts of Deccan Plateau of Maharashtra, eastern Madhya Pradesh and Kathiawar Peninsula of

Table 9.2 Selected physico-chemical properties of series representing arid ecosystem

Horizon	Depth	Colour	Sand (2–0.05)	Silt (0.05–0.002)	Clay (<0.002)	Textural class	pH (1:2 water)	CEC (cmol (+) kg^{-1} of soil)
7.2.1 Hot arid desert and saline soils—Masitawali series: *Torrifluventic Haplustepts*								
Ap	0–14	Pale brown 10YR 6/3 D	65.0	18.4	16.6	sl	8.4	9.7
Bw1	14–28	Brown 10YR 4/3 M	64.5	21.6	13.9	sl	8.7	8.0
Bw2	28–49	Brown 10YR 4/3 M	70.2	18.0	11.9	sl	8.7	8.0
Bw3	49–78	Brown 10YR 4/3 M	75.9	12.5	11.6	sl	8.9	8.7
2 Bw4	78–104	Dark yellowish brown 10YR 4/3 M	72.9	18.0	9.1	sl	9.3	9.0
3Bw5	104–138	Dark yellowish brown 10YR 4/3 M	75.9	15.9	8.2	sl	9.4	7.7
4Bw6	138–160+	Brown 10YR 5/3 M	61.9	28.8	9.3	sl	9.4	7.0
7.2.2 Hot arid red and black soils—Teligi series: *Sodic Haplusterts*								
Ap	0–14	Very dark greyish brown 10 YR 3/2 M	30.9	19.1	50.0	c	8.2	40.0
Bss1	14–40	Very dark greyish brown 10 YR 3/3 M	33.8	17.0	49.2	c	8.6	39.1
Bss2	40–69	Very dark greyish brown 10 YR 3/2 M	36.1	19.5	44.4	c	9.0	33.9
Bss3	69–98	Very dark grey 10 YR 3/1 M	31.5	17.4	51.1	c	9.1	43.0
Bss4	98–117	Very dark greyish brown 10 YR 3/2 M	32.3	23.7	44.0	sc	8.7	35.6

Gujarat, parts of Kota in Rajasthan, Bidar, Gulbarga, Bijapur, parts of Belgaum and Bagalkot in Karnataka.

Description of Soils

Clayey, calcareous, moderately to strongly alkaline, swell shrink black cotton soils occur on nearly level to gently sloping pediplains, associated with alluvial soils on very gently sloping plains. The black cotton soils have high potential for crop production, however, constrained with management problems. These soils belong to *Haplusterts* great group of the Vertisols soil order.

The major soil series identified are Sarol, Kamliakheri, Jambha, Linga, Barsi, Jamkhandi and Dandi series (Fig. 9.9).

Sarol Series: *Typic Haplusterts*

These soils occur in Malwa Plateau and are developed on basaltic alluvium. They are moderately deep, clayey, moderately to well-drained, calcareous and moderately to strongly alkaline. The CEC ranges from 40 to 54 cmol (+) kg^{-1} of soil. These soils have dark greyish brown to very dark greyish brown, clayey, moderately alkaline A horizon and dark greyish brown to very dark greyish brown, clayey, strongly alkaline B horizons. These soils have high AMC and medium to high productivity potential. These soils are mainly cultivated to sorghum, pigeon pea, wheat and Bengal gram (Bhattacharyya et al. 2014).

9.2.8 Hot Semi-arid with Mixed Red and Black Soils

Area: 25.9 Mha (7.9% of total geographical area of India).

Location and AER

Agro-ecological Region 8

These soils occur in hot semi-arid regions in Kurnool, Prakasam, parts of Cuddpah, Guntur, Krishna, West and East Godavari in Andhra Pradesh, Adilabad, Nizamabad, Karimnagar, Medak, Warangal, Khammam, Nalgonda, Mahabubnagar, Rangareddi in Telangana, Gadag, parts of Koppal, Raichur, parts of Belgaum, Dharwad, parts of Chitradurga and Shimoga in Karnataka.

Fig. 9.6 Hot semi-arid agro-ecoregion with coarse-loamy alluvial soils (AER-4). *Source* NBSS & LUP ARCHIVES 2017

Description of Soils
Deep, clay loam to clayey, neutral to slightly acidic soils with moderate permeability are associated with deep, clayey soils on gently sloping plains in the region. These are affected with severe erosion. These soils are agriculturally important due to high water-holding capacity. These are classified as *Paleustalfs*, *Rhodustalfs* and *Haplusterts* great groups of *Alfisols* and *Vertisols*. The major soils identified, falling in this classification, are Adesar, Achmatti (Fig. 9.10), Hungund, Kagalgomb, Raichur, Kadirabad, Chinnaloni, Kasireddipalli, Patancheru and Lakhpat series.

Raichur Series: *Typic Haplusterts*
These soils have dark grey to very dark grey, moderately alkaline clay A and B horizons overlying lithologically discontinuous sandy C horizons. These soils are well-drained with slow permeability (Murthy et al. 1982).

Kasireddipalli: *Typic Haplusterts*
These soils are developed in basaltic alluvium and occur in depressions in Medak and adjoining districts of Andhra Pradesh. These soils are deep, very dark greyish brown, clayey, moderately well-drained, calcareous and moderately alkaline. The CEC ranged between 42 and 58 cmol (+) kg^{-1} of soil, and the AWC is medium to high. These soils are mainly cultivated to sorghum and pigeon pea. The major limitations include salinity and sodicity (Bhattacharyya et al. 2014).

9.2.9 Hot Semi-arid with Red Loamy Soils

Area: 18.7 Mha (5.7% of total geographical area of India).

Location and AER

Agro-ecological Region 9

These soils occur in hot semi-arid regions covering the parts of south Deccan Plateau, Tamil Nadu and Karnataka plateau. The districts in Karnataka include Chikmagalur, Chitradurga, Hassan, Tumkur, Bangalore (Rural and Urban), Kolar, Chamarajnagar, Mysore, Kodagu and Dakshin Kannada. In Tamil Nadu, these soils occur in Vellore, Dharmapuri, Salem, Namakkal, Erode, Nilgiri, Coimbatore, Theni, Madurai, Dindigul, Karur, Tiruchirappalli, Perambalur, parts of Pudukkottai, Sivaganga and Virudhunagar.

Description of Soils
Deep to moderately deep, occasionally shallow, gravelly sandy loam to sandy clay loam soils, having moderate to strong alkalinity constitutes the soil-scape. These soils are known for low available water-holding capacity, and their susceptibility to soil erosion and surface crusting the important soils series in this class are Tyamagondalu, Palathurai, Coimbatore (Fig. 9.11), Channasandra and Vijayapura series.

Channasandra Series: *Oxic Rhodustalfs*
These soils are deep, clayey-skeletal, well-drained and neutral to slightly acid. The CEC ranges from 4 to 12 cmol (+) kg^{-1} of soil with medium to high AWC. The productivity potential of these soils is medium (Murthy et al. 1982).

Vijayapura Series: *Oxic Haplustalfs*
These are deep, clayey, well-drained and slightly to strongly acid. The CEC is 3–4 cmol (+) kg^{-1} of soil. The soils have medium available water capacity (Murthy et al. 1982).

Selected physico-chemical properties of series representing semi-arid ecosystem are represented in Table 9.3.

Fig. 9.7 Chirai series—*Typic Haplocambids* (AER-5). *Source* NBSS & LUP ARCHIVES 2017

9.2.10 Hot Sub-humid with Moderately Deep Black Soils

Area: 15.2 Mha (4.6% of total geographical area of India).

Location and AER

Agro-ecological Region 10

These soils occur in hot sub-humid regions covering eastern plateau of Satpura, Mahanadi Basin and fringes of Maharashtra plateau. These soils are distributed in the districts of Chhindwara, Seoni, Betul, Balaghat, Mandla, Dindori, Jabalpur, parts of Damoh, Sagar, Raisen, Narsinghpur in Madhya Pradesh; Gadchiroli, Chandrapur, Bhandara, Gondia, parts of Nagpur and Yavatmal in Maharashtra; Rajnandgaon, Kawardha, Bilaspur, Janjgir–Champa, Raipur and Durg in Chhattisgarh.

Description of Soils

Medium to deep black soils and red soils belonging to *Ustorthents, Haplustepts, Haplustalfs* and *Haplusterts* constitute the dominant soil-scape. Deep black soils are associated with the moderately deep, light brown to dark brown soils, which are characterized with calcareousness; slight alkalinity and high swell shrink potential. Red soils occur on eastern Dandakaranya plateaus which are shallow to moderately deep, clayey, neutral to slightly acidic. The main soil series found in this class is Kheri series (Fig. 9.12).

Kheri Series: *Typic Haplusterts*

These soils are formed on the basaltic alluvium in the Satpura ranges of Deccan Plateau, in Khajri Kheri, Jabalpur, Madhya Pradesh. These soils are moderately deep, clayey, mildly alkaline and moderately well-drained. These soils have greyish brown, clayey, mildly alkaline A horizon, very dark greyish brown, clayey, mildly to moderately alkaline B horizons. The CEC ranges from 47 to 54 cmol (+) kg^{-1} of soil. The AWC is medium to high. These soils are mainly cultivated to soybean, wheat, paddy with medium productivity potential (Ray et al. 2014b).

Fig. 9.8 Hot semi-arid to dry sub-humid alluvial and Tarai soil; Typic *Haplustepts* (AER-6). *Source* NBSS & LUP ARCHIVES 2017

9.2.11 Hot Sub-humid Red and Yellow Soils

Area: 21.7 Mha (6.6% of total geographical area of India).

Location and AER

Agro-ecological Region 11

These soils occur in hot sub-humid regions covering a part of eastern Chota Nagpur Plateau, north Chhattisgarh, and Bundelkhand region of Madhya Pradesh, Uttar Pradesh and Jharkhand. These soils occur in the districts of Sidhi, Rewa, Shahdol, Satna, Panna, Chhatarpur, Tikamgarh in Madhya Pradesh; Korba, Surguja, Jashpur, parts of Bilaspur and Raigarh of Chhattisgarh; Garhwa, Palamu, Chatra, Hazaribagh, Koderma, Lohardaga, Gumla, Ranchi, Bokaro, Giridih, Jharkhand, Sonbhadra, Mirzapur, parts of Chandauli, Allahabad, Chitrakoot, Banda, Jhansi, Maoba, Hamirpur, Lalitpur in Uttar Pradesh; and parts of Kaimur and Rohtas, Bihar.

Description of Soils

The dominant soil-scape in the region is moderately to gently sloping uplands. Soils are dark red to reddish brown having sandy clay loam to sandy clay texture and have moderate to strong acidity. Soils are gravelly, very shallow to shallow and slightly acidic at places. The soils are classified in great groups *Ustorthents, Haplustepts, Haplustalfs, Rhodustalfs* and *Plinthustalfs* (Fig. 9.13).

Marha Series: *Entic Chromusterts*

These soils occur on gently sloping flood plains of Bundelkhand region of Madhya Pradesh. They are deep, clayey, moderately well-drained and moderately alkaline. The CEC ranges from 45 to 55 cmol (+) kg^{-1} of soil. The AWC is high. These soils are cultivated to rainfed sorghum, wheat, pigeon pea and Bengal gram. Their productivity potential is very high (Murthy et al. 1982).

9.2.12 Hot Sub-humid Red and Lateritic Soils

Area: 26.6 Mha (8.1% of total geographical area of India).

Location and AER

Agro-ecological Region 12

These soils occur in hot sub-humid regions covering Chota Nagpur Plateau, Jharkhand, western parts of West Bengal, Eastern Ghats of Odisha and Bastar region of Chhattisgarh. The districts include Dantewada, Bastar, Kanker, Dhamtari, parts of Raipur and Mahasamund in Chhattisgarh; Malkangiri, Koraput, Rayagada, Gajapati, Nabarangpur, Balangir, Sundergarh, Deogarh, Boudh, Phulbani, Ganjam, Nayagarh, Cuttack, Angul, Dhenkanal, Jajpur, Mayurbhanj, Kendujhar, Khorda in Odisha; parts of Gadchiroli District in Maharashtra; parts of Puruliya, Bankura and Medinipur districts in West Bengal; Srikakulam and Visakhapatnam districts in Andhra Pradesh.

Description of Soils

Soils are brown to reddish brown, fine textured and developed over very gently sloping pediment surfaces of weathered granite–gneiss complex. These soils are deep, slight to strongly acidic, having low cation exchange capacity. The other soils on gently sloping side slopes have dark reddish brown colour and strong acidity. Soils of Garhjat hills and eastern plateau are sandy loam to sandy clay loam with reddish to dark reddish colour and moderately acidic reaction. Other soils developed over granitic gneiss are

Fig. 9.9 Kauta B soils of Aurad block, Bidar District, Karnataka; *Typic Haplusterts* (AER-7). *Source* NBSS & LUP ARCHIVES 2017

fine-loamy dark reddish brown to dark yellowish and neutral to slightly acidic reaction on gentle slopes. The dominant soil great groups are *Haplustepts, Haplustalfs, Plinthustalfs, Paleustalfs, Rhodustalfs* (Fig. 9.14), *Ustorthents* and *Endoaquepts*. The soil series identified in this group are Chougel series, Mrigindihi series and Bhubaneswar series.

Mrigindihi Series: *Ultic Paleustalfs*

These soils are developed in old alluvium of the undulating interfluve plain in the western part of Midnapore and the southern part of Bankura District in West Bengal. Typically, Mrigindihi soils have yellowish red, very strongly acid, sandy loam A horizons, yellowish red, strongly acid, sandy loam BA horizons and yellowish red to red, strongly to very strongly acid, sandy clay loam Bt horizons. The CEC ranged from 5.8 to 9.6 cmol (+) kg^{-1} of soil. These soils are somewhat excessively drained with moderate permeability and are cultivated to rainfed vegetables and direct sown rice, wheat, potato and winter vegetables grown under irrigation (Ray et al. 2014b).

9.2.13 Hot Sub-humid Alluvial Soils

Area: 5.3 Mha (1.6% of total geographical area of India).

Fig. 9.10 Achmatti series of Belwataki, Dharwad, *Sodic Haplusterts* (AER-8). *Source* NBSS & LUP ARCHIVES 2017

Location and AER

Agro-ecological Region 13

These soils occur in hot sub-humid regions covering the lower Gangetic plains in south western and central parts of Bihar. It is distributed in the districts of Buxar, Bhojpur, Patna, Nalanda, Jamui, Banka, Bhagalpur, Munger, Paschim and Purbi Champaran, Muzaffarpur, Vaishali, Saran, Siwan, Samastipur, Begusarai and Khagaria in Bihar.

Description of Soils

The soil-scapes in the region are level to very gently sloping plain. These occur in association with imperfectly drained. The other soils in this region are greyish brown to yellowish brown fine-loamy, imperfectly to poorly drained and slightly acidic to neutral in soil reaction. The soils on the old alluvial plains are brown, fine textured, poorly drained and slightly acidic (Fig. 9.15).

Fig. 9.11 Vijayapura series from Tumkur District, Karnataka; Typic Haplustalfs (AER-9). *Source* NBSS & LUP ARCHIVES 2017

Hathiapathar Series: *Typic Ochraqualfs*

These soils are developed on weathered granite–gneiss and occur on toeslopes of peneplaned plateaus in Santal Parganas of Bihar. They are deep, clayey, imperfectly drained and strongly acid. The CEC ranges from 12 to 18 cmol (+) kg^{-1} of soil. The AWC is medium to high. These soils are mostly cultivated to rice, and the productivity potential is medium (Murthy et al. 1982).

9.2.14 Warm Sub-humid to Humid Sub-montane Shallow and Skeletal Soils

Area: 15.6 Mha (4.7% of total geographical area of India).

Location and AER

Agro-ecological Region 14

These soils are confined to Western Himalayas of Jammu and Kashmir, Himachal Pradesh and north-western hilly areas of Uttarakhand. These soils occur in Kathua, Udhampur, Jammu, Reasi, Punch, parts of Jammu/Kashmir, Chamba, Lahul and Spiti, Kangra, Una, Hamirpur, Bilaspur, Manali, Kullu, Kinnaur, Shimla, Son and Sirmaur in Himachal Pradesh; Uttarkashi, Dehradun, Tehri Garhwal, Pauri Garhwal, Rudrapur, Chamoli, Pithoragarh, Bageshwar, Almora, Nainital, Champawat and Udham Singh Nagar in Uttarakhand.

Description of Soils

The major soils occurring in the hill and mountain region are shallow to deep, medium to high in organic carbon, classified as brown forest and podzolic soils. The soils are characterized as medium deep, loamy-skeletal, mixed calcareous with high stoniness (>80%) in the sub-soils. The soils of Shiwaliks zone are deep fine-loamy with good water-holding capacity, nutrient retention capacity and capable to produce good yield for climatically adapted crops. The soils of piedmont *tarai* zone are fine sand, silt and clay occurring on nearly level to gently sloping lands.

The major soil series falling in this class are Gogji Pather, Wahthora, Shinwali and Haldi series (Fig. 9.16).

Haldi Series: *Typic Haplustalfs*

These soils are formed in the recent alluvium and occur in the sub-montane *Tarai* region of Uttar Pradesh. These soils are deep, coarse-loamy, well-drained and neutral in reaction. These soils have greyish brown to dark greyish brown, loamy, neutral A horizon, very dark greyish brown to brown, silt loamy, mildly alkaline Bt horizon, dark brown to brown, sandy loam, neutral to mildly alkaline BC horizons over greyish brown to dark grey to grey, loamy sand, mildly to strongly alkaline C horizon. The CEC ranges between 3.4 and 17.4 cmol (+) kg^{-1} of soil. The soils have medium to high available moisture capacity. These soils are cultivated to paddy, maize, soybean and wheat. These soils are highly productive with good air–water relationship (Ray et al. 2014b).

9.2.15 Hot Sub-humid Loamy to Clayey Alluvial Soils

Area: 21.7 Mha (6.6% of total geographical area of India).

Table 9.3 Selected physico-chemical properties of series representing semi-arid ecosystem

Horizon	Depth	Colour	Sand (2–0.05)	Silt (0.05–0.002)	Clay (<0.002)	Textural class	pH (1:2 water)	CEC (cmol (+) kg^{-1} of soil)
9.2.6 Hot semi-arid coarse-loamy alluvial soils—Fatehpur series: *Inceptic Haplustalfs*								
Ap1	0–10	Pale brown 10YR 4/3 M	79.4	14.8	5.8	ls	7.3	6.1
Bw1	10–21	Brown 10YR 4/3 M	80.6	12.7	6.7	ls	7.6	6.1
Bw2	21–36	Strong brown 7.5YR 4/6 M	75.8	15.6	8.5	ls	7.8	7.8
Bt1	36–53	Brown 7.5YR 4/4 M	69.8	19.8	10.4	sl	7.8	10.4
Bt2	53–73	Brown 7.5YR 4/4 M	68.0	19.9	12.1	sl	7.9	11.3
Bt3	73–95	Brown 7.5YR 4/4 M	65.2	23.4	11.4	sl	8.2	11.3
Bt4	95–115	Brown 7.5YR 4/4 M	63.5	24.8	11.6	sl	8.2	11.4
Bt5	115–137	Very dark grey to dark grey 7.5YR3.5/4 M	66.3	22.5	11.2	sl	8.3	11.3
Bt6	137–155+	Brown to strong brown 7.5 YR 4/5 M	66.3	22.6	11.2	sl	8.4	10.5
9.2.7 Hot semi-arid old alluvial soils—Singpura series: *Vertic Haplustalfs*								
Ap	0–13	Dark yellowish brown 10 YR 4/6 M	50.6	32.1	17.3	l	8.4	10.0
Bt1	13–37	Dark yellowish brown to brown 10 YR 4/3 M	37.8	32.6	29.6	cl	8.6	15.0
Bt2	37–68	Dark yellowish brown 7.5 YR 4/3 M	32.2	33.9	33.9	cl	8.7	17.0
Bt3	68–102	Brown 7.5 YR 4/4 M	38.5	34.3	27.2	cl	8.5	18.0
Bt4	102–146	Dark yellowish brown 7.5YR 4/4 M	31.7	37.6	30.7	cl	8.4	18.0
Bt5	146–158	Dark yellowish brown to brown 10 YR 4/4 M	31.7	39.8	28.5	cl	8.8	16.0
9.2.8 Hot semi-arid to dry sub-humid alluvial and *Tarai* soils—Basiaram series: *Udic Haplustepts*								
Ap	0–13	Dark greyish brown 10 YR 4/2 M	24.2	65.5	10.3	sil	7.2	6.3
AB	13–28	Brown 10 YR 5/3 M	24.5	60.5	15.0	sil	7.4	7.5
Bw1	28–44	Dark yellowish brown 10 YR 4/4 M	20.5	58.0	21.5	sil	7.4	8.5
Bw2	44–82	Yellowish brown 10 YR 5/4 M	16.2	58.5	25.3	sil	7.3	10.6
Bw3	82–110	Yellowish brown 10 YR 5/4 M	21.7	57.0	21.3	sil	7.3	9.8
BC1	110–136	Brown 10 YR 5/3 M	11.6	65.8	22.6	sil	7.2	10.1
BC2	136–170	Brown 10 YR 5/3 M	8.3	62.7	29.0	sicl	7.1	10.2
9.2.9 Hot semi-arid moderately deep black soils—Sarol Series: *Typic Haplusterts*								
Ap	0–12	Very dark greyish brown 2.5 Y 3.5/2 M	3.3	42.2	54.5	c	8.4	51.3
Bw	12–34	Very dark greyish brown 2.5 Y 3.5/2 M	2.8	44.2	53.0	c	8.6	47.8
Bss1	34–64	Very dark greyish brown 2.5 Y 3.5/2 M	3.6	37.7	58.7	c	8.6	53.9
Bss2	64–101	Very dark greyish brown 2.5 Y 3.5/2 M	3.7	35.5	60.8	c	8.6	44.3
Bss3	101–129	Very dark greyish brown 2.5 Y 3.5/2 M	2.3	40.1	57.6	c	8.5	40.0
Bss4	129–150	Olive brown to light olive brown 2.5 Y 4.5/4 D	2.2	37.2	60.6	c	8.5	54.7

(continued)

Table 9.3 (continued)

Horizon	Depth	Colour	Sand (2–0.05)	Silt (0.05–0.002)	Clay (<0.002)	Textural class	pH (1:2 water)	CEC (cmol (+) kg^{-1} of soil)
9.2.10 Hot semi-arid with mixed red and black soils—Kasireddippalli series: *Typic Haplusterts*								
Ap	0–8	Very dark greyish brown 10 YR 3/2 D	23.3	29.6	47.1	c	8.0	42.6
Bw	8–18	Very dark greyish brown 10 YR 3/2 D	22.6	25.6	51.8	c	8.1	49.2
Bss1	18–32	Very dark greyish brown 10 YR 3/2 D	21.2	25.9	52.9	c	8.2	50.4
Bss2	32–44	Very dark greyish brown 10 YR 3/2 M	20.4	25.0	54.6	c	8.2	48.7
Bss3	44–65	Very dark greyish brown 10 YR 3/2 M	17.0	24.4	58.6	c	8.1	57.4
Bss4	65–87	Very dark greyish brown 10 YR3.5/2 M	14.3	24.9	60.8	c	8.2	54.7
Bss5	87–115	Very dark greyish brown 10 YR4.5/4 M	13.0	22.4	64.6	c	8.1	54.7
9.2.11 Hot semi-arid with red loamy soils—Vijayapura series: *Oxic Haplustalfs*								
Ap1	0–7	Yellowish red 5 YR 4/6 M	72.5	11.0	16.5	sl	5.6	3.0
Ap2	7–21	Yellowish red 5 YR 4/6 M	69.7	12.5	17.8	sl	5.2	3.0
B21t	21–46	Yellowish red 5 YR 4/6 M	57.8	12.7	29.5	scl	5.6	4.2
B22t	46–87	Yellowish red 5 YR 4/6 M	52.0	9.5	38.5	sc	5.6	4.5
B23t	87–104	Dark red 2.5 YR 3/4 M	49.6	12.8	37.6	sc	6.1	5.2
IIC1	104–135	Dark red 2.5 YR 3/6 M	57.9	12.0	30.1	gscl	6.5	4.7
IIC2	135–152+	Dark red 2.5 YR 3/6 M	45.3	17.0	37.7	sc	6.5	5.0

Location and AER

Agro-ecological Region 15

These soils cover parts of Bihar and West Bengal and southern parts of Assam. These soils are distributed in the districts of Sitamarhi, Madhubani, Kishanganj, Madhepura, Kosi, Samastipur and Darbhanga of Bihar, 24 Parganas (North and South), Howrah, Nadia, parts of Bankura, Barddhaman, Birbhum, Baharampur and Malda in West Bengal.

Description of Soils

Soils are slight to strongly acidic and have low to moderate base saturation which occurs on nearly level to gently sloping plains. In the upper part of soil-scape, soils are reddish yellow to red, strongly to slightly acidic sandy loam developed over weathered granitic gneiss and are prone to erosion. In the lower part of the soil-scape, the soils are greyish brown to grey, deep to very deep in the flood plains. The soils have aquic moisture regimes, slight to strongly acidic and imperfectly to poor drainage in the low-lying plains. These are subjected to seasonal floods during rainy season. At places, deep, sandy loam and dark yellowish brown soils have low water-holding capacity and suffer from drought in summer. Soils belonging to *Ustifluvents, Endoaquepts, Endoaqualfs, Plinthustalfs* and *Ustipsamments* great groups occur in level to very gently sloping plains. The major soil series in this class are Kanagarh, Madhpur, Hanrgram and Jagdishpur series.

Madhpur Series: *Chromic Vertic Endoaqualfs*
These soils are formed in alluvium and occur on very gently sloping lower plains in Barddhaman District of West Bengal (Fig. 9.17). These soils are very deep, fine textured, moderately acidic, imperfectly drained with moderate to slow permeability. These soils have pale brown to brown, moderately acid, loam A horizons, and olive brown to light yellowish brown, slightly to strongly alkaline, clay loam to clay Bt horizons, with distinct to prominent mottles of yellowish brown to brownish yellow colour. The CEC ranges between 17 and 37 cmol (+) kg^{-1} of soil. Madhpur soils are mainly cultivated to paddy, wheat and mustard with high productivity potential (Ray et al. 2014b).

Selected physico-chemical properties of series representing sub-humid ecosystem are given in Table 9.4.

9.2.16 Warm Humid to Per-Humid Loamy to Clayey Alluvial Soils

Area: 9.7 Mha (3.0% of total geographical area of India).

Fig. 9.12 Hot sub-humid with moderately deep black soils; Typic Haplusterts (AER-10). *Source* NBSS & LUP ARCHIVES 2017

Location and AER

Agro-ecological Region 16

These soils are present in northern hilly parts of West Bengal and Brahmaputra valley. These are distributed in Katihar and Purnia districts of Bihar, parts of North Cachar Hills, Karbi Anglong, Golaghat, Jorhat, Sibsagar, Dirbrugarh, Tinsukia, Dhemaji, Lakhimpur, Sonitpur, Nagaon, Tezpur, Barpeta, Bongaigaon, Dhubri, Morigaon, Kamrup, Goalpara, Kokrajhar, Nalbari in Assam, west Tripura, North Tripura, South Tripura, Dhalia districts of Tripura and Dinajpur (Uttar and Dakshin) Jalpaiguri, Cooch Behar and parts of Darjeeling in West Bengal.

Description of Soils

The dominant soils in the region are rich in organic matter with moderate to low base status. The soil-scape of the area is gently sloping chaur lands with soils of *Typic/Aeric Fluvaquents, Typic/Mollic Udifluvents* and *Aeric/Vertic*

Fig. 9.13 Hot sub-humid red and yellow soils; Typic Haplustalfs (AER-11). *Source* NBSS & LUP ARCHIVES 2017

Epiaquepts and *Endoaquepts*. These are grey to dark grey in colour and poorly drained with sandy loam to sandy clay loam soils. The other soils on the chaur lands are grey to light grey in colour developed over alluvium on nearly level flood plain, subjected to overflow during rainy season (Fig. 9.18).

9.2.17 Warm Per-Humid Shallow and Skeletal Soils

Area: 8.7 Mha (2.6% of total geographical area of India).

Location and AER

Agro-ecological Region 17

These soils occur in hilly states of Arunachal Pradesh and Sikkim. These soils are mainly distributed in East Kameng,

Fig. 9.14 Hot sub-humid red and lateritic soils, Ganjam District, Odisha; *Typic Rhodustalfs* (AER-12). *Source* NBSS & LUP ARCHIVES 2017

Papum Pare, Upper and Lower Subansiri, West and East Siang, Upper Siang, Dibang Valley, Lohit, Changalang, Tawang and West Kameng in Arunachal Pradesh.

Description of Soils

These soils are fine-loamy to loamy-skeletal, deep, dark brown to yellowish brown, moderately to strongly acidic and excessively drained with high organic carbon occur on steeply sloping hills and valleys. These soils are susceptible

Fig. 9.15 Hot sub-humid alluvial soils; Typic *Ustifluvents* (AER-13). *Source* NBSS & LUP ARCHIVES 2017

to severe erosion. The soils are suitable for forest plantations. The dominant great groups are *Dystrudepts, Hapludolls* (Fig. 9.19), *Argiudolls, Entrudepts* and *Udorthents* on steeply sloping hills and valleys. The major soil series in this class are Jaihing series and Gemotali series.

Jaihing Series: *Typic Dystrochrepts*

These soils occur on gently sloping piedmont plains in Assam valley of Lakhimpur District of Assam. These soils are deep, well-drained and strongly to extremely acid. The CEC ranges between 13 and 16 cmol (+) kg^{-1} of soil. The AWC is medium. These soils are mainly cultivated to tea (Murthy et al. 1982).

9.2.18 Warm Per-Humid Red and Yellow Soils

Area: 10.0 Mha (3.0% of total geographical area of India).

Fig. 9.16 Warm sub-humid to humid sub-montane shallow and skeletal soils; Typic *Hapludalfs* (AER-14). *Source* NBSS & LUP ARCHIVES 2017

Fig. 9.17 Hot sub-humid loamy to clayey alluvial soils Madhpur series (West Bengal), *Chromic Vertic Endoaqualfs* (AER-15). *Source* NBSS & LUP ARCHIVES 2017

Location and AER

Agro-ecological Region 18

These soils occur in the Purvanchal region of northeast comprising states of Meghalaya, Manipur, Nagaland and Mizoram.

Description of Soils

The soils in the area are shallow to deep loamy red and yellow soils. *Lithic Hapludalfs* and *Humic Dystrudepts* are some of the extensive soils occurring in the Meghalaya and Mizoram region. These soils are red dark greyish brown to strong brown in colour and are strongly acidic with pH < 5.0 and contain fairly good amount of organic carbon at the surface and in the subsurface. These are well-drained. Other soils in the valley are *Haplaquepts* and *Humaquepts* in the state of Manipur, occurring on nearly level-to-level inter-hilly area. These are dark greyish to yellowish brown in colour, slight to strongly acidic.

The major soil series in this class are Phullen, Dialong and Selsekgiri series (Fig. 9.20).

Dialong Series: *Ultic Hapludalfs*

These soils occur on steep middle hill slopes. They are moderately deep to deep, loamy-skeletal, well-drained and strongly to extremely acidic. The CEC ranges from 14 to

Table 9.4 Selected physico-chemical properties of series representing sub-humid ecosystem

Horizon	Depth	Colour	Sand (2–0.05)	Silt (0.05–0.002)	Clay (<0.002)	Textural class	pH (1:2 water)	CEC (cmol (+) kg^{-1} of soil)
9.2.16 Hot sub-humid moderately deep black soils—Kheri series: *Typic Haplusterts*								
Ap	0–14	Light olive brown 2.5 Y 4.5/3 M	18.4	30.5	51.1	c	7.5	47.9
Bw1	14–32	Very dark greyish brown 10 YR 3/2 M	16.6	29.7	53.7	c	7.6	47.9
Bw2	32–61	Very dark greyish brown 10 YR 3/2 M	16.8	36.9	46.3	c	7.6	53.5
Bss1	61–82	Very dark greyish brown 10 YR 3/2 M	6.0	40.4	53.6	c	7.6	49.3
Bss2	82–112	Very dark greyish brown 10 YR 3/2 M	14.8	38.6	46.6	c	7.8	49.3
Bss3	112–138	Very dark greyish brown 10 YR 3/2 M	16.4	39.0	44.6	c	7.7	49.3
Bss4	138–156	Very dark greyish brown 10 YR 3/2 M	15.9	37.4	46.7		8.0	52.1
9.2.17 Hot sub-humid red and yellow soils—Marha series: *Entic Chromusterts*								
Ap	0–12	Dark greyish brown 2.5 Y 4/2 M	11.7	44.1	44.2	sic	8.0	48.0
A12	12–44	Dark greyish brown 2.5Y3/2 M	9.2	43.6	47.2	sic	8.0	52.2
A13	44–70	Dark greyish brown 2.5 Y 4/2 M	5.5	43.4	51.1	sic	8.0	54.4
A14	70–105	Dark greyish brown 2.5 Y 5/2 M	7.0	45.1	47.9	sic	8.0	47.4
A15	105–137	Dark greyish brown 2.5 Y 4/2 M	5.3	49.4	45.3	sic	8.0	46.4
AC	137–180	Dark greyish brown 2.5 Y 4/2 M	10.6	49.0	40.4	sicl	8.1	47.5
9.2.18 Hot sub-humid red and lateritic soils—Mrigindihi series: *Ultic Paleustalfs*								
A	0–15	Yellowish red 5YR 4/8 M	69.9	16.0	14.1	sl	4.9	5.8
AB	15–33	Yellowish red 5YR 4/6 M	67.9	14.7	17.4	sl	4.8	6.2
BA	33–59	Yellowish red 5YR 4/6 M	64.8	15.2	20.0	sl	5.2	7.0
Bt1	59–87	Yellowish red 5YR 4/6 M	60.0	16.9	23.1	scl	5.0	8.0
Bt2	87–114	Yellowish red 5YR 4/6 M	62.4	14.1	23.5	scl	5.1	8.8
Bt3	114–141	Dark red 2.5 YR3.5/6 M	60.6	16.4	23.0	scl	4.9	9.2
Bt4	141–165	Dark red 2.5 YR3.5/6 M	57.4	17.4	25.2	scl	4.4	9.6
9.2.19 Hot sub-humid alluvial soils—Hathiapathar series: *Typic Ochraqualfs*								
Ap	0–13	Dark greyish brown to greyish brown 2.5 Y 4.5/2 M	31.9	52.5	15.6	sil	5.3	12.5
A3	13–24	Greyish brown 2.5 Y 5/2 M	27.7	53.1	19.2	sil	5.4	12.5
B21t	24–47	Light brownish grey 2.5 Y6/2 M	26.4	34.5	39.1	cl	5.5	17.5
B22t	47–71	Light grey 2.5 Y 6.5/2 M	19.4	40.9	39.7	cl	5.5	16.4
B23tg	71–101	Light grey 2.5 Y 7/2 M	25.0	44.3	30.7	cl	5.5	14.8
B24tg	101–127	Light grey 5 Y 7/2 M	15.8	55.5	28.7	sicl	5.3	16.2

(continued)

Table 9.4 (continued)

Horizon	Depth	Colour	Sand (2–0.05)	Silt (0.05–0.002)	Clay (<0.002)	Textural class	pH (1:2 water)	CEC (cmol (+) kg^{-1} of soil)
9.2.20 Warm sub-humid to humid sub-montane shallow and skeletal soils—Haldi series: *Typic Haplustalfs*								
Ap	0–18	Dark greyish brown 10 YR 3.5/2 D	34.4	50.2	15.4	sil	7.4	14.2
Bt1	18–36	Brown 10 YR 4/3 M	23.9	55.9	20.3	sil	7.6	17.5
Bt2	36–58	Brown 10 YR 4/3 M	26.4	54.9	18.7	sil	7.4	16.7
Bt3	58–76	Brown 10 YR 4/2 M	43.1	43.9	13.0	l	7.6	10.3
BC1	76–95	Greyish brown 10 YR 4/2 M	70.1	22.0	7.9	sl	7.7	7.5
BC2	95–117	Greyish brown 2.5 Y 5/2 M	82.4	13.1	4.5	ls	7.8	5.4
BC3	117–146	Greyish brown 2.5 YR 5/2 M	76.4	18.6	5.0	gsl	8.1	5.0
BC4	146–164	Greyish brown 2.5 YR 5/2 M	87.6	8.5	3.8	ls	8.6	6.8
9.2.21 Hot sub-humid loamy to clayey alluvial soils—Madhpur series: *Chromic Vertic Endoaqualfs*								
Ap	0–9	Brown 10 YR 4/3 M	34.9	44.0	21.2	l	5.7	8.7
Bt1	9–27	Light Olive brown 2.5 Y 5/4 M	30.7	40.6	28.7	cl	7.8	12.2
Bt2	27–44	Dark greyish brown to olive brown 2.5 Y 5/3 M	22.2	40.9	36.9	cl	8.3	21.7
Bt3	44–56	Olive brown to light olive brown 2.5 Y 4.5/4 M	20.8	37.5	41.6	c	8.3	24.4
Bt4	56–80	Light olive brown 2.5 Y 5/4 M	20.9	38.2	40.9	c	8.4	23.5
Bt5	80–99	Light olive brown 2.5 Y 5/3 M	20.5	38.0	41.4	c	8.4	23.5
Bt6	99–109	Light olive brown 2.5 Y 5/4 M	22.1	36.9	41.1	c	8.4	20.9
Bt7	109–150+	Light Yellowish brown 2.5Y6/4 D	21.7	38.0	40.3	c	8.5	22.6

18 cmol (+) kg^{-1} of soil. Dialong soils are under subtropical moist deciduous forests, occasionally under Jhum cultivation (Murthy et al. 1982).

Selected physico-chemical properties of series representing humid/per-humid ecosystem are given in Table 9.5.

9.2.19 Hot Sub-humid Coastal and Deltaic Alluvial Soils

Area: 11.3 Mha (3.5% of total geographical area of India).

Location and AER

Agro-ecological Region 19

These soils occur in coastal plains of Odisha and West Bengal, Subarnarekha and lower Gangetic delta, coastal Andhra Pradesh, Krishna and Godavari delta, coastal region of Tamil Nadu and Puducherry, Andaman and Nicobar Islands.

Description of Soils

The major soils in this region include the alluvium derived from soils of coastal and deltaic plains. The soils (*Vertic Endoaquepts*, *Vertic Fluvaquents* and *Typic Endoaquepts*) developed on deltaic plain (Mahanadi) are deep, susceptible to severe flooding, slight to severely acidic and have wide cracks during non-rainy season. The soils of alluvial plains (Utkal Plains and Bengal Plains) are deep, slightly acidic to neutral, grey to dark grey and are imperfectly to poorly drained (Fig. 9.21). The soils (*Typic Ustipsamments, Rhodic Paleustalfs* and *Fluventic Haplustepts*) in the semi-arid part of the region are deep, fine, neutral to moderately alkaline and sandy at places.

The major soil series in this class are Motto series and Kalathur series.

Kalathur Series: *Sodic Haplusterts*

These soils are formed in alluvium occurring in the Cauvery delta plain, Thanjavur District of Tamil Nadu. These soils have dark grey to very dark grey, very strongly alkaline clay loam to clay A horizons, and dark grey, very strongly alkaline, clay B horizons over dark greyish brown, very strongly alkaline clay C horizons. These soils are moderately well-drained with slow to very slow permeability and are mainly cultivated to rice (Ray et al. 2014c).

Fig. 9.18 Warm humid to per-humid loamy to clayey alluvial soils, *Udifluvents* (AER-16). *Source* NBSS & LUP ARCHIVES 2017

Fig. 9.19 Warm per-humid shallow and skeletal soils, Mangan block, North Sikkim District; *Typic Hapludolls* (AER-17). *Source* NBSS & LUP ARCHIVES 2017

9.2.20 Hot Humid/Per-Humid Red and Lateritic and Alluvial Soils

Area: 9.5 Mha (2.9% of total geographical area of India).

Location and AER

Agro-ecological Region 20

These soils occur in hot humid/per-humid regions in the north central and south Sahyadris falling in the states of Maharashtra, Goa, Karnataka and Kerala.

Description of Soils

The major soils in this region are red and lateritic soils (Fig. 9.22). Soils of the hills are yellowish red to dark reddish brown, moderately to strongly acidic, clayey-skeletal. The soils of the interhill valleys and low-lying areas in the region are dark brown to dark reddish brown in colour, moderate to strongly acidic and formed on alluvium with low CEC and low base saturation. The soils of Lakshadweep Island, on the other hand, are highly calcareous and sandy in nature. The major soil series in this class are Thekkady, Trivandrum, Kunnamangalam and Ambalapuzha series.

Trivandrum Series: *Oxic Dystropepts*

These soils occur on laterite mounds and level lands in Trivandrum District of Kerala. These soils are moderately deep, clayey-skeletal, well-drained and strongly to very strongly acidic in reaction. The CEC is between 4 and

Fig. 9.20 Warm per-humid red and yellow soils; Typic *Dystrudepts* (AER-18). *Source* NBSS & LUP ARCHIVES 2017

7 cmol (+) kg^{-1} of soil. The AWC is medium to high. Trivandrum soils are mainly cultivated to tapioca, ginger and pepper along with coconut and cashew plantations, with medium productivity potential (Murthy et al. 1982).

Selected physico-chemical properties of series representing coastal ecosystem are presented in Table 9.6.

9.3 Benchmark Soils and Soil Taxonomy: Correlation

The benchmark soils with corresponding soil series and classification (soil orders and sub-orders) in different locations are presented in Tables 9.7 and 9.8. Soil classification can also be used in transfer of technology. Altogether seven major soil orders occurring in India are Inceptisols (36.8%), Entisols (20.01%), Alfisols (13.75), Vertisols (10.09%), Ultisols (3.25%), Aridisols (3.23%) and Mollisols (0.48%). These soil orders cover 18 benchmark sites (series) under Alfisols, 16 benchmark sites under Inceptisols and 13 benchmark sites under Vertisols. In addition, Entisols, Aridisols, Ultisols, Mollisols and Histosols cover 7, 2, 1, 1 and 1 benchmark sites, respectively.

In fact, there are twenty benchmark soils identified which have sixty-four soil series sites throughout India qualifying seven soil orders, viz. Entisols, Alfisols, Vertisols, Ultisols, Aridisols and Mollisols (Table 9.8). Out of total, 18 benchmark sites (series) comes under Alfisols covering Gogji Pather, Dialong, Selsekgiri, Zarifa Viran, Sakit, Itwa, Jagadishpur, Gemotali, Chougel, Jamakhandi, Tyamagondalu, Vijayapura, Channasandra, Patancheru, Palathurai, Pusaro, Hathiapathar and Mrigindihi, whereas 16 benchmark

Table 9.5 Selected physico-chemical properties of series representing humid/per-humid ecosystem

Horizon	Depth	Colour	Sand (2–0.05)	Silt (0.05–0.002)	Clay (<0.002)	Textural class	pH (1:2 water)	CEC (cmol (+) kg^{-1} of soil)
7.2.17. Warm humid to per-humid loamy to clayey alluvial soils—Jaihing series: *Typic Dystrochrepts*								
A1	0–16	Yellowish brown 10 YR5/4 M	54.4	21.3	24.3	l	4.2	13.6
B1	16–61	Yellowish brown 10YR5/4 M	51.2	22.8	26.0	cl	4.5	14.4
B2	61–92	Strong brown 7.5 YR 5/6 M	49.4	24.6	26.0	cl	4.5	16.0
C	92–150	Yellowish brown 10YR 5/4 M	61.0	18.1	20.9	l	4.6	13.2
7.2.18. Warm per-humid red and yellow soils—Dialong series: *Ultic Hapludalfs*								
A1	0–15	Dark brown 10YR 4/3 M	44.6	28.0	27.4	si cl	4.5	14.4
B2t	15–29	Brown to dark brown 7.5 YR4/4 M	34.6	29.0	36.4	si cl	4.4	17.6
B3	29–61	Strong brown 7.5 YR 5/6 M	44.6	26.0	29.4	gsicl	4.4	14.6
C3	61+	Weathered grey and pink						

Fig. 9.21 Coastal and deltaic alluvial soils, Gosaba block, South 24 Parganas District, West Bengal; *Typic Haplaquepts* (AER-19). *Source* NBSS & LUP ARCHIVES 2017

sites (Wahthora, Kanjli, Nabha, Sadhu, Hirapur, Bijaipur, Basiaram, Madhupur, Hanrgram, Jaihing, Kamliakheri, Singpura, Chinnaloni, Trivandrum, Coimbatore and Motto) under Inceptisols and 13 benchmark sites include Sarol, Kheri, Marha, Jambha, Linga, Barsi, Achmatti, Hungund, Kagalgomb, Raichur, Kadirabad, Kasireddipalli and Kalathur under Vertisols. Likewise, there are seven sites (Chinwali, Phullen, Fatehpur, Kanagarh, Thar, Masitawali and Chomu) under Entisols, five sites (Chirai, Pali, Adesar, Dandi and Lakhpat) under Aridisols and two sites (Thekkady and Kunnamangalam) under Oxisols. Besides, each one benchmark series is found in Ultisols (Bhubaneswar), Mollisols (Haldi) and Histosols (Ambalapuzha) belonging to Orissa, Uttar Pradesh and Kerala, respectively (Table 9.8). The benchmark sites (series) are presented in location numbers on soil order map of India following the Table 9.8. Soil order map showing benchmark sites (series) following location number (Fig. 9.23).

Fig. 9.22 Hot humid/per-humid red and lateritic and alluvial soils, Santhanpara, Idukki, Kerala, *Ustic Kandihumults* (AER-20). *Source* NBSS & LUP ARCHIVES 2017

Table 9.6 Selected physico-chemical properties of series representing coastal ecosystem

Horizon	Depth	Colour	Sand (2–0.05)	Silt (0.05–0.002)	Clay (<0.002)	Textural class	pH (1:2 water)	CEC (cmol (+) kg^{-1} of soil)
7.2.19. Hot sub-humid coastal and deltaic alluvial soils—Kalathur series: *Sodic Haplusterts*								
Ap	0–14	Dark grey to very dark grey 10 YR3.5/1 M	43.4	17.9	38.7	c	9.1	38.9
Bw	14–49	Dark grey to very dark grey 10 YR 3.5/1 M	38.6	16.3	45.1	c	9.7	43.9
Bss	49–87	Dark grey 10 YR 4/1 M	27.1	16.5	56.4	c	9.4	54.1
BC	87–104	Dark grey 10 YR 4/1 M	22.6	18.7	58.7	c	9.2	57.9
Ck	104–120	Dark greyish brown 10 YR 4/2 M	24.1	17.6	58.3		9.2	52.8
7.2.20 Hot humid/per-humid red and lateritic and alluvial soils—Trivandrum series: *Oxic Dystropepts*								
Ap	0–9	Yellowish red 5YR 4/6 M	38.5	9.9	51.6	gc	4.5	6.7
A12	9–25	Yellowish red 5YR 4/6 M	39.9	8.7	51.4	gc	4.5	5.8
B	25–52	Dark red 2.5 YR 3/6 M	37.9	9.1	53.0	gc	4.8	4.0
C	52–84+	variegated blue, white, yellow and dark red	34.1	13.2	52.7		5.0	4.5

Table 9.7 Benchmark soils and their taxonomic grouping (USDA Soil Taxonomy)

AER	Benchmark soils	Soil series	Subgroup	Soil orders	Sub-orders
1	Cold arid shallow skeletal soils	–	*Typic Cryorthents*	*Entisols*	*Orthents*
2	Hot arid desert and saline soils	Masitawali series Thar series	*Torrifluventic Haplustepts* *Typic Torripsamments*	*Inceptisols* *Entisols*	*Ustepts* *Psamments*
3	Hot arid red and black soils	Teligi series	*Sodic Haplusterts*	*Vertisols*	*Usterts*
4	Hot semi-arid coarse-loamy alluvial soils	Fatehpur series Kanjili series Nabha series Zarifa Viran series Sadhu Khoh series	*Inceptic Haplustalfs* *Typic Haplustepts* *Typic Haplustepts* *Typic Natrustalfs* *Vertic Haplustepts* *Typic Ustipsamments*	*Alfisols* *Inceptisols* *Inceptisols* *Alfisols* *Inceptisols* *Entisols*	*Ustalfs* *Ustepts* *Ustepts* *Ustalfs* *Ustepts* *Psamments*
5	Hot semi-arid old alluvial soils	Chirai series Pali Chomu Singpura series	*Typic Haplocambids* *Lithic Calciorthids* *Typic Ustipsamments* *Vertic Haplustalfs*	*Aridisols* *Aridisols* *Entisols* *Alfisols*	*Cambids* *Orthids* *Psamments* *Ustalfs*
6	Hot semi-arid to dry sub-humid alluvial and *Tarai* soils	Hirapur Sakit Itwa Bijaipur Basiaram series	*Vertic Natrustalfs* *Vertic Haplustalfs* *Typic Endoaqualfs* *Udic Ustochrepts* *Udic Haplustepts*	*Alfisols* *Alfisols* *Alfisols* *Inceptisols* *Inceptisols*	*Ustalfs* *Ustalfs* *Aqualfs* *Ochrepts* *Ustepts*
7	Hot semi-arid moderately deep black soils	Sarol series Kamliakheri Jambha Linga Barsi Jhamkhandi Dandi	*Typic Haplusterts* *Lithic Vertic Ustochrepts* *Typic Chromusterts* *Udic Chromusterts* *Typic Chromusterts* *Typic Paleustalfs* *Typic Salorthids*	*Vertisols* *Inceptisols* *Vertisols* *Vertisols* *Vertisols* *Alfisols* *Aridisols*	*Usterts* *Ochrepts* *Usterts* *Usterts* *Usterts* *Ustalfs* *Orthids*
8	Hot semi-arid with mixed red and black soils	Adesar Hungund Kagalgomb Kadirabad Chinnaloni Patancheru Lakhpat series Achmatti series Raichur series Kasireddipalli series	*Typic Paleargids* *Typic Chromusterts* *TYpic Chromusterts* *Typic Chromusterts* *Paralithic vertic ustropepts* *Udic Rhodustalfs* *Typic Natrargids* *Sodic Haplusterts* *Typic Haplusterts* *Typic Haplusterts*	*Aridisols* *Vertisols* *Vertisols* *Vertisols* *Inceptisols* *Alfisols* *Aridisols* *Vertisols* *Vertisols* *Vertisols*	*Argids* *Usterts* *Usterts* *Usterts* *Tropepts* *Ustalfs* *Argids* *Usterts* *Usterts* *Usterts*
9	Hot semi-arid with red loamy soils	Tyamagondalu, Palathurai, Coimbatore Channasandra Vijayapura series.	*Oxic Paleustalfs* *Typic Haplusterts* *Oxic Rhodustalfs* *Oxic Haplustalfs* *Oxic Haplustalfs*	*Alfisols* *Vertisols* *Alfisols* *Alfisols* *Alfisols*	*Ustalfs* *Usterts* *Ustalfs* *Ustalfs* *Ustalfs*
10	Hot sub-humid with moderately deep black soils	Kheri series	*Typic Haplusterts*	*Vertisols*	*Usterts*
11	Hot sub-humid red and yellow soils	Marha series	*Entic Chromusterts*	*Vertisols*	*Usterts*
12	Hot sub-humid red and lateritic soils	Chougel Mrigindihi Bhubaneswar series.	*Plinthustalfs* *Ultic Paleustalfs* *Typic Haplustults*	*Alfisols* *Alfisols* *Ultisols*	*Ustalfs* *Ustalfs* *Ustults*
13	Hot sub-humid alluvial soils	Hathiapathar series	*Typic Ochraqualfs*	*Alfisols*	*Aqualfs*

(continued)

Table 9.7 (continued)

AER	Benchmark soils	Soil series	Subgroup	Soil orders	Sub-orders
14	Warm sub- humid to humid sub-montane shallow and skeletal soils	Gogji Pather series Wahthora series Shinwali series Haldi series	*Typic Haplustalfs* *Mollic Haplaquepts* *Lithic Ustorthents* *Typic Haplustalfs*	*Alfisols* *Inceptisols* *Entisols* *Alfisols*	*Ustalfs* *Aquepts* *Orthents* *Ustalfs*
15	Hot sub-humid loamy to clayey alluvial soils	Kanagarh Madhpur Hanrgram Jagdishpur series.	*Chromic Vertic Endoaqualfs* *Vertic Endoaqualfs* *Chromic Vertic Endoaqualfs*	*Alfisols*	*Aqualfs* *Aqualfs* *Aqualfs*
16	Warm sub- humid to per-humid loamy to clayey alluvial soils		*Aquic Ustifluvents*	*Entisols*	*Fluvents*
17	Warm per-humid shallow and skeletal soils	Jaihing series Gemotali series	*Typic Dystrochrepts* *Typic Haplustalfs*	*Inceptisols* *Alfisols*	*Ochrepts* *Ustalfs*
18	Warm per-humid red and yellow soils	Phullen, Dialong Selsekgiri series.	*Lithic Udorthents* *Ultic Hapludalfs* *Typic Hapludalfs*	*Entisols* *Alfisols* *Alfisols*	*Orthents* *Udalfs* *Udalfs*
19	Hot sub-humid coastal and deltaic alluvial soils	Motto series Kalathur series	*Vertic Halaquepts* *Sodic Haplusterts*	*Inceptisols* *Vertisols*	*Aquepts* *Usterts*
20	Hot humid/per-humid red and lateritic and alluvial soils	Thekkady Trivandrum Kunnamangalam Ambalapuzha	*Tropeptic Eutrorthox* *Oxic Dystropepts* *Typic Haplustox* *Salidic Sulfihemists*	*Ultisols* *Inceptisols* *Oxisols* *Histosol*	*Humults* *Tropepts* *Ustox* *Hemists*

Table 9.8 Soil series, classification and location of benchmark soils

Sl. No.	Soil series	Soil sub-order	Soil classification	State
1	Gogji Pather	Ustalfs	Typic Haplustalfs	Jammu and Kashmir
2	Wahthora	Aquepts	Mollic Haplaquepts	Jammu and Kashmir
3	Chinwali	Orthents	Lithic Ustorthents	Uttar Pradesh
4	Phullen	Orthents	Lithic Udorthents	Mizoram
5	Dialong	Hapludalfs	Ultic Hapludalfs	Manipur
6	Selsekgiri	Udalfs	Typic Hapludalfs	Meghalaya
7	Fatehpur	Psamments	Typic Ustipsamments	Ludhiana, Punjab
8	Kanjli	Ochrepts	Typic Ustochrepts	Punjab
9	Nabha	Ochrepts	Udic Ustochrepts	Punjab
10	Zarifa Viran	Ustalfs	Typic Natrustalfs	Haryana
11	Sadhu	Ochrepts	Vertic Ustochrepts	Patiala, Punjab
12	Hirapur	Aquepts	Aeric Halaquepts	Uttar Pradesh
13	Sakit	Ustalfs	Typic Natrustalfs	Uttar Pradesh
14	Itwa	Aqualfs	Aeric Ochraqualfs	Uttar Pradesh
15	Bijaipur	Ochrepts	Udic Ustochrepts	Uttar Pradesh
16	Basiaram	Ochrepts	Typic Eutrochrepts	Uttar Pradesh
17	Kanagarh	Aquents	Aeric Fluvaquents	West Bengal
18	Madhupur	Ochrepts	Fluventic Eutrochrepts	West Bengal
19	Hanrgram	Ochrepts	Vertic Eutrochrepts	West Bengal
20	Jagadishpur	Aqualfs	Vertic Ochraqualfs	West Bengal
21	Jaihing	Ochrepts	Typic Dystrochrepts	Assam
22	Gemotali	Udalfs	Typic Hapludalfs	Arunachal Pradesh

(continued)

Table 9.8 (continued)

Sl. No.	Soil series	Soil sub-order	Soil classification	State
23	Haldi	Udolls	Typic Hapludolls	Uttar Pradesh
24	Chirai	Orthids	Typic Camborthids	Rajasthan
25	Pali	Orthids	Lithic Calciorthids	Rajasthan
26	Thar	Psamments	Typic Torripsamments	Rajasthan
27	Masitawali	Fluvents	Typic Torrifluvents	Rajasthan
28	Chomu	Psamments	Typic Ustipsamments	Rajasthan
29	Sarol	Usterts	Typic Chromusterts	Madhya Pradesh
30	Kamliakheri	Ochrepts	Lithic Vertic Ustochrepts	Madhya Pradesh
31	Singpura	Ochrepts	Typic Ustochrepts	Madhya Pradesh
32	Kheri	Usterts	Typic Chromusterts	Madhya Pradesh
33	Marha	Usterts	Entic Chromusterts	Madhya Pradesh
34	Jambha	Usterts	Typic Chromusterts	Maharashtra
35	Linga	Usterts	Udic Chromusterts	Maharashtra
36	Barsi	Usterts	Typic Chromusterts	Maharashtra
37	Adesar	Argids	Typic Paleargids	Gujarat
38	Achmatti	Usterts	Typic Pellusterts	Karnataka
39	Hungund	Usterts	Typic Chromusterts	Karnataka
40	Kagalgomb	Usterts	Typic Chromusterts	Karnataka
41	Raichur	Usterts	Typic Pellusterts	Karnataka
42	Chougel	Ustalfs	Plinthustalfs	Madhya Pradesh
43	Jamakhandi	Ustalfs	Typic Paleustalfs	Karnataka
44	Tyamagondalu	Ustalfs	Oxic Paleustalfs	Karnataka
45	Vijayapura	Ustalfs	Oxic Haplustalfs	Karnataka
46	Channasandra	Ustalfs	Oxic Rhodustalfs	Karnataka
47	Kadirabad	Usterts	Typic Chromusterts	Andhra Pradesh
48	Chinnaloni	Ustropepts	Paralithic Vertic Ustropepts	Andhra Pradesh
49	Kasireddipalli	Usterts	Typic Pellusterts	Andhra Pradesh
50	Patancheru	Ustalfs	Udic Rhodustalfs	Andhra Pradesh
51	Thekkadi	Orthox	Tropeptic Eutrorthox	Kerala
52	Trivandrum	Dystropepts	Oxic Dystropepts	Kerala
53	Kunnamangalam	Ustox	Typic Haplustox	Kerala
54	Palathurai	Ustalfs	Typic Haplustalfs	Tamil Nadu
55	Coimbatore	Ustropepts	Vertic Ustropepts	Tamil Nadu
56	Pusaro	Ustalfs	Ultic Paleustalfs	Bihar
57	Hathiapathar	Aqualfs	Typic Ochraqualfs	Bihar
58	Mrigindihi	Aqualfs	Ultic Paleustalfs	West Bengal
59	Bhubaneswar	Ustults	Typic Haplustults	Orissa
60	Motto	Aquepts	Vertic Halaquepts	Orissa
61	Kalathur	Usterts	Udorthentic Pellusterts	Tamil Nadu
62	Ambalapuzha	Hemists	Salidic Sulfihemists	Kerala
63	Dandi	Orthids	Typic Salorthids	Gujarat
64	Lakhpat	Argids	Typic Natrargids	Gujarat

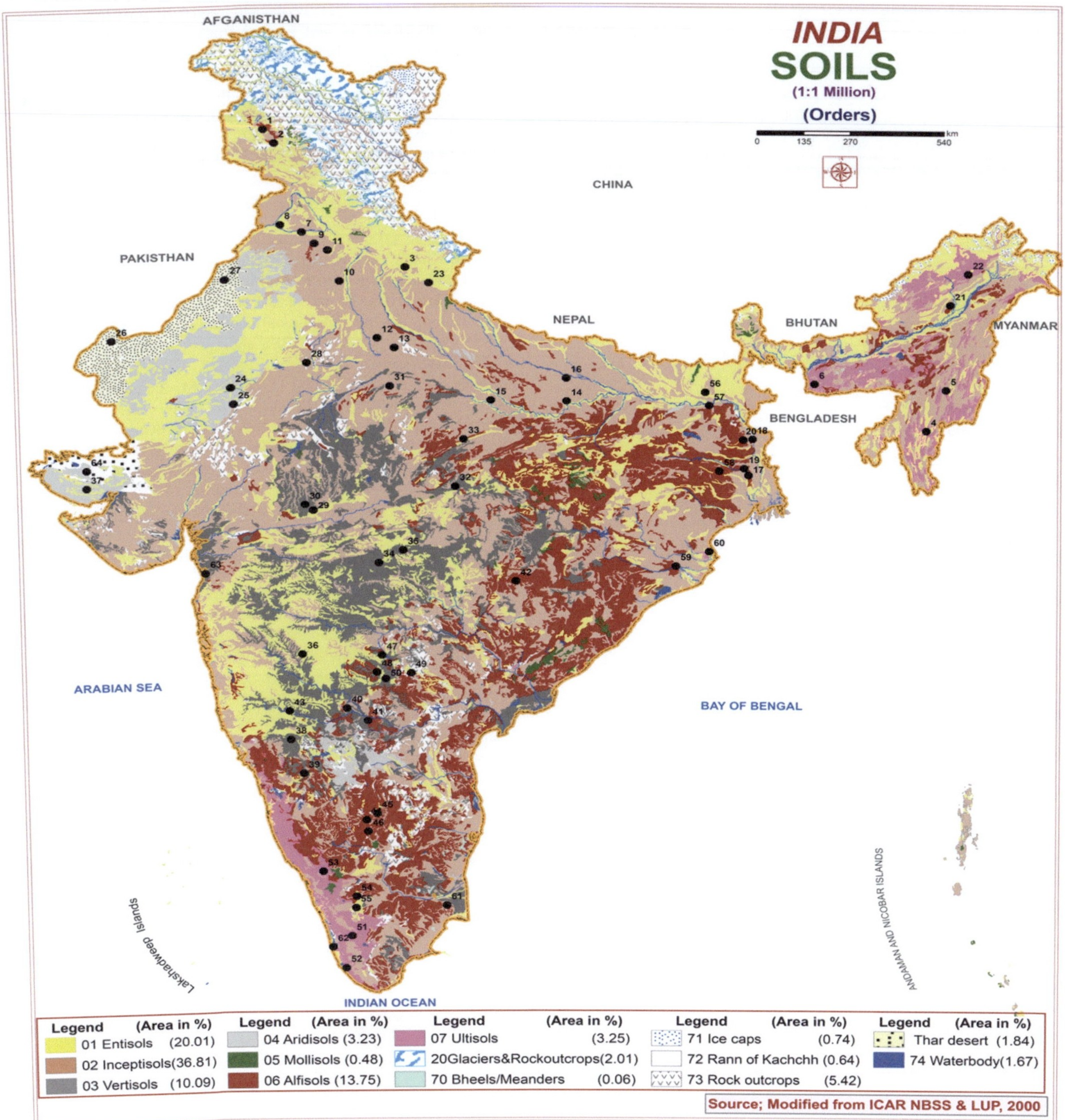

Fig. 9.23 Soil order map of India showing the benchmark sites (series) in numbers. *Source* NBSS & LUP ARCHIVES 2017

9.4 Conclusions

Benchmark soils in twenty agro-ecological regions of the country are recognized and presented following the established soil series. Such basic information would support to interpret the respective soils, their pedogenic features as well as physical and chemical characteristics together with associated constraint and potential. Benchmark soils would help to develop the soil resource inventory as well as classification options and land use plans. However, there is need to harmonize soil management issues with benchmark soils in order to formulate a soil-based agriculture road map in India. Further refinement and purposeful linking of benchmark soils with GIS tools are of high national priorities.

References

Bhattacharyya T, Sarkar D, Ray SK, Chandran P, Pal DK, Mandal DK, Prasad J, Sidhu GS, Nair KM et al (2014) Soil datasets of the hot spots black soil region (BSR). Working Report No. 4, NAIP component-4 Project on georeferenced soil information system for land use planning and monitoring of soil and land quality for agriculture". NBSS Publ. No. 1065, NBSS and LUP, Nagpur, 225 pp

GOI (2019) Benchmark soils of Kerala. ENVIS Centre: Kerala, State of Environment and Related Issues; Kerala State Council for Science, Technology and Environment; Ministry of Environment, Forests and Climate Change, Govt of India, 18th Jan 2019

Mandal DK, Mandal C, Singh SK (2016) Agro-ecological regions of India (Revised), NBSS and LUP Publ. No. 170. ICAR-NBSS and LUP, Nagpur, India, p 73

Murthy RS (1982) Benchmark soils of India: morphology, characteristics, and classification for resource management. National Bureau of Soil Survey and Land Use Planning, ICAR, New Delhi

Murthy RS, Hirekerur LR, Deshpande SB, Venkat Rao BV (1982) Benchmark soils of India. National Bureau of Soil Survey and Land Use Planning, Nagpur, 440010 (MS)

NBSS and LUP (2015) Agro ecological region map of India (Revised edition). ICAR-NBSS and LUP, e-Publication: ICAR-NBSS and LUP Technologies. https://www.nbsslup.in/assets/uploads/clinks/Delineating%20Agro-Ecological%20Regions.pdf

NBSS & LUP Archives (2017). Compilation of legacy soil information from different projects of ICAR-NBSSLUP, Nagpur

Ray SK, Bhattacharyya T, Sarkar D, Pal Chandran P, Tiwary P, Mandal DK, Prasad J, Sidhu GS, Nair KM, Sahoo AK, Das TH, Singh RS, Mandal C, Srivastava R, Sen TK, Chatterji S, Obireddy GP, Patil NG, Mahapatra SK, Anil Kumar KS, Das K, Singh AK, Reza SK, Dutta D, Srinivas S, Karthikeyan K, Venugopalan MV, Velmourougane K, Srivastava A, Raychaudhuri M, Kundu DK, Mandal KG and others (2014a) Baseline data Indo-Gangetic plains (IGP). Working Report No. 1, NAIP component—4 Project on "georeferenced soil information system for land use planning and monitoring soil and land quality for agriculture" NBSS Publ. No. 1062 (b), NBSS and LUP, Nagpur, pp 601–1165

Ray SK, Bhattacharyya T, Sarkar D, Pal DK, Chandran P, Tiwary P, Mandal DK, Prasad J, Sidhu GS, Nair KM, Sahoo AK, Das TH, Singh RS, Mandal C, Srivastava R, Sen TK, Chatterji S, Obireddy GP, Patil NG, Mahapatra SK, Anil Kumar KS, Das K, Singh AK, Reza SK, Dutta D, Srinivas S, Karthikeyan K, Venugopalan MV, Velmourougane K, Srivastava A, Raychaudhuri M, Kundu DK, Mandal KG and others (2014b) Baseline data black soil region (BSR). Working Report No.2, NAIP component—4 Project on "georeferenced soil information system for land use planning and monitoring soil and land quality for agriculture" NBSS Publ. No. 1063 (b), NBSS and LUP, Nagpur, pp 425–658

Ray SK et al (2014c) Baseline data: black soil region. Working Report No.2, Part II. NAIP component—4 Project on "georeferenced soil information system for land use planning and monitoring of soil and land quality for agriculture". NBSS Publ. No. 1063 (b), NBSS and LUP, Nagpur, pp 425–658

Sehgal J (1994) Soil series: concept and criteria. In: Sohan L, Deshpande SB, Sehgal J (eds) Soil series of India. Soil Bulletin 40, National Bureau of Soil Survey and Land Use Planning, Nagpur, India, 684 pp

Sehgal JL, Mandal DK, Mandal C, Vadivelu S (1992) Agro-ecological regions of India, 2nd edn, Technical Bulletin. NBSS Publ. 24, Nagpur, India

USDA (2017) Benchmark soils Part 630. Title 430: National Soil Survey Handbook (430-630-H, 1st ed, Amend. 25 Nov 2017) 630-A.1, United States Department of Agriculture

Land Evaluation and Land Use Planning

10

V. Ramamurthy, O. Challa, L. G. K. Naidu, K. S. Anil Kumar, S. K. Singh, D. Mamatha, K. Ranjitha, and Bipin B. Mishra

Abstract

In order to restore the declining trend of qualities of different soils of India, there has been an emerging need for soil as well as land evaluation and land use planning. In fact, land evaluation is a prerequisite for land use planning. This chapter is aimed at correlation of soil survey information, climate, vegetation and other aspects of land with the specific use for which land is evaluated. In this process, the suitability of the land is assessed and classified. Data set requirements for land evaluation are described in relation to Indian context. Updates of different land evaluation approaches like quantitative and qualitative approaches as adopted and practiced in India are briefly described with case studies. For the Second Green Revolution, India wants successful adoption of land evaluation and land use planning under strong vision, mission and overall goal. Being the foundation base of production functions, the mode for soil evaluation needs to be shifted in accordance with a wide range of objectives. Scientific approaches in quantifying the land evaluation would ensure targeted production of the best suitable crop in a well-defined land use planning system. There is further scope to link the land use planning system with supply chain process integrating the farming activities from point of origin i.e. soil and land to the point of consumption, i.e. market. India wants prime land and needs soil-based efforts to alleviate poverty through profitable production on sustainable frameworks. Profitable production with sustainable management could be enhanced considerably to ensure even more than double of the farmer's economic growth merely by improving the correctable limitations with a given land.

Keywords

Land evaluation • Land use planning • Soil suitability rating • Prime lands

10.1 Introduction

The demand for the finite land resources is increasing exponentially due to the growing population at the current rate of almost 1.67%. The population growth is leading to unfavourable man-to-land ratio. In India, per capita cultivable landholding has been declining from 0.48 ha in 1951 to 0.16 ha in 1991 and it is likely to decline further to 0.11 ha in 2025 and less than 0.09 ha in 2050 (NAAS 2009). Although the food production has increased from 52 m ton in the 1950s to almost 256 m ton in 2014 (GOI 2015), this increase has been large as a result of expansion in cultivated and irrigated area and high chemical (fertilizer) inputs. The significant growth of agriculture has been at the cost of decline in soil quality and risk of soil degradation. We are now facing the serious threat of ensuring sustainability in our production systems. In many of the so-called First Green Revolution areas, a whole range of second-generation problems are posing serious challenges to the sustainable agricultural production. About 57% of soils are under different kinds of degradation and these are getting further deteriorated with risk of jeopardizing our food

V. Ramamurthy (✉) · L. G. K. Naidu · K. S. Anil Kumar · D. Mamatha · K. Ranjitha
ICAR-National Bureau of Soil Survey and Land Use Planning, Hebbal, Bangalore 560024, India
e-mail: ramamurthy20464@gmail.com

O. Challa · S. K. Singh
ICAR-National Bureau of Soil Survey and Land Use Planning, Amravati Road, University P.O., Nagpur 440033, India

B. B. Mishra
Bihar Agricultural University, Sabour, Bhagalpur 813210, India

B. B. Mishra
Pedology and Land Use Planning, School of Natural Resources Management and Environmental Sciences, Haramaya University, P O Box 138 Dire Dawa, Ethiopia

B. B. Mishra
International Union of Soil Sciences, Vienna, Austria

B. B. Mishra (ed.), *The Soils of India*, World Soils Book Series,
https://doi.org/10.1007/978-3-030-31082-0_10

security (Sehgal and Abrol 1994). In addition to this, many issues concerning environmental sustainability, carrying capacity of our land resources, etc. are also cropping up and adversely affecting soil and human health. These problems demand a systematic appraisal of our soil and climatic resources to recast and implement an effective and appropriate land use plan at regional level and local level. The land use or land produce is a subject that tends to ensure profitability. Thus, the concept of land economics is very relevant to make the land use profitable by applying economic principles within the control of farmers by applying the key components of supply chain process beginning from the point of origin (soil) to the point of consumption, i.e. market (Mishra 2017).

The purpose of land evaluation is to understand the relationships between the conditions of the land and the manner in which it is utilized. The close relationship between physiography and soils has been reported by many workers (Chamuah et al. 1996; Tamgadge et al. 1996). Ideally, it should also predict the effects (output), both positive and negative, resulting from the use of land in a particular manner. Land evaluation procedures involve the interpretation of biophysical resource inventories in relation to their use (Sehgal 1993).

Land evaluation has been defined by FAO (1976) as the process of assessment of land performance when used for specified purposes, involving the execution and interpretation of surveys and studies of landforms, soils, vegetation, climate and other aspects of land in order to identify and make a comparison of promising kinds of land use in terms applicable to the objectives of the evaluation. The evaluation is a vital link in the chain leading to the sustainable management of land resources (Sharma et al. 1994). This includes productive uses such as farming, livestock production and forestry and other uses such as catchments area protection, recreation, tourism and wild life conservation. The land possesses a definite set of capabilities for supporting different crops (Dent and Deshpande 1993). Performance of any crop is largely dependent on soil topography and climatic conditions and the levels of management. It is because of the fact that each plant species requires a different set of conditions for optimum growth, and there is a need to evaluate the land resources for sustainable crop growth and to delineate major and efficient crop zones.

10.2 Aims of Land Evaluation

The framework for land evaluation as structured and elaborated by FAO (1976) is the international pillar for reference and is recommended to follow with regional procedures. Land evaluation may be concerned with present land performance and takes into consideration the economics of the proposed enterprises and trade of goods, the social consequences for the people of the area and the region concerned, and the consequences, beneficial or adverse, for the environment. Mishra (2017) emphasized that land economics must be the integral part of land evaluation. Velayutham (1997), however, forwarded the detailed strategic planning based on advantage of spatial and temporal soil variability that needs to be integrated with land evaluation. Thus, land evaluation should answer at least the following questions:

- How is the land currently managed and what will happen if present practices remain unchanged?
- What improvements in management practices, within the present use, are possible?
- What other uses of land are physically possible and economically and socially relevant?
- Which of these uses offer possibilities of sustained production or other benefits?
- What adverse effects, physical, economic or social are associated with each use?
- What recurrent inputs are necessary to bring about the desired production and minimize the adverse effects?
- What are the benefits of each form of use?

If the introduction of a new land use involves a significant change in the land itself, as for example in irrigation schemes, then the following additional questions should be answered:

- What changes in the condition of the land are feasible and necessary, and how can they be brought about?
- What non-recurrent inputs are necessary to implement these changes?

The evaluation process does not in itself determine the land use changes that are to be carried out, but provides data on the basis of which such decisions can be taken. To be effective in this role, the output from an evaluation normally gives information on two or more potential forms of use for each area of land, including the consequences, beneficial and adverse effects of each.

10.3 Land Evaluation and Land Use Planning

Land evaluation is part of the process of land use planning. Its precise role varies in different circumstances that also consider the functions and services of soils in the landscape as well as management options such as supply chain process (Mishra 2017). It is sufficient to represent the land use planning process by following generalized sequence of activities and decisions:

1. Recognition of a need for change;
2. Identification of aims;
3. Formulation of alternative forms of land use and recognition of their main requirements;
4. Recognition and delineation of the different types of land present in the area;
5. Comparison and evaluation of each type of land for the different uses;
6. Selection of a preferred use for each type of land;
7. Detailed analysis of a selected set of alternatives for distinct parts of the area; this, in certain cases, may take the form of a feasibility study.
8. Decision to implement;
9. Implementation;
10. Monitoring of the operation.

Land evaluation plays a major part in stages 3, 4 and 5 of the above sequence and contributes information to the subsequent activities. Thus, it is preceded by the recognition of the need for some change in the use to which land is put; this may be the development of new productive uses, such as agricultural development schemes, irrigation and flood control dams, forestry plantations or the provision of services, such as the designation of a national park or recreational area, etc.

The evaluation process itself includes description of a range of promising kinds of uses and the assessment and comparison of these with respect to each type of land identified in the area. This leads to recommendations involving one or a small number of preferred kinds of use. These recommendations can then be used in making decisions on the preferred kinds of land use for each distinct part of the area. Later stages will usually involve further detailed analysis of the preferred uses, followed, if the decision to go ahead is made, by the implementation of the development project or other form of change, and monitoring of the resulting systems.

10.4 Principles of Land Evaluation

Certain principles are fundamental to the approach and methods employed in land evaluation. These basic principles are as follows:

1. Land suitability assessment and classification with respect to specified kinds of use (Ramamurthy et al. 2018)
2. Evaluation requires a comparison of the benefits obtained and the inputs needed on different types of land
3. A multidisciplinary approach is required (Niranjan et al. 2011; Das and Sudhakar 2014; Gautam et al. 2017)
4. Evaluation is made in terms relevant to the physical, economic and social context of the area concerned (Niranjan et al. 2011; Das and Sudhakar 2014; Gautam et al. 2017; Ramamurthy et al. 2018)
5. Suitability refers to use on a sustained basis (Ramamurthy et al. 2018)
6. Evaluation involves a comparison of more than a single kind of use (Narayanaswamy 2017)

10.5 Land Evaluation Procedures

The land evaluation activities undertaken and the order in which the work is done depend on the type of approach adopted, whether parallel or two stages.

The main activities in a land evaluation are as follows:

- Initial consultations, concerned with the objectives of the evaluation, and the data and assumptions on which it is to be based
- Description of the kinds of land use to be considered and establishment of their requirements
- Description of land mapping units and derivation of land qualities
- Comparison of kinds of land use with the types of land present
- Economic and social analysis
- Land suitability classification (qualitative or quantitative)
- Presentation of the results of the evaluation.

It is important to note that there is an element of iteration, or a cyclic element, in the procedures. Although the various activities are here of necessity described successively, there is, in fact, a considerable amount of revision to early stages consequent upon findings at later periods. Interim findings might, for example, lead to reconsideration of the kinds of land use to which evaluation is to refer or to changes in boundaries of the area evaluated.

10.6 Data Set Requirements for Land Evaluation

The land units and their homogeneity form the basic requirement for proper land evaluation. The land units selected for land evaluation have no scale limitation. The information on the land units is generated through different kinds of soil surveys.

The land characters and land qualities considered in defining the land units are as under:

Land characters Land characteristics used in the land evaluation are measurable properties of the physical environment directly related to land use and are available from the soil survey. These characteristics are

Biophysical characteristics: factors like topography (t)—slope length and gradient; wetness (w)—drainage and flooding
Physical soil characteristics: texture, soil depth and intensity of acid sulphate layer and gypsum or *kankar* layer
Fertility characteristics (f): cation exchange capacity of the clay as an expression of weathering stage, base saturation and organic matter content
Salinity and alkalinity (n): salinity status and alkalinity status

Climatic database Factors, such as temperature, potential evaporation, the temporal and spatial variability of rainfall, specific to an area are considered as database for estimation of growing period.

There are a number of other important properties, which co-vary with changes in the property; however, these properties are of great value in interpreting the various uses. Soil classification systems very much rely extensively on quantitative composition of soils and these compositions are selected on their assumed importance in understanding the genesis of the soil.

Land qualities It is a complex attribute of land which acts in a distinct manner, its influence on the suitability of land for a specific kind of use. They may be positive or negative. They are in fact practical consequences of land characteristics. They could be segregated into two groups: FAO (1976) suggests three comprehensive land qualities:

Internal qualities Water holding capacity; oxygen availability; availability of foothold to roots; tolerance to iron-induced chlorosis; nutrient availability; resistance to structural degradation of top soil; absence of salinity and alkalinity.

External qualities Correct temperature regime; resistance against erosion; ability for layout of farm plan and workability.

As per the land use, these qualities can be grouped as;

A. **Land qualities related to productivity for crops or other plant growth**: Crop yields; moisture availability; nutrient availability; oxygen availability in root zone; adequacy of foothold for roots; conditions for germination; workability of the land; salinity and alkalinity; resistance to soil erosion; pests and diseases incidence related to land; flooding hazard; temperature regime; radiation energy and photoperiod; climatic hazard (Wind, hail and frost); air humidity; drying period for ripening of crops.
B. **Land qualities related to domestic animal productivity**: productivity of grazing land; climatic hardships affecting animals, endemic pests and diseases; nutritive value of grazing land; toxicity of grazing land; resistance to degradation of vegetation; resistance to soil erosion under grazing condition; and availability and quality of drinking water.
C. **Land qualities related to forest productivity**: mean annual increments of timber species; types and qualities of indigenous species; site factors affecting establishment of young trees; pests and diseases; and fire hazards.
D. **Land qualities related to management and inputs:** These qualities may refer to arable use, animal production or forestry. These include trafficability (mechanization); accessibility (laying roads); size of potential managements units; (forest block, farms and fields); location in relation to markets and supplies of inputs.

The land qualities (both internal and external) are practical consequences on plant growth, performance and producing yield.

10.6.1 Spatial Database

(i) **Land units**

The land unit covers an area of land, usually mapped with specified characteristics, employed as units for land evaluation. Example: major climate, growing period and agro-climatic zones.

- Soil series, soil associations and other soil mapping units.
- Land system and land facets.

Two kinds of land units are employed at different stages, e.g. agro-climatic region—major part of land evaluation at regional level; detailed units—landforms and soils at local level.

Criteria for Land Units in land evaluation

- Land units should be as homogenous as possible.
- The grouping should have a practical value in relation to proposed land use.
- It should be possible to map the units consistently.

- Should be defined as simply as possible based on properties which are readily observable in the field—soils and land surface.

(ii) **Present land use**

Current land use in an area is being practiced over the years as per the needs of the people. It has the spatial concept. The present land use or the existing land use is of prime importance in land evaluation. Land evaluation principle emphasizes on the existing land use and its details. It indicates that before considering a particular land unit for evaluation towards alternative land use, it is important to look into the improvements possible in the existing land use through management. Even after the management improvements are made in the present land use, if the expected production levels are not achieved, then the evaluation is done for alternative land use.

(iii) **Land Utilization Types (LUTs)**

Land utilization type is the subdivision of land use which is an important component of land evaluation. Land utilization type deals with specific land use and management.

Produce: The produce of a crop or crop rotation can be subdivided into four groups such as arable land (annual crops); permanent crops (fruits and tropical perennial crops); grassland and forests.

(iv) **Land Use Requirements**

Land use requirements are important components in land evaluation since these parameters help in different phenological phases of the crop with a better biomass production. These requirements differ with the type/variety of crop and stage of the crop. Broadly, these land use requirements can be grouped into agro-climatic requirements, soil-associated crop requirements and additional requirements.

In land evaluation, defining the land use requirements is a key issue. For finalizing the land use requirement's criteria specific to a crop/variety/locality/region, it is necessary to take the help of existing literature, individual's experience and the experimental data. In the present approach of obtaining crop growth requirements criteria and their ratings, the basis of anticipated yields or experimental yields is also considered. The proper definition of the crop growth requirements needs a good knowledge of the phenology and ecology of the crops/plants under consideration and requires the input of crop specialists. The procedure of defining the crop growth requirements includes two main phases, respectively, linked to the listing of criteria which are of relevance and rating in terms of the optimal and marginal growth conditions. With respect to the identification of criteria, good care should be taken not to duplicate the impact of factors by introducing them at several levels. As such, soil texture affects rooting conditions, aeration and workability, but it should be included only at its most relevant level, e.g. under the rooting conditions.

Agro-climatic crop needs are mainly related to the moisture and energy conditions during the growing season, e.g. during the time that the crop is effectively on the land. It is therefore important to know the length of the growth cycle and of the different phenological stages and to rate ultimately the optimal and marginal conditions for each of these stages. Requirements in terms of moisture supply depend on the groundwater table and related capillary rise, the soil moisture retention capacity and the evapotranspirative demands. The energy regime is determined by the current temperature, insolation and day length. Secondary climatic requirements may refer to the sensitivity to extreme levels of air humidity (too high levels promote the development of diseases, particularly in flowering and maturation periods; too low levels may require too high evaporative demands and disruptions in the plant physiological system), extreme temperatures, etc.

Soil-associated crop requirements refer mainly to rooting and aeration conditions, the availability of nutrients and moisture and the sensitivity or tolerance to toxic elements. Rooting and aeration conditions affect the penetration and development of the plant root system in search for water and nutrients. The soil depth, the texture and/or the eventual presence of coarse fragments in the profile, drainage conditions and flooding hazards mainly influence these qualities. The crop nutrient supply is in the first place determined by the cation exchange capacity and the base status, as expressed per 100 g of soil. In most cases, the sum of bases may be a sufficient parameter to express the total amount of bases available to the plant; for certain crops, however, it may also be worthwhile to include additional parameters, such as Ca/Mg or K/Mg ratios as those relations affect directly crop behaviour (bananas, oil palm, etc.). The introduction of pH as a crop parameter is meant to be an expression of the overall base saturation within the soils medium and to indicate solubility and uptake potential of nutrients in the root zone.

The evaluation of the NPK status in the soil is not a common practice in soil survey reports, but should be promoted. In this respect, minimum and optimal levels of these elements should become a criterion to be taken care of in the evaluation procedures for arable cropping. The sensitivity to specific chemical and other toxic components in the root zone can directly be quantified by the introduction of threshold figures for $CaCO_3$ or gypsum contents in the case of arid zone crops, salinity and alkalinity (expressed through electrical conductivity and ESP values), or sensitivities to exchangeable aluminium for tropical crops. The tolerance to heavy metals may become of steadily increasing importance under conditions where pollution phenomena occur.

Additional requirements related to seedbed preparation and harvesting procedures refer to workability, trafficability and erosion hazards. These depend largely on the combined effects of the soil moisture status (rainfall, groundwater depth, internal drainage and lateral water movement), slope and soil surface characteristics. In terms of management practices, slope and surface properties (stoniness and rockiness) may have an influence on the potential use of machinery for tillage and harvesting and may hence determine the appropriate management system.

Finally, it should be emphasized in term of economical and ecological viability assessment, a number of additional requirements may be added as per local need, it acts as a checklist of requirements to be taken into consideration for land evaluation.

Once the factors affecting crop or land use requirements have been selected and listed, the ratings have to be given to all criteria taken into consideration. If, for example, optimal crop development requires a minimum growth cycle of 180 days, this figure is given the highest rating, and the critical time below which the crop will fail is given the lowest rating. Because the land use requirements are different, factor ratings vary from one crop to another and obviously from one utilization type to another.

Factor ratings are usually expressed in degrees of limitations (or constraints according to the following rating scale (FAO 1983; Verheye 1992)

- No limitation or constraint: the specific characteristic is considered (almost) optimal for plant growth;
- Slight limitations: the characteristic/quality is nearly optimal for the given utilization type and affects productivity for not more than 20% with regard to optimal yields;
- Moderate limitations: the characteristic/quality has a moderate influence on the yield decrease, which may reach up to 50–60% of optimal yield; nevertheless, benefits can still be made and the use of the land remains profitable.

In some studies, reference is made to the fifth type of constraint, corresponding to very severe limitations. The difference between a severe and very severe limitation is not considered relevant in the present context, but could become so when a differentiation will be made between correctable and non-correctable types of constraints.

10.6.2 Socio-economic Data

Socio-economic data of any land unit may not directly influence the land evaluation process. However, the following socio-economic data sets are available, and they may help to draw practical conclusions.

I. **Management**

It covers five aspects such as
Size of the farms: levels of size used are

- Marginal holding (<1 ha)
- Small holding (1–2 ha)
- Semi-medium (2–4 ha)
- Medium (4–10 ha)
- Large (10 ha and above)

II. **Levels of Inputs**

For land evaluation, five levels of production inputs are proposed: They are

a. Low; can, in general, be borne by the landowners (stone cleaning and simple levelling);
b. Medium: can be borne by the landowner with credit facilities (grading and open drains);
c. High: Govt., funds or long-term credit to landowner (simple land reclamation works);
d. Very high: with normal recurring costs and
e. Very high: with high recurring costs where large Govt. funds are required.

III. **Labour intensity and availability**: labour; animal power; labour intensive; labour extensive

IV. **Sources of farm power**: Heavy mechanized with crawler tractors; fully mechanized with four wheel tractors; light mechanization with two-wheel or one-wheel-operated machinery; animal power; hand-operated tools–man power.

V. **The technical know-how of the farmer**: Low technical know-how limits the ambitious planning for land use, land management and improvement practices.

- Use of capital and its availability
- Non-recurring requirements or development cost such as land reclamation and installation of irrigation, drainage and erosion structures.

10.7 Land Evaluation Approaches

Land evaluation is the ranking of soil units on the basis of their capabilities (under given circumstances including levels of management and socio-economic conditions) to provide highest returns per unit area and conserving the natural resources for future use (Van Wambeke and Rossiter 1987). Several systems of land evaluation have been recognized (Storie 1954; Riquier et al. 1970; Sys 1985; Sehgal et al. 1980). The FAO (1976) panel for land evaluation suggested the classification of land in different categories: orders, classes, sub-classes and units. The soil-site characteristics are expressed in terms of degree of limitation (0, 1, 2, 3 or 4); the limitation of 2 is considered critical at which the expected yield declined significantly and the cultivation is considered marginally economical. The final soil-site evaluation/suitability is based on the number and degree of limitation (s). Modern approaches involve simulation model predicting yield as a measure of suitability. Although very well refined, yet these approaches are largely based on local experience of farmers or of the researchers. Naidu et al. (1986) stressed the need of geo-referenced information on the location, extent and quality of soil and land resources. For land evaluation, the soil database in conjunction with climatic data may be used in which the input data included land use requirements, land and soil characteristics, climatic data and soil map, wherein output was in the form of suitability the purpose specified. Naidu et al. (1986) further warned that the manual methods of soil data handling are insufficient in managing the huge volumes of diversified types of information. So, computer-based data management is recommended.

Since crop performance reflects the integrated effect of the environmental and soil characteristics, it would be appropriate to study the relationships, through regression analysis, between the crop performance and yield-influencing parameters (Gbadegesin and Areola 1987). In order to construct a knowledge base by which deductive reasoning may lead to ranking of land units, the present attempt is made to interpret the black cotton soils in terms of their characteristics and qualities for developing soil-site suitability models for different crops through a multivariate regression yield model (Sehgal et al. 1989)

The model may need further refinement by having a large number of test sites. The yield and soil-site parameters were compared through a linear equation of the following form based on collected yield data under similar management practices from different locations varying in rainfall and covering the entire black soil region (Table 10.1).

$$Y = a + b_1x_1 + b_2x_2 + b_3x_3 + b_4x_4 + b_5x_5 + \mu.$$

where

Y yield of the crop, q ha^{-1};
X_1 rainfall, mm;
X_2 soil depth, cm;
X_3 growing period, weeks;
X_4 clay, per cent;
X_5 calcium carbonate, per cent;
μ random error;

Table 10.1 Relationship of crop performance to soil-site parameters in Vertisols

Name of crop	Number of observations	Intercept	Regression coefficients					R^2
			X_1	X_2	X_3	X_4	X_5	
Sorghum	15	−14.4091	0.0074* (0.0025)	0.0432 (0.0204)	0.2423* (0.1024)	0.0746 (0.0957)	−0.0606 (0.0923)	0.8505
Cotton	10	−12.0748	0.0172* (0.0045)	−0.0313 (0.0295)	−0.2043 (0.1465)	0.1803 (0.2255)	0.7278* (0.2243)	0.8583
Pigeon pea	12	−4.3625	0.0023* (0.0008)	0.0059 (0.0020)	0.1646* (0.0756)	0.0671 (0.0982)	0.0926 (0.1154)	0.4199
Chick pea	13	4.1164	0.0065* (0.0015)	−0.0478* (0.0291)	−0.0347 (0.0686)	−0.0931 (0.0311)	−0.3878* (0.0735)	0.7515
Pearl millet	8	12.9200	0.0073 (0.0075)	0.0193* (0.0019)	−0.0049 (0.1475)	−0.2310 (0.0917)	−0.0755 (0.0735)	0.9477
Groundnut	9	−8.6524	0.0071 (0.0057)	0.0038 (0.0372)	0.3702* (0.0633)	−0.0093 (0.1179)	0.1089 (0.1047)	0.8181

* significant at 5% level
X_1 rainfall (mm)
X_2 soil depth (cm)
X_3 growing period (weeks)
X_4 clay %
X_5 $CaCO_3$%
( ) standard error of variable

a	intercept;
b_1	partial regression coefficient.

In order to find the optimum range of any parameter for crop performance, a quadratic equation was fitted in the following form:

$$Y = a + bx + cx^2 + \mu.$$

where

Y	yield of the crop (q ha^{-1})
X	explanatory variable (rainfall, calcium carbonate)
μ	random error
a	intercept
b, c	are regression coefficients

The Vertic intergrades (Inceptisols and Entisols) occurring in geographic association with the Vertisols, are mainly cultivated for sorghum and cotton. Yield of sorghum and cotton crops from six experiments and three locations near Nagpur area under similar rainfall pattern was also compared with the soil parameters.

Land evaluation involves the assessment of land and soils for their potential for different uses involving matching the land qualities and requirements for the given land use. Both qualitative and quantitative approaches are in vogue.

A. **Qualitative evaluation**
   - (i) Land capability classification (Klingbiel and Montgomery 1961).
   - (ii) Land irrigability classification (Soil Survey Staff 1951; USBR 1953).
   - (iii) Fertility capability classification
   - (iv) Soil suitability classification (FAO 1976; Sys 1985; Sys et al. 1993)
   - (v) Prime land classification (Ramamurthy et al. 2012)

B. **Quantitative evaluation**
   - (i) Soil index rating (Shome and Raychaudhari 1960; Storie 1978)
   - (ii) Actual and potential productivity (Riquier et al. 1970)
   - (iii) Soil suitability classification-statistical approach (Sehgal et al. 1989)
   - (iv) Land use planning and analysis system (LUPAS) (Laborte et al. 2002):
   - (v) Land suitability assessment by parametric approach (Rabia and Terribile 2013)
   - (vi) Land suitability by fuzzy AHP and TOPSIS methods (Mukhtar Elaalem et al. 2010)

10.7.1 Land Capability Classification

It is an interpretative grouping of the soils based on inherent soil characteristics, land features and environmental factors that limit land use or impose risk of erosion. Soils are grouped into eight capability classes from I to VIII on the basis of their ability to produce commonly cultivated crops. The classes from I to IV indicate arable and V to VIII indicates non-arable lands that can be used for non-agriculture uses. The risk of soil damage progressively increases from Class I to Class VIII. There is a provision to assign sub-class on the basis of kind of predominant hazard, limitation or conservation problem. A sub-class may be further divided into capability units according to the similarity in potential and response to management.

While land capability classification system is useful for relatively broad level planning, it needs to be supplemented by more precise evaluation for micro-level planning. Further, the land capability classification is conservation-oriented which considers the negative aspects. Yet this system is still widely used because of its simplicity and ease of comprehension. This classification gives a general idea about the capability of the soils but does not explain specific crop performance unless supplemented by additional information. This method could be followed effectively for highlighting the conservation-oriented limitations which need immediate attention and for broad grouping of soils into agricultural and non-agricultural lands.

Class: Groups of land units that have the same degree of limitation is denoted as 'Class'. The risk of soil damage or limitation becomes progressively greater from Class I to Class VIII. The classes show the general suitability of a land unit for agricultural use. Class I to IV—arable and V to VIII —non-arable

Sub-classes: These are based on major conservation problems such as 'e'—erosion and runoff; 'w'—excess water or waterlogging; 's'—root zone limitation and 'c'—climatic limitations.

The definitions of the different classes are given in Tables 10.2 and 10.3.

The disadvantage of this method is that it is (i) highly subjective; (ii) it is a limitation approach; based on one parameter, the land is brought under a lower class.

Martin and Saha (2009) used quantitative land evaluation procedures, namely USDA land capability classification (LCC) and FAO land evaluation procedure for soil-site

Table 10.2 Principles for the definitions of the arable classes

Arable land classes				
Parameters	Class I	Class II	Class III	Class IV
Definition	Few limitations restrict their use	Moderate limitations	Severe limitations	Very severe limitations
Range of crops	All crops give optimal yields	Most crops give nearly optimal yields	Limited crops do not yield satisfactorily	Yield marginal
Slope erosion (e)	Level no or low erosion	Gentle slope, moderate susceptibility to wind or water erosion	Moderately steep slope Wind and water erosion	Steep slope, very high wind and water erosion
Wetness (w) Flooding/drainage	Not subject to waterlogging or overflow Well drained	Occasional overflow Moderate permeability limitation	Frequent overflow Waterlogging, very slow permeability	Frequent overflow Excessive waterlogging
Physical soil condition(s)	Hold water well, good workability deep (+100 cm)	Unfavourable workability, less ideal depth (50–100 cm)	Low moisture holding-capacity Shallow depth (25–50 cm)	Low moisture <25 cm
Fertility	Well supplied with plant nutrients	Responsive to fertilizers	Low fertility	Low fertility
Salinity and alkalinity	No or slight	Slight to moderate easy to correct	Moderate salinity/sodium hazard	Severe salinity, sodium hazard
Management requirement	Ordinary	Careful management	Very careful	Very careful

Table 10.3 Principles for the definitions of the non-arable classes

Parameters	Pastures		Forest	Recreation and wild life
	Class V	Class VI	Class VII	Class VIII
Definition	Not suited to cultivation	Severe limitations	Very severe limitations	Unsuitable for any crop
Range of crops	Pastures	Pasture or range	Woodland	Recreation and wildlife
Slope and erosion (e)	Nearly level no erosion	Very steep severe erosion	Very steep severe erosion	Erosion hazard
Wetness (w) flooding	Frequent overflow	–	–	–
Drainage	Drainage feasible	–	Too wet soils	Too wet soils
Soils (s) conditions	Stony or rocky	Stoniness, low moisture-holding capability. Too shallow	Stoniness. Too shallow	Low moisture-holding capacity, stoniness. Too shallow
Salinity and Alkalinity	–	Severe salinity and sodicity hazard	–	–
Management requirement	Pasture	Pasture	–	–

suitability for various land utilization types to assess the land suitability for different crops and for generating cropping pattern for *kharif* and *rabi* seasons in a watershed (Table 10.4). The database on soil, land use/land cover rainfall and temperature was generated from data derived from Landsat TM remote sensing satellite and soil survey to perform an integrated analysis in the geographic information system environment. Arable and non-arable lands were delineated in the watershed using the USDA LCC and non-arable areas were masked for removal from future analysis. Different land quality parameters, viz. soil texture, depth, erosion, slope, flooding and coarse fragments under various land units were evaluated for a number of crops. Subsequently, all of them were integrated using a sequence of logical operations to generate the land suitability maps for various crops. *Kharif* and *rabi* season cropping patterns were

Table 10.4 Land evaluation based on USDA land capability classification and FAO procedure

Mapping unit	Rabi season				Kharif season			
	Wheat	Mustard	Sugarcane	Suitability	Paddy	Maize	Sugarcane	Suitability
M11	N	N	N	N	N	N	N	N
M12	N	N	N	N	N	N	N	N
P11	S3e	S3e	N	Wheat, mustard	S3e	S3ecf	N	Paddy, maize
P12	S2tse	S3e	S3et	Wheat, mustard, sugarcane	S3e	S2ts	S3et	Paddy, maize, sugarcane
P21	S3t	S2se	S3e	Mustard, wheat	S2tse	S2ts	S3e	Paddy, maize, sugarcane
P22	S3t	S3e	S3e	Mustard, wheat	S2ts	S2tse	S3e	Paddy, maize, sugarcane
T1	S2te	S2te	S2e	Mustard, wheat	S2ts	S2te	S2e	Paddy, maize, sugarcane
T2	N	S3dse	N	Mustard	S3d	S3ds	N	Paddy, maize
AT1	S2d	S2ef	S2ed	Wheat, mustard, sugarcane	S2ds	S2de	N	Paddy, maize, sugarcane
AT2	S2t	S2ef	S2e	Wheat, mustard, sugarcane	S2tse	S2tds	S2e	Paddy, maize, sugarcane
AT3	S2t	S2ef	S2e	Wheat, mustard, sugarcane	S2tse	S2tds	S2e	Paddy, maize, sugarcane

E—erosion, *cf*—coarse fragments, *t*—texture, *d*—drainage, *s*—slope, *f*—flooding, *N*—not suitable

developed by integrating crop suitability maps for the winter and summer seasons separately. Finally, cropping system maps for the watershed were obtained by integrating the two season cropping sequences within the crop calendar. Results indicated that the present agricultural area of 47% could be increased to 71% by adopting scientific land evaluation methods for watershed development. It was also found that better land use options could be implemented in different land units as the conventional land evaluation methods suffer from the limitation of spatial analysis for the suitability of various crops.

10.7.2 Land Irrigability Classification

In this system (USBR 1953), soils are first categorized according to physical factors (topography, drainage and water quality) and socio-economic factors (location and size of farms, characteristics of landownership, cultural patterns, and the skill and resources of individual operators development costs, etc.). Separation of land irrigability classes is made on specified limits of soil properties and other physical parameters. Land irrigability system may be used for selection of irrigable lands, estimation of water requirements, development costs and benefits from irrigation. Such information will help in land use planning decisions. Soil Survey and Land Use Planning Scheme, Sabour in Bihar also used similar classification (Singh and Mishra 1997, Singh et al. 1996).

This system also provides six suitability classes (Table 10.5) for irrigation based on the soil and land characteristics and the repayment capacity. The sub-classes provided are based on deficiencies or problems with respect to topography (t), soil (s) and drainage (w).

- Soils are categorized based on their suitability for sustained use under irrigation.
- Physical factors (topography, drainage and water quantity).
- Socio-economic factors

Sub-class: Groups of land units with some dominant limitations like soil (s), topography (t) and drainage (d).
Irrigability units: Grouping of lands that are nearly alike in suitability for irrigation.

10.7.3 Land Suitability Classification

Land suitability classification refers to the fitness of a given type of land for a defined use. This classification is arrived at on the basis of soil survey information, economic and social analysis, kinds of land use and need for change. Separate classifications are made with respect to each kind of land use

Table 10.5 Characteristics of land irrigability classification

Class	1	2	3	4	5	6
Arability	Arable	Arable	Arable	Limited arable	Temporarily non-arable	Non-arable
Repayment capacity	High	Intermediate	Intermediate	Low	–	Not repayable
Crop suitability	Wide	Restricted	Restricted	Few specific crops	Not suitable	Not suitable
Yield	High and sustained	Moderate and sustained	Moderate and sustained	Low	Very low	Extremely low
Water use	Efficient	Moderately efficient	Moderately efficient	Little efficient	Very low efficient	Inefficient
Physical parameters						
Slope %	<1	1–3	3–5	5–10	>10	>10
Soil depth (cm)	>90	45–90	22.5–45.0	7.5–22.5	<7.5	<7.5
Permeability (mm/hr)	5.0–50	50–130	130–250	>250	>250	>250
Texture	Sandy loam, clay loam	Sandy loam, sandy clay loam	Clay, loamy sand	Sandy, clay	Any	Any
WHC (cm)	12	9–12	6–9	2–6	<2	<2

that appears to be relevant for the area (FAO 1976). The land evaluation proposed by FAO (1976) defines the basic concepts and principles followed universally. The basic concepts include the land and its major use, utilization type, characteristics, qualities and diagnostic criteria. Land includes soil, vegetation, hydrology, landform and climate. The framework suggested classification of land into different categories. viz. orders, classes, sub-classes and units. There are two orders namely S for suitable lands and N for non-suitable lands; further three classes (S1, S2, S3) within the order S and two classes (N1, N2) under the order N depending on degree of limitations with respect to specific land use. The appraisal of the classes, within the order is done according to the land limitations. The sub-classes reflect the kind of limitations that are the major kinds of improvement measures required within these classes. They are indicated by the symbols using lower case letter following the Arabic numeral.

In the land evaluation, there are four steps namely (i) characterization of existing soil, climatic and land use conditions (ii) development of soil-site criteria or crop requirements (iii) matching of crop requirements with existing soil and climatic conditions and (iv) choosing of the best fit among the crops and the selecting the same as the alternative crop strategy.

Among the above four steps, the formulation of the soil-site criteria to meet the crop requirements forms a vital and important step. For the development of crop requirements, one has to do either experimentation at each well-characterized growing environment or take the help of published literature. Naidu et al. (2006) have compiled the soil-site requirement of major crops of India by reviewing published literature and consulting crop-specific researcher teams.

Matching of crop requirements consists of comparing existing climate, soil and physiographic conditions with the soil-site criteria with respect to individual crop. On the basis of the degree and the number of limitations identified, the suitability class is established, viz. highly suitable (S1), moderately suitable (S2), marginally suitable (S3) and unsuitable land (N1 and N2) for specific kind of land use. The S1 classes correspond to areas, which have a yield potential above 80% of the maximal attainable harvest within the climatic region of the area. This figure drops to 60 and 40% for classes S2, and S3, respectively.

An ideal method to decide adoption of a cropping pattern (land use) on a particular soil unit is to have prior knowledge of the yield performance. Yields are the integrated end products of interactive processes of all factors and inputs and are, therefore, the best indices of productivity potentials. It is neither possible to obtain such information for all soil units in all the areas in view of the cost, nor it is necessary. Soil survey and classification aid in transfer of technology and, therefore, form the basis for evolving rational land use and management methods. Analysis of crop yields obtained by farmers over the years in relation to management levels on known soils (soil series) in surveyed area or field experimental data should help in deciding cropping pattern and transfer of technology to similar areas.

The land suitability can be determined by three methods.

(i) Simple limitation method
(ii) Limitation method with criteria of number and intensity of limitation.
(iii) Parametric method

(i) **Simple limitation method**: In this limitation method, the suitability classes and sub-classes are directly assigned to land units based on suitability criteria.

In this method, with criteria of number and intensity of limitation, the first step is to assign the limitation to each of the parameter.0—no limitation

1—slight limitation
2—moderate limitation
3—severe limitation
4—very severe

(ii) **Based on number and intensity of limitations, classes are assigned.**

S1—Very suitable → 3 or 4 slight limitation

S2—Moderately suitable → More than 3 or 4 slight limitation
S3—Marginally suitable → 2 or 3 moderate or one or more severe limitation
N1—Actually unsuitable, but potentially suitable
N2—Unsuitable

In India, land evaluation is mostly the land suitability classification method (Fig. 10.1). It is being carried out after land resource inventory of an area at different scales like 1:10 K (watershed/panchayat/taluka/mandal), 1:50 K (district) and 1:250 K (State). Land resource data of an area will be integrated with landform/physiography/sub-physiography and climate (agro-climatic zones/agro-ecological sub-regions/agro-ecological regions) to delineate land management units (LMUs), which are homogenous with respect to major soil and climatic parameter and require similar management for particular crop. LMUs will be evaluated for crop suitability and management strategies will be recommended for particular crop and cropping systems.

(iii) **Parametric method**: This method suggests the calculation of productivity index (Riquier et al. 1970) considering nine factors, viz. moisture (H), drainage (D), effective depth (P), texture/structure (T), base saturation (N), soluble salt concentration (S), organic matter content (O), mineral exchange capacity/nature of clay (A) and mineral reserve (M). Each of the above parameters is given numerical value between 1 and 100 and resultant index obtained by a multiplication of nine factors are positioned in one of the five different classes.

Productivity index = $H \times D \times P \times T \times N/S \times O \times A \times M$

Based on overall rating by multiplicative method, classes are assigned as under:

S1: 100–75
S2: 75–50
S3: 50–25
N1: 25–12
N2: 12–0

Under each method, the suitability of the soil units for soybean is assessed and presented in Tables 10.6, 10.7, 10.8 and 10.9. Similar method was extensively used elsewhere (Singh and Mishra 1995, 1996).

10.7.4 Fertility Capability Classification

The soil survey report provides information on the relative suitability of soils for alternative uses, and however, its utility can be enhanced if the taxonomic units are grouped into

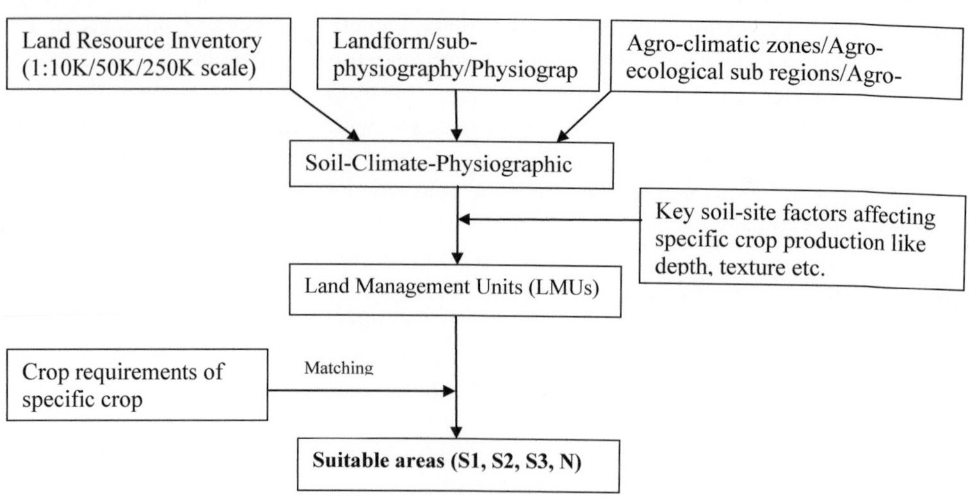

Fig. 10.1 Stepwise methodology for land evaluation. *Source* Author

Table 10.6 Climatic and soil-site suitability criteria for soybean

Land use requirement			Rating			
Soil-site characteristic		Unit	Highly suitable S1	Moderately suitable S2	Marginally suitable S3	Not suitable **N**
Climatic regime	Mean temperature in growing season	°C	25–28	29–32	33–36	>36
	Mean RH in growing season	%	70–75	60–65	50–60	<50
	Total rainfall	mm	600–750	500–600	400–500	<400
Land quality						
Moisture availability	Length of growing period	Days	>120	100–120	85–100	<85
	AWC	mm/m	>200	150–200	150–50	
Oxygen availability to roots	Soil drainage	Class	Well drained	Moderately well drained	Poorly drained to imperfectly drained	
Nutrient availability	Texture	Class	cl, scl, l, sil	sl, c, sic, sicl	c + (SS), ls	s
	pH	1:2.5	6.5–7.5	7.6–8.5 6.4–6.0	>8.5	
	OC	%	Medium	High	Low	
Rooting conditions	Effective soil depth	cm	>75	50–75	50–25	<25
	Stoniness	%	>15	15–25	25–35	
	Coarse fragments	vol%				
Soil toxicity	Salinity (EC saturation extract)	dS/m	<1.0	1–2	2–4	>4
	Sodicity (ESP)	%	<5	5–10	10–15	>15
Erosion hazard	Slope	%	<3	3.5	5–8	>8

Table 10.7 Land suitability classification for soybean by simple limitation method

Soil units	AmB	AmB2	AmC2	Af B	Af B2	Bm B2	CmC2	DmB	DmB2	DmC2	DkB2
Climatic characters											
Total rain (mm)	S_3	S_3	S_3	S_3	S_3	S_3	S_3	S_3	S_3	S_3	S_3
Rainfall during growing period	S_2	S_2	S_2	S_2	S_2	S_2	S_2	S_2	S_2	S_2	S_2
LGP	S_1	S_1	S_1	S_1	S_1	S_1	S_1	S_1	S_1	S_1	S_1
Mean temp. GP	S_1	S_1	S_1	S_1	S_1	S_1	S_1	S_1	S_1	S_1	S_1
Site characteristics											
Slope %	S_1	S_1	S_2	S_1	S_1	S_1	S_2	S_1	S_1	S_2	S_1
Drainage	S_1	S_1	S_1	S_1	S_1	S_1	S_1	S_1	S_1	S_1	S_1
AWC	S_1	S_1	S_1	S_1	S_1	S_3	S_1	S_1	S_1	S_1	S_1
Soil characters											
Texture	S_2	S_2	S_2	S_1	S_1	S_2	S_2	S_2	S_2	S_2	S_2
Depth	S_1	S_1	S_1	S_1	S_1	S_3	S_2	S_1	S_1	S_1	S_1
pH	S_2	S_2	S_2	S_2	S_2	S_1	S_3	S_2	S_2	S_2	S_2
EC	S_1	S_1	S_1	S_1	S_1	S_3	S_1	S_1	S_1	S_1	S_1
ESP	S_1	S_1	S_1	S_1	S_1	S_3	–	S_1	S_1	S_1	S_1
Suitability class	S_3	S_3	S_3	S_3	S_3	S_3	S_3	S_3	S_3	S_3	S_3

Table 10.8 Land suitability classification for soybean based on number and intensity of limitations

Soil units	AmB	AmB2	AmC2	AfB	AfB2	BmB2	CmC2	DmB	DmB2	DmC2	DkB2
Total rainfall	3	3	3	3	3	3	3	3	3	3	3
Rainfall during growing period	2	2	2	2	2	2	2	2	2	2	2
GP	0	0	0	0	0	0	0	0	0	0	0
Mean temp. GP	0	0	0	0	0	0	0	0	0	0	0
Site characteristics											
Slope (%)	1	1	2	1	1	1	2	1	1	2	1
Drainage	1	1	1	1	1	1	1	1	1	1	1
AWC	0	0	0	0	0	3	1	0	0	0	0
Soil											
Texture	2	2	2	2	2	2	2	2	2	2	2
Depth	0	0	0	0	0	3	2	0	0	0	0
Soil fertility											
pH	2	2	2	2	2	2	3	2	2	2	2
EC	0	0	0	0	0	0	1	0	0	0	0
ESP	1	1	1	1	1	1	4	1	1	1	1
Suitability class	S_3	S_3	S_3	S_3	S_3	S_3	S_3	S_3	S_3	S_3	S_3

Table 10.9 Land suitability classification for soybean by parametric method

Soil unit	AmB	AmB2	AmC2	AfB	AfB2	BmB2	CmC2	DmB	DmB2	DmC2	DkB2
Climatic											
Total rainfall	50	50	50	50	50	50	50	50	50	50	50
Rainfall during growing period	50	50	50	50	50	50	50	50	50	50	50
LGP	97.5	97.5	97.5	97.5	97.5	97.5	97.5	97.5	97.5	97.5	97.5
Mean temp. GP	97.5	97.5	97.5	97.5	97.5	97.5	97.5	97.5	97.5	97.5	97.5
Site characteristics											
Slope	90	90	72.5	90	90	90	72.5	90	90	72.5	90
Drainage	90	90	90	90	90	90	90	90	90	90	90
AWC (mm/m)	97.5	97.5	97.5	97.5	97.5	50	90	97.5	97.5	97.5	97.5
Soil											
Texture	72.5	72.5	72.5	72.5	72.5	72.5	72.5	72.5	72.5	72.5	72.5
Depth	97.5	97.5	97.5	97.5	97.5	50	72.5	97.5	97.5	97.5	97.5
Fertility											
pH	72.5	72.5	72.5	72.5	72.5	72.5	50	72.5	72.5	72.5	72.5
EC	97.5	97.5	97.5	97.5	97.5	97.5	90	97.5	97.5	97.5	97.5
ESP	90	90	90	90	90	90	20	90	90	90	90
Overall suitability class	8.4	8.4	6.8	8.4	8.4	2.2	0.66	8.4	8.4	6.8	8.4
	N_2	N_2	N_2	N_2	N_2	N_2	N_2	N_2	N_2	N_2	N_2

management units which can easily indicate the potentials and constraints of an area in terms of its fertility, tillage and irrigation management. The fertility capability classification (FCC) system is a technical soil classification system that focuses quantitatively on the physical and chemical properties of the soil that are important to soil fertility management (Buol et al. 1975). The FCC lays maximum emphasis on the component of soil fertility within 50 cm layer from the surface; however soil taxonomy puts more emphasis on sub-surface soil properties because of their more permanent nature.

The system consists of three categorical levels **'type'** (texture of plough layer or surface 20 cm) at the highest category.

S	Sandy topsoils: loamy sands and sands (by USDA definition)
L	Loamy topsoils: <35% clay but not loamy sand or sand
C	Clayey topsoils: >35% clay
O	Organic soils: >30% O.M. to a depth of 50 cm or more

The **substrata type**, which is the next lower category of the system, refers to the texture of the subsoil that occurs within 50 cm from the surface and is used only when there is any marked textural change from the surface or if a hard root restricting layer is encountered within 50 cm. The substrata types are as follows:

S	Sandy subsoils: texture as in type
L	Loamy subsoil: texture as in type
C	Clayey subsoil: texture as in type
R	Rock or other hard root restricting layer
R	Weathered parent material (murrum) as proposed by Jagdish Prasad (2000)

Modifiers: Condition modifier is the lowest category of FCC system, which is determined after assessing the physical and chemical properties of the surface and sub-surface soils, where more than one criterion is listed for each modifier, only one need to be met. The criterion listed first is the most desirable one and should be used if data are available. Subsequent criteria are presented for use where data are limited.

Waterlogging (g)	(gley): soil or mottles with < 2 chroma within 60 cm of the soil surface and below all A horizons, or soil saturated with water for > 60 days in most years;
Soil moisture stress (d)	(dry): ustic, aridic or xeric soil moisture regimes (subsoil dry > 90 cumulative days per year within 20–60 cm depth);
Low cation exchange capacity (e)	(low cation exchange capacity): applies only to plough layer or surface 20 cm, whichever is shallower: CEC < 4 meq./100 g soil by Σ bases + KCl-extractable Al (effective CEC), or CEC < 7 meq/100 g soil by Σ cations at pH 7, or CEC < 10 meq./100 g soil by $\sum$ cations + Al + H at pH 8.2;
Aluminium toxicity (a)	(aluminium toxicity): > 60% Al saturation of the effective CEC within 50 cm of the soil surface, or > 67% acidity saturation of CEC by Σ cations at pH 7 within 50 cm of the soil surface or > 86% acidity saturation of CEC by Σ cations at pH 8.2 within 50 cm of the soil surface or pH < 5.0 in 1:1 H_2O within 50 cm, except in organic soils where pH must be less than 4.7;
Acidity (h)	(acid): 10–60% Al saturation of the effective CEC within 50 cm of soil surface, or pH in 1:1 H_2O between 5.0 and 6.0;
High P-fixation by iron (i)	(high P-fixation by iron):% free Fe2O3 to % clay > 0.15 and more than 35% clay, or hues of 7.5 YR or redder and granular structure. This modifier is used only in clay types; it applies only to plough layer or surface 20 cm of soil surface, whichever is shallower;
Amorphous clays (x)	(X-ray amorphous): pH > 10 in 1 N NaF, or positive to field NaF test or other indirect evidence of allophane dominance in the clay fractions;
Low nutrient reserves (k)	(low K reserves): < 10% weatherable minerals in silt and sand fraction within 50 cm of the soil surface, or exchangeable $K < 0.20$ meq./100 g, or $K < 2\%$ of Σ bases; if bases < 10 meq/100 g;
Calcareousness (b)	(basic reaction): free $CaCO_3$ within 50 cm of soil surface (effervescence with HCl) or pH > 7.3;
Salinity (s)	(salinity): >4 S m^{-1} of electrical conductivity of saturated extract at 25 °C within 1 m of the soil surface;
Alkalinity (n)	(natric): >15% Na saturation of CEC within 50 cm of the soil surface;
Sulphides (c)	(cat clay): pH in 1:1 H_2O is <3.5 after drying and jarosite mottles with hues of 2.5 Y or yellower and chroma 6 or more are present within 60 cm of the soil surface;
Gravel (r)	(gravel): a prime (′) denotes 15–35% gravel or coarser (>2 mm) particles by volume to any type or substrata type texture (example: S′L = gravelly, sand over loamy; SL ′ = sandy over gravelly loam); two prime marks (″) denote more than 35% gravel or coarser particles (>2 mm) by volume in any type or substrata type (example: LC ″ = loamy over clayey skeletal; L′C ″ = gravelly loam over clayey skeletal);
Slope (s)	(slope): where it is desirable to show slope with the FCC, the slope range percentage can be placed in parenthesis after the last condition modifier (e.g. Sb (1–6%) = uniformly sandy soil calcareous in reaction, 1–6% slope.

(continued)

Management measures to overcome the limitations

Waterlogging (g)	Denitrification frequently occurs in anaerobic subsoil; tillage operations and growth of certain crops may be adversely affected by excess rain unless tilling or other drainage procedures improve drainage; good soil moisture regime for rice production
Soil moisture stress (d)	Moisture is limiting during the dry season unless soil is irrigated; planting date should take into account the flush of N at onset of rains; germination problems are often experienced if first rains are sporadic

(continued)

High leaching potential	Low ability to retain nutrients against leaching, mainly K, Ca and Mg; heavy applications of these nutrients and of N fertilizers should be split; potential danger of over-liming
Aluminium toxicity (a)	Plants sensitive to Al toxicity will be affected unless lime is applied; extraction of soil water below the depth of lime incorporation will be restricted; lime requirements are high unless a modifier is also indicated; this modifier is desirable for rapid dissolution of phosphate rocks and for good latex flow in rubber; Mn toxicity may occur in some of these soils
Acidity (h)	Low to medium soil acidity; requires liming for Al-sensitive crops, such as cotton and alfalfa
High P-fixation (i)	High P-fixation capacity; requires high levels of P fertilizer or special P management practices; sources and methods of P fertilizer application should be considered carefully; with C texture, these soils have granular soil structure
Amorphous clays (x)	High P-fixation capacity; amount and most convenient source of P to be determined; low organic N mineralization rates
Vertic character (v)	Clayey textured topsoil with shrink and swell properties; tillage is difficult when too dry or too moist, but soils can be highly productive; P deficiency common
Low nutrient reserves	Low ability to supply K; availability of K should be monitored and K fertilizers may be required frequently; potential K–Mg–Ca imbalances
Calcareousness (b)	Calcareous soils; rock phosphate and other non-water-soluble phosphates should be avoided; potential deficiency of certain micro-nutrients; principally iron and zinc
Salinity (s)	Presence of soluble salts; requires drainage and special management for salt-sensitive crops or the use of salt-tolerant species and cultivars
Alkalinity (n)	High levels of sodium; requires special soil management practices for alkaline soils, including use of gypsum amendments and drainage
Sulphides (c)	Potential acid sulphate soil; drainage is not recommended without special practices; should be managed with plants tolerant to high water table level

How to Arrive at FCC Units

The soil is first classified based on the presence or absence of soil constraints. Most of the quantitative limits used in FCC are the criteria already used in the soil. Types and substrata types are represented in capital letters, modifier in the lower case as prime and slope are in the parenthesis

For example: In typical Ultisols, the FCC unit 'Lcaei' indicates

L Loamy surface texture
c Clayey subsoil texture
a Aluminium toxicity
e Low CEC
i High P-fixation by iron

Mathan et al. (1994) studied 21 soils of Kamarajar District of Tamil Nadu state and grouped into 8 FCC units based on type, substrata type and conditional modifiers. These FCC units can be used for conducting fertility-based experiments. The condition modifiers identified are 'd', 'b', 'm', 'v', 'm', 'n', 'k', 'i' and 'e', whereas in acid soils of Nilgiris District, condition modifiers observed are 'a', 'h', 'i' and 'e' (Mathan 1990). Additionally, as a local modifier, Mg (m) deficiency was identified.

10.7.5 Prime Lands Classification

Heterogeneity is a basic characteristic of lands and this heterogeneity means the capacity of lands to support various functions simultaneously. Land use is the key activity which determines the performance of lands such as land-based production, infrastructure and housing (Wiggering et al. 2003). The recent trend in land use in some areas resulted in loss of some prime farm lands to industrial and urban uses (Sturdevant et al. 2001). The loss of prime farm lands to other uses puts pressure on marginal lands which generally are more fragile, erodible, less productive and cannot be easily cultivated. The significant reduction in the area under culturable waste lands indicates clearly the conversion of even marginal lands for agricultural purposes. The multi-functionality of crop production system refers to the fact that crop production activity is not limited to produce food and fibre and may also have other functions. For that purpose, land has been classified into different classes according to their potential for raising crops or other purposes. According to USDA, prime farm lands are defined as lands that have the best combination of physical and chemical characteristics for producing food, feed, fibre, forage, oilseed and other agricultural crops with minimum inputs of fuel, fertilizer, pesticides and labour, with tolerable soil erosion. It has the combination of favourable soil properties, growing season and moisture supply needed to produce sustained high yields of crops in an economic manner if it is treated and managed according to acceptable farming methods. The classification will identify the potential to use a given piece of land for different purposes, based on biophysical limitations of the land such as soils, climate and topography that cannot be removed or improved by acceptable level of management. Land with higher capability has more options for use and also likely to have reasonable resilience to adaptation *vis-a-vis* climate change. Further, climatic constraints also have equal importance for land capability classification by restricting plant growth rate, ploughing, sowing and harvesting. Procedure and criteria given by Read (1988) and Giles and Koeln (1983) for classification of prime lands in

Table 10.10 Criteria used in delineation of prime lands

Particulars	Prime irrigated	Prime rainfed	Marginal lands
	Indicators		
Soil depth	–	>50 cm	<50 cm
Landform	–	Level to gently sloping	Moderate to highly sloping lands
Gravelliness	Non-gravelly	Non-gravelly to slightly gravelly	Gravelly
Length of growing period	–	>120 days	<120 days
Erosion	–	Nil to slight	Moderate to severe
Productivity of crops	<30% deviation from attainable yield	<30% deviation from attainable yield	>30% deviation from attainable yield

Fig. 10.2 Methodology followed for delineation of prime agricultural land. *Source* Author

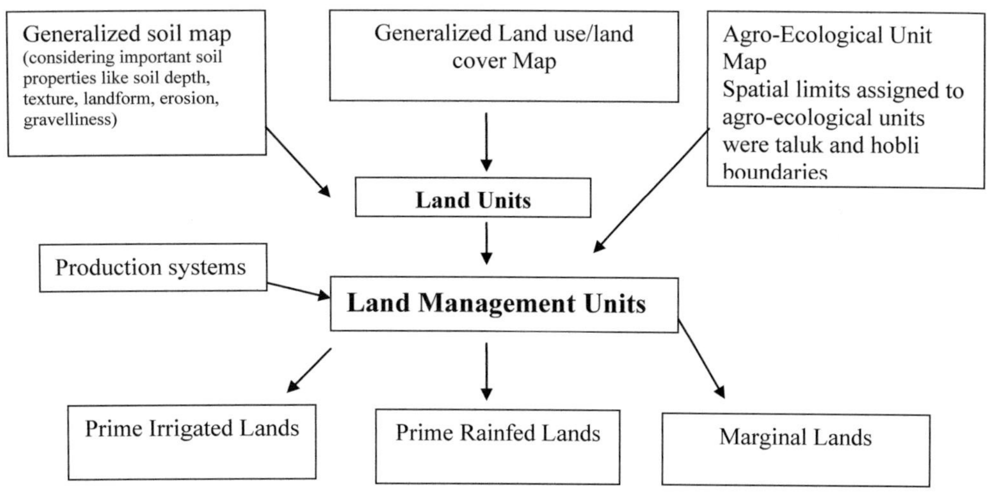

Mysore District have been suitably modified by considering landform, soil depth, gravelliness, erosion, LGP and productivity of crops (Ramamurthy et al. 2012).

Methodology followed to delineate prime irrigated lands (PIL), prime rainfed lands (PRL) and marginal lands (ML) is presented in Table 10.10 and Fig. 10.2 and prime lands map of Mysore District of Karnataka is presented in Fig. 10.3 (Ramamurthy et al. 2015).

Distribution and per cent area under prime irrigated, rainfed and marginal lands in different *taluks* is presented in Table 10.11.

T. Narsipura *Taluk* has the highest prime irrigated lands followed by K. R Nagar and Hunsur. Minimal prime irrigated lands are in Piriyapatna *Taluk*. Nanjangud *Taluk* has the highest prime rainfed area followed by Piriyapatna, Mysore, H.D. Kote and Hunsur *taluks*. There is a need for government agencies to consider the potentials of land before earmarking land for other than agricultural uses. Hunsur and Piriyapatna *taluks* have comparatively more marginal lands, which can be diverted to non-agricultural uses.

10.7.6 Storie Index Rating

Soil evaluation is the determination of productivity ratings, which is defined as capacity of the soil to produce crops. Soil texture, climate, soil management, drainage, soil salinity or alkalinity, nutrient status are some of the important factors that govern the productivity of the soil. Storie (1950) used the four parameters for evaluation of soil productivity. It is a parametric approach. Parameters are not based on requirement of any crop, only soil and site characteristics are taken into consideration.

Four groups of factors are considered.

SIR = A · B · C · X or D

A = Cracter of soil profile

B = Texture of surface soil

C = Slope

X or *D* = Miscellaneous

Ratings are given based on the soil parameters and overall rating is calculated by the multiplicative method. Based on this rating, the classes are allotted.

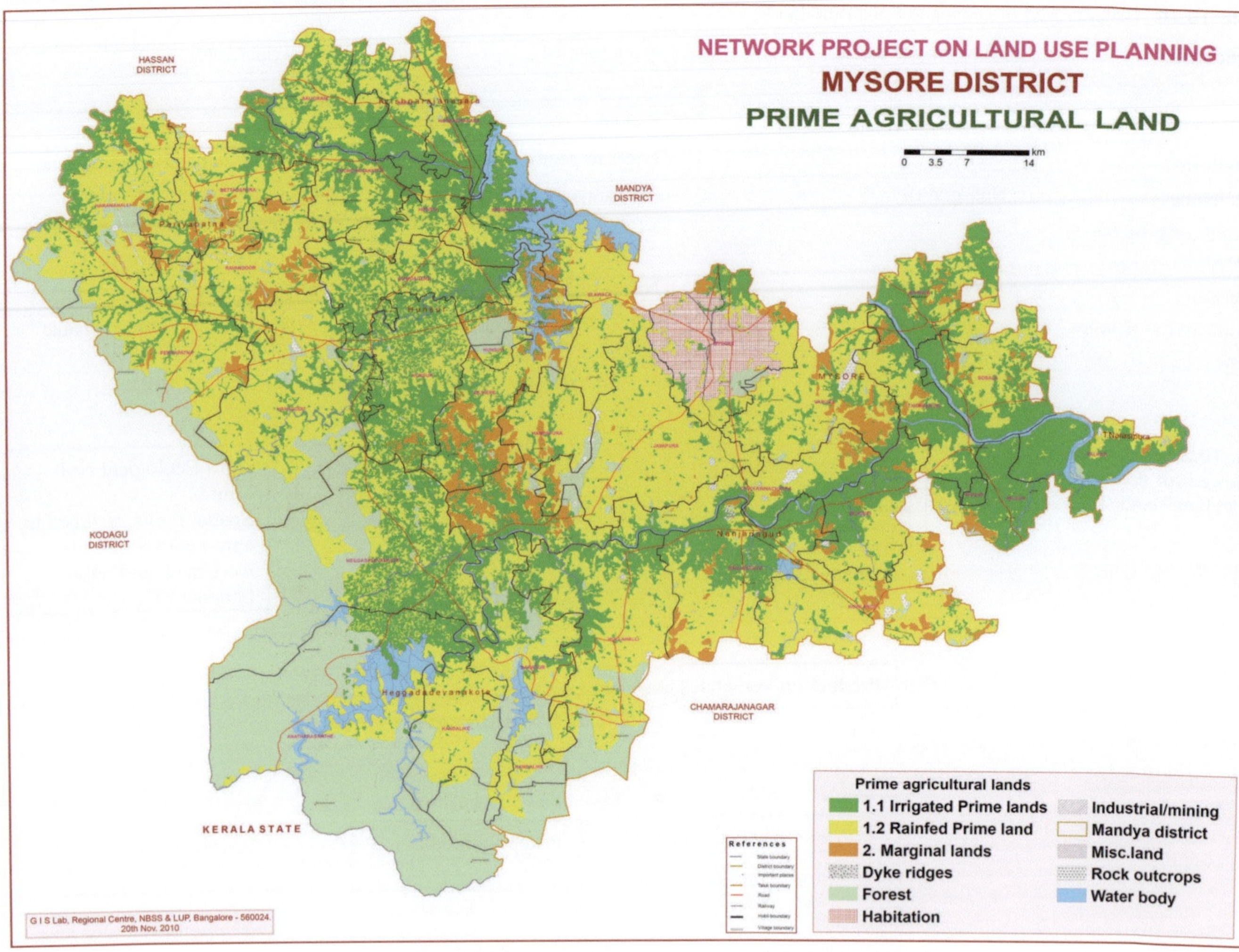

Fig. 10.3 Prime agricultural lands in Mysore district (Ramamurthy et al. 2015). *Source* Author

Table 10.11 Area under different categories of prime lands in Mysore District

Taluks	Per cent area		
	Prime irrigated lands	Prime rainfed lands	Marginal lands
H.D. Kote	4.25	6.91	0.63
Hunsur	4.50	6.46	0.95
K.R. Nagar	4.81	3.50	0.14
Mysore	1.27	7.70	0.68
Nanjanagud	4.24	8.71	0.74
Piriyapatna	2.08	7.77	0.91
T. Narsipura	5.72	2.47	0.41

Ratings	Grade	Class
80–100	1	Excellent
60–79	2	Good
40–59	3	Fair
20–39	4	Poor
10–19	5	Very poor
0-9	6	Not suitable

Shome and Raychoudhuri (1960) by using three soil parameters, viz. A:soil profile characteristics; B:topography, texture and structure and C:degree of climate suitability, salinity, stoniness and tendency to erode and assessed soils of 294 Indian districts by using soil index rating. They concluded that soil index rating is a permanent feature of the soil, whereas Storie's productivity ratings can be varied by adopting more and more improved management practices.

10.7.7 Actual and Potential Productivity

The earliest approach in soil productivity indices and rating was initiated by Shome and Raychoudhuri (1960). Later on, Riquier et al. (1970) have evolved a system of soil appraisal in terms of actual and potential productivity. It is a modified version of Storie's Index. Nine factors, viz. moisture, drainage, depth, texture, base saturation, soluble salts, organic matter, CEC and mineral reserves are rated on a scale 0–100 and the percentages cumulatively multiplied to obtain productivity index (*P*). In a similar manner, the potentiality index (*P*′) is calculated after affecting the management measures. The ratio *P*:*P*′ indicating the extent to which productivity can be improved is called the coefficient of improvement. Soils with rating index 65–100 are excellent, 35–64 good, 20–34 average, 8–10 poor and below 8 extremely poor. Maps showing productivity and potentiality index can also be prepared. It is evident that the land evaluation system of Riquier et al. (1970) does not explain the variability in the yield. Like Storie's Index, this system of land evaluation has the limitation in that one limiting factor reduces the index of productivity. Also, assigning values to factors like drainage is difficult. Perhaps, factors should be chosen according to the operating limitations affecting the crop growth within a particular region to obtain a more realistic productivity rating. However, there is a need to improve this technique under diversified Indian conditions for reliable applications (Mishra 2015, 2017).

Naidu et al. (1986) assessed productivity and potentiality of eight extensively occurring soil series of Delhi (UT) through Riquier's method. Out of these eight series, three were graded as good class and another three series graded as average and while two were graded as poor class due to their inherent physico-chemical limitations. It was observed that soil texture and moisture properties are the prime factors which influenced the productivity and potentiality ratings. Vishalakshi Devi and Naidu (2016) evaluated major sugarcane-growing soils of Chittoor District of Andhra Pradesh by using qualitative and quantitative methods. The qualitative methods employed were USDA land capability classification and land suitability classification, while the quantitative evaluation method includes Riquier's parametric approach. The soils of the study area were classified into land capability classes IV, V and VI. However, the land suitability evaluation suggested that these sugarcane-growing soils were moderate to marginally suitable for growing sugarcane crop. Riquier's parametric approach was found to be a good indicator for identification of production potential of sugarcane-growing soils. The land evaluation study revealed that, characteristics and suitability of these soils for sugarcane crop were highly variable, and hence, their management must be site-specific. Singh and Mishra (1996, 1997), Singh et al. (1996), Choudhary et al. (2009) and Mishra (2016) applied Riquier's approach for soil evaluation in Bihar.

Mishra (2016) quantified the extent of potential productivity through improvement in Kosi Zone of Bihar. The Soil Survey and Land Use Planning Scheme, Sabour in collaboration with NBSS and LUP, Regional Centre, Kolkata, has established altogether 10 soil series in Kosi Zone with their land capability and irrigability sub-classes (Tables 10.12 and 10.13). The soil series is a group of soil horizons, similar in differentiating characteristics and arrangement within the series control section, except for the surface soil, developed under comparable climatic and geomorphic environments, and is a key to any landscape unit.

Table 10.12 Agro-climatic zone II: location, area, districts and major land use choice

AC zone II	Latitudes	Longitudes	Total geographical area (mha)	Net sown area (mha)	Main crop grown	Districts
North-east alluvial plain	25° 10′–26° 32′ N	86°21′–88°19′ E	2.08	1.21	Rice, wheat, maize, jute, moong, millets, sugarcane, kalai, barley, potato, grasses, vegetables, oilseeds and spices	Purnea, Katihar, Saharsa, Supaul, Madhepura, Araria and Kisanganj

Table 10.13 Soil series established in Kosi region of Bihar

Physiographic region	Soil series	Soil taxonomy	Land capability sub-class	Irrigability sub-class	Land use pattern (major)
Indo-Gangetic alluvial plain *(Alluvial cone of Kosi river)* Saharsa, Madhepura, Supaul, Araria and Purnea districts	Arraha	Typic Ustifluvent	IIw	2d	Rice, gram and khesari
	Baruari	Typic Psammaquents	IVws	3ds	Rice
	Bhargaon	Aeric Fluvaquents	IIIws	3ds	Rice, wheat and mustard
	Hanuman Nagar	Typic Ustipassements	IVws	3ds	Rice and potato
	Keskata	Aeric Endoaquents	IIIsw	3 sd	Rice, wheat and linseed
	Madhepura	Typic Ustifluvents	IIIs	3s	Rice and mustard
	Madhuban	Aeric Fluvaquents	IIIws	3d	Rice and khesari
	Nirpur	Aeric Endoaquepts	IIw	2d	Rice and lentil
	Paina	Typic Haplustepts	IIIw	3d	Rice, wheat and khesari
	Tikapatti	Aeric Endoaquents	IIIw	3d	Rice, wheat and khesari

Actual and potential productivity for 10 soil series computed following the methods outlined by Riquier and associates in 1970. It is apparent that proper management/improvement of soils could enhance the productivity even up to 1.34–1.77 times (coefficient of improvement), although such figures (Tables 10.14 and 10.15) could further be authenticated with agronomic yields. However, there is a need for refinement of this technique so that productivity of a soil could be computerized/authenticated.

10.7.8 Crop Growth Models for Land Evaluation

The FAO framework is basically a classification system working with classified land data, inferring land qualities and resulting in suitability classes. Initially, this made sense as data were collected over map units. But with the advent of computer and multi-temporal/continuous sampling methods (remote sensing), it is possible to collect large quantities of data in space and time. This permits us to model the response of the land to various land uses, thus fulfilling the fundamental purpose of land evaluation, i.e. assessment of land performance when land is used for specified purposes.

Crop growth modelling is a procedure through which yield can be simulated and used for land evaluation. The obvious advantage of this method is its ability to provide probabilistic estimates of yield and to include evaluation of spatial and temporal variations. Crop growth models can be empirical or mechanistic. Empirical models are developed by regressing a sample of yield with sample of input variable (s). Commonly, simple linear, nonlinear or multivariate analysis are used to fit historical yield data to climate variables like mean temperature or precipitation. They are useful to predict yield potentials based on climate. However, they cannot be used directly for land evaluation. They do not account for dynamic changes of state variables. Moreover, the regression coefficients need to be calibrated when these models are applied to new circumstances.

Mechanistic- or process-oriented models are mathematical simulation of physiological, chemical and physical processes which govern crop growth. There is an unlimited potential to expand a simulation model-based land evaluation system from the traditional biophysical modelling (Varcoe 1990).

A major use of modelling for land evaluation is to predict yields (either average yields or time sequence). The value of the land is directly reflected by its productivity. The modelled yield, along with a price for the product, gives the gross return. Since yields vary with management level (e.g. type and level of inputs and timeliness of operations), modelling yield requires a careful specification of the input levels of farming systems.

Currently, models are used at three distinct levels in land evaluation

- As a scientific tool for the investigation of processes
- As a predictor of yield and
- As a standalone evaluation or classification system

Table 10.14 Actual soil productivity levels with associated parameters

Parameter	Series1	Series2	Series3	Series 4	Series 5	Series 6	Series 7	Series 8	Series 9	Series 10
Moisture (H)	H4c	H3c	H4c	H4c	H4c	H4c	H4c	H4c	H4c	H4c
Drainage (D)	D2a	D2a	D2a	D2a	D3a	D2a	D2a	D2a	D2a	D2a
Depth (P)	T6	T2	T4	T4	T4	T4	T6	T6	T6	T6
Base saturation (N)	N5	N5	N5	N5	N5	N5	N5	N5	N5	N5
Salinity (S)	S1	S1	S1	S1	S1	S1	S1	S1	S1	S1
Organic matter (O)	O1	O1	O1	O1	O1	O1	O1	O1	O1	O1
Clay CEC, cmolkg^{-1}(A)	A0	A0	A0	A0	A0	A0	A0	A0	A0	A0
Mineral reserve (M)	M1	M1	M1	M1	M1	M1	M1	M1	M1	M1
Parent material	Alluvium	Alluvium	Alluvium	Alluvium	Alluvium	Alluvium	Alluvium	Alluvium	Alluvium	Alluvium
Permeability	Moderate	Rapid	Moderate	Rapid	Rapid	Rapid	Slow	Slow	Slow	Slow
Soil pH	Neutral	Neutral	Neutral	Neutral	Neutral	Neutral	Neutral	Neutral	Neutral	Neutral
Stratification due to sand	–	15 cm	67 cm	30 cm	46 cm	78 cm	20–25 cm	–	–	–

Table 10.15 Actual soil productivity ratings, index of potentiality (PI) and coefficient of improvement (CI)

Parameter	Series 1	Series 2	Series 3	Series 4	Series 5	Series 6	Series 7	Series 8	Series 9	Series 10
Moisture (H)	100	70	100	80	100	100	100	100	100	100
Drainage (D)	80	80	80	80	90	80	80	80	80	80
Depth (P)	100	100	100	100	100	100	100	100	100	100
Base saturation (N)	100	100	100	100	100	100	100	100	100	100
Salinity (S)	100	100	100	100	100	100	100	100	100	100
Organic matter (O)	85	85	85	85	85	85	85	85	85	85
Clay CEC, $cmol_c kg^{-1}$(A)	85	85	85	85	85	85	85	85	85	85
Mineral reserve (M)	85	85	85	85	85	85	85	85	85	85
Actual productivity index (PI)	49	3.4	24	19	27	24	49	49	49	49
Rating class of PI	Good	Extremely poor	Average	Poor	Average	Average	Good	Good	Good	Good
Index of potentiality rating (IP) after improvement	85	8.5	42.5	25.5	42.5	42.5	85	85	85	85
IP rating class	Excellent	Poor	Good	Average	Good	Good	Excellent	Excellent	Excellent	Excellent
Coefficient of Improvement (CI)	1.73	2.5	1.77	1.34	1.57	1.77	1.73	1.73	1.73	1.73
Associated correctable limitations	D-O-A	H-O-A	D-O-A	H-D-O-A	O-A-D	D-O-A	D-O-A	D-O-A	D-O-A	D-O-A
Limitations un-correctable at farmer's level	M	T-M	T-M	T-M	T-M	T-M	M	M	M	M

Note H—period of soil moisture between field capacity and wilting point in a year, *D*—flooding and drainage condition, *P*—effective depth of profile, *T*—texture and structure of soil, *N*—base saturation as an index of nutrient access, *S*—salinity hazard, *O*—soil organic matter, *A*—nature of clays and their CEC and *M*—mineral reserve

Modelling can also be used to predict some land qualities that are important components of yield, e.g. moisture supply, nutrient supply, radiation balance, as well as land qualities important for the land use but not directly affecting yield, e.g. trafficability and workability.

Advantages of crop growth models in agricultural land evaluation

(i) Simulation of yield provides a quantifiable method of classifying land
(ii) A large number of variables and complex interactions can be analysed.
(iii) It provides a single, accessible, organized and standardized body of reference.
(iv) Simulation of result is a rapid and cheap method of investigation, particularly when time frames or money do not permit data collection.
(v) The modelling process identifies particularly important inputs to each modelled system.
(vi) Simulation is often claimed to be scale-neutral in concept.
(vii) Simulation is useful for extrapolating experimental results to other sites where climate and soil conditions may differ.
(viii) Production assessments can be made for crops not previously grown in a region or to compare a range of crops.
(ix) Models can be continuously updated and modified.
(x) Models can be extended to incorporate economic and/or social constraints.
(xi) Estimation of the production levels can be used for valuation and taxation purposes

Disadvantages of crop growth models in agricultural land evaluation

(i) Even in mechanistic models, many relationships are empirical.
(ii) There may be a tendency to accept simulated results without adequate validation.
(iii) Many models are very complex (at least initially) and may require detailed data.
(iv) Access to data is becoming increasingly difficult/costly.
(v) There are seldom comprehensive records of the required data at region or district level, particularly in less developed countries
(vi) Certain degree of expertise is needed to use simulation models.

Some commonly used crop simulation models in India

There are many dynamic simulation models to predict crop yield.

DSSAT: Decision support system for agro-technology transfer (DSSAT) was developed by ICRISAT to estimate crop production, resource use and risks associated with different crop production practices. The software package contains crop simulation models, databases for weather, soils and crop and strategy evaluation programme integrated with a shell programme which is the main user interface. It contains the following families of models

The CERES family for simulating wheat, maize, barley, sorghum and millets
The CROPGRO family models to simulate grain legumes–soybean, groundnut and dry bean and
ROOT crop models to simulate potato, cassava and aroids.
OTHER crops—tomato, sunflower, sugarcane and pasture.

These crop simulation models simulate the effects of weather, soil, water, cultivar and N on crop growth and yield for well-drained soils.

INFOCROP: Developed by Aggarwal et al. (2004); it is a generic simulation model for annual crops in tropical environments. It uses weather (radiation, rainfall, temperature, wind speed, humidity and frost), soil (texture, pH, depth and fertility), variety (physiology, phenology and morphology), management (planting date, fertilization, irrigation and residue management) and pests (type, population and severity) as inputs. The outputs include economic and biomass yield, crop duration, water stress, N stress, yield loss due to pests, soil C and N dynamics and greenhouse gas emissions. Currently, it simulates chickpea, cotton groundnut, maize, mustard, pearl millet, pigeon pea, potato, rice, sorghum, soybean, sugarcane and wheat.

Aggarwal et al. (2001) adopted a systems' approach combining simulation models and DSS for land use analysis and planning for sustainable food security in Haryana. Roetter et al. (2004) provided a detailed account of the Systems research Network (SysNet) for land use planning in tropical Asia with a focus on its main scientific–technical output: the development of the land use planning and analysis system (LUPAS) and its component models. These include crop simulation models, expert systems, GIS and multiple goal linear programming (MGLP) model for land evaluation and optimization.

Crop growth models are emerging as new tools for land evaluation to quantify production possibilities and constraints under different land use systems. There is no single land evaluation modelling approach. The choice of model affects the reliability and scope of application of the land evaluation. The models generally use data on land unit characteristics and land utilization attributes and generate estimates of production potentials. However, these models need to be calibrated and validated to local conditions before using. The paucity of accurate data to run simulation models is still a major obstacle in their widespread use as tools for quantitative land evaluation.

10.8 Conclusions

Among the different approaches of land evaluation, the land capability classification is applicable for grouping the lands into arable and non-arable lands, while the irrigability classification attempts to group the lands into irrigable and non-irrigable classes. Both these methods are qualitative and are based on inherent limitations of the lands. Categorization of prime and marginal lands is an approach which considers the limitations of the lands for general use, whereas fertility capability classification (FCC) deals with fertility aspects of the lands for better management. The Storie index and Riquier's approach are multiplicative approaches used to arrive at productivity and potentiality of the lands. These methods though quantitative may not be useful for specific crops. FAO and Sys methods are used for land suitability for different crops. These methods are crop-specific since they consider the crop requirements, land characteristics and land qualities. In present context, prime land classification assumes greater importance to conserve these lands for present and future needs of people. Second Green Revolution or Evergreen Revolution will be a reality in the near future. Therefore, integration of prime lands and land suitability for specific crops helps in conserving the land resources and also enhances the resource use efficiency. The application of models in land evaluation has limitations since they are mostly theoretical and need validation for wider applicability. Among the above approaches, the methods which consider land characteristics—qualities of a location and crop requirements are found to be better for crop planning and delineating efficient crop—zones. However, Riquier's approach needs to be refined under Indian situations to quantify the production levels of a given land unit. So, more systematic- and location-specific works are desired to authenticate the practical relevance with coefficient of improvement, so that profitable production could be enhanced considerably to ensure even more than double of the farmer's economic growth merely by improving the correctable limitations with a given land.

References

Aggarwal PK, Kalra N, Chander S, Pathak H (2004) INFOCROP. A generic simulation model for annual crops in tropical environments. IARI, New Delhi, pp 132

Aggarwal PK, Roetter RP, Kalra N, Van Keulen H, Hoanh CT, Van Laar (2001) Land use analysis and planning for sustainable food security: with an illustration for Haryana, India, New Delhi IARI, Los Bonos IRRI, Wageningen WURC, pp 167

Buol SW, Sanchez PA, Cate RB, Granger MA (1975) Soil fertility capability classification. In: Bornemisza E, Alvarado A (eds) Soil management in tropical America. North Carolina State University, Raleigh, pp 126–141

Chamuah GS, Gangopadhyay SK, Walia CS, Baruah U (1996) Soils of Jorhat district: physiography relationship. Agropedology 6(2): 29–36

Choudhary BC, Chandra SK, Prasad SN, Mohan K, Mishra BB (2009) 25th years' perspective plan of land use planning of Bihar (2000–2025). Soil Survey and Land Use Planning Scheme, Bihar Agricultural College, Sabour

Das PT, Sudhakar S (2014) Land suitability analysis for orange and pineapple: a multi criteria decision making approach using geo spatial technology. J Geogr Info Syst 6:40–44

Dent DL, Deshpande SB (eds) (1993) Land evaluation for land use planning (papers of Indo-UK Workshop). NBSS Publ. 42, NBSS and LUP, Nagpur, India, 167 pp

FAO (1976) A framework for land evaluation. Soils Bull 32. Food and Agriculture Organization of the United Nations, Rome, Italy

FAO (1983) Guidelines: land evaluation of rained agriculture. Soil Bull 52, FAO Rome AEZ—Agro-ecological Regions of India Tech. Bull NBSS Publ. 24

Gautam D, Parul P, Kalubarme KH (2017) Agricultural land suitability evaluation for wheat cultivation using geomatics for Patan district, India. Int J Agric Res Gov Ecol 13(1):91–108

Gbadegesin S, Areola O (1987) Soil factors affecting maize yields in the south-western Nigerian savanna and their relation to land suitability assessment. Soil Surv Land Eval 7:167–175

Giles RH Jr, Koeln GT (1983) Land and cropland primeness: concepts and methods of determination. Environ Manage 7(2):129–142

GOI (2015) Directorate of economics and statistics. Government of India, New Delhi

Jagdish Prasad J (2000) Application of fertility capability classification system in soils of a Watershed in semi-arid tropics. J Indian Soc Soil Sci 48(2):329–338

Klingbiel AA, Montgomery PH (1961) Land capability classification. USDA Agriculture Handbook 210, pp 16

Laborte AG, Roetter R, Hoanh CT (2002) The land use planning and analysis system of the systems research network in Asia. In: Lawrence, J. Robinson (eds) Proceedings of the 2nd international conference on multiple objective decision support systems for land, water and environmental management (MODSS'99), 1–6 Aug 1999, Australia, P.A

Martin D, Saha SK (2009) Land evaluation by integrating remote sensing and GIS for cropping system analysis in a watershed. Curr Sci 96(4):569–575

Mathan KK (1990) Application of soil fertility capability classification to acid soils. J Indian Soc Soil Sci 38:469–473

Mathan KK, Natarajan S, Mahendra PP (1994) Application of fertility capability classification concept in major soil groups of Kamarajar district, Tamil Nadu. J Indian Soc Soil Sci 42(30):416–420

Mishra BB (2015) Soil science and land use planning: myth, reality, evidence and challenge. EC Agric 1.3:140–148

Mishra BB (2016) Report on sustainable agricultural production in Kosi region of Bihar using soil based evaluation and management. Soil Survey and Land Use Planning Scheme, Bihar Agricultural University, Sabour, Bhagalpur

Mishra BB (2017) Land economics versus land use planning. Agric Res TechOpen Access J 10(4):555791. https://doi.org/10.19080/artoaj.2017

Mukhtar E, Alexis C, Pete F (2010) Land evaluation techniques comparing fuzzy AHP with TOPSIS methods. In: 13th AGILE international conference on geographic information science, Guimaraes, Portugal, pp 1–8

NAAS (2009) State of agriculture. National Academy of Agricultural Sciences (NAAS) publication, New Delhi

Naidu LGK, Verma KS, Jain SP, Rana KPC, Sidhu GS (1986) An appraisal of productivity and potentiality of dominant soils of Delhi territory. J Indian Soc Soil Sci 34:558–567

Naidu LGK, Ramamurthy V, Challa O, Rajendra H, Krishnan P (2006) Manual soil site suitability criteria for major crops. NBSS Publ. No. 129, NBSS and LUP, Nagpur, 118 pp

Narayanaswamy YA, Inayathulla M, Thabrez M, Shashishankar A (2017) Land suitability evaluation of soils for crop production. Int J Innovative Res Sci Eng Technol 6(9):18187–18196

Niranjana KV, Ramamurthy V, Hegde Rajendra, Srinivas S, Koyal Arti, Naidu LGK, Sarkar Dipak (2011) Characterization, classification and suitability evaluation of banana growing soils of Pulivendla Region, Andhra Pradesh. J Indian Soc Soil Sci 59 (1):1–5

Rabia AH, Terribile F (2013) Introducing a new parametric concept for land suitability assessment. Int J Environ Sci Dev 4(1):15–19

Ramamurthy V, Nair KM, Naidu LGK, Dipak S (2012) Delineation of prime agricultural land for land use planning—a case study of Mysore district, Karnataka. In: National workshop on "integrated land use planning for sustainable agriculture and rural development" at NIRD, Hyderabad from 18–20th June 2012

Ramamurthy V, Naidu LGK, Nair KM, Ramesh Kumar SC, Srinivas S, Thayalan S, Sarkar D, Chaturvedi A, Singh SK (2015) District land use planning, Mysore, Karnataka. NBSS Publ. No.169, 95 pp

Ramamurthy V, Mamatha D, Niranjan KV, Vasundhara R, Ranjitha K, Chandrakala M, Singh SK (2018) Suitability evaluation for pigeon pea in southern transition zone of Karnataka Plateau, India. Legume Research. https://doi.org/10.18805/lr-4047

Read VT (1988) The concept of prime agricultural land: a Western Australian perspective. Department of Agriculture and Food, Western Australia, Perth. Report 72

Riquier J, Bramo D, Cornel JF (1970) A new system of soil appraisal in terms of actual and potential productivity. FAO—AGL-TESR/70/6, Rome

Roetter RP, Hoanh CT, Laborte AG, Van Keulen H, Van Ittersun MK, Dreiser C, van Laar (2004) Environmental modelling and software (in press)

Sehgal J (1993) Land use planning based on soils and agro-ecological zones (papers of Indo-UK Workshop). ICAR-National Bureau of Soil Survey and Land Use Planning, Nagpur, NBSS Publ. 42, pp 1–14

Sehgal JL, Medi T, Rehman T (1980) The soils of middle tigris project (Mesopotamian Plain) for land use planning. Soils Bull 376

Sehgal JL, Abrol IP (1994) Soil degradation in India: status and impact. Bull. No. 38

Sehgal JL, Challa O, Gajja BL, Yadav SC (1989) Suitability of swell-shrink soils of India for crops growth. In: Clumput O (ed) Soils for development. Pub. ITC, Gent, Belgium, pp 29–53

Sharma KR, Sharma PK, Sawhney JS (1994) Soil suitability for rice in different agroclimatic zones of Punjab. Agropedology 4:91–98

Shome KB, Raychaudhuri SP (1960) Rating of soils of India. In: Proc Nat Inst Sci India 26(Part-A):260–289

Singh VN, Mishra BB (1995) Land evaluation of some Alfisols in toposequence affected by sodiumization. J Indian Soc Soil Sci 43:112–116

Singh VN, Mishra BB (1996) Pedogenic characterization of some typical soils of Gandak command area of Bihar for evaluation of land suitability. J Indian Soc Soil Sci 44:136–142

Singh VN, Mishra BB (1997) Irrigability and productivity classification of some typical soils of sedentary and alluvial origins. J Indian Soc Soil Sci 45:805–809

Singh VN, Tiwary JR, Mall J, Mishra BB (1996) Irrigability and productivity classification of some soils of Chotanagpur plateau. J Indian Soc Soil Sci 44:350–353

Soil Survey Staff (1951) Soil survey manual. USDA Handbook 18. U. S. Govt. Printing Office, Washington, D.C

Storie RE (1950). Rating soils for agricultural forest and grazing use. Trans IV Amsterdam Int Congr Soil Sci I:336–339

Storie RE (1954) Land classification as used in California for the appraisal of land for taxation purposes. Leopoldsville (Congo). Trans V Int Congr Soil Sci VI-10

Storie RE (1978). Storie index soil rating (Revised, 1978). Special Publication Division of Agriculture Science No. 3203, University of California

Sturdevant GW, Moore MJ, Preston JD (2001) Soil survey of Laclede County, Missouri. USDA-NRCS. U.S. Gov. Print Office, Washington, DC

Sys C (1985). Land evaluation. State university of Ghent, International Training Centre for Postgraduate Soil Scientists, Parts I, II, and III, Ghent

Sys C, Van Ranst E, Debaveye J (1993) Land evaluation, Part I & II

Tamgadge DB, Ghosh SK, Raja P, Gajbhiye KS (1996) Impact of landform and different parent materials on intensity of soil mineral weathering with reference to sand fractions. Agropedology 6(2):37–42

USBR (1953) Bureau of reclamation manual. Volume V: irrigated land use, part 2: Land classification Denver, Colorado. U.S. Dept. Interior

Valayutham M (1997) Rural communities and sustainable land management and soil survey information dissemination: a common ground to forge. J Indian Soc Soil Sci 45(4):626–636

Van Wambeke A, Rossiter D (1987) Automated land evaluation systems as a focus for soil research. IBSRAM News Lett 6

Varcoe VJ (1990). A note on the computer simulation of crop growth in agricultural land evaluation. In: Catt CA (ed) Soil use management, vol 6, pp 157–160

Verheye W (1992) Soil survey interpretation. Land Evaluation and land resource management. Agropedology 1:17–32

Vishalakshi Devi PA, Naidu MVS (2016) Land evaluation for alternate land use planning of sugarcane growing soils of Chittoor District in Andhra Pradesh. J Indian Soc Soil Sci 64(1):13–24

Wiggering H, Mueller K, Werner A, Helming K (2003) The concept of multifunctionality in sustainable land development. In: Helming K, Wiggering H (eds) Sustainable development of multifunctional landscapes. Springer, Berlin, Germany, pp 3–18

Soil Health and Fertility

11

Suresh K. Chaudhari, P. P. Biswas, and Hemlata Kapil

Abstract

Soil as a natural resource performs various critical functions, out of which the production function is the basis for food and nutritional security sustaining human and animal life on earth. It is the soil biological, chemical and physical properties that interact to provide a soil with capacity to function. Deterioration of soil health is considered as one of the main reasons of declining nutrient use efficiencies vis-à-vis stagnation of agricultural productivity in the country. Soil health is conceptualized on the ecological attributes of the soil. These attributes are chiefly those associated with the soil biota which is playing a critical role in maintaining soil health, fertility, ecosystem functions and productivity. Government of India is implementing a number of schemes for judicious use of soil resources in order to ensure maximum agricultural productivity and profitability for farming communities based on the technology backstopping provided by Indian Council of Agriculture Research (ICAR) and State/Central Agricultural Universities. Some of the notable government initiatives are Soil Health Management, component of National Mission for Sustainable Agriculture (NMSA), National Mission on Soil Health Card, Biogas and Manure Management Schemes, *Paramparagat Krishi Vikas Yojana*, Watershed Management component of *Pradhan Mantri Krishi SinchayeeYojana* and Nutrient-Based Subsidy scheme.

Keywords

Soil health • Soil fertility • Soil quality indicators • Policy initiatives

S. K. Chaudhari (✉) · P. P. Biswas · H. Kapil
Indian Council of Agricultural Research,
New Delhi, 110 012, India
e-mail: surchaudhari@hotmail.com

S. K. Chaudhari
Indian Society of Soil Science, New Delhi, India

B. B. Mishra (ed.), *The Soils of India*, World Soils Book Series,
https://doi.org/10.1007/978-3-030-31082-0_11

11.1 Introduction

The major issues confronting soil users and researchers include stagnating productivity and simultaneous decline in soil health and quality (Sehgal 1995). Caring soil is extremely important for sustaining healthy life on earth. The root cause of deteriorating human/animal health actually lies on poor health of soil. Only a healthy soil can support healthy plant growth to provide nutritious produce to keep us healthy. From agricultural point of view, soil health is the capacity of a soil to function within ecosystem boundaries to sustain biological productivity, maintain environmental quality and promote plant, animal and human health. A healthy soil would ensure proper retention and release of water and nutrients, promote and sustain root growth, maintain soil biotic habitat, respond to management, resist degradation and act as a buffer for environmental pollution (Brevik and Sauer 2015).

India presently supports 17.5% of the world's human and 15% livestock population with only 2.4% of the land mass. The country is presently home to about 175 million food-insecure population out of the world's total of about 795 million (2015), who on average consume less than 2100 calories a day. To feed the teeming millions, India has to increase food grain production @ of more than 6 million tonnes (MT) annually. With tremendous demographic pressure on land, the per capita availability of agriculture land in India has already decreased sharply from 0.48 ha in 1951 to 0.13 in 2011, which is likely to decrease further to 0.08 in 2035. This has resulted in intensive use of soil resource to feed the population growing @1.2% per annum. PostGreen Revolution concerns have been raised about non-sustainability arising from the deterioration of soil chemical, physical and biological health through accelerated erosion, salinization (primary and secondary), depletion of soil organic matter (SOM), elemental imbalance, deficiencies of some essential nutrients, soil compaction, surface sealing, improper crop husbandry practices and scalping of top soil for brick-making and other non-farming uses.

In fact, agricultural sustainability depends to a large extent on the maintenance or enhancement of soil quality. The terms "soil health" and "soil quality" are functional concepts and interchangeable, which can be viewed in two ways: (i) as inherent properties of a soil and (ii) as the dynamic nature of soils as influenced by climate and human use and management (Doran et al. 1996). It is indeed a real challenge to restore soil resource for achieving food, nutritional, environmental and livelihood security, conserving this vital natural resource base for future generations without any deterioration.

However, soil fertility is a very specific term that refers to the ability of a soil to sustain agricultural plant growth by providing plant habitat and result in sustained and consistent yields of optimal quality through supplying the essential plant nutrients in adequate amounts. The nutrient balance is the difference between the nutrient inputs, entering the soil primarily as manures and fertilizers, and the nutrient outputs either through uptake of nutrients by plants or through losses, fixation or leaving system by other means. A nutrient deficit, i.e. negative value of nutrient balance indicates the declining soil fertility. The Leibig law states that growth is not directed by total resources available, but by the limiting factor. The United Nations Millennium Development Task Force on hunger has made soil health enhancement as one of the five recommendations for increasing agricultural productivity and fighting hunger. Poverty does not prevent the generation, but is extremely difficult and unfavourable for caring of children. Similarly, seed is sown in soil, the tender plant is produced, but in a very hot or cold soil with surrounding unfavourable climate, plant soon withers and dies. So, India needs healthy soils that must be fertile as well as nourishing. Obviously, soil types and their classification in accordance with associated climate, geology, geomorphology, mineralogy, micromorphology, biogeochemistry and biodiversity are of immense importance and integrated with overall quality of the soil.

11.2 Soil Resources

India is endowed with diverse soil types covering different soil orders in the US Soil Taxonomy classification with distinct sub-orders and great-groups. The extent and distribution of different soil classes of India have been given in Table 11.1, though the details are discussed elsewhere in Chap. 5.

Table 11.1 Extent and distribution of the different soil classes of India with their equivalence in USDA Soil Taxonomy

Major soils (traditional name)	Soil orders US soil taxonomy	Extent		Distribution in states
		'000 ha	Percentage	
Alluvial	Inceptisols, entisols, alfisols, aridisols	100,006	30.4	J&K, HP, Punjab, Haryana, Delhi, UP, Gujarat, Goa, MP, MS, AP, Karnataka, TN, Kerala, Puducherry, Bihar, Odisha, WB, ArP, Assam, Nagaland, Manipur, Mizoram, Tripura, Meghalaya, A&N
Coastal alluvial	Aridisols, inceptisols, entisols	10,049	3.1	AP, Karnataka, TN, Kerala, WB, Gujarat, Odisha, Puducherry, Lakshadweep, A&N
Red	Alfisols, ultisols, entisols, inceptisols, mollisols, aridisols	87,989	26.8	AP, Karnataka, Kerala, TN, Puducherry, Rajasthan, MP, MS, Gujarat, Goa, ArP, Assam, Manipur, Meghalaya, Nagaland, Mizoram, Tripura, Delhi, UP, HP, A&N
Laterites	Alfisols, ultisols, inceptisols	18,094	5.5	AP, Karnataka, Kerala, TN, Puducherry, MS, Odisha, WB
Brown forest	Mollisols, inceptisols	540	0.2	Karnataka, Maharashtra
Hill	Inceptisols, entisols	2262	0.7	Manipur, Odisha, WB, Tripura, Nagaland
Terai	Mollisols	326	0.1	UP, Sikkim
Mountain meadow	Mollisols	60	–	J&K
Sub-montane	Alfisols	104	–	J&K
Black	Vertisols, mollisols, inceptisols, entisols, aridisols	54,682	16.6	MP, MS, Rajasthan, Puducherry, TN, UP, Bihar, Odisha, AP, Gujarat
Desert	Aridisols, inceptisols, entisols	26,283	8.0	Rajasthan, Gujarat, Haryana, Punjab
Others[a]		28,305	8.6	–
Total		328,700	100	–

[a]Includes glaciers (0.4%), sand dunes (0.01%), mangrove swamps (0.04%), salt waste (0.01%), water bodies (0.1%), rock land (0.25%) and rock outcrops (7.8%)

MP Madhya Pradesh; *MS* Maharashtra; *UP* Uttar Pradesh; *J&K* Jammu and Kashmir; *TN* Tamil Nadu; *AP* Andhra Pradesh; *ArP* Arunachal Pradesh; *WB* West Bengal; *HP*, Himachal Pradesh; *A&N* Andaman and Nicobar Islands (*Source* Bhattacharyya et al. 2013)

The soils of India are represented by red and laterite soils (Alfisols, Oxisols, Ultisols, Incepitisols, etc.; 117.2 M ha), black soil (Vertisols and their associations; 73.5 M ha), alluvial soil (Entisols and Inceptisols; 58.4 M ha), desert soils (mostly Aridisols and Entisols; 30 M ha) and the rest other soil types. The major alluvial track of the country lies along the Indo-Gangetic plains and Brahmaputra and Barak valleys. Alluvial soils are also located in coastal and deltaic region and along the important rivers of the country. The soils are generally deep, loamy and alluvium-derived having moderate to high available water holding capacity. The soils have low organic matter content and moderate to medium exchange capacity. The Indo-Gangetic plains of north India are the most fertile lands and are largely irrigated, contributing 65% of the total food basket. Red and lateritic soils occur on gently sloping to undulating lands. The physical constraints to crop production in red and lateritic soils are soil erosion, hardening of soil, low water holding capacity, reduced soil volume due to concertinos, occurrence of hard plinthite or petroplinthite, drought-related stress, low cation exchange capacity, low organic matter, high acidity, iron and aluminium toxicity, high phosphorous fixations and poor fertility status. Black soils occur mostly in the peninsular region on gently sloping lands at low elevations. These soils have high cation exchange capacity (CEC) with advantage of nutrient retention and swelling for moisture retention. Moisture shortage under rainfed conditions and waterlogging, salinity and alkalinity under irrigated conditions adversely affect the productivity of the soils. On saturation, they become impervious; as a result they are more erodible. Since wet Vertisols are difficult to cultivate, cropping on them is concentrated more during Rabi season (post-rainy season). Soils of arid regions (mostly Aridisols and Entisols) occur in Rajasthan, Haryana and Gujarat. The track is, in general, sandy and in many parts saline or alkaline in nature with unfavourable physical conditions and high pH value. Extent of area affected by various soil physical constraints in India is shown in Table 11.2.

11.3 Soil Health Issues

11.3.1 Chemical Health

Indian soils are, in general, low in organic carbon content. The continuous mining of nutrients from soil @ of 34 MT per annum against the external supplement of only 26 MT coupled with low addition of organic manures, secondary and micronutrients resulted in the emergence of multi-nutrient deficiencies in several districts in the country (Table 11.3). The nutrient deficiency in the country is in the order of 95, 94, 48, 25, 41, 20, 14, 8 and 6% for N, P, K, S, Zn, B, Fe, Mn and Cu, respectively. The limiting nutrients do not allow the full expression of other nutrients, lower the fertilizer response and crop productivity. The fertilizer response ratio (kg grain per kg nutrient) decreased nearly by five times (from 13.4 in 1970 to around 2.7 in 2015) in irrigated areas of the country (Fig. 11.1). While only 54 kg fertilizer nutrients were required per ha during 1970 to maintain the yield level around 2.0 t ha, nearly five times fertilizer nutrients (254 kg in 2015) is being required to sustain the same yield level, which is a matter of concern (Chaudhari et al. 2015).

The nutrient use efficiency in type-specific Indian soils is also low ranging from 30 to 50% (N), 15 to 20% (P), 60 to 70% (K), 8 to 10% (S) and 1 to 2% (micronutrients). While the unutilized nitrogen is subjected to leaching and denitrification or volatilization loss polluting groundwater and atmosphere, respectively, considerable amount of P and K nutrients are lost through the process of soil erosion. It has been estimated that over 5.3 billion tonnes of soil is lost annually through water erosion with a loss of $\sim$8 MT of plant nutrients (NPK).

It is further reported that around 17.36 million ha (Mha) of arable land of the country is affected by various kinds of chemical degradation with very low productivity (<1 t/ha). This includes about 10.72 Mha suffering from acute soil acidity (pH < 5.5), around 6.64 Mha are under

Table 11.2 Extent of area affected by various soil physical constraints in India

Physical constraints	Area (Mha)	Major states affected
Shallow depth	26.40	Andhra Pradesh, Maharashtra, West Bengal, Kerala and Gujarat
Soil hardening	21.57	Andhra Pradesh, Maharashtra and Bihar
High permeability	13.75	Rajasthan, West Bengal, Gujarat, Punjab and Tamil Nadu
Sub-surface hard pan	11.31	Maharashtra, Punjab, Bihar, Rajasthan, West Bengal and Tamil Nadu
Surface crusting	10.25	Haryana, Punjab, West Bengal, Odisha, Gujarat
Temporary water logging	6.24	Madhya Pradesh, Maharashtra, Punjab, Gujarat, Kerala and Odisha

Table 11.3 Districts deficient in soil available nitrogen (N), phosphorus (P), potassium (K), sulphur (S), zinc (Zn), iron (Fe), manganese (Mn) and boron (B) in different states

State/UTs	Nutrient	Districts
Andhra Pradesh including Telangana	N	Adilabad, Chittoor, Cuddapah, East Godavari, Guntur, Karimnagar, Khammam, Krishna, Kurnool, Mahbubnagar, Nizamabad, Visakhapatnam, Vizianagaram, Warangal, West Godavari
	P	Adilabad, Anantapur, Chittoor, Cuddapah, Guntur, Hyderabad, Karimnagar, Khammam, Krishna, Kurnool, Mahbubnagar, Medak, Nalgonda, Nellore, Nizamabad, Rangareddi, Srikakulam, Warangal
	K	Nil
	S	Kurnool, Mehboobnagar, Karimnagar, Kadappa, Guntur, Anantpur, Nizamabad Nalgonda
	Zn	Kurnool, Mehboobnagar, Karimnagar, Guntur, Anantpur, Rangareddy, Krishna West Godavari, Adilabad, Parakasham, Srikakulam
	Fe	Kurnool, Anantpur, Nizamabad, Adilabad, Parakasham, Vishakhapatnam
	Mn	Rangareddy, West Godavari, Nizamabad, Medak
	B	Mehboobnagar, Karimnagar, Rangareddy, Krishna, West Godavari, Nalgonda, Adilabad, Vishakhapatnam, Srikakulam, Medak
Assam	N	Bongaigaon, Bopeta, Chirang, Darrang, Kokrajhar, Morigaon, NC Hills, Nalbari
	P	Jorhat, Karbi angling, Udalguri
	K	Bongaigaon, Cachar, Chirang, Golaghat, Hailakandi, Jorhat, Karimganj, Kokrajhar, N C Hills, Nagaon, Sivsagar, Udalguri
	S	Jorhat, Sibsagar, Kamrup
	Zn	Jorhat, Golaghat, Barpeta, Kamrup, Sonitpur, Dibrugarh, Darang, Tinsukia
	Fe	Nil
	Mn	Nil
	B	Jorhat, N Lakhimpur, Dibrugarh
Chhattisgarh	N	Bastar, Dantewara, Dhamtari, Durg, Kanker, Kawardha, Mahasmund, Raipur, Rajnandgaon
	P	Bastar, Dantewara, Dhamtari, Kanker, Korba, Mahasmund, Raipur
	K	Bastar, Dantewara, Kanker
Gujarat	N	Amreli, Banaskantha, Bharuch, Gandhinagar, Jamnagar, Kutch, Mahesana, Narmada, Patan, Sabarkantha, Surat, Surendranagar, Vadodara
	P	Banaskantha, Bharuch, Bhawnagar, Dahod, Mahesana, Narmada, Navsari, Panchmahal, Patan, Porbandar, Surendranagar, Valsad
	K	Nil
	S	Banaskantha, Anand, Kheda, Panchmahal, Vadodara, Ahmedabad, Dahod
	Zn	Patan, Bharuch, Ahmedabad, Sabarkantha, Mehsana, Banaskantha, Kutch
	Fe	Anand, Kheda, Patan, Vadodara, Mehsana, Banaskantha, Gandhinagar Kutch
	Mn	Nil
	B	Panchmahal, Patan, Sabarkantha, Mehsana, Gandhinagar
Haryana	N	Bhiwani, Faridabad, Fatehbad, Gurgaon, Hisar, Jhajjar, Jind, Kaithal, Karnal, Kurukshetra, Mahendragarh, Panchkula, Panipat, Rewari, Sirsa, Sonipat, Yamuna Nagar
	P	Bhiwani, Faridabad, Fatehbad, Gurgoan, Hisar, Jhajjar, Jind, Kaithal, Karnal, Kurukshetra, Panchkula, Panipat, Rewari, Sirsa, Sonipat, Yamuna Nagar
	K	Nil
	S	Kurukshetra, Mohindergarh, Jhajjar, Rewari, Ambala, Palwal, Bhiwani, Rohtak, Fatehabad
	Zn	Mohindergarh, Bhiwani
	Fe	Sirsa, Hisar, Mohindergarh, Fatehabad, Bhiwani, Rohtak
	Mn	Hisar, Karnal, Fatehabad
	B	Nil

(continued)

Table 11.3 (continued)

State/UTs	Nutrient	Districts
Himachal Pradesh	N	Nil
	P	Hamirpur, Kangra, Mandi, Shimla, Una
	K	Chamba, Hamirpur, Kangra, Kinnaur, Lahaulspiti, Una
	S	Nil
	Zn	Hamirpur, Una, Chamba, Mandi
	Fe	Nil
	Mn	Nil
	B	Bilaspur, Una, Kangra, Shimla, Solan
Karnataka	N	Kolar
	P	Bellari, Bijapur, Hassan, Norht Kannada, South Kannada, Udupi
	K	South Kannada, Udupi
Kerala	N	Kasaragod, Kollam, Thiruvananthapuram
	P	Nil
	K	Nil
Madhya Pradesh	N	Bhind, Chhattarpur, Daria, Gwalior, Indore, Jabalpur, Mandsaur, Neemuch, Morena, Panna, Ratlam, Sheopu, Shivpuri, Sidhi
	P	Ashok Nagar, Betul, Bhind, Chhattarpur, Damoh, Daria, Dewas, Gwalior, Japlpur, Jhabua, Katni, Panna, Shivpuri, Ujjain, Umaria
	K	Dhar, Anuppur, Betul, Gwalior, Morena, Sagar, Sidhi
	S	Chattarpur, Satna, Chindwara,Narsinghpur, Reewa, Dewas, Panna, Morena
	Zn	Balaghat, Seoni, Shahadol, Mandala, Bhopal, Raisen, Tikamgarh, Chattarpur, Satna, Chindwara, Jabalpur, Narsinghpur, Reewa, Dewas, Panna, Morena
	Fe	Bhopal, Panna, Narsinghpur
	Mn	Nil
	B	Nil
Maharashtra	N	Akola, Amaravati, Aurangabad, Beed, Bhandara, Buldhana, Gondiya, Hingoli, Jalgaon, Jalna, Latur, Nagpur, Nanded, Nashik, Parbhani, Pune, Raigad, Ratnagiri, Sangali, Satara, Solapur, Usmanabad, Wardha, Washim, Yeotmal
	P	Akola, Amaravati, Aurangabad, Bhandara, Bhuldhana, Dhule, Gondiya, Hingoli, Jalgaon, Jalna, Kolhapur, Latur, Nagpur, Nanded, Nandurbar, Nashik, Parbhani, Pune, Raigad, Ratnagiri, Sangali, Satara, Sindhudurg, Solapur, Usmanabad, Wardha, Washim, Yeotmal
	K	Raigad, Sindhudurg
	S	Akola, Aurangabad, Washim, Nanded, Gondia, Nagpur, Parbhani, Latur
	Zn	Akola, Bhandara, Jalna, Yavatmal, Amravati, Buldhana, Chandrapur, Aurangabad, Wardha, Hingoli, Nanded, Nagpur, Parbhani, Latur
	Fe	Akola, Jalna, Amravati, Aurangabad, Washim, Wardha, Beed, Parbhani
	Mn	Nil
	B	Nil
Orissa	N	Bhadrak, Boudh, Cuttuck, Dhenkanal, Gajapati, Ganjam, Jagatsinghpur, Kalahandi, Kendrapada, Khurda, Mayurbhanj, Naupada, Nayagarh, Bhulbani, Puri, Sundargarh
	P	Balasore, Bhadrak, Cuttack, Gajapati, Ganjam, Jharsuguda, Keonjhar, Mayurbhanj, Nawrangpur, Phulbani, Sambalpur
	K	Cuttack, Ganjam, Nayagarh
	S	Bargarh, Bhadrak, Dhenkanal, Kalahandi, Nayagarh, Nuapada, Sambalpur, Sonepur
	Zn	Angul, Bhadrak, Boudh, Puri, Sonepur
	Fe	Nil
	Mn	Nil
	B	Angul, Bargarh, Bhadrak, Boudh, Dhenkanal, Kandhmal, Kendrapada, Nayagarh, Nuapada, Puri, Sambalpur, Sonepur

(continued)

Table 11.3 (continued)

State/UTs	Nutrient	Districts
Punjab	N	Bhatinda, Faridkot, Ferozepur, Gurdaspur, Hoshiarpur, Jalandhar, Kapurthala, Ludhiana, Mansa, Moga, Muktsar
	P	Nil
	K	Nil
	S	Nil
	Zn	Gurdaspur
	Fe	Bhatinda
	Mn	Bhatinda, Faridkot, Gurdaspur, Tarantaran
	B	Nil
Rajasthan	N	Alwar, Banswara, Baran, Bharatpur, Barmer, Bundi, Churu, Dausa, Dholpur, Durgapur, Hanumangarh, Jaisalmer, Jalore, Jhunjhun, Jodhpur, Karauli, Kota, Nagpur, Pali, Rajsamand, SawaiMadhopur, Sikar, Sirhi, Sriganganagar, Tonk
	P	Bharatpur, Barmer, Churu, Dausa, Dholpur, Durgapur, Hanumangarh, Jaisalmer, Jalore, Karauli, SawaiMadhopur, Sikar, Sirohi, Sirhi, Sriganganagar
	K	Nil
Tamil Nadu	N	Coimbatore, Cuddalore, Dharmapuri, Dindigul, Erode, Fudukkottai, Kanchipuram, Kanyakumari, Karur, Madurai, Nagapattinam, Namakkal, Peerambalur, Ramanathapuram, Salem, Sivagangai, Thanjavur, Theni, Thiruallur, Thiruvarur, Thoothukudi, Tiruvannamalai, Tiruvarur, Trichiraplli, Vellore, Villupuram, Virudhunagar
	P	Kanchipuram, Sivagangai, Thoothukudi, Trichirapalli
	K	Ariyalur
	S	Nagapattinam, Coimbatore, Virudhunagar, Theni, Krishnagiri, Pudukkotai
	Zn	Thanjavur, Cuddalore, Villupuram, Virudhunagar, Theni, Krishnagiri Kanyakumari, Toothukudi, Pudukkotai
	Fe	Trichy, Erode, Villupuram, Virudhunagar, Krishnagiri
	Mn	Villupuram
	B	Erode, Salem, Krishnagiri, Kanyakumari, Toothukudi, Pudukkotai
Uttar Pradesh	N	Aazamgarh, Agra, Aligarh, Allahabad, Ambedkarnagar, Auraiya, Badanyu, Baghpat, Baharaich, Baliya, Balrampur, Banda, Barabanki, Bareli, Basti, Bijnaur, BulandShahar, Chandouli, Chitrkut, Devariya, Eta, Etahwa, Faizabad, Farukkhabad, Fatehabad, Firozabad, Gautambudhnagar, Gazipur, Ghaziabad, Gorakhapur, Hameerpur, Hardoi, Hathras, Jalaun, Jaunpur, Jhansi, Jyotishaphoolnagar, Kannauj, KanpurDehat, Kanpur Nagar, Kashiram Nagar, Kaushambee, Kushinagar, Lalitpur, Lucknow, Lukhimpur, Maharajganj, Mahowa, Mainpuri, Mathura, Mau, Meerut, Muradabad, Muzzafar Nagar, Peelibhit, Pratapgarh, Rampur, Raybareli, Saharanpur, Santkabeer Nagar, Shahjahapur, Sidhrth Nagar, Sitapur, Sonebhadra, Sribasti, Sultanpur, Unnav, Varanasi
	P	Aazamgarh, Agra, Aligarh, Allahabad, Ambedkarnagar, Auraiya, Badanyu, Baghpat, Baharaich, Baliya, Balrampur, Banda, Barabanki, Bareli, Basti, Bijnour, Buland Shahar, Chandouli, Chitrakut, Devariya, Eta, Etawa, Faizabad, Farukhabad, Fatehabad, Firozabad, Gautambudh nagar, Gazipur, Ghaziabad, Gorakhpur, Hameerpur, Hathras, Jalaun, Jaunpur, Jhansi, Jyotishaphool nagar, Kannauj, Kanpur Dehat, Kanpur Nagar, Kashiram Nagar, Kaushambee, Kushinagar, Lalitpur, Lucknow, Lukhimpur, Maharajganj, Mahowa, Mainpuri, Mathura, Mau, Meerut, Mirzapur, Muradabad, Muzzafarnagar, Peelibhit, Pratapgarh, Rampur, Raybareli, Saharanpur, Santkabeer nagar, Santravidasnagar, Shahjahapur, Sidhrath Nagar, Sitapur, Sonebhadra, Sribasti, Sultanpur, Unnav, Varanasi
	K	Nil
	S	Allahabad, Etawah, Farrukhabad, Lakhimpur, Pilibhit, Raibareli, Ramabainagar, Unnao
	Zn	Agra, Allahabad, Farrukhabad, Gorakhpur, Kannauj, Kanpur, Lakhimpur, Pilibhit, Raibareli, Ramabainagar, Sitapur
	Fe	Nil
	Mn	Etawah, Farrukhabad, Kannauj, Kanpur, Sitapur, Varanasi
	B	Nil

(continued)

Table 11.3 (continued)

State/UTs	Nutrient	Districts
Uttarakhand	N	Dehradun, Tehari Gadwal, Udhamsingh Nagar, Uttarkashi
	P	Bageswar, Chamoli, Champawat, Dehradun, Haridwar, Paudi, Rudraprayag, Udhamsingh Nagar, Uttarkashi
	K	Nil
	S	Champawat, Dehradun
	Zn	Udham Singh Nagar
	Fe	Nil
	Mn	Rudraprayag
	B	Pithoragarh, Uttarkashi, Tehri
West Bengal	N	Midnapore E, Midnapore W, North 24- Parganas, South 24- Parganas
	P	Midanpore E, Prakama, Purulia
	K	Jalpaiguri
	S	Nil
	Zn	Jalpaiguri, North Dinajpur, N 24 Pargana, Bardhaman, Coochbehar
	Fe	Nil
	Mn	Nil
	B	Hooghly, Murshidabad, Bardhaman, Nadia, Coochbehar, S 24 Pargana

Source agricoop.nic.in/dacdivision/Comsoilhealth28612.pdf and IISS, Bhopal

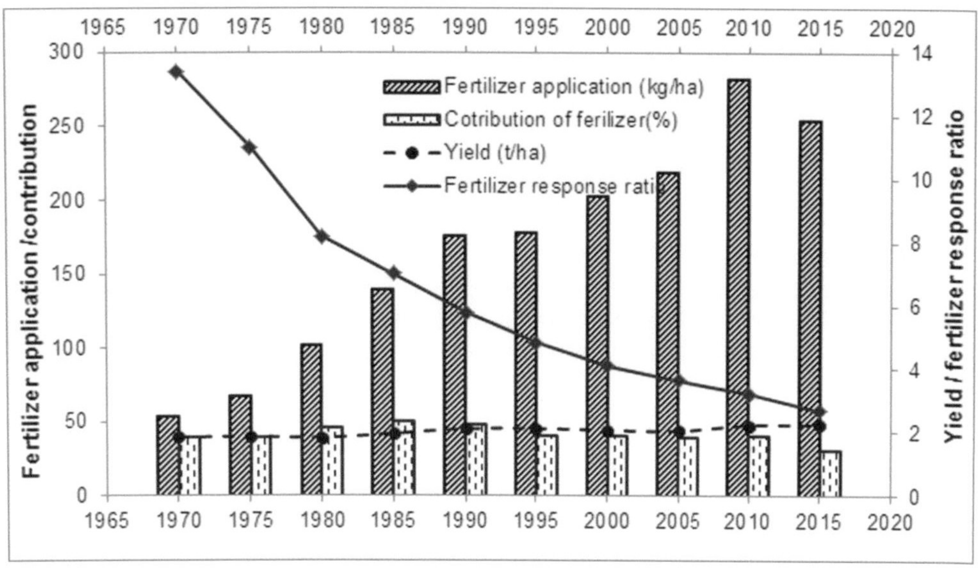

Fig. 11.1 Response and contribution of fertilizer in foodgrain production in irrigated areas over the years in India

salt-affected soils comprising of 3.64 Mha under high sodicity (pH > 9.5), and about 3 Mha (including 1.25 Mha coastal salinity) are under salinity. The crop production loss due to salinity and alkalinity at the national level has been estimated to be 5.66 and 11.18 MT, respectively. In economic terms, this is equivalent to the annual monetary loss of Rs. 80,000 million and Rs. 1,50,000 million due to salinity and alkalinity problems, respectively, assuming 2014–15 as base year (ICAR-CSSRI 2016a, b).

11.3.2 Physical Health

Understanding and managing soil physical properties and processes are essential for sustainable management of soil and water resources. Physical environment of soil has a significant role in water and nutrient uptake and losses, pollutant transport and also emission of greenhouse gases from soil (Acharya et al 2016). It also influences its chemical and biological properties and processes through movement of water and gases in soil and soil thermal regime. McKenzie et al. (2011) defined the soil physical quality as the ability of a given soil to meet plant and ecosystem requirements for water, aeration and strength over time and to resist and recover from processes that might diminish that ability.

It is estimated that 104 Mha of arable land is degraded in one way or the other producing less than 20% of its potential capacity (Maji et al. 2010). Out of which, around 73.27 Mha suffer from water erosion and 12.24 Mha from wind erosion. Erosion-induced loss in crop production in rainfed areas under major cereal, oilseed and pulse crops has been estimated as 13.4 million tons ($\sim$16%), which in economic terms is equivalent to Rs. 162.8 billion (Sharda et al. 2010). Besides, nearly 89.52 Mha (Table 11.2) suffer from one or the other form of physical constraints like shallow depth, soil hardening, slow and high permeability, sub-surface compacted layer, surface crusting, temporary waterlogging, etc. (Painuli and Yadav 1998). In addition, around 0.9 Mha area of land is affected due to permanent surface inundation. Mechanization of farm operations, frequent tillage in intensive cropping systems, unscientific and indiscriminate use of inputs and decline in soil organic matter, etc., are adding new areas with new problems to the existing degraded area.

11.3.3 Biological Health

Soil is a unique biological laboratory depending on soil types, organic inputs and surrounding climates. Attention to soil biological health is as essential as that to chemical and physical health for sustaining higher crop productivity. Biological properties of soil also influence soil's chemical and physical properties. Therefore, three are interactive and not mutually exclusive. This demands wider recognition of soil biological health than what it has received so far. Soil health is conceptualized on the ecological attributes of the soil, which have implications beyond its quality or capacity to produce a particular crop. These attributes are chiefly those associated with the soil biota: its diversity, its food web structure, its activity and the range of functions it performs (Sharma and Adhya 2016). By soil health, we generally mean to soil physical and chemical health (forgetting biological health) both largely rely on soil biodiversity and soil biological processes. Soil and its living organisms are an integral part of agricultural ecosystems, playing a critical role in maintaining soil health, ecosystem functions and productivity. These services not only facilitate functioning of natural ecosystems but constitute an important resource for sustainable management of agricultural systems. The value of "ecosystem services" provided each year by soil biota in agricultural systems worldwide may exceed US$1542 billion (FAO 2002). Almost 40–48 million tons N per year is biologically fixed in agricultural crops compared to 83 million tons N per year fixed industrially for the production of fertilizer. Despite recognition of the fundamental role of soil biota in maintaining sustainable and efficient agricultural systems, it is still largely neglected in the majority of agricultural development initiatives. The consequences of neglecting or abusing soil life will weaken the soil functions and contribute to greater loss of fertile lands and subsequently, an over-reliance on chemical fertilizers for maintaining agricultural production.

Organic manures/compost is an eco-friendly source of soil organic carbon providing energy to soil biota which act as the primary driving agents of nutrient cycling, regulating the dynamics of soil organic matter, soil carbon sequestration and greenhouse gas emission, modifying soil physical structure and water regimes, enhancing the amount and efficiency of nutrient acquisition by the vegetation and enhancing plant health (Benbi 2016). Crop residues returned to the soil provide substrate to soil organisms, which help in SOM turnover. It is estimated that burning of one ton of rice straw accounts for loss of 5.5 kg nitrogen, 2.3 kg phosphorus, 25 kg potassium and 1.2 kg sulphur, besides organic carbon. Crop residues burning is a potential source of greenhouse gases (GHGs) such as CH_4, CO, N_2O and NO_x causing global warming. Despite being aware of the beneficial effects of residue recycling on SOC and overall soil health enhancement, farmers resort to in situ burning of crop residues. In India, an estimated 500–550 Tg of crop residues are produced annually. After accounting for multiple competitive uses about 141 Tg are surplus most of which are burnt in situ. The crop residues on an average contain 45% C, and assuming a humification rate 10% the incorporation of surplus crop residues can result in C sequestration of 6.3 Tg C annually (NAAS 2012).

11.3.4 Soil Pollution

Soil pollution is not the integral part of a soil, but is imposing a threat to soil health. This can be defined as the build-up of pollutants in soils like persistent toxic compounds, chemicals, salts, radioactive materials or disease-causing agents with adverse effects on plant growth and human/animal health (Sanyal et al 2016). The pollutants are released into the environment, including soil, from both natural (geogenic) and anthropogenic sources. The main factors of soil pollution are the high state of soil erosion, excessive use of chemical fertilizers and pesticides, use of urban and industrial wastes and irrigation with poor quality water (sodic/saline water, arsenic, selenium and fluoride contaminated ground water, sewage water, etc.). In India, about 100 MT of pollutants are being added to the atmosphere annually through burning of fossil fuel and industrial emissions causing considerable air pollution. Coal combustion in thermal power plants releases 100–110 t Hg/year, which finally gets precipitated on soil and water body. Polluted surface water and groundwater add several harmful chemicals into the soil body when used for irrigation. More than 35 billion litres of urban waste water and 25 billion litres of industrial waste water are released every day, a significant part of which enters into agricultural land as irrigation carrying different pollutants. Some of the pollutants are constituents of extensively used agrochemicals like Cd through phosphatic fertilizer and pesticides (organic pollutants) and enter into the rhizosphere when these are used for higher production and economic return.

11.4 Climate Change Impact on Soil

Climate change or variability can impact the overall soil quality through its influence on physical (porosity, maximum water holding capacity, bulk density, mean weight diameter, water stable aggregates, hydraulic conductivity), chemical (clay content, cation exchange capacity, total N, available N, P, K, Zn Fe Mn, S, P and K fixing capacities) and biological (SOC, MBC, metabolic quotient, potentially mineralizable N, soil respiration, dehydrogenase activity, phosphatase activity) attributes (Pathak et al 2016). It will impact the soil organic matter dynamics, including soil organisms and the multiple soil properties that are tied to organic matter, soil water and soil erosion. This has already been discussed elsewhere in Chap. 3.

11.5 Opportunities for Evaluation of Soil Health

11.5.1 Soil Health Assessment

Soil health and quality represent a composite of physical, chemical and biological attributes. Soil quality cannot be measured directly, but essentially needs to be inferred from measuring changes in its attributes of the ecosystem, referred to as indicators (Sharma et al 2016). The changes in these indicators are used to determine whether soil quality is improving, stable or declining with changes in management, land use or conservation practices (Brejda and Moorman, 2001). The predominant indicators for physical, chemical and biological qualities of soil at micro- and macro-farm scale have been suggested by Singer and Ewing (2000). The indicators should (i) correlate well with natural processes in the ecosystem (this also increases their utility in process-oriented modelling), (ii) integrate soil physical, chemical and biological properties and processes, and serve as the basic inputs needed for estimation of soil properties or functions, which are more difficult to measure directly, (iii) be relatively easy to use under field conditions, so that both specialists and producers can use them to assess soil quality, (iv) be sensitive to variations in management and climate and (v) be the components of the existing soil database wherever possible (Doran et al. 1996; Doran and Parkin 1996; Chen 1998). Rapid measurements of soil quality attributes are possible using hyper-spectral remote sensing data as well as visible near infrared diffuse reflectance spectroscopy (VisNIR DRS).

Another approach of soil quality assessment called "a comparative assessment technique" is based on (i) selection of a minimum data set (MDS) of indicators that best represent the soil function, (ii) scoring of the MDS indicators based on their performance of soil functions, (iii) corroboration of the MDS indicators with functional goals set by the land manager or grower and (iv) integration of the indicator score into a comparative index of soil quality (Andrews and Carroll 2001; Andrews et al. 2002a, b). Positive effects of green manuring, INM, manure, crop residue recycling, legume-based crop rotations and balanced fertilization have been observed on predominant soil quality indicator, overall soil quality indices and SYI of crops.

11.5.2 Soil Testing

Soil testing is a basic necessity to determine the quantities of nutrients to be applied to ensure balanced nutrition to plants. The country has about 700 static soil-testing laboratories with an analysing capacity of over seven million soil sample per annum. Given the vast cultivated area and number of farm holdings, the soil-testing facilities are grossly inadequate. The ICAR has developed two portable digital quantitative soil test kits, namely mini lab (*Mridaparikshak*) by Indian Institute of Soil Science, Bhopal and STFR (soil test and fertilizer recommendation) meter by Indian Agricultural Research Institute (IARI) New Delhi, with GPS facilities to supplement soil-testing service in the country. The kits are useful in analysing soil samples for the purpose of distributing soil health cards to farmers along with fertilizer recommendations and developing geo-referenced soil fertility maps at village level.

11.5.3 Balanced and Integrated Nutrient Management Versus Soil Sustainability

It has been amply demonstrated under long-term fertilizer experiments that balanced application of nutrients along with organic manures facilitates better soil quality and yield sustainability (Figs. 11.2 and 11.3) compared to chemical

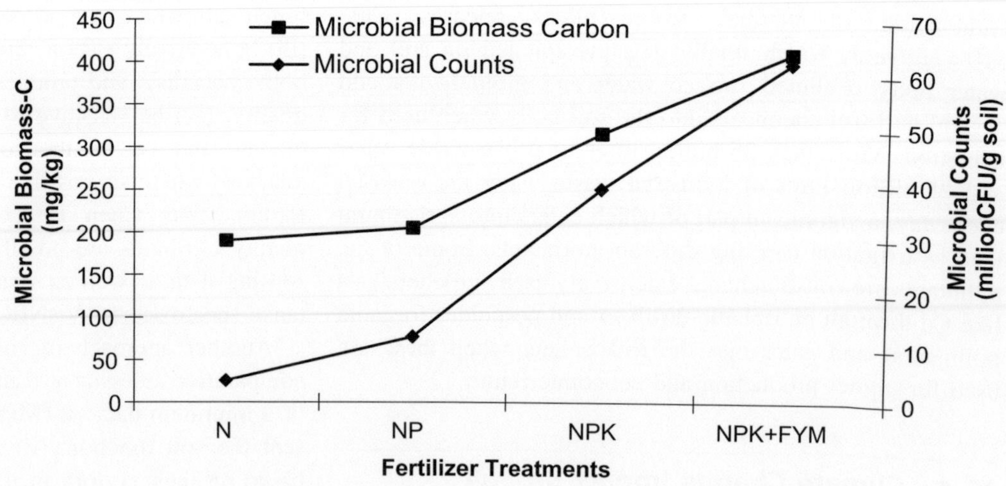

Fig. 11.2 Soil biological attributes under long-term fertilizer experiments. *Source* Sharma et al. (2002)

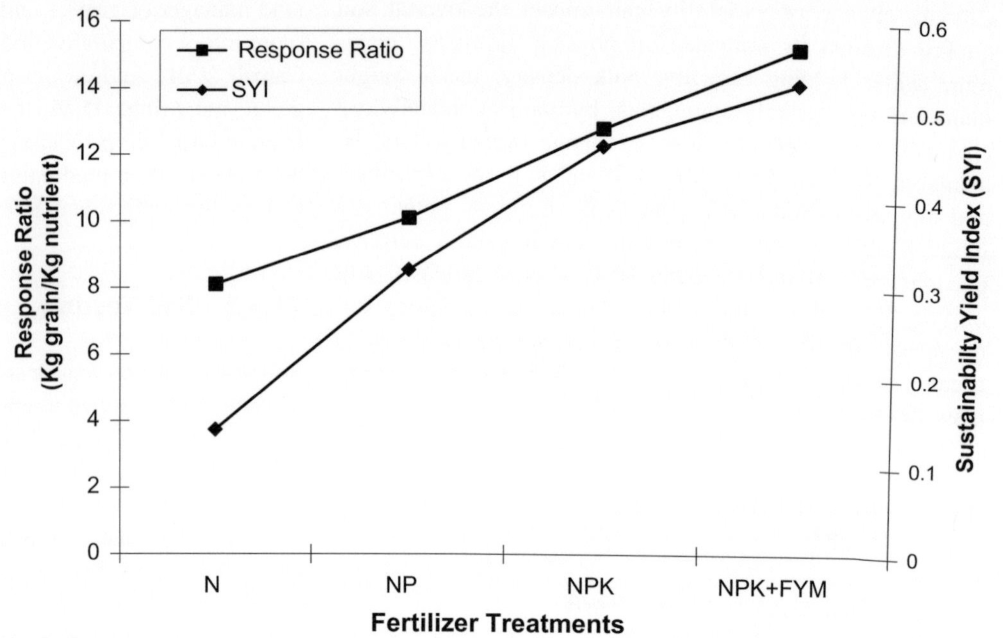

Fig. 11.3 Average fertilizer response ratios (kg grain/kg nutrient) and Sustainable Yield Index (SYI) under long-term fertilizer experiments in India. *Source* Biswas and Sharma (2008); Wanjari et al. (2004)

fertilization alone. Such efforts must be tested on type-specific as well as benchmark soils to generate reliable technologies for sustainable production.

It has been found that soils with relatively higher organic matter content are better in performing functions that are critical for crop production and environmental conservation. Indian Council of Agricultural Research (ICAR) has developed technologies to prepare various types of organic manures such as phosphocompost, vermincompost, bio-enriched compost, municipal solid waste compost, etc., from various organic wastes. Effective biofertilizer strains of *Rhizobium* for promoting nodulation and nitrogen fixation in legumes and of plant growth promoting rhizobacteria (PGPR) consisting of *Azotobacter*, phosphate solubilizing bacteria (*Bacillus*, *Pseudomonas*) and other PGPR for cereals, legumes, millets, oilseeds, vegetables and horticultural crops were developed which resulted in the increase in productivity by 10%, saving of 20–25% chemical fertilizers, improvement of nutrient use efficiency by 15–25%, produce quality and soil health. When biofertilizers are applied along with compost @ 5 t/ha or vermicompost @ 2 t/ha, fertilizer saving is almost 50%. The ICAR is recommending soil test-based balanced and integrated nutrient management through conjunctive use of both inorganic and organic sources (manure, biofertilizers, etc.) of plant nutrients to prevent deterioration of soil health. It is concluded, based on several biological, chemical and physical indicators of soil quality that conjunctive nutrient use as well as sole organic nutrient treatments was superior to even balanced NPK application with higher crop productivity and soil quality index (Chaudhary et al. 2005; Sharma 2009). The viable INM practices of dominant cropping systems in different agro-climatic regions of the country have been summarized (Table 11.4).

11.5.4 Assessment of Soil Organic Matter

Soil organic matter (SOM) is a unique indicator which exerts major influence on a number of soil physical, chemical and biological attributes. The soil physical properties, most commonly influenced by SOM, include bulk density, aggregate stability and moisture retention. Increase in SOM lowers soil bulk density, improves aggregation and the proportion of water stable macroaggregates and favourably impacts soil water retention and transmission properties (Benbi 2016). The increase in aggregate formation and aggregate stability is mainly due to the production of organic macromolecules by microorganisms that bind primary particles and macroaggregates.

The chemical properties of soil that are mainly influenced by SOM include nutrient availability, exchange capacity, reaction with metals and contaminants and its capacity to act as proton buffer. It contributes 25–90% of the CEC of surface layers of mineral soils depending on the nature and content of organic matter. The soil biological properties or processes influenced by SOM include mineralization, microbial biomass and enzyme activities. A number of studies in India have shown that SOC is the most important parameter for soil quality index formulation indicating that soils with relatively higher organic matter content are better in performing functions that are critical for crop production and environmental conservation. Indeed, an increase of SOC stock by 1 Mg C ha^{-1} in the root zone can raise the crop yield by 15–33 kg ha^{-1} for wheat (Benbi and Chand 2007), 160 for kg ha^{-1} for rice, 170 kg ha^{-1} for pearl millet, 13 kg ha^{-1} for groundnut, 18 kg ha^{-1} for lentil, 90 kg ha^{-1} for sorghum, 101 kg ha^{-1} for finger millet and 145 kg ha^{-1} for soybean, (Srinivasarao et al. 2013). Therefore, greater SOC content can result in higher foodgrain production restoration in the country.

11.5.5 Carbon Sequestration Potential

Carbon sequestration potential of different nutrient management practices is estimated to range between 2.1 and 4.8 Mg C ha^{-1} with a total potential of 300–620 MT (Pathak et al. 2011). In India, balanced application of fertilizers can enhance SOC concentration by 6–100% and C sequestration by 20–600 kg ha^{-1} year^{-1}, depending on soil, crop and climatic conditions. The same potential of integrated nutrient management practices is estimated at 100–1200 kg C ha^{-1} year^{-1} with an enhance SOC concentration of 17–132% under various soil, crop and climatic conditions (Benbi 2013). Carbon sequestration potential of rainfed production systems under different nutrient management practices ranges between 0.08 and 0.45 Mg ha^{-1} year^{-1} for groundnut (*Arachishypogaea*) on Alfisols, 0.04 and 0.038 for finger millet (*Eleusivecoracana*) on Alfisols, 0.1 and 0.2 for winter sorghum on Vertisols and up to 0.33 for soybean on black soils (Srinivasarao et al. 2014). However, for agroforestry systems, such potential varied widely (1.3–173.0 Mg C ha^{-1}) depending on tree species, climatic conditions and age of plantation (Nair et al. 2009). No-till agriculture can enhance soil C sequestration by reducing the degree of soil disturbance.

Methane is produced in soil during microbial decomposition of organic matter under strictly anaerobic continuous submergence. Improved water management such as alternate wetting and drying, direct seeding of rice (DSR) and System of Rice Intensification (SRI) crop do not require continuous soil submergence, and therefore reduce or totally eliminate methane emission when rice is grown as an aerobic crop. The DSR and SRI have potential to reduce the global warming potential (GWP) by about 25–50% compare to the conventional puddled transplanted rice. The incorporation of rice residue in soil instead of burning can lead to C accrual of 0.80% Tg C year^{-1} over an area of 2.28 Mha in Punjab, which is equivalent to about 350 kg C ha^{-1}. Still higher rates of C sequestration, up to 745 kg C ha^{-1}, can be achieved if both animal manure and rice residue are applied annually in the rice–wheat cropping sequence (Benbi et al. 2012). Details are already presented elsewhere in Chap. 3.

11.5.6 Emission of N_2O and Nitrate Pollution in Ground Water

Enhancing the efficiency of fertilizer N with 5R approach, i.e. use of right kind of fertilizer, right rate of application, right time of application, right place of application, right method of application, use of leaf colour chart (LCC) and nitrification inhibitors such as coated calcium carbide and dicyandiamide is the key solution to reduce the nitrous oxide emission and nitrate pollution in groundwater by about 10–15% (Bhatia et al. 2012; Jain et al. 2014). There are some plant-derived organics such as *neem* oil, *neem* cake and *karanja* seed extract, which can also act as nitrification inhibitors. Chapter 3 is, however, devoted to the details of N_2O emission in soils.

Table 11.4 Soil fertility versus integrated plant nutrient supply system (IPNS packages) for dominant cropping system in different agro-climatic regions

Cropping system	IPNS package
Western Himalayan Region	
Rice–wheat	**Rice**: 40 kg N + FYM/Green Manure @ 15 t/ha + 20 kg Zinc sulphate (in Zn deficient soils) **Wheat**: 120 kg N + 80 kg P_2O_5 (through SSP) + 40 kg K_2O
Maize–wheat	**Maize**: 60 kg N + 30 kg P_2O_5 (through SSP) + 20 kg K_2O + 10 t FYM + fresh Eupatorium/Lantana Mulch @ 10 t/ha **Wheat**: 80 kg N + 30 kg P_2O_5 (through SSP) + 15 kg K_2O
Eastern Himalayan Region	
Rice–rice	**Rice**: 20 kg N + 20 kg P_2O_5 + 15 kg K_2O + FYM/GM @ 10t/ha + Azolla @ 10t/ha + 20 kg Zinc Sulphate once in 3 years + 5 kg borax + 1 kg ammonium molybdate + 5 kg copper sulphate Rice: 60 kg N + 40 kg P_2O_5 25 kg K_2O + Azolla @ 10 t/ha
Rice–wheat	**Rice**: 40 kg N + 20 kg P_2O_5 + 40 kg K_2O + FYM @ 5 t/ha/GM + Azolla @ 10 t/ha + 20 kg Zinc Sulphate once in 3 years + 5 kg borax + 1 kg ammonium molybdate + 5 kg copper sulphate **Rice**: 50 kg N + 20 kg P_2O_5 + FYM @ 5 t/ha
Rice–mustard	**Rice**: 40 kg N + 30 kg P_2O_5 (through SSP) + 40 kg K_2O + FYM/GM @ 10 t/ha + Azolla @ 10 t/ha + 20 kg Zinc Sulphate once in 3 years + 5 kg borax + 1 kg ammonium molybdate + 5 kg copper Sulphate **Mustard**: 20 kg N + 10 kg P_2O_5 (through SSP) + 25 kg K_2O
Rice–potato	**Rice**: 40 kg N + 20 kg P_2O_5 + 15 kg K^2O + Azolla/GM @ 10 t/ha + 20 kg Zinc Sulphate once in 3 years + 5 kg borax + 1 kg ammonium molybdate + 5 kg copper sulphate **Potato**: 50 kg N + 50 kg P_2O_5 + 30 kg K_2O + FYM @ 10 t/ha + seed treatment with Azotobacter and PSB
Lower Gangetic plain	
Rice–rice	**Rice**: 60 kg N + 40 kg P_2O_5 + 30 kg K_2O + FYM/GM @ 10 t/ha + 20 kg Zinc Sulphate **Rice**: 90 kg N + 80 kg P_2O_5 + 60 kg K_2O + Azolla @ 10 t/ha
Rice–wheat	**Rice**: 40 kg N + 45 kg P_2O_5 + 30 kg K_2O + FYM/GM @ 10 t/ha + Azolla @ 10 t/ha/BGA @ 10 kg/ha + kg Zinc Sulphate **Wheat**: 90 kg N + 45 P_2O_5 + 45 kg K_2O
Jute–rice–potato	**Jute**: 30 kg N + FYM @ 5 t/ha **Rice**: 30 kg N + 30 kg P_2O_5 + 30 kg K_2O + Azolla @ 10 t/ha/BGA @ 10 kg/ha + 20 kg Zinc sulphate **Potato**: 150 kg N + 40 kg P_2O_5 + 100 kg K_2O + FYM @ 5 t/ha + seed treatment with Azotobacter and PSB
Middle Gangetic plain	
Rice–wheat	**Rice**: 50 kg N + 30 kg P_2O_5 + 20 kg K_2O + Green Manure (greengram/stover) 20 kg Zinc Sulphat (in calcareous soils) **Wheat**: 90 kg N + 60 P_2O_5 + 30 kg K_2O + FYM@ 10 t/ha OR **Rice**: 75 kg N + 45 kg P_2O_5 + 30 kg K_2O + BGA @ 15 kg/ha + FYM @ 10 t/ha + 20 kg Zinc Sulphate (in calcareous soils) **Wheat**: 100 kg N + 65 kg P_2O_5 + 30 kg K_2O
Maize–wheat	**Maize**: 90 kg N + 60 P2O5 (through SSP) + 30 kg K_2O + GM + 16 kg borax (in calcareous soil) **Wheat**: 90 kg N + 60 kg P_2O_5 + 30 kg K_2O + FYM @ 10 t/ha
Groundnut–Pigeonpea intercropping	100% RDF + lime @2 t/ha + FYM @ 2 t/ha + Soil water conservation measure (furrows between groundnut and pigeon pea rows)
Upper Gangetic plain	
Rice–wheat	**Rice**: 90 kg N + 30 kg K_2O + FYM/GM (Sesbania/Leucaena Lopping) @ 10 t/ha **Wheat**: 90 kg N + 60 kg P_2O_5 (through SSP) + 30 kg K_2O
Maize–wheat/mustard	**Maize**: 50 kg N + 20 kg K_2O + FYM @ 10 t/ha **Wheat**: 120 kg N + 60 kg P_2O_5 (through SSP) + 40 kg K_2O **Mustard**: 60 kg N + 40 kg P_2O_5 (through SSP) + 30 kg K_2O
Sugarcane–potato	**Sugarcane (Autumn planting)**: 100 kg N + 45 kg P_2O_5 + 30 + Sulphitation press mud/GM + Incorporation of Potato foliage **Potato (Intercropping)**: 135 kg N + 20 kg P_2O_5 + 60 kg K_2O + FYM 2 10 t/ha + seed treatment with Azotobacter and PSB (In case of ratoon crop, incorporate sugarcane fresh along with only 75 kg N)[a]

(continued)

Table 11.4 (continued)

Cropping system	IPNS package
Sugarcane–wheat	**Sugarcane (Autumn planting)**: 135 kg N + 45 kg P_2O_5 + 30 t (FYM/Sulphitationp ressmud)/GM (Sesbania/Sunhemp/cowpea @ 10 t/ha **Wheat (Intercropping)**: 80 kg N + 40 kg P_2O_5 + 40 kg K_2O (In case of ratoon crop, incorporates sugarcane fresh along with only 75 kg N)[a]
Trans Gangetic plain	
Rice/cotton/maize/bajra–wheat	**Rice**: 60 kg N + 30 kg K_2O + FYM/poultry manure/GM @ 10 t/ha **Maize**: 70 kg N + FYM/GM (sesbania/cowpea) @ 10t/ha **Cotton**: 120 kg N **Bajra**: 60 kg N + 30 kg P_2O_5 FYM @ 10 t/ha **Wheat:** 150 kg N + 30 kg P_2O_5 (through SSP) + 30 kg K2O + Azotobactor/Azospirillium + PSB
Eastern Plateau and Hills	
Rice–winter maize/wheat/pulses	**Rice**: 30 kg N + 15 kg P_2O_5 (through SSP) + 15 kg K_2O + FYM/GM @ 10 t/ha + 15 kg BGA **Winter maize**: 100 kg N + 45 kg P_2O_5 (through SSP) + 20 kg K_2O **Wheat**: 90 kg N + 45 kg P_2O_5 (through SSP) + 30 kg K_2O **Pulses**: 10 kg N + 20 kg P_2O_5 (through SSP) + FYM @ 2.5 t/ha + Rhizobium + 500 g PSB
Rice–wheat/mustard	**Rice**: 75 kg N + FYM/Green Manure @ 5 t/ha **Wheat**: 90 kg N + 45 kg P_2O_5 + 30 kg K_2O Mustard: 30 kg N + 15 kg P_2O_5 + 10 kg K_2O FYM @ 10 t/ha
Soybean–wheat	**Soybean**: 10 kg N + 25 kg P_2O_5 (through Boronated SSP) + 4 t FYM + Rhizobium + 25 kg Zinc Sulphate in alternate years **Wheat**: 90 kg N + 45 kg P_2O_5 (through SSP)
Rice–gram	**Rice**: 25 kg N + 15 kg P_2O_5 + pulse crop residue incorporation + BGA @ 10 kg/ha/Azolla @ 10 t/ha **Gram**: 10 kg N + 20 kg P_2O_5 + Rhizobium + 5 t FYM + 500 PSB
Soybean–chickpea	**Soybean-Chickpea** (Rainfed system): 100% RDF + 2 t/ha FYM to soybean and 50% RDF to chickpea
Western Plateau and hills	
Soybean–wheat	**Soybean**: 10 kg N + 25 kg P_2O_5 (through SSP) + 4 t FYM + Rhizobium + 25 kg Zinc Sulphate in alternate year **Wheat**: 90 kg N + 45 kg P_2O_5 (through SSP)
Cotton–fallow/pigeon pea/wheat	**Cotton**: 50 kg N + 25 kg P_2O_5 + 25 kg K_2O + seed treatment with Azotobacter + 4 t FYM/in situ Green manuring (cowpea) followed by mulching with subabulloppings **Pigeon pea**: 10 kg N + 20 kg P_2O_5 (through SSP) + 10 kg K_2O + FYM @ 2.5 t/ha + Rhizobium + 500 g PSB **Wheat**: 90 kg N + 30 kg P_2O_5 (through SSP) + 30 kg K_2O + Azotobactor/Azospirillium + PSB
Green gram-Safflower	**Safflower** (Rainfed): Incorporation of green gram stalk before sowing of safflower along with 75% RDF of safflower + soil moisture conservation measure (Summer ploughing and inter-culture with blade hoe)
Fallow-Sunflower	**Sunflower** (Rainfed): 100% RDF + FYM @ 2 t/ha +SMC measure (Opening furrow after every 6 rows)
Southern plateau and hills, East Coast Plains and Ghats and West Coast Plains Regions	
Rice–rice	**Rice**: 75 kg N +15 kg P_2O_5 + 15 kg K_2O + FYM/Green Manure @ 5 t/ha **Rice**: 90 kg N + 60 kg P_2O_5 + 40 kg K_2O + 40 kg K_2O + Azolla @ 10 t/ha BGA @ 10 kg/ha + 20 kg Zinc Sulphates
Rice–pulses	**Rice**: 25 kg N + 15 kg K_2O + pulse crop residue incorporation + BGA @ 10 kg/ha/Azolla @ 10 t/ha **Pulses**: 10 kg N + 20 kg P_2O_5 + 10 kg K_2O + Rhizobium + 2.5 T FYM + 500 g PSB
Fallow-sunflower	**Sunflower** (Rainfed): 100% RDF + FYM @ 2 t/ha +SMC measure (opening furrow after every 2 rows)
Castor	**Castor monocropping (Rainfed)**: Cowpea incorporation after first picking and 75% RDF of castor
Gujarat plains and hills regions	
Groundnut/wheat/mustard	**Groundnut**: 15 kg N + 30 kg P_2O_5 (through SSP) + 45 kg K_2O + Gypsum @ 250 kg/ha in furrow + 25 kg Zinc Sulphate + 1 kg Boron **Wheat**: 70 kg N + 30 kg P_2O_5 (through SSP) + 20 kg K_2O + Azotobactor/Azospirillium + PSB **Mustard**: 30 kg N + 15 kg P_2O_5 (through SSP) + 10 kg K_2O + FYM @ 10 t/ha
Cotton–castor	**Cotton**: 50 kg N + 25 kg P_2O_5 + 25 kg K_2O + seed treatment with Azotobacter + 4 t FYM **Castor (irrigated)**: 25 kg N + 50 kg P_2O_5 (through SSP) + 1 t castor seed cake/FYM @ 5 t/ha + seed treatment with Azospirillium and PSB @ 5 kg/ha

(continued)

Table 11.4 (continued)

Cropping system	IPNS package
Western dry regions	
Kharif pulses	**Pulses–fallow**: 10 kg N + 20 kg P_2O_5 + 10 kg K_2O + Rhizobium + 2.5 t FYM
Pearmri millet–mustard	**Pearl mrimillet**: 25 kg N + 20 kg P_2O_5 (through SSP) + 10 kg K_2O + Azotobacter/Azospirillium **Mustard**: 45 kg N + 20 kg P_2O_5 (through SSP) + 15 kg K_2O + FYM @ 5 t/ha
Fallow–mustard	**Mustard (Rainfed)**: Green manuring with Sesbania and FYM @ 2 t/ha + 75% RDF (80 kg N + 40 kg P_2O_5 + 20 kg S)
Maize-raya (rainfed)	Maize-Raya (Rainfed): 100% RDF + S @ 20 kg/ha + SMC Measures (summer ploughing + maize residue application on surface)

[a]Liming @ 3–4 q/ha in furrows at the time of sowing for soils having pH < 5.5 may be practiced except for submerged rice

11.5.7 Extent of Soil Degradation

In order to prevent loss of top fertile soil and deterioration of soil physical conditions due to soil erosion, the Indian Institute of Soil and Water Conservation (IISWC) has developed location-specific bio-engineering measures. Similarly, Central Arid Zone Research Institute, Jodhpur, has developed sand dune stabilization and shelter belt technology to check wind erosion. The council through Central Soil Salinity Research Institute, Karnal, and All India Coordinated Research Project (AICRP) on salt-affected soils has developed reclamation technology, salt-tolerant varieties of rice (CSR-30, CSR-36), wheat (KRL-210, KRL-213) and mustard (CS-52, CS-54) and agroforestry interventions for rehabilitation of lands affected by salinity and sodicity. Also, sub-surface drainage and bio-drainage technologies have been developed to improve the productivity of saline waterlogged soils in the country. Similarly, the council has developed cost-effective amelioration techniques, i.e. liming @ 3–4 q/ha for managing acid soils too.

11.6 Government's Policy Initiatives

The Government of India, based on the recommendations of task force on balanced and integrated use of fertilisers has taken several initiatives, namely strengthening of soil-testing facilities, distribution of soil health cards, promotion of integrated nutrient management, production of organic fertilizers, development of new value-added fortified/customized fertilisers, fertigation, nutrient-based fertiliser subsidy, etc., to improve soil health. The Indian Council of Agricultural Research (ICAR) through Indian Institute of Soil Science (IISS) and All India Coordinated Research Projects (AICRPs) on soil test crop response (STCR), micro- and secondary-nutrients and pollutant elements (MSNP) and plants, long-term fertilizer experiments (LTFE) and Network Project on Soil Biodiversity-Biofertilizers are addressing researchable issues related to soil health in the country and providing requisite technology backstopping to a variety of programmes/schemes being implemented by different Ministries/Departments ranging from single component/commodity-based scheme to area-based integrated approach. Some of the notable schemes, viz. National Mission of Sustainable Agriculture, National Mission on Soil Health Cards, Nutrient-Based Subsidy scheme, *Param paragat Krishi Vikas Yojana* have direct relevance to soil health management.

India is a country of diverse soil, crop and climatic conditions. Under such situations, the sustainability of high crop productivity can be assured through site-specific nutrient management. Accordingly, a National Mission on Soil Health Card has been launched to provide soil test-based fertilizer recommendation to all the farmers in the country based on the twelve soil parameters. The Government under the component of soil health management of National Mission on Sustainable Agriculture (NMSA) is promoting soil test-based balanced and integrated nutrient management in the country through setting up as well as strengthening of soil-testing laboratories, establishment of biofertilizer and compost unit, use of micronutrients, trainings and demonstrations. A number of value-added fertiliser materials fortified with secondary- and micro-nutrients have been enlisted in Fertilizer Control Order (FCO) to promote balanced and efficient use of fertilisers. Also, the customized fertilisers which are crop, soil and area-specific show a good promise to maintain soil health by ensuring balanced fertilization. The Govt. of India took a historical policy decision of introduction of Nutrient-Based Subsidy (NBS) on N, P, K and sulphur containing fertilizers with effect from the 1 April 2010. Additional subsidy for fertilizers fortified with zinc and boron was paid at the rate of Rs. 500 and Rs. 300 per tonne, respectively. It well help in soil health enhancement through balanced and efficient use of plant nutrients including secondary- and micro-nutrients.

The Department of Fertilizers, Ministry of Chemicals and Fertilizers has declared subsidy on city compost @ Rs. 1500 per tonne to serve twin objectives, viz. (i) supporting government's *Swachh Bharat Abhiyan* and (ii) providing manures to farmers. The Ministry of New and Renewable Energy is implementing National Biogas and Manure Management Programme, which is a Central Sector Scheme of Biogas Technology Development Division of the Ministry aiming at setting up of family type biogas plants at rural and semi-urban/households level for recycling of rural wastes linking sanitary toilets with biogas plants (http://mnre.gov.in/schemes/decentralized-systems/schems-2/).

The Government of India through various schemes like National Centre of Organic Farming, National Horticulture Mission is promoting organic farming and thereby improving soil health in the country. Cultivated area under certified organic farming has grown almost 17-fold in the last one decade (42,000 ha in 2003-04 to 7.23 lakh ha in 2013–14) covering 27 states. Recently, dedicated schemes, namely *Param paragat Krishi Vikas Yojana* (PKVY) and Mission Organic Value Chain Development for North Eastern Region (MOVCDNER) under National Mission for Sustainable Agriculture (NMSA) have been launched. This will encourage farmers to adopt eco-friendly concept of cultivation and reduce their dependence on fertilizers and agricultural chemicals to improve yields. Under this programme, organic farming is promoted through adoption of village by Cluster Approach and Participatory Guarantee System (PGS) certification.

The Department of Land Resources, Ministry of Rural Development had been implementing an area development programme, i.e. Integrated Watershed Management Programme (IWMP) w.e.f. 26.02.2009, for development of rainfed/degraded areas. The major activities taken up under IWMP *inter-alia* include ridge area treatment, drainage line treatment, soil and moisture conservation, rain water harvesting, nursery raising, afforestation, horticulture, pasture development and livelihood for resource poor persons. In 2015–16, the IWMP has been amalgamated as the Watershed Development Component of the "*PradhanMantriKrishiSinchayeeYojana* (WDC-PMKSY)". The funding pattern under WDC-PMKSY is 60: 40 between Centre and State Government except in the States of North Eastern Region and Hill States where the funding pattern between Centre and State Government is 90:10.

The Government implemented a Centrally Sponsored Scheme "Reclamation and Development of Alkali and Acid Soils (RADAS)" through Macro Management of Agriculture (MMA) Scheme in seven states. Since inception up to March 2013, almost 9.0 lakh ha area has been developed. This programme has been subsumed in National Mission for Sustainable Agriculture (NMSA) with effect from April 2014 as component of Reclamation of Problem soils (viz. saline, alkali and acid soils). The assistance under this programme is 50% of cost to the limit of Rs. 60,000 per ha for reclamation of alkali/saline soils and Rs. 15,000 per ha for amelioration of acid soil.

Till now, nearly 1.8 Mha of sodic soils and 0.75 Mha saline soils have been reclaimed. Salt-tolerant varieties of rice, wheat and mustard have been spread in more than 2 lakhs ha areas. Benefit–cost ratio (BCR) of the technology is 1.43 with internal rate of return of 25%. Besides, about 1.0 lakh ha waterlogged (water table < 1.5 m below ground level) saline soils have been reclaimed through sub-surface drainage technology in Haryana, Rajasthan, Maharashtra, Karnataka, Gujarat, Punjab and Andhra Pradesh. The technology resulted in 40% improvement in cropping intensity and yields of different crops leading to 2–3-fold increase in farmer's income at a benefit–cost ratio of 1.5 and 20% internal rate of return, besides generating 128-man days of employment per ha annually. Nearly 2.00 Mha in Indo-Gangetic Plains have been brought under RCTs mainly zero tillage and bed planting through National Food Security Mission (NFSM) and NMSA schemes.

11.7 Conclusions

India needs growth of about 4% annually in agriculture sector to meet our food requirements. The agriculture will not be sustainable unless soil health is managed scientifically to meet present and future needs. The real challenge before the scientists is to provide a viable farming system based on resource-efficient evaluation measures that may influence soil chemical, physical and biological properties and processes in such a way as to improve and sustain land productivity. Viable farming system needs efficient crop planning based on soil characterization and classification, climate, water availability/irritability, land use pattern, socio-economic conditions, infrastructure, marketing facilities, etc., to enable farmers to harness full potential of their land choosing the right crop/cropping system suitable for the region. Land use has to be harmonized with the soil quality and carrying capacity of the soil resources. The all-round initiatives and commitments to restore and sustain the soil health lead to promote the national framework for more transparent soil evaluation procedures and land use planning at grass root level in order to achieve the national target for total food security.

References

Acharya CL, Bandyopadhyay KK, Chaudhary RS (2016) Management of soil physical quality. In: Pathak H, Sanyal S, Takkar PN (eds) State of the Indian agriculture—soil. National Academy of Agricultural Sciences, New Delhi, pp 65–84

Andrews SS, Carroll CR (2001) Designing a soil quality assessment tool for sustainable agro-ecosystem management. Ecol Appl 11 (6):1573–1585

Andrews SS, Karlen DL, Mitchell JP (2002a) A comparison of soil quality indexing methods for vegetable production systems in northern California. Agric Ecosyst Environ 90:25–45

Andrews SS, Mitchell JP, Mancineelli R, Karlen DL, Hartz TK, Horwath W, Pettygrove GS, Scow KM, Munk DS (2002b) On-farm assessment of soil quality in California's central valley. Agron J 94:12–23

Biswas PP, Sharma PD (2008) A new approach for estimating fertilizer response ratio-the Indian scenario. Indian J Fert 4(7):59–62

Benbi DK (2013) Opportunities for carbon sequestration in soil. J Res (PAU) 49:186–192

Benbi DK (2016) Organic matter turnover in Soil. In: Pathak H, Sanyal S, Takkar PN (eds) State of the Indian agriculture—soil. National Academy of Agricultural sciences, New Delhi, pp 36–50

Benbi DK, Chand M (2007) Quantifying the effect of soil organic matter on indigenous soil N supply and wheat productivity in semiarid sub-tropical India. Nutr Cycl Agroecosyst 79:103–112

Benbi DK, Toor AS, Kumar S (2012) Management of organic amendments in rice-wheat cropping system determines the pool where carbon is sequestered. Plant Soil 360:145–162

Bhatia A, Aggarwal PK, Jain N, Pathak H (2012) Greenhouse gas emission from rice and wheat-growing areas in India: spatial analysis and upscaling. Greenhouse Gas Sci Technol 2:115–125

Bhatia A, Jain N, Pathak H (2013) Methane and nitrous oxide emissions from Indian rice paddies, agricultural soils and crop residue burning. Greenhouse Gas Sci Technol 3:196–211

Bhattacharyya T, Pal DK, Mandal C, Chandran P, Ray SK, Sarkar D, Velmourougane K, Srivastava A, Sidhu GS, Singh RS, Sahoo AK, Dutta D, Nair KM, Srivastava R, Tiwary P, Nagar AP, Nimkhedkar SS (2013) Soil of India: historical perspective, classification and recent advances in knowledge; a review. Curr Sci 104:1308–1323

Brejda JJ, Moorman TB (2001) Identification and interpretation of regional soil quality factors for the Central High Plains of the Midwestern USA. In: Stott DE, Mother RH, Steinhardt GC (eds) Sustaining the global farm-selected papers from the 10th International Soil Conservation Organization Meeting (ISCO99), USDA and Purdue University, pp 535–540

Brevik EC, Sauer TJ (2015) The past, present and future of soils and human health studies. Soil 1:35–45

Chaudhari SK, Adlul Islam, Biswas PP, Sikka AK (2015) Integrated soil, water and nutrient management for sustainable agriculture in India. Indian J Fertil 11(10):51–62

Chaudhary J, Mandal UK, Sharma KL, Ghosh H, Mandal B (2005) Assessing soil quality under long-term rice based cropping system. Commun Soil Sci Plant Anal 36(9/10):1141–1161

Chen ZS (1998). Selecting indicators to evaluate the soil quality of Taiwan soils and approaching the national level of sustainable soil management. In: Proceedings of international seminar on soil management for sustainable agriculture in the tropics, held during December 14–19, 1998 at Taichung, Taiwan, pp 131–171

Doran JW, Parkin TB (1996). Quantitative indicators of soil quality: a minimum data set. In: Doran JW, Jones AJ (eds) Methods for assessing soil quality. Soil Sci Am Special publication, No. 49, Madison, pp 25–37

Doran JW, Sarrantonio M, Liebig MA (1996) Soil health sustainability. Adv Agron 56:1–54

FAO (2002) Soil biodiversity and sustainable agriculture. Background paper for the Ninth Regular Session of the Commission on Genetic Resources for Food and Agriculture (CGRFA) FAO-Rome, pp 14–18

ICAR—CSSRI (2016a) Reclamation of waterlogged saline soils through subsurface drainage technology. ICAR-Central Soil Salinity Research Institute, Karnal-132001(Haryana). ICAR-CSSRI/Carnal/ Technology Folder/ 2016/02

ICAR—CSSRI (2016b) Reclamation of alkali soils through gypsum technology. ICAR-Central Soil Salinity Research Institute, Karnal-132001(Haryana). ICAR-CSSRI/Karnal/Technology Folder/ 2016/01

Jain N, Dubey R, Dubey DS, Singh J, Khanna M, Pathak H, Bhatia A (2014) Mitigation of green house gas emission with system of rice intensification in the Indo-Gangetic Plains. Paddy Water Environ 12:355–363

Katyal JC, Chaudhari SK, Dwivedi BS, Biswas DR, Rattan RK, Majumdar K (eds) (2016) Soil health: concept, status and monitoring. Bull Indian Soc Soil Sci 30:1–98

Maji AK, Obi Reddy GP, Sarkar D (2010) Degraded and wastelands of India: status and spatial distribution. ICAR and NAAS, New Delhi

McKenzie BM, Tisdall JM, Vance WH (2011) Soil physical quality. In: Glinskin J et al (eds) Encyclopaedia agro-physics. Springer, The Netherlands, pp 770–777

NAAS (2010) Degraded and wastelands of India, status and spatial distribution. ICAR, New Delhi

NAAS (2012) Management of crop residues in the context of conservation agriculture. Policy Paper No. 58, National Academy of Agricultural Sciences, New Delhi, p 12

Nair PKR, Mohan Kumar B, Nair VD (2009) Agroforestry as a strategy for carbon sequestration. J Plant Nutria Soil Sci 172:10–23

Painuli DK, Yadav RP (1998) Tillage requirements of Indian soils. In: Singh GB, Sharma BR (eds) 50 years of natural resource management research division. Of Natural Resource Management. ICAR, New Delhi, pp 245–262

Pathak H, Benbi DK, Srinivasa Rao Ch (2016) Soil and climate change. In: Pathak H, Sanyal S, Takkar PN (eds) State of the Indian agriculture—soil. National Academy of Agricultural Sciences, New Delhi, pp 267–293

Pathak H, Byjesh K, Chakrabarti B, Aggarwal PK (2011) Potential and cost of carbon sequestration in Indian agriculture: estimates from long-term field experiments. Field Crops Res 120:102–111

Sanyal SK, Gupta SK, Kukal SS, Jeevan Rao K (2016) Soil degradation, pollution and amelioration. In: Pathak H, Sanyal S, Takkar PN (eds) State of the Indian agriculture—soil. National Academy of Agricultural Sciences, New Delhi, pp 234–266

Sehgal J (1995) Land resources appraisal for land use planning to meet the challenges of 21st century. J Indian Soc Soil Sci 43(4):504–528

Sharda VN, Dogra P, Prakash C (2010) Assessment of production losses due to water erosion in rainfed areas of India. J Soil Water Conserv 50(2):79–91

Sharma KL (2009) Assessing soil quality key indicators for development of soil quality index using latest approaches under predominant management practices in rainfed agro ecology. Consolidated Report (2005–09) of National Fellow Project, submitted to the ICAR, New Delhi. Central Research Institute for Dryland Agriculture, p 419

Sharma KL, Srinivasa Rao Ch, Sammi Reddy K (2016) Soil quality and its management. In: Pathak H, Sanyal S, Takkar PN (eds) State of the Indian agriculture—soil. National Academy of Agricultural Sciences, New Delhi, pp 51–64

Sharma MP, Adhya TK (2016) Management of soil biological quality. In: Pathak H, Sanyal S, Takkar PN (eds) State of the Indian agriculture—soil. National Academy of Agricultural Sciences, New Delhi, India, pp 85–99

Sharma SP, Subehia SK, Sharma PK (2002) Long term effects of chemical fertilizer on soil quality, crop productivity and sustainability. Research Bull 33 pp. Department of Soil Science, CSKHPKV, Palampur

Singer MJ, Ewing S (2000) Soil quality. In: Sumner ME (ed) Handbook of soil science. CRC Press, Boca Raton, pp G-271-G-298

Srinivasarao Ch, Lal R, Kundu S, Prasad Babu MBB, Venkateswarlu B, Singh AK (2014) Soil carbon sequestration in rainfed production systems in the semiarid tropics of India. Sci Total Environ 487:587–603

Srinivasarao Ch, Venkateswarlu B, Lal R, Singh AK, Kundu S (2013) Sustainable management of soils of dryland ecosystems of India for enhancing agronomic productivity, sequestering carbon. Adv Agron 121:253 329

Wanjari RH, Singh MV, Ghosh PK (2004) Sustainability yield index: an approach to evaluate the sustainability of long term intensive cropping systems in India. J Sustain Agric 24(40):39–56

Soil Resilience

12

K. Sammi Reddy, K. L. Sharma, K. Srinivas, A. K. Indoria, M. Vassanda Coumar, Pushpanjali, and V. Girija Veni

Abstract

Following an understanding on soil health assessment as well as land evaluation to a particular soil, this chapter highlights the concept of soil quality and soil resilience, and further to develop a minimum data set of soil quality indicators for deriving soil quality index (SQI), and methods for evaluating it through indicator approach to various soil functions. Though the framework for SQI is put in use for monitoring soil quality, efforts are made to use such a framework for resilience purpose to assess ease and degree of recovery of degraded soil. Based on the experiments conducted on farmers' fields in Madhya Pradesh, it was observed and recorded that the integrated use of wood charcoal (5.4 t/ha) along with balanced fertilization to soybean and wheat crops improved soil resilience index. Further studies on soil resilience were conducted elsewhere across the country in degraded soils. The chapter may thus answer what integration of inputs could achieve effective soil resilience for soil sustainability.

Keywords

Soil quality • Soil resilience • Soil resilience index • Soil sustainability

12.1 Introduction

The production function of a soil is the basis for food and nutritional security of the Indian population crossing 1.25 billion marks. Soil, however, is a highly complex, multi-component system of interacting materials such as minerals, rock, water, air, organic materials and living biota. It has biological, chemical and physical properties that interact to provide a soil and its capacity to function. As the soil resources are finite, their intensive uses are inevitable to meet the demands for food and fibre. India has to produce about 345 mt of food grains in 2030, requiring to be increased at rate of more than 5 mt annually which is difficult to achieve without identifying the emerging threats to soils for sustainable development. Soil of the pedosphere is in dynamic equilibrium with the biosphere, hydrosphere, atmosphere and lithosphere (Patiram 2003) and thus a critically established soil evaluation procedure needs to be developed. Soil degradation, by and large, is a major threat to agricultural sustainability and environmental quality both in irrigated and rainfed agroecosystem. Consequently, the decline in crop yield, partial factor productivity of inputs and quality of produce particularly under intensive agriculture is widely common (Abrol and Sehgal 1994; Sunanda Biswas 2016; Vittal et al. 1990). The partial factor productivity of fertilizers decreased from 20 kg grain/kg N + P_2O_5 + K_2O at the end of Sixth Five Year Plan to 14 in Eight Plan, to 9.2 at the end of 11th Plan (2009–10). The decline in rate of response of crops to added fertilizers under intensive cropping systems has possibly resulted from deterioration in soil quality (Muralidharudu et al. 2012).

Decline in soil organic matter and its associated nutrient supply in soil is the major factor for yield decline under intensive cropping systems (Manna and Ganguly 2003). For decreasing trend of the partial factor productivity of input (fertilizers), it is essential to identify location-specific soil physical, chemical and biological constraints of crop production and thereafter to implement measures for alleviating these constraints for sustaining higher crop productivity over a longer period of time. In addition, stresses due to acidity, salinity, alkalinity, waterlogging, etc., are also there for a considerable land area in different parts of the country. These degradation forces and processes impair soil's essential ecosystem functions and ultimately its quality. Therefore, for sustainable use of soil and its protection against degradation,

K. S. Reddy (✉) · K. L. Sharma · K. Srinivas · A. K. Indoria · M. V. Coumar · Pushpanjali · V. G. Veni
ICAR-Central Research Institute for Dryland Agriculture, Santoshnagar, Saidabad Post, Hyderabad, Telangana 500 059, India
e-mail: k.sammireddy@icar.gov.in

B. B. Mishra (ed.), *The Soils of India*, World Soils Book Series, https://doi.org/10.1007/978-3-030-31082-0_12

the soil quality assessment, its resilience capacity and identification of diagnostic recovery modules are the only options available to address this critical issue.

The concept of soil quality and the methods for evaluating it with respect to various soil functions have been evolved to a considerable extent for identifying major soil quality indicator. Though the indicators are put in use for monitoring soil quality (Andrews et al. 2004), such indicators for resilience purpose are developed in a very few studies to assess ease and degree of recovery of degraded soil system in India (Kundu et al. 2015). Soil resilience and soil resistance are affected by both inherent and dynamic soil quality characteristics, and thus, will vary substantially from one area to the next and will change over time and management practices (Lal 1997). Therefore, the issues of soil quality, the concept of soil resilience and framework developed for quantification of soil resilience using soil quality index are discussed in this chapter with a reference to Indian soils.

12.2 Soil Resilience: Concept and Background

Soil resilience has been defined as the capacity of a soil to recover its functional and structural integrity after a disturbance. Functional and structural integrity are defined as a soil's capacity to perform vital soil functions such as those proposed by Karlen et al. (1997). Structural integrity is linked to soil function and deals with the physical arrangement of primary soil particles and their aggregation. A disturbance is broadly defined as any event that causes a significant change from the normal pattern of functioning of an ecosystem. A wide variety of disturbances are included in this broad definition, including those that are primarily natural in origin and others that are largely or wholly anthropogenic. Natural disturbances and causes of disturbances include fires, earthquakes, floods, landslides and high-intensity storms. Nearly, all human activities associated with land evaluation, improvement and use can be classified as "disturbances" including logging, grazing, urban and industrial development, recreation and annual cropping. Agriculture itself is one of the greatest sources of stress, risk and disturbance of the environment. Common disturbances or stresses associated with agriculture include heavy load as a result of vehicular traffic, tillage, application of fertilizers and pesticides and removal or exclusion of competing plant species, *e.g.* monoculture, shifting cultivation, etc. (Indoria et al. 2017; Benjamin et al. 2007; Halvorson et al. 2002; Dalal and Bridge 1996).

Since the soil resilience is the soil's inherent ability to restore its quality after disturbance that influences the soil health and quality, the sustainability thus depends on resilience and resistance of soil by conservation and improvement measures in time and space. The soils of north-eastern hill states, for example, are affected by type-specific degradation (loss of resilience), as caused by over exploitation of forest for timber, fuel and fodder surrounding human settlements, shifting agriculture (jhum) on hill slopes, improper land use/land cover, population explosion including livestock, infrastructure development and mining activities. These practices result in poor soil quality and loss of soil productivity for sustaining agricultural productivity and hill agroecosystem (Patiram 2003). Recently, a protocol for quantifying resilience of degraded Vertisols of Madhya Pradesh with field-level validation has been developed (Kundu et al. 2015). Meanwhile, Mandal et al. (2017) have attempted to establish soil resilience in rainfed soils.

The soil's capacity to recover has two components, the rate of recovery and the degree of recovery. The rate of recovery is the amount of time it takes for soil to recover its potential after a disturbance. The magnitude of recovery to some stabilized potential in relation to its pre-disturbance state defines the degree of recovery. If the disturbance is too drastic or if the soil is inherently fragile, the soil can undergo irreversible degradation in which its capacity to function will not recover within any reasonable time frame (*e.g.* human lifespan). In this case, the soil's resilience capacity has been exceeded, resulting in permanent damage or the need for very costly restoration. The greater the rate and/or degree of recovery, the more resilient the soil system is to a specific disturbance. The soils of India being widely diversified suffer greatly from multiple types of disturbances.

12.3 Soil Resilience and Pedogenesis

Soil resilience may be looked at in terms of soil formation and development, collectively refers to pedogenesis, which is a continuous process taking thousands of years and so the soil is treated as a non-renewable resource. Soil misuse and the extreme climatic conditions can damage self-regulating capacity and give way to regressive pedogenesis (Pal et al. 2013), and thus, might lead to the soil to regress from higher to lower usefulness and/or drastically diminished productivity. However, humans have extensively exploited, intensively utilized and frequently changed the soil as a natural resource, since they depend directly on soil. Pedogenesis is the result of different factors and processes, wherein parent material and topography are passive and contribute to soil mass and position, while climate and the biosphere are active and supply energy in soil formation (Paton 1978).

In soils of humid tropical climate of India, the dominance of kaolin indicates that inspite of prolonged weathering for millions of years, the weathered products of Ca-rich minerals of the parent materials have not reached even the pure kaolinitic mineral stage. The formation of kaolin clay mineral suggests that the formation of acidic Alfisols, Mollisols and Ultisols, and their pedogenic threshold at this time supports the supposition that steady state may exist in soils developed over long periods of time spanning not just thousands of years (Smeck et al. 1983; Yaalon 1975) but also millions of years (Bhattacharyya et al. 1993, 1999; Chandran et al. 2005). Similarly, the organic carbon (OC)-rich Ultisols have less Al saturation in surface horizons due to the downward movement of Al as organo-metal complexes or chelates, but have higher base saturation than the sub-surface horizons due to addition of alkaline and alkaline metal cations (Pal et al. 2014) through litter fall (Nayak et al. 1996), and there is no desilication and transformation of kaolin to gibbsite. In Ultisols, Alfisols and Mollisols of tropical climate, C sequestration is relatively high as indicated by their OC concentration ranging from 1 to 5% (Pal et al. 2014). Soils in a wet climate under forest have high OC content, sufficient to qualify as Mollisols. The OC addition to Ultisols and Alfisols has been possible as a result of favourable soil temperature and moisture regime. It is known that 2:1 expanding clay minerals provide higher surface area for OC accumulation. But a high positive balance of OC in kaolin-dominated soils clearly indicates a positive role of this mineral and other hydroxy-interlayered clay minerals because kaolin often shows a relatively high value of CEC $\sim 30\ \text{cmol}_{(p+)}\text{kg}^{-1}$(Ray et al. 2001) than that of well crystalline kaolinite. Therefore, besides the dominating effect of humid climate in cooler winter months with profuse vegetation, the soil substrate quality in terms of larger surface area is of fundamental importance in OC sequestration in soils, an important issue in maintaining soil health and sustainability (Velayutham et al. 2000; Pal et al. 2014). This has been possible due to considerable chemical reactivity of soil kaolin that helps provide an effective substrate to support agriculture and other land uses in the tropics. Soil kaolin provides the intrinsic ability to regenerate the productivity of these acid soils, thus making them resilient. Therefore, to sustain the crop productivity at an enhanced level, large tracts of lands dominated by acid soils can be brought under improved soil, water and nutrient management to help meet the food needs of ever increasing Indian population and elsewhere of the tropical world (Pal et al. 2014: Gilkes and Prakongkep 2016).

12.4 Framework to Measure Soil Resilience

12.4.1 Factors Affecting the Soil Resilience

The interacting factors that affect soil resilience can be grouped into two broad categories: (a) endogenous and (b) exogenous. Endogenous factors are related to inherent soil properties and micro- and meso-climate. Endogenous factors include rooting depth, soil texture and mineralogy, parent material type, landscape position and terrain, moisture regime, micro- and meso-climate, soil biodiversity, etc., whereas exogenous factors include land use and farming system, technological innovations, inputs and management, etc. Factors that enhance soil resilience are sufficient rooting depth, loam to clay loam texture, structurally active soils containing high activity clay minerals, a high proportion of stable micro-aggregates, gentle to rolling terrain, good internal and other characteristics of fertile soil depend on these conditions (Lal 1997).

Soil resilience includes such important properties as a buffering capacity of soil in respect of chemical, physical and biological impacts. Physical buffering capacity plays an important role as far as the hazard of erosion is concerned; whereas, chemical buffering capacity comes into play in the case of soil acidification or alkalization. It is somewhat more difficult to define the biological buffering capacity of a soil which concerns several biological processes as well as the soil biota. Potential indicators of soil resilience are soil structure, micro-aggregation, soil water, retention and transmission characteristics, CEC, exchangeable cations, soil organic matter content and transformation, nutrient supplying capacity, rooting depth, soil biodiversity, microbial activity, etc. (Lal 1997).

Soil resilience is not the same as soil resistance, because resilience refers to "elastic" attributes that enable soil to regain its quality upon alleviation of any perturbation or destabilizing influence (Lal 1997). Sound rhizospheric processes are essential for soil resilience against anthropogenic and natural perturbations. Being a dominant site of microbial metabolism, it is pertinent to identify evaluation and improvement systems that stimulate soil microbiotic activity and related microbial processes. In this context, an "eco-physiological index" has been proposed to assess the impact of soil resilience (Lynch 2002) on soil processes. Managing the quantity and quality of soil organic carbon (SOC) pool is once again a crucial guiding principle in identifying appropriate management practices that will strengthen resilience and reduce risks for soil degradation (Syers 1997). The SOC pool size is strongly related to the quantity of both above and belowground biomass-C inputs. It is the assured, continuous input of the biomass-C that moderates MBC, provides a reservoir of plant nutrients (e.g. N, P, S), influences nutrient cycling and improves/stabilizes soil structural morphology and geometry (Lal 1997; Lakaria et al. 2012). Biochar is a C-rich soil amendment derived from biomass by pyrolysis and used to improve soil resilience while mitigating climate change (Venkatesh et al. 2013).

12.4.2 Computation of Soil Quality Index

The soil resilience is an accumulated soil quality and an inherent key element of sustainability. Factors that affect the soil quality also affect the soil resilience. Fundamental approaches for evaluating and assessing the soil resilience may be the measurement of recovery directly after a disturbance, quantifying the integrity of recovery mechanisms after a disturbance and measuring soil quality parameters that serve as indicators of recovery mechanisms. Critical scientific monitoring followed by reliable research is essentially desirable in the development of indicators for quantitative measures of the ability of soils to recover from specific or even multiple disturbances. Of late, sincere attempts have been made in India to quantify soil resilience by quantifying the soil quality index at different stages of degradation of soil quality that too before and after imposing interventions.

Systematic research has been conducted in India to develop minimum data set (MDS) of soil quality indicators for different soils under both irrigated and rainfed conditions and to study the effect of different nutrient management interventions/practices on the crop productivity and soil quality (Chaudahry et al. 2005; Sharma et al. 2005; Mandal et al. 2011; Kundu et al. 2015). In general, the minimum data set of soil quality indicators included mean weight diameter (MWD), bulk density, plant available moisture (Physical properties); pH, EC, soil organic carbon (SOC), total organic carbon (TOC), available N, P, K, Cu, Zn, Mn, Fe, Zn, S, B, total N, total organic P, total inorganic P, non-exchangeable K, different carbon pools (chemical properties); microbial biomass carbon, alkaline phosphatase, dehydrogenase enzyme, (biological properties) which vary from soil to soil. After developing minimum data set (MDS) of soil quality, research had been conducted to quantify the soil quality index as influenced by different nutrient and soil management practices. From those studies, in general, it was found that the integrated/balanced nutrient management options were found to be a viable option for maintaining better soil quality, crop productivity and yield sustainability over longer period of time on different soils and cropping systems. Sharma et al. (2005) assessed the effect of different nutrient management options such as inorganic fertilizers alone, organic manures alone and integrated nutrient management practices on sustainable yield index (SYI) and soil quality of rice–blackgram–horsegram system under dryland conditions of Phulbani, Orissa. Integrated use of farmyard manure and inorganic fertilizer nitrogen not only produced higher SYI of rice–blackgram–horsegram system but also maintained the highest soil quality index. Similarly, Sharma et al. (2005) also found that at different rates of N application, the soil quality was better with conventional tillage with crop residue practices as compared to without crop residue under sorghum–caster system on Alfisols. Best nutrient management practices that were found to maintain better soil quality and sustain higher crop yields in different soils under various cropping systems have been compiled and furnished (Table 12.1). But in all these studies, the resilience index was not quantified and restricted to soil quality index.

12.4.3 Quantification of Soil Resilience Index

Although Lal (1997) observed that there is no standard method of assessing or quantifying soil resilience, there has been a demand to develop practical methods of assessing soil resilience, as this would provide a means of predicting the long- and short-term consequences of soil disturbance on a given site. So, before taking reliable decision for improvement and/or management of a disturbed or degraded soil, assessment and quantification of soil resilience seem to be of scientific relevance. Different researchers have, however, used different yardsticks to measure soil resilience. Biological resilience in soil was quantified by measuring changes in the short-term mineralization of plant residues, dissolved organic carbon (DOC) and catabolic function in response to disturbance. Kuan et al. (2007) have studied the biological and physical resilience of selected Scottish soils. However, in this study, they have measured the biological resilience by quantifying CO_2 evolution from soil amended with plant residues after either a transient (heat) or a persistent (copper) stress.

Recently, a protocol for quantifying resilience of degraded Vertisols of Madhya Pradesh in central India with field-level validation has been developed (Kundu et al. 2015). Steps to be followed in determining the soil resilience index are given below:

Steps for Quantification of Soil Resilience Index (Kundu et al. 2015)

- Stratified multistage random sampling method was followed for selecting farmers' fields in identified districts (Sehore and Vidisha), where tehsil/block was considered as strata.
- Twenty-five villages were selected from each district covering all blocks.
- From each village, six farmers' fields have been selected based on high, medium and low resource use following participatory rural appraisal (RRA) technique.
- From each site, composite soil sampling has been carried out using core sampler as per standard soil test method.
- Soil samples have been analysed for various soil properties, viz. mean weight diameter (MWD), bulk density,

Table 12.1 Predominant nutrient management practices which proved effective for maintaining higher productivity and soil quality indices (Kundu et al. 2015)

S. no.	Soil type	Cropping system	Best practice that produced the highest SQI	Corresponding SQI recorded with the best practice	References
1	Indo-Gangetic alluvial soils (Inceptisols)	Rice–wheat–jute	100% NPK + FYM, 100% NPK	2.25 1.75	Chaudhury et al. (2005)
2	Alfisols (red soil)	Sorghum–castor	Conventional tillage + fresh gliricidia loppings @2 t ha^{-1} + N @90 kg ha^{-1} (CTGLN90) Conventional tillage + fresh gliricidia loppings @2 t ha^{-1} + N @60 kg ha^{-1} (CTGLN60)	1.27 1.19	Sharma et al. (2005)
3	Alfisols (red soil)	Sorghum–moong bean	4t Compost + 2t gliricidia loppings) + reduced tillage (RT) 2t gliricidia loppings + 20 kg N as urea + reduced tillage (RT)	0.89 0.87	Sharma et al. (2008)
4	Alfisols (red soil)	1. Agroforestry system, 2. Agrihorticulture system	Agroforestry system Agrihorticulture system	0.92 0.86	Sharma et al. (2009)
5	Alfisols (red soil)	Groundnut Groundnut Castor + groundnut	10-20-20 NPK kg ha^{-1} + FYM@ 4 t ha^{-1}	1.96	Sharma et al. (2010)
			Low tillage (LT) + 100% organic	2.77	
			LT + Herbicide + 100% organic	2.56	
			50% N (Gliricidia) + 50% N (Inorganic)	1.52	
6	Vertisols	Cotton + greengram intercropping (1:1) system	25 kg P_2O_5 ha^{-1} + 25 kg N ha^{-1} through leuceana green biomass	2.10	Sharma et al. (Sharma et al. 2011)
			25 kg N + 25 kg P_2O_5 + 25 kg N ha^{-1} through FYM	2.01	
7	Alfisols , Vertisols (Mixed type)	Rice–rice (Irrigated)	FYM: 4–5 Mg ha^{-1} N: 100–120 kg ha^{-1} P: 26–30 kg ha^{-1}	2.97	Mandal et al. (2011)
	Alfisols, Vertisols (Mixed Type)	Vegetable-based system (Irrigated)	FYM: 2–4 Mg ha^{-1} N: 100–150 kg ha^{-1} P: 26–30 kg ha^{-1}	3.06	Mandal et al. (2011)
	Alfisols, Vertisols (Mixed Type)	Cotton Supplementary irrigation and rainfed	FYM: 1–2 Mg ha^{-1} N: 60–100 kg ha^{-1} P: 20–25 kg ha^{-1}	2.68	Mandal et al. (2011)

plant available moisture (physical properties); pH, EC, SOC, TOC, available N, P, K, Cu, Zn, Mn, Fe, Zn, S, B, total N, total organic P, total inorganic P, non-exchangeable K, different carbon pools (chemical properties); microbial biomass carbon, alkaline phosphatase, dehydrogenase enzyme, (biological properties) following standard procedures.

- Significant variables were chosen for minimum data set (MDS) formation through principle component analysis (PCA) (Andrews et al. 2002a; Shukla et al. 2004).
- After determining the MDS indicators, every observation of each MDS indicator was transformed using linear scoring method (Andrews et al. 2002b).
- Once transformed, the MDS variables for each observation were weighted using the PCA results and then summed up the weighted MDS variable's scores for each observation using the following equation:

$$\text{SQI} = \sum_{i=1}^{n} \text{Wi Si}$$

- where S is the score for subscripted variable, and W is the weighing factor derived from the PCA. Here, the assumption is that the higher index scores meant better soil quality or greater performance of soil function. Based on these PCA derived SQI value, ten farmers' fields having varying SQI values (gradient of SQI value ranging from low to high) were selected for the field experiment to validate the protocol developed for resilience study.
- Based on the quantitative value of the key indicators, SQI values were computed as described above for all the interventions of ten sites.
- The soil resilience index (SRI) of the soils for all management interventions tested at ten different sites was computed using the following formula.

$$\text{Soil Resilience Index} = \frac{\text{SQI(i)}-\text{SQI(d)}}{\text{SQI(p)}-\text{SQI(d)}} \times 100$$

where,

SQI (*i*) The SQI value of soil after management intervention (*i*)
SQI (*d*) The computed SQI value of soil before the management intervention (*i*)
SQI (*p*) The computed SQI value of pristine soil nearer to the corresponding site

Kundu et al. (2015) computed soil resilience index (SRI) using SQI values derived from quantitative values of the key soil indicators for all management interventions tested at ten different sites using the above formula. From these experiments conducted on farmers' fields, they found that the integrated use of wood charcoal (5.4 t/ha) along with balanced fertilization to soybean and wheat crops which qualifies as an appropriate remedial option to improve soil resilience index. Further, it was observed that the farmers' field soils which had initially low soil quality index values showed maximum improvement in both resilience index as well as soil quality index values due to balanced fertilization with wood charcoal addition (Table 12.2).

Soil resilience depends on inherent soil properties and the net balance between soil formative and degradative processes following the equation: $S_r = S_a + f(S_n - S_d \pm \text{Im})$ d*t* where S_r is the soil resilience, S_a is the antecedent soil condition, S_n is the rate of new soil formation, Sd is the rate of degradation and depletion, 1m is the management inputs, and t is the time (Lal 1995). The magnitude and sign of the term ($S_n - S_d \pm$ 1m) are critical in determining the soil resilience. The equation may be applied to specific soil properties, viz. rooting, depth, topsoil thickness, soil organic

Table 12.2 Effect of addition of 5.4 t/ha charcoal along with balanced fertilization on soil quality index and resilience index of Vertisol at different sites in Vidisha and Sehore districts of Madhya Pradesh state (Kundu et al. 2015)

Site number	Soil Quality Index (SQI)			Soil Resilience Index (SRI)
	SQI (p)	SQI (d)	SQI (i) after intervention of wood charcoal with balanced fertilization	
1	1.546	0.959	1.067	17.85
2	1.641	0.991	1.116	19.23
3	1.512	1.097	1.266	40.27
4	1.541	1.117	1.363	58.01
5	1.593	1.298	1.373	25.42
6	1.564	1.393	1.452	34.50
7	1.678	1.462	1.514	24.07
8	1.805	1.564	1.593	12.03
9	1.809	1.695	1.728	28.94
10	1.982	1.745	1.771	10.97

matter, available nutrients and water, etc., as long as the "time-dependent" relationship of the specific property is established. At the time of disturbance, soil quality becomes the function of soil resilience, and soil resilience is the component of soil quality after disturbance (Patiram 2003). Thus, sustainability depends on soil resilience and resistance of the system. The improper or imbalanced management or even misuse of soil over the time may result into reduction of soil quality due to degradation or loss of soil resistance and so poor resilience.

In north-eastern hill (NEH) region of India, shifting cultivation, valley rice-based cropping, horticulture and mixed farming (agriculture and livestock) are the common land uses in the states of Arunachal Pradesh, Manipur, Meghalaya, Mizoram, Nagaland, Sikkirn and Tripura (Patiram 2003). Soil degradation is resulting into persistent decrease of soil potential productivity and loss of environmental regulatory capacity. The land has been degraded mainly due to over exploitation of forest for fuel, timber and fodder surrounding human settlement, shifting cultivation *(jhum)* with 3–5 years abundant cycle on hill slopes, improper land-use practices, population explosion including livestock, infrastructure development and mining activities (Patiram 2001). Acid soil and low temperature especially at higher altitude retard the mineralization of organic P into inorganic P, which in turn, minimize the soil availability of P to crops (Patiram 2003). The aluminium toxicity in acidic soils by farming organo-Al complexes and favours the increase of soil pH, OC, available nutrients and soil CEC (Patiram 2003). Sharma et al (1998) found that continuous application of FYM + NPK increased the soil organic carbon 50% over the initial value after 20 years with higher microbial biomass-C, bacterial and fungal population by creating more favourable good soil health in acid sub-humid temperate agro-climate of Western Himalayas.

Mandal et al. (2017) have attempted to assess the soil resilience in rainfed soils. They subjected the samples to the heat stress in the laboratory, and the effects of heat stress on biological functions were monitored after relieving the stress as well as under addition of substrate (dry gliricidia leaf: C% 44.1; N% 3.73). As a result, heat stress reduced the respiration up to sixth day of incubation to the extent of 17%. There was hardly any reduction in respiration under heat stress condition after 13th and 15th day of incubation. Out of four conditions, (normal soil or control, soil under heat treatment, soil under heat treatment with gliricidia leaf and normal soil with gliricidia leaf) maximum respiration rate was noted under normal soil with gliricidia leaf, followed by heat-treated soil with gliricidia, normal soil and lowest in heat-treated soils. The resilience and stability index were developed based on relative respiration rate under heat stress, normal soil, normal soil with substrate and heat-treated soil with substrate. Overall, stability index was relatively more than resilience index. They reported that under different land-use systems, maximum resilience index was found under fallow followed by paddy, redgram, cotton, maize, intercrop and lowest under castor system. The trend for stability index was fallow > paddy > maize > red gram > cotton > intercrop > castor. The irrigated system also showed higher resilience than rainfed system. A regression equation was developed between soil quality index, resilience index and rice equivalent yield.

Working with Inceptisols (Delhi), Vertisols (Bhopal) and Alfisols (Ranchi), Saha et al. (2015) observed that the soil physical resilience is more acute and problematic in nature as it requires long time duration for bouncing back to its original state. Some index properties like plasticity, swell–shrink potential, compaction, maximum dry density and strength characteristics of soil are very important for estimation of soil physical resilience. There was a significant correlation between soil strength characteristics and clay content of soil. These studies indicated that Alfisols have better resistance and resilience than Vertisols and Inceptisols. Soil amendments such as fly ash, farmyard manure, poultry manure and biochar application had significant favourable effect on soil resilience. Kumar et al. (2014) found that the heat stress decreased soil respiration and dehydrogenase activity by 20–80% in alluvial soils. But the integrated use of NPK at recommended rates and farmyard manure for continuously over 36 years enhanced the resistance and resilience (stability) of soil microbial activity against heat stress. Thus, it can be applied in compacted degraded soil for better resilience and sustainable crop production.

12.5 Conclusions

Soil resource inventory is of paramount importance to study the soil health, quality and resilience. Few attempts have been made to assess the soil resilience using integrated soil quality indices computed from data on soil physical, chemical and biological properties. These studies showed that the integrated use of soil amendments such as wood charcoal and balanced fertilization enhanced soil resilience of Vertisols under soybean–wheat system. However, systematic studies under field conditions at different locations are desired to assess the impact of improvement interventions on soil resilience.

References

Abrol IP, Sehgal JL (1994) Degraded lands and their rehabilition in India. In: Soil resilience and sustainable land use. pp 129–144

Andrews SS, Karlen DL, Cambardell CA (2004) The soil management assessment framework: a quantitative soil quality evaluation method. Soil Sci Soc Am J 68:1945–1962

Andrews SS, Karlen DL, Mitchell JP (2002a) A comparision of soil quality indexing methods for vegetable system in northern California. Agric Ecosyst Environ 90:25–45

Andrews SS, Mitchell JP, Mancinrlli R, Karlen DL, Hartz TK, Horwarth WR, Pettygrove GS, Scow KM, Munk DS (2002b) On farm assessment of soil quality in California's Central Valley. Agron J 94:12–23

Benjamin JG, Mikh M, Nielsen DC, Vigil MF, Calderon F, Henry WB (2007) Cropping intensity effects on physical properties of a no-till silt loam. Soil Sci Soc Am J 71:1160–1165

Bhattacharyya T, Pal DK, Deshpande SB (1993) Genesis and transformation of minerals in the formation of red (Alfisols) and black (Inceptisols and Vertisols) soils on Deccan Basalt in the Western Ghats, India. J Soil Sci 44:159–171

Bhattacharyya T, Pal DK, Srivastava P (1999) Role of zeolites in persistence of high altitude ferruginous Alfisols of the humid tropical Western Ghats, India. Geoderma 90:263–276

Chandran P, Ray SK, Bhattacharyya T, Srivastava P, Krishnan P, Pal DK (2005) Lateritic soils of Kerala, India: their mineralogy, genesis and taxonomy. Aust J Soil Res 43:839–852

Chaudhury J, Mandal UK, Sharma KL, Ghosh H, Mandal B (2005) Assessing soil quality under long-term rice-based cropping system. Commun Soil Sci Plant Anal 36:1141–1161

Dalal RC, Bridge BJ (1996) Aggregation and organic matter storage in subhumid and semi-arid soils. In: Carter MR, Stewart BA (eds) Advances in soil science. CRC Lewis Publishers, Boca Raton, FL, pp 263–307

Gilkes RJ, Prakongkep N (2016) How the unique properties of soil kaolin affect the fertility of tropical soils. Appl Clay Sci 131:100–106

Halvorson AD, Peterson GA, Reule CA (2002) Tillage system and crop rotation effects on dry land crop yields and soil carbon in the Central Great Plains. Agron J 94:1429–1430

Indoria AK, Srinivasa Rao Ch, Sharma KL, Sammi Reddy K (2017) Conservation agriculture—a panacea to improve soil physical health. Curr Sci 112(1):52–61

Karlen DL, Mausbach MJ, Doran JW, Cline RG, Harres RF, Schuman GE (1997) Soil quality: a concept definition, and framework for evaluation. Soil Sci Soc Am J 61:4–10

Kuan HL, Hallett PD, Griffiths BS, Gregory AS, Watts CW, Whitmore AP (2007) The biological and physical stability and resilience of a selection of Scottish soils to stresses. Eur J Soil Sci 58:811–821

Kumar S, Patra AK, Singh D, Purakayastha TJ (2014) Long-term chemical fertilization along with farmyard manure enhances resistance and resilience of soil microbial activity against heat stress. J Agron Crop Sci 200(2):21–23

Kundu S, Vassanda Coumar M, Saha JK, Adhikari T, Rajendran S, Patra AK (2015) Protocol for resilience of degraded Vertisol with field level validation. J Soil Water Conserv 14(4):310–316

Lakaria BL, Singh M, Sammi Reddy K, Biswas A K. Jha P, Chaudhary R S, Singh A B and Subba Rao A (2012) Carbon addition and storage under integrated nutrient management in soybean–wheat cropping sequence in a vertisol of Central India. Natl Acad Sci Lett India (May–June 2012) 35(3):131–137. https://doi.org/10.1007/s40009-012-0040-z

Lal R (1995) Global soil erosion by water and carbon dynamics. In: Lal R, Kimble JM, Livine E, Stewart BA (eds) Soil and global change. CRC Press, Boca Raton, Florida, pp 131–141

Lal R (1997) Degradation and resilience of soils. Philosophical transactions of the royal society of London B 352:997–1010

Lynch J (2002) Resilience of the rhizosphere to anthropogenic disturbance. Biodegradation 13:21–27

Mandal UK, Ramachandran K, Sharma KL, Satyam B, Venkanna K, Udaya Bhanu M, Mandal M, Masane RN, Narsimlu B, Rao KV, Srinivasarao Ch, Korwar GR, Venkateswarlu B (2011) Assessing soil quality in a semiarid tropical watershed using a geographic information system. Soil Sci Soc Am J 75(3):1144–1160

Mandal UK, Sharma KL, Venkanna K, Solomon Raju AJ, Pushpanjali Srinivasa Rao, Ch Rahul N M, Adake RV, Prasad JVNS, Venkatesh G, Venkatravamma K (2017) Sustaining soil quality, resilience and critical carbon level under different cropping systems. Curr Sci 112(9):1882–1894

Manna MC, Ganguly TK (2003) Impact of integrated nutrient management practices and land use on soil organic matter dynamics, soil biological processes and sustainable productivity in semi-arid and tropical India. In: Acharya CL, Sammi Reddy K, Sharma AK (eds) Proceedings of the ICAR-ACIAR workshop on Farmyard manure: life cycle analysis and management, Indian Institute of Soil Science, Bhopal, 18–20 Feb 2003

Muralidharudu Y, Subba Rao A, Sammi Reddy K (eds) (2012) District-Wise soil test based fertilizer and manure recommendations for balanced nutrition of crops. Indian Institute of Soil Science, Bhopal, pp 1–270

Nayak DC, Sen TK, Chamuah GS, Sehgal JL (1996) Nature of soil acidity in some soils of Manipur. J Indian Soc Soil Sci 44:209–214

Pal DK, Sarkar D, Bhattacharyya T, Datta SC, Chandran P, Ray SK (2013) Impact of climate change in soils of semi-arid tropics (SAT). In: Bhattacharyya T et al (eds) Climate change and agriculture. Studium Press, New Delhi, pp 113–121

Pal DK, Wani SP, Sahrawat KL, Srivastava P (2014) Red ferruginous soils of tropical Indian environments: a review of the pedogenic processes and its implications for edaphology. Catena 121:260–278. https://doi.org/10.1016/j.catena2014.05.023; Panda N (1998) Fertiliser News 43(5):39–50

Patiram (2001) Proceedings of national seminar on water and land management for social and economic upliftment of NE region, Guwahati, pp 11–19

Patiram (2003) Soil resilience in North Eastern Hills. Indian J Hill Farm 16 (1, 2): 1–8

Paton TR (1978) The formation of soil material. George Allen and Unwin. LondPratibha G, Sharma KL (2016) Conservation agriculture and organic farming for adaptation and mitigation of climate change. In: Venkateswarlu B, Ravindra Chary G, Gurbachan Singh YS (eds) Shivay climate resilient agronomy , The Indian Society of Agronomy, New Delhi, pp 220–231

Ray SK, Chandran P, Durge SL (2001) Soil taxonomic rationale: kaolinitic and mixed mineralogy classes of highly weathered ferruginous soils. In: Abstract, 66th Annual convention and national seminar on "Developments in Soil Science" of the Indian Society of Soil Science, Udaipur, Rajasthan, pp 243–244

Saha R, Hati KM, Mohanty M, Jha Pramod, Somasundaram J, Chaudhary RS (2015) Characterization of soil physical resilience by index properties and strength characteristics of selected Indian soils. J AgriSearch 2(3):195–199

Sharma KL, Mandal UK, Srinivas K, Vittal KPR, Mandal Biswapati, Grace JK, Ramesh V (2005) Long-term soil management effects on crop yields and soil quality in a dryland Alfisol. Soil Till Res 83:246–259

Sharma KL, Raju KR, Das SK, Rao BR, Prasad C, Kulkarni BS, Srinivas K, Kusuma GJ, Madhavi M, Gajbhiye PN (2009) Soil fertility and quality assessment under tree-, crop-, and pasture-based

land-use systems in a rainfed environment. Commun Soil Sci Plant Anal 40(9), 1436–1461
Sharma KL, Grace JK, Mishra PK, Venkateswarlu B, Nagdeve MB, Gabhane VV, Sankar GM, Korwar GR, Chary GR, Rao C, Srinivasa G, Pravin N, Madhavi M, Mandal UK, Srinivas K, Ramachandran K (2011) Effect of soil and nutrient-management treatments on soil quality indices under cotton-based production system in rainfed semi-arid tropical Vertisol. Commun Soil Sci Plant Anal 42(11):1298–1315
Sharma KL, Balaguruvaiah D, Babu MVS, Reddy BR, Srinivasa Rao Ch, Mishra PK, Grace JK, Ramesh G, Madhavi M, Srinivas K, Mandal UK, Korwar GR, Maruthi SG, Ravindra CG (2010) Long-term impact of soil and nutrient management practices on soil quality in rainfed Alfisols at Anantapur in Andhra Pradesh. Indian J Dryland Agric Res Dev 25(1):74–85
Sharma KL, Grace JK, Mandal UK, Gajbhiye PN, Srinivas K, Korwar GR, Bindu VH, Ramesh V, Ramachandran K, Yadav SK (2008) Evaluation of long-term soil management practices using key indicators and soil quality indices in a semi-arid tropical Alfisol. Aust J Soil Res 2008(46):368–377
Sharma SP, Sharma J, Subehia SK (1998) Proceedings of long term fertility management through integrated plant nutrient supply. IISS, Bhopal, pp 125–138
Shukla MK, Lal R, Ebinger M (2004) Soil quality indicators for reclaimed mine soils in Southeastern Ohio. Soil Sci 169(2):133–141
Smeck NE, Runge ECA, Mackintosh EE (1983) Dynamics and genetic dealing of soil system. In: Wilding LP, Smeck NE, Hall GF (eds) Pedogenesis and soil
Sunanda B (2016) Soil resilience: a new approach to recover degraded soil. Biotech Articles; Views 1333. www.biotecharticles.com/Agriculture-Article/Soil
Syers JK (1997) Managing soils for long-term productivity. Philos Trans Royal Soc London (B) 352:1011–1021
Velayutham M, Pal DK, Bhattacharyya T (2000) Organic carbon stock in soils of India. In: Lal R, Kimble JM, Stewart BA (eds) Global climate change and tropical ecosystems. Lewis Publishers, Boca Raton, Fl, pp 71–96
Venkatesh G, Venkateswarlu B, Gopinath KA, Srinivasarao Ch, Korwar GR, Sanjeeva Reddy B, Prasad JVNS, Grover M, Raju BMK, Sasikala Ch, Venkanna K (2013) Biochar production technology for conservation of cotton stalk bioresidue into biochar and its characterization for soil amendment qualities. Indian J Dryland Agric Res Dev 28(1):48–57
Vittal KPR, Vijayalakshmi K, Rao UMB (1990) The effect of cumulative erosion and rainfall on sorghum, pearl millet and castor bean yields under dry farming conditions in Andhra Pradesh, India. Exp Agric 26:429–439
Yaalon DH (1975) Conceptual models in pedogenesis. Can soil forming functions be solved? Geoderma 14:189–205

Soil and Industry

13

Bipin B. Mishra, Sanjay K. Choudhary, and Richa Roy

Abstract

Soil has strong linkage with industries as well as industrial opportunities. This chapter highlights different aspects that link soil and its mineral as well as organic fractions including parent rocks and minerals with industries. As raw material for earthenware, brick, tile, mudwall, crockery, idol, doll, toy, etc., the type-specific soils are widely used in different parts of India. In constructions (building, highway, road, dam, embankment or other infrastructures), soil as the raw material plays vital role, but seldom caring for environmental impacts, and thus needs policy intervention. Efforts are made to address each soil function to its relevance in specific industrial framework. The modern Agriculture in India is somehow partially organized industry since the introduction of the first green revolution in 1960. Even the soil, being a huge laboratory for microbiological, photochemical and biogeochemical interactions can be translated into industrial terms covering soil functions including nutrient recycling, biodiversity, humification, carbon sequestration, molecular biotechnology, etc. Emphasis is given on certain emerging as well as future scopes that would enable type-specific soils and clays to be medically used for human health even. Even the electric power as the energy source for industries is necessarily connected to electric earthing that is controlled by type, texture, fineness, moisture-holding capacity and depth of soil.

Keywords

Type-specific soils • Industry • Raw material • Constructions • Human health • Environment impacts

B. B. Mishra (✉)
Bihar Agricultural University, Sabour, Bhagalpur, 813210, India
e-mail: bbmsoil@rediffmail.com
URL: http://www.iuss.org

B. B. Mishra
Pedology and Land Use Planning, School of Natural Resources Management and Environmental Sciences, Haramaya University, P O Box 138 Dire Dawa, Ethiopia

B. B. Mishra
International Union of Soil Sciences, Vienna, Austria

S. K. Choudhary
TNB College, TM Bhagalpur University, Bhagalpur, 812007, India

R. Roy
Department of Biotechnology, TNB College, TM Bhagalpur University, Bhagalpur, 812007, India

B. B. Mishra (ed.), *The Soils of India*, World Soils Book Series,
https://doi.org/10.1007/978-3-030-31082-0_13

13.1 Introduction

The functions of a soil are multi-dimensional viz. (i) Foundation base for plant growth (ii) Restoration and enrichment of biodiversity (iii) Recycling of organic residues, nutrient elements and waste materials (iv) Filtration, storage and detoxification of water and nutrients (v) Raw materials either in the form of sand, silt, clay or peat or even whole soil and parent materials (vi) Construction, conservation and land fillings (vii) Buffering features and resilience (viii) Human health and clinical therapy (ix) Ecosystem and climate change mitigation, and (x) Carrier of ancient plant parts for synthesis of fossil fuels from buried organic sources of geologic time (paleosols). The farming systems that significantly contribute to the domestic GDP are the subsistence, organic, commercial or industrial farming. By and large, Indian agriculture is moving towards industrial farming. The humans manipulate the soil and its fractions in order to utilize them for economic gains through certain defined processes of industrial relevance. Today's agriculture follows a systematic planning based on soil evaluation. Organically planned conservation agriculture is an organized agriculture having industrial essence that can restore not only the sustainability, productivity and profitability, but also promote the steady rate of carbon sequestration (Friedrich 2008).

The distinct colours, varying fineness, chemical and physical behaviours, mineralogical suites as well as corresponding parent materials, optical or ceramic values, microbiological potential, organic carbon stock, etc. of the soil are some traits that open industrial relevance of a type-specific soil including parent material. If a man finds something benefitting, he tries to do so without caring for its consequences in the long run. Such habit of a man is the serious cause of disturbances, if it reflects the imbalance in physical, chemical or even biological stability of the soil body. Can we correct such wrongdoings already made in past as well as present on different industrial uses of soils? In fact, soil has immense uses of industrial relevance, but a soil scientist must appreciate only those uses within the framework of sustainability. The present chapter is intended to cover various industrial uses of soils and related soil fractions for economic growth particularly in rural areas of the country.

13.2 Soil as Raw Material in Construction Industry

Soil plays vital role for human shelter, livelihood and desired infrastructures through constructions of houses, buildings, schools, colleges, playgrounds, picnic spots, roads, highways, embankments, dams including land fillings.

13.2.1 Sand

Sand being an integral soil fraction is widely used for a variety of industrial purposes viz. water filtration, foundry, painting, water treatment and plastering. The Bureau of Indian Standards prescribed time to time different grades suitable for industrial uses of sands. High purity silica sand is used for glass industry with 99% SiO_2 content and impurities of iron, aluminium oxides and lime (IS 488 1963). Quartz, quartzite, and silica sand are the various forms of silica. The chemical composition of silica is SiO_2. These forms of silica are used in a number of industries, the important being glass, foundry, sodium and potassium silicate, ferro-silicon, refractory, iron and steel, cement, fertilizer, ceramic, silicon carbide and other abrasives, chemical industry, asbestos products industry, insecticide, electrode, paint industry, rubber, water filtration, sand/lime bricks industry and plastering mortars. India is a land of rivers with sandy riverbeds, surrounded from three sides with Bay of Bengal, Indian and Arabian seas loaded with abundant coastal sands.

Sand is classified as a minor mineral by the Ministry of Mines (Indian Bureau of Mines 2015) along with clay, marble and a few others. This accounts for over 12% of the total mineral production in the country. Silica sand is used for the manufacture of glass. Other sand is used for all grades of construction from buildings to roads. So, as economic activity expands, the demand for sand too rises. Unsustainable sand mining from riverbeds and water bodies can have huge social, environmental, geomorphic and disastrous impacts for rivers and water bodies. The demand for sand is so high that there are illegal sand mining operations by *Baloo Mafia* across the country (Fig. 13.1).

Fig. 13.1 Illegal sand mining in Sone river of Bihar. *Source* (https://sandrp.in)

However, sand, the backbone of construction industry, has become a problematic resource to procure in India. Accepting that sand was essential for the health of Rivers, Prakash Javedkar, Minister, MoEF&CC stated 'Sand is for river, what RBC is for blood'. Introduction of 'sustainable sand mining policy' draft notification by Ministry of Environment Forest and Climate Change (MoEF&CC) was the most significant development pertaining to sand mining in 2015. In March 2015, MoEF&CC team conducted a field inspection of Haridwar and reported of large scale illegal mining of sand and stones going on in Ganga River. The violation of the mining Act is prohibited under 379 (punishment for theft) and the Minerals Act 1957 of India.

13.2.2 Silt

Silt as the soil fraction and particulate matter in running water does settle at the bottom of standing water and often create a problem for de-siltation. In Karnataka, silts are gathered in village tanks and lakes and applied to fields in order to improve soil fertility. There are a report of about 36,000 tanks in 26,000 villages in Karnataka state, which are built with centuries-old expertise, that act as water storage systems for communities. They often trap the silt in run-off, and such silts, rich in plant nutrients, are used as a viable means of increasing soil fertility. Silts collected from tanks and lakes are spread (normally 20–25 tractor loads) on the field evenly before sowing is made. The adhesive properties of silt allow it to mix with soil in the main fields during the first monsoon rains. Usually, silt is applied to the soil once in 3 years in order to improve soil conditions.

In Ganga and Kosi rivers, siltation has been a challenging problem. One of the alternative measures may be to utilize such huge silt loads to high clay *Tal* land soils spreading in a long strip along south of the Ganga river (Tiwary and Mishra 1990) as well as in major Vertisols in other states. Silt as a soil conditioner in such high clay soils needs to be popularized based on some unique features of silt. Silt as the fine soil fraction is slippery and soft, but easy to make it compact. It is also easier to keep nutrients and moisture in place for appreciably long periods of time. Thus, silt may be a good compromise between clay and sand. Use of silt as the soil conditioner needs to be encouraged extensively at large scale.

13.2.3 Clays and Clay Minerals

The term clay is derived from 'Argil' in Latin and is applied to those fine-grained earthy materials whose most prominent properties are plastic when wet, capable of retaining shape when dried and formation of hard rock-like mass without losing the original contour when fired at red heat (Bose 1948). Ghosh (1997) undertook fundamental and basic research on electrometric investigations, nature of clay acid, ion-exchange reactions, clay constituents, clay-organic interactions and research on different industrial uses of clays. Comprehensive reviews on clay mineral research in India by a number of researchers (Mukherjee et al. 1971; Raman and Ghosh 1974; Ghosh 1997) provide valuable information on uses of clay minerals in India. The rheological properties of clays are made uses in its large scale utilization in ceramic industries. Clays find multiple uses such as in construction of roads, dams, irrigation canals, drainage canals, mud houses, in manufacture of bricks, tiles, stoneware pipes. Bentonite linings are used for reservoir and in landfills for waste disposal. Since bentonite has swelling and shrinkage properties, mixtures of illite and bentonite were found to be better than either illite or bentonite alone for use as clay liners (Sivapullaiah and Savitha 1996; Das 2003). Clays, the natural material either as found or after processing to purified forms are tested in different industries such as petroleum, pharmaceutical, pesticide formulation, textile and ceramic industries (Bhat and Sidheswaran 1996). Decontamination and decolorization of wastewater from textile industry are burning problems. The property of adsorption of textile dye by organo-clay complex is used for this purpose. The dye molecules have high affinity for organo-clays. The organo-clays prepared with Rajasthan bentonite have been found to have capacity to adsorb Remazol Blue from aqueous solution (Bhatt and Pandya 1998). Das (2003) in a review work reported that type-specific clays are used for manufacture of soaps, detergents, catalysis and in cosmetic preparation, wherein kaolin and attapulgite are used in pharmaceutical preparation (calamine lotion).

In developed countries, industrial uses of clays have many applications such as coating and filler pigment for paper, filler for paint, rubber and plastics, formulation additives in food, insecticides, cosmetics, pharmaceutics, fertilizers and soil correctors, and also as a major component in ceramics (Manoharan et al. 2012).

(i) Kaolin or china clay, is a natural clay predominantly kaolinite ($Al_2Si_2O_5(OH)_4$), associated with other clay minerals like dickite, halloysite, nacrite and anauxite. China clay is a white powdery mineral with specific gravity of 2.6 and fusion point of 1785 °C. The term 'Kaolin' is derived from the two Chinese words 'Kao-Liang,' meaning 'high ridge,' a local designation for the area where white china clay is found. The kaolins are known for unique resistance to heat, whiteness when fired, and moderate plasticity. It is extracted from felspathic rocks having no iron oxide. It has virtually no or least shrinkage. The major use of kaoline in the country is in cement industry and its processed form in ceramic industry. Besides, processed kaolin is used in other industries like sealants, paper coatings, as extender in fibre glass, paint and as the filler for paper, rubber, plastic, cosmetics, pharmaceuticals and textiles. Crude china clay also finds use in Insecticide and Refractory Industries. Some uses are in ink, ultramarine, synthetic zeolite, catalyst, water filter candles, soaps and detergents and explosives and pyrotechnic industries. In areas where use of kaolin is gaining importance are in the manufacture of plastic film, video and audio tapes where clays are used as anti-blocking agents, and in the field of biotechnology, where ceramics are widely in use for its lightweight and high strength properties. High growth is expected both for hydrous and calcined clay, particularly in paint, cables, plastics, rubber and ceramics. The apparent demand of china clay is estimated at 4.61 million tonne by 2016–17 and that of ball clay at 1.82 million tonne by 2016–17 at 9% growth rate by the Planning Commission of India (Indian Minerals Yearbook 2013).

(ii) According to Indian Minerals Yearbook (2013), ball clay is another clay type that consists of 20–80% kaolinite, 10–25% mica and 6–65% quartz. Ball clays differ from kaolin in having more degree of plasticity. It is sedimentary white burning clay of fine particle size, excellent plasticity and dry strength. It is thus utilized after mixing with non-plastic clay to impart the desired plasticity in pottery, porcelain and refractory materials. Compared to china clay, ball clay is richer in silica and powder alumina. It contains a larger proportion of alkalies, iron oxide and carbonaceous matter. The ball clay is so-called because it is usually mined in the form of balls. The main characteristic of ball clay is that it can be put into use directly after extraction from the mines. It also helps in the preparation of glaze, enamels and for imparting a dense vitrified body. Ball clay is used in making of pottery articles, tablewares, white wall tiles, electrical equipments, iron enamelling clays and fillers for paints. The total resources of ball clay as on 1.4.2010 in the country are placed at 83.39 million tonne. More than 62% resources are in Andhra Pradesh, followed by Rajasthan with 38%. Resources in Gujarat are nominal. Out of the total resources, ceramic/pottery grade constitutes 89%. The production of ball clay at 1856 thousand tonne in 2012–13 increased by 13% as compared to that in the previous year. Rajasthan continued to be the leading state in production accounting for 89% of the total production followed by Andhra Pradesh with 10%. Consumption of ball clay increased from 579,600 tonne in 2011–12 to 585,000 tonne in 2012–13. About 97% consumption was accounted for by the Ceramic Industry. The remaining consumption was reported by the Refractory and Abrasive industries (Indian Minerals Yearbook 2013).

(iii) Shale is a fine-grained, clastic sedimentary rock comprised of mud that is a mixture of flakes of clay minerals and tiny fragments of minerals like quartz and calcite. The ratio of clay to other minerals is variable. Shale which occurs with limestones as parting is rich in alumina content. Total shale resource in India is 15.9 million tonne as on 1.4.2010 and is located in Andhra Pradesh. Its uses are confined to manufacturing of cement (Indian Minerals Yearbook 2013).

(iv) Fireclays that include nearly all clays that have a fusion point above 1600 °C and are not white burning are the backbone of refractories industry. They may be described as an earthy plastic, detrital material with low percentage of iron oxide, lime, magnesia and alkali in order to enable the material to withstand a temperature of at least 1500 °C and preferably over 1600 °C (Rao 1966). Fire clay may be regarded as a variety of impure Kaolin. Fire clay is generally greenish, grey in colour, compact and dense in structure and varies in degree of hardness. Fireclay is generally found beneath the coal seams. It varies widely in composition and properties. It can be roughly divided into three types (Rao 1966).

(v) Brick and tile clays comprise wide varieties of clays of varying composition, the clay mineral being of the kaolinitic or illitic type. It is invariably high in iron content and often contains gross amounts of other impurity, notably calcium compounds. Because of the high impurity content, fluxing additions are not normally necessary and the clay can be fired at a relatively low temperature. Some deposits are high in organic matter, which ignites on firing and reduces the amount of fuel necessary to fire the ware. For a judicious selection of brick and tile clay, more importance is usually given to its physical properties rather than to its chemical composition. The clay employed in the manufacture of bricks and tiles should be sufficiently plastic for being moulded into the required shape. It is generally found that clay containing mixed grains has better plasticity. There are two main types of tile clay deposits, viz. the lacustrine type and the flood plain type deposits. The clay is generally fine, plastic, dull, white or variegated. Certain organic substances like horse dung can increase the plasticity of the clay. Such clay shall retain its shape in both wet and dry states and that it may be capable of being sufficiently vitrified at 950 °C/1100 °C to form hard bricks without excessive shrinking or deformation (Rao 1966). A typical brick and tile clay may have approximate chemical compositions as given (Table 13.1).

Table 13.1 Chemical composition of typical tile clay

Chemical constituent	Amount (%)
Silica (SiO_2)	59.23
Alumina (Al_2O_3)	18.92
Iron oxide (Fe_2O_3)	5.61
Lime (CaO)	1.39
Magnesia (MgO)	1.03
Potash (K_2O)	1.61
Soda (Na_2O)	0.92
Loss on ignition	11.29

Source Rao (1966)

13.2.4 Laterite

The term "laterite" was originally used for highly ferruginous deposits first observed in Malabar Region of coastal Kerala and Dakshin Kannad and other parts of Karnataka. It is a highly weathered material, rich in secondary oxides of iron, aluminium or both. It is either hard or capable of hardening on exposure to moisture and drying. Laterite and bauxite show a tendency to occur together. Aluminous laterites and ferruginous bauxites are quite common. Laterite with Fe_2O_3:Al_2O_3 ratio more than one, and SiO_2:Fe_2O_3 ratio less than 1.33 is termed as ferruginous laterite, while that having Fe_2O_3:Al_2O_3 ratio less than one and SiO_2:Al_2O_3 ratio less than 1.33 is termed as aluminous laterite (Indian Bureau of Mines 2013).

Laterite can be considered as polymetallic ore as it is not only the essential repository for aluminium, but also a source of iron, manganese, nickel and chromium. Furthermore, it is the home for several trace elements like gallium and vanadium which can be extracted as by-products. Laterite occurrences are widespread in the country (Indian Bureau of Mines 2013). Almost all Indian bauxite deposits are associated with laterite, except those in Jammu & Kashmir. Laterite normally occurs on the hills and plateaus of Madhya Pradesh and in some states of the Deccan peninsula. Andhra Pradesh is the leading state in laterite production contributing 77% of the total production, followed by Madhya Pradesh (11%), Karnataka (5%) and Gujarat (3%). The remaining 4% is contributed by Kerala, Maharashtra and Jharkhand.

The compact and ferruginous variety of laterite is used widely as a road metal and as a local stone for culverts and buildings. It cannot withstand heavy pressure and as such it is used for construction of light structures, partition walls, boundary walls, etc. Laterite as a building stone possesses one advantage that it is soft when quarried and can be easily cut and dressed into blocks and bricks which on exposure to air become hard. The industrial use of laterite is in the cement industry. Laterite is capable of removal of phosphorus from solutions and percolating columns of laterite remove cadmium, chromium and lead to very low concentrations. The consumption of laterite in cement has increased due to increased demand for cement in the country. In future, laterite could be used as a source of metallic minerals like iron, aluminium, chromite and of trace elements like gallium and vanadium (Indian Bureau of Mines 2013).

13.2.5 Zeolite

In India, occurrence of zeolites was reported by Kapoor et al. (1981) in sodic soils and Bhattacharya et al. (1999) in ferruginous soils of Western Ghats. Mining zeolites from shrink-swell soils of semi-arid part of Western ghats and addition in humid ferruginous soils to prevent chemical degradation and preserve biodiversity hold great industrial promise. Natural zeolites are used as building stone. Zeolites are also used as NH_3 filters in kidney dialysis units (Das 2003). Steady rate of slow release of nutrients in the environments could be achieved by using zeolites that are a group of naturally occurring minerals having a honeycomb-like layered crystal structure. Zeolites are often seen in the cavity of basalt. In nanoscale, zeolite acts as a reservoir for nutrients that are slowly released 'on demand.' Fertilizer particles can be coated with nanomembranes that facilitate slow and steady release of nutrients. Nano-fertilizer technology is very innovative but scantily reported in the literature (Manjunatha et al. 2016).

13.3 Soil Based Construction and Products

13.3.1 Mudwall

Mud is a mixture of soil and water. It is a natural building material that is found in abundance. In India, mud wall is very advantageous, when building materials such as bricks, stone or wood are scarce due to affordability and or availability. The mud architecture is a great resource that focuses on architecture constructed of mud brick, rammed earth,

Fig. 13.2 Typical mud wall in a village. *Source* http://blog.aaobihar.com/the-story-of-mud-houses-of-bihar/

compressed earth block and other methods of earthen construction. It is a cheap and local material available in most parts of the world. It provides excellent heat insulation, so the inside of a mud building is cooler in summer and hotter in winter than a building made with steel and concrete. It is strong in compression and so makes good walls. It can also be made strong in shear and tension through additives and reinforcement. In such construction, locally available materials are used and good amount of money can be saved. Poor people mostly in rural areas can afford such a type of house building (Fig. 13.2).

13.3.2 Bricks

Clay-based bricks are mostly used in India for building (Fig. 13.3). At present, India has the production capacity to manufacture over 1000 crore bricks through 45,000 local kilns in the unorganized sectors (Manoharan et al. 2012). Brick making is the most common soil-based industry commonly on small-scale. The number of kilns and their bricks are quite large and correct figures are difficult to present as the industries are highly unorganized and dispersed. Bricks are the true indicator of how soil is the shelter for humans (Fig. 13.4). As per some projected estimates, brick production is increasing annually in India by 5–10% due to emergence of a large number of development sectors, buildings, urbanization and infrastructure. India is thus the second largest producer of bricks in the world after China.

13.3.3 Clay Tiles

Bricks and clay roof tiles are made from clay and clayey mixtures after being fired at high temperatures. The power of fire transforms the soil into ceramic products and strengthens the unique properties of bricks and tiles.

The origin of clay roofing tile is traced to Indus valley civilization and Babylonian civilizations (Krishnamurthy et al. 2015). Then, the use of clay tile has spread throughout Asia and Europe. Clay roofing tiles fell out of fashion again for a short time at the end of the nineteenth century, but once more gained acceptance in the twentieth century, in which clay tile roofs featured prominently. With the availability of

Fig. 13.3 Brick making with clay in Begusarai and Bhagalpur, Bihar

machines capable of extruding clay in a variety of forms in large quantities, clay tiles became more readily available across the nation (Figs. 13.5 and 13.6). A typical clay tile called 'Mangalore tiles' has its own place in the clay tile industry. These tiles are native to Mangalore in Karnataka on the western coast of India.

13.3.4 Earthenwares

Until 50 years before, most of the household materials and utensils used in kitchen (*tawa* for frying *chapati*, *haria* for cooking grains/rice), for storing water (*ghara*, *surahi*, *matka*), milk and milk products (*metia*, *berua*), grinding device for grains (*jata-chakki*), tea cup (*kullhar*) and for storing agricultural products (*kothi*) were all made of earthen materials with distinct regional names. Soil made furnace (*chulha and tandoor*) are very common in every part of India. During Hindu's marriage, uses of different forms of soil made structures like *mrida kalash* are mandatory. Even with death in Hindu, soil made structures (*pind*) and candles (*deep*) are often used to pay homage to the departed soul. Earthen pots (*gamla*) for planting materials are extensively used in nursery as well as landscaping and buildings almost everywhere. Every Hindu festival in India is marked with soil made specific structures including idols (*murti* of gods or goddesses), lightening candle (*deep* or *deeya*) and often pitcher (*kalash*). The 'light festival' of India (*deewali*) is a sign of global prosperity through lightening with soil made candles (*deepmala*) in every house in India and some other countries also (Fig. 13.7).

Fig. 13.4 Bricks ready for construction Bihar

Fig. 13.5 A mud house with clay tiles on roof

Fig. 13.6 Common form of tiles on roof in Bhagalpur

Fig. 13.7 Potter's wheel for making earthen pots and *Deep*. *Source* (https://www.shutterstock.com/image-photo/indian)

13.3.5 Soil Stabilization

Indian population, particularly in rural areas, lives, shelters or works in buildings made of earth, including mud, tiles, and bricks. In such construction, we need sand and silt besides clays of specific types, while others are built using soil-dependent plant materials, such as wood, straw, bushes etc. The relevance of soil in industries lies primarily with fact on its uses in making toys, utensils, tiles, bricks and mudwalls, and in such processes, soil stabilization is often of importance. As Kulkarni (1973) has described, there are two types of soil stabilization. In the first type, the soil to be used as foundation of a structure, say for example a building or a dam, has to be stabilized in its in situ condition, without removing it. This aspect of soil stabilization did not progress much in India. However, the second aspect of soil stabilization is the one in which the soil from one place can be removed, stabilizing agent added in it and then it may be utilized at some other place. In fact, a large amount of information is available in India on soil stabilization (Kulkarni 1973; Majumdar 1957). The stabilized soil was used for making or fabricating bricks, for applying a waterproof plaster to wall and often for preparing statues of Gods and Goddesses and also for applying waterproof plaster to these idols (Kulkarni 1973). During Indus civilization, the use of burning bricks and utensils made of soil or earth was made perfect to a very high degree. The utensils prepared in Maharashtra during *Chalcolithic* period were made of soil which was stabilized by adding sand in the naturally available clayey soils. Indians thus in prehistoric time had developed the art of mechanical stabilization of soil by adding soil particles of desired sizes. *Yejurveda Samhita* gives information about preparation and burning the utensil made of soil, which was called '*Ukha*'. This seems to be the first authentic information of soil stabilization. Thus, Sankalia (1970) concluded that the art of soil stabilization was first perfected by Indians from ancient time. The soil could be stabilized by adding fine powder of coarse sand, stone and rock. This method is mentioned in *Satapatha Brahmana*. Addition of ash of burnt coal is another method of stabilizing soil and this method was known to Indians. Any soil to be utilized for fabrication of bricks or for construction of mudwall was stabilized by beating it thoroughly for a long time under the feet of men. The construction of the walls of Harappa civilization was very old Indian tradition (Majumdar 1957). Different methods of soil stabilization followed by Indians from prehistoric and Vedic periods (up to 600 BC) to the period of *Silpa Sastra* (1500 AD) are indicative of the facts that the soils in India have been extensively used for industrial purposes since ancient past.

13.4 Pottery and Ceramic

The pottery in its widest sense includes all objects fashioned from clay and then hardened by fire (Encyclopaedia Britannica 1947). The ceramic product is an article made from clay with or without the addition of other materials. Ceramic is the art and technology of making objects of clay and similar materials treated by firing (Ceramics 1988). It is largely synonymous with pottery and other articles made of burned clay. It includes cement, refractories, glass, whitewares and fired building materials and abrasives (Poornam 1966). Clay products as roofing material began to replace straw and leaf thatching only in a much later era of human civilization. Tile is a thin, flat slab, usually of burnt clay, glazed or unglazed, used either structurally or decoratively in building (Encyclopaedia Britannica 1947). The Educator Encyclopaedia (1962) defines tile as a slab of baked clay used in covering roofs, paving floors decorating walls and carrying of drainage. Tile industry embraces a wide variety of terracotta clay products.

The origin of pottery in India can be traced back to the neolithic age, with coarse handmade pottery—bowls, jars, vessels—in various colors such as red, orange, brown, black and cream. The real beginning of Indian pottery is with the Indus Valley Civilization. There is proof of pottery being constructed in two ways, handmade and wheel-made. *Harrappan* and *Mohanjodaro* cultures heralded the age of wheel-made pottery, characterized by well-burnt black painted red wares.

Pottery and ceramics have been the integral part of human culture in India. In art history, ceramics and ceramic arts cover figures, tiles, and tableware made from type-specific soil, clay or other materials through pottering. Decorative ceramics are often known as 'art pottery', which is widely popular. In general, pottery is considered to be containers made from clay. 'Pot' is a word used for container in any form. Bihar has a rich history of clay pottery work. Since the time of Mauryan and Gupta, this art has been in practice in Bihar. The archeological excavations at places near Nalanda and Rajgir had confirmed the existence of this artistic craft in Bihar. Beautiful earthen utensils and tiles are made by potters throughout the country. They have the abilities and skill to do artistic and beautiful paintings on earthen pots. Kolkata and Patna are very famous for making earthen statutes of various gods and goddesses.

In rural areas, children are fond of making toys, idols, pottery and typical landscape (*dharaondha*) using type-specific soils in their houses. Besides, soil-based toy business is very common across the country since ancient era. There is a community called *kumhar* or *kumbhkar*

Fig. 13.8 Soil made *Deep* for Deewali in Bihar

Fig. 13.9 Soil made kitchen furnace (*Chulha*) in a village

(potter), who is known for this profession using type-specific soils (Figs. 13.8, 13.9, 13.10 and 13.11). However, they are more traditional and need to be updated with modern knowledge in order to compete global markets. *Kumhar* or *Kumbhkar* also deserve association with activities in soil science in India.

India has been engaged in manufacturing as well as export of pottery. However, the share of global market of India in pottery production needs improvement. The pottery production at Khurja units (Uttar Pradesh) is popular covering the major area in Khurja, Chinhar, Chunar, Phulpur, Mathura, Agra, Ghaziabad, Jhansi, Moradabad, Kanpur, Rampur and Aligarh. The products of pottery made in the village are sold around restricted local markets. There is thus need for all round encouragement in such soil-based industries in India.

Fig. 13.10 Soil made pitcher (*Ghara*) store water

Fig. 13.11 *Surahi* to store water cool

13.5 Soil in Art, Painting, Doll and Idol Making

In art history, ceramic art includes art objects such as dolls, idols, figures, tiles, and tableware made from clay and other raw materials by the process of pottery. Certain ceramic products grouped under fine art, while others are either decorative or industrial or applied art, and artifacts in archaeology. Decorative ceramics are often called art pottery. Bihar has a rich history of clay pottery work. Beautiful earthen utensils and tiles are made by potters of Bihar. They have the abilities and skills to do artistic and beautiful paintings on earthen pots, toys and dolls. Patna is very famous for such work. Patna is also famous for making earthen statutes and idols of gods and goddesses. In Gujarat, mud work is a traditional art which is made by mixing sugar lid with mud and made a perfect piece which can be a great decorating ways for walls. This mudwork is made on mudwork frame. Affectionately called Panditji, Bihar-born Brahmadeo Ram Pandit, was conferred with the prestigious 'Padma Shri'—one of the highest civilian award in the country—in recognition of his excellent craftsmanship and outstanding contribution in the field of ceramic pottery. The soils used for making these toys following their painting and other crafts come mainly from river beds, lake and tank sediments and other areas. The famous Indian artist from Kerala, Raja Ravi Varma had been using indigenous paints made from leaves, flowers, tree bark and soil.

13.6 Peat as Fuel and in Horticultural Industries

Peat deposits in India are confined to localized pockets in Kerala, Tripura and Andaman and Nicobar islands (Sehgal 2006). Deposits are also found in the Himalayan regions. These soils are locally called *Kari* in Kottayam and Alleppey districts of Kerala. Such marshy soils are developed as a result of waterlogging and anaerobic conditions, which leads to partial decomposition of organic matter. They occur in the areas of heavy rainfall and high humidity, where vegetative growth is abundant round the year. It often occurs in the northern part of Bihar, the southern part of Uttaranchal and the coastal areas of West Bengal, Orissa and Tamil Nadu besides Kerala. The soils are generally heavy and black in colour characterized by a humus-rich organic matters. Normally they often contain iron and varying amounts of organic matter (10–40%), which may go even up to 40–50%.

Peat has been found to be a key bed in the Late Quaternary sedimentary deposit of Kerala and has essentially been derived from the tropical evergreen forests and freshwater swamps of lowland and midland during early Middle Holocene (Kumaran et al. 2014). Peats are also reported to be in coastal wetlands associated with the lagoons, paleo-delta and creeks (Rajendran et al. 1989; Narayan et al. 2007) as well as in the form of reworked Neogene deposits (Kumaran et al. 2014). Peat forms in wetlands or peat lands, often called bogs, swamp and marsh (Sehgal 2006).

13.7 Nanotechnology and Nano-Clay

Efforts are made to increase the efficiency of applied fertilizer with the help of nano clays and zeolites and restoration of soil fertility by releasing fixed nutrients. Studies indicate that the use of nanofertilizers causes an increase in nutrients to use efficiency, reduces soil toxicity, minimizes the potential negative effects associated with overdosage and reduces the frequency of the application (Manjunatha et al. 2016). In fact, 'Nano' means one-billionth of a particle, thus nanotechnology deals with materials measured in a billionth of a meter. A nanometer is 1/80,000th the diameter of a human hair or approximately ten hydrogen atoms wide (Manjunatha et al. 2016). Adhikari et al. (2010) have opined that nanofertilizers for slow release and efficient use of water and fertilizers by plants are the prime potential application of the technology. Encapsulation of fertilizers within a nanoparticle is one of these new facilities which are done in three ways (i) the nutrient can be encapsulated inside nanoporous materials, (ii) coated with thin polymer film, or (iii) delivered as particle or emulsions of nanoscales dimensions (Rai et al. 2012). Research on the controlled release pattern of nutrients using clay nanoparticle at Tamil Nadu Agricultural University, Coimbatore is encouraging (Chinnamuthu and Boopathi 2009). Fertilizers encapsulated in nanoparticles may increase the uptake of nutrients (Tarafdar et al. 2012). Zeolite acts as a reservoir for nutrients that are slowly released 'on demand.' Fertilizer particles can be coated with nanomembranes that facilitate slow and steady release of nutrients (Chinnamuthu and Boopathi 2009). Besides, nanoclays are nanoparticles of layered mineral silicates with specific properties of chemical resistance, mechanical strength and thermal stability so as to be a preferred material in industries such as automotive, pharmaceuticals, and coatings amongst others. The organo-clays are used in the manufacture of polymer-clay nanocomposites.

13.8 Human Health, Cosmetics and Medical Treatments

A man working in soil, walking on soil and playing with soil is healthier than a man keeping his hands, feet and body far off the soil. People around Ganga river do use clayey soil as tooth powder or paste and often apply mud externally on whole body before bath. Such examples need to be translated in clinical and medical terms after due validation medically.

13.8.1 Evidences in Medical Uses

The human civilizations in past collapsed due to one or more reasons, wherein neglect of soil was common. Why one should worship the ground he walks on? People have been using mud therapies to help with dermatitis, diabetes, and arthritis. The antibiotics such as streptomycin was discovered by soil microbiologists to fight infection. Kaolin and pectin provide the base ingredient for anti-diarrheal drugs and antacids that help with stomach aches. Soil minerals, water, air, and organic matter under given temperature provide a habitat for a diversity of microbes, animals, and the underground parts of most plants and rhizosphere. Some of the commonly used antibiotics and anticancer drugs have their origins in soil microbes. Based on reports of Tony de Morais (2009), Mishra and Roy (2015a, b) stated that clay may be used externally as well as internally in different forms. It is antiseptic to prevent decay or putrefaction, promotes wound healing, relieves and prevents inflammation, cleans cancer cells (anti-carcinogenic), softens and soothes the skin (emollient), refrigerant cools and reduces body temperature (refrigerant), and improves skin texture (cosmetic). When used internally, clay-like bentonite acts as an effective detoxifier, which can absorb heavy metals such as mercury, arsenic, lead, and tin. Besides, it provides minerals and trace elements. Being colloidal in nature, it reduces or even eliminates toxins and harmful ingredients from body. Medicinal values of bentonite or montmorillonite clays are historically experienced particularly on (i) skin to heal eczema, dermatitis and psoriasis (ii) bath as a soaking material to remove toxins, alkalizes the body, boost immunity by keeling harmful bacteria and viruses (iii) teeth to improve strength and boost probiotics, and (iv) for relieving certain digestive problems like constipation, nausea etc. (https://draxe.com/10-bentonite-clay-benefits-uses/).

13.8.2 Detoxification

Today's lifestyle and food habits in India are turned to readymade fast food following the health problems causing ailments and disorders such as blood pressure, blood sugar, constipation, gastric, weight gain, mental strain, etc. Type-specific soils and clays are tested at different levels for recovery from such ailments. However, many of such tests are not validated medically, though evidences, being the witnesses of truth, necessitate systematic investigations of type-specific clays and soils in India and elsewhere (http://www.toothwisdom.net/detox_mobilization.html) for their clinical relevance in curing ailments and disorders. Clay eating followed by detoxifying potential of clays deserves attention for clinical validation. This will open a new industrial avenue with rigorous scientific soil research in collaboration with medical and other sciences. Taking a pinch of well-tested clay or soil may be a benefit to the immune system. Some reports indicate that normal children of one and three years of age often eat soil, while aged children may continue to eat soil if there is delay in their growth. A dose of 500 mg a day of soil or clay consumption is considered normal in children up to 3 years old as reviewed by Mishra and Roy (2015a, b). Clay is antiseptic to prevent decay or putrefaction, promotes wound healing, relieves and prevents inflammation, cleans cancer cells (anti-carcinogenic), softens and soothens the skin (emollient), refrigerant cools and reduces body temperature (refrigerant) besides improving skin texture (cosmetic). When used internally, the clay such as bentonite acts as a detoxifier, which can absorb heavy metals like mercury, arsenic, lead, and tin. Besides, it provides minerals and trace elements. Being colloidal in nature, it reduces or even eliminates toxins and harmful ingredients from body. Metallic ions of silver, copper, and zinc have inhibitory and

Fig. 13.12 *Multani mitti* (Fuller's earth) in situ and prepared for use as face pack and other skin applications. *Source* Internet

bactericidal impact. However, there is need to undertake collaborative research in India with medical teams, since it has high medical and industrial values too.

13.8.3 Multani Mitti (Fuller's Earth)

Multani mitti in India is commonly used for cleaning hair. Also known as fuller's earth, *multani mitti* has been used to get rid of skin problems and achieve radiant and blemish-free skin since ages. It is a powerful means to clean and nourish the skin. It has active elements that effectively absorb oil, dirt, sweat and impurities, leaving the skin clean, soft and supple (Fig. 13.12).

In the eighteenth century, a clay-rich lime bearing soil was identified around the city of Multan, and the inhabitants were surprised by its amazing cleansing properties. And now, the *multani mitti* (Fuller's earth) has become a part of every household for being used for cooling the skin and making it glow. It is rich in minerals like aluminium silicate, and offers high absorbing properties that keep the skin and hairs fresh, glowing and radiant. This traditional skin care ingredient is highly beneficial particularly for oily and acne-prone skin. Its lime content is beneficial to destroy or kill the harmful bacteria and capable of controlling excess oil and dirt in order to keep the skin clean and soft. It is believed that *multani mitti* has quality of a cooling effect to the skin. It is also believed that its application could relieve the inflammation as caused by acne and so. Besides, it also helps to tighten the skin and thus minimizes the wrinkles and linings on skin. Currently, this typical soil has high industrial values and marketed in different trade names in India.

13.8.4 Soil and Clay Eating

As a mark of the World Soil Day on 5th December, 2015, the Department of Soil Science at Bihar Agricultural University, Sabour identified Karu Paswan of more than 100 years of age at a village of Babupur (Bakharpur) in Pirpainti Block of Bhagapur District in Bihar (Fig. 13.13), who has been eating a type specific soil daily for the last 60 years (Mishra and Roy 2015a, b). However, he has normal food diet, but additionally, he eats almost 200 g of soil daily. He has two daughters and two sons. At this old age, he has black hairs and walks on foot for 10–12 km daily. Evidence makes the truth, but intervention of medical

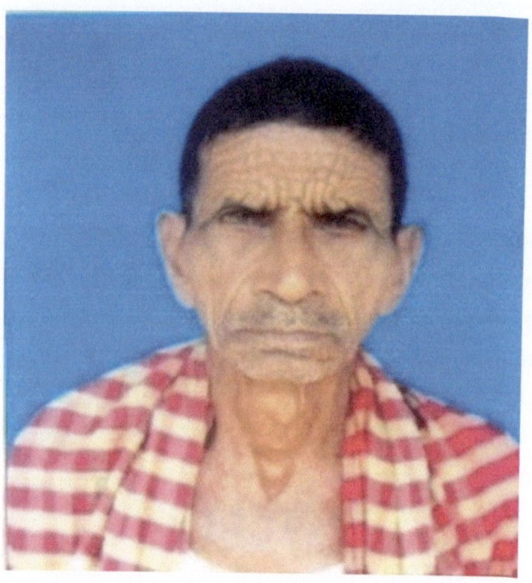

Fig. 13.13 Karu Paswan, Bhagalpur (Mishra and Roy 2015a)

verification may help to authenticate the type specific clay or soil eating. Soil of specific type and nature may be for protective medical treatment and its industrial values need to be established in days to come.

A review work of Mishra and Roy (2015a, b) indicates that clay tablets were used widely across the Mediterranean as well as European territories in certain religious cause besides curing the poisoning and the plague. However, the clay tablet was used by Roman Catholic Church and was listed in pharmacopeia as late as 1848. The use of eating clay has been studied in America, Sweden, Africa, Indonesia and Australia. In India and many other countries, however, knowledge of soil and clay eating is scanty. The rates of pregnant women eating soil or clay in African countries range approximately from 28% in Tanzania to 65% in Kenya, where clay is selectively identified and sold in markets. They collect it from termite mounds being rich in minerals and eat at an average of 30 g daily.

13.9 Emerging Issues and Opportunities

13.9.1 Soil Based Industry in Agriculture

Conservation agriculture is a concept for resource-saving crop production that strives to achieve acceptable profits with high sustained production while conserving the environment. This is centred on enhancing natural biological processes above and below the ground. It consists of three simple principles—*disturb the soil as little as possible*, *keep the soil covered round the year*, and *mix and rotate crops* (IIRR and ACT 2005). Farmers have successfully put the elements of healthy soils (surface residues/crop covers to increase organic matter inputs, minimal soil erosion, in situ soil moisture storage, conjunctive use of organic and inorganic nutrient inputs and use of deep-rooted crops—nitrogen fixers, minimal disturbance of soil through No till) into practice by developing crop-production systems which satisfy three important conditions favourable to biotic activity in the soil, viz. (a) permanent cover of the soil with organic matter provided by a mulch of retained residues from the previous crop or cover crops; (b) minimal soil disturbance by tillage, and preferably no-tillage once the soil has been brought to good condition; (c) rotation of crops, (to include N-fixing legumes) which contribute to maintaining biodiversity above and below the soil and to avoid build-up of pest-populations (FAO 2008). Organically managed conservation agriculture has immense industrial relevance.

The concept of supply chain process in land uses or produces is neither fully understood among rural farmers nor accepted in most parts of India, though Food Industries are running around urban areas in different cities (Mishra 2017). The supply chain starts from soil or land and decides the fate of distribution and consumption of the produce in the markets. Farmers, in general, are not aware of the merit of supply chain process for their produces. The corporate sectors may take a lead to implement a planning system to judge the existing land use scenario and make management for decisions based upon various situations including storage of raw materials (land uses), processing and finished goods from point-of-origin (soil resource) to point-of-consumption (markets) in order to ensure the best competitive price to the farmers (Mishra 2017). Such planning approach through supply chain process in agriculture will ensure 'poverty alleviation' and make the soil science as "employment provider' in days to come. As stated by Hartemink (2006), the future of soil science is brighter than ever and this seems to be more relevant to Indian soils too.

13.9.2 Soil Evaluation and Soil Testing Laboratory

The Government of India has launched many notable schemes including National Mission on Soil Health Cards. With around 120 million farm holdings in India, soil analyzing capability for about 40 million soil samples annually is expected to cover analysis for almost each landholding once in three years. Obviously, there is a need for a massive expansion of the soil testing programme across the country. There is already a centrally sponsored scheme 'National Project on Management of Soil Health and Fertility' launched in 2008–09 in this mission. Other identical programmes launched by States and Agricultural Universities are also functioning. In fact, the soil testing prior to planning for growing crops, based on soil and land evaluation, has become a strong choice of the most of farmers. Apart from these soil testing laboratories of the governments or agricultural universities, other organizations are being flourishing in this sector and gaining recognition for providing soil test services and ensuring the quality and productivity of the soil. If the soil testing programme runs as an industry through private or public sectors, it would be more acceptable and reliable because of market competition. If soil and land evaluation are reliably followed, the production from the given soil could be enhanced considerably and thus, land evaluation and soil testing would develop purposefully on industrial framework, wherein corporate sector would expand.

13.9.3 Soil and Electricity

Indian Physicist Dr. Sunil Kumar, working as a Professor at Dongguk University in South Korea, has developed a 24-h electricity-generating hybrid green energy module from soil.

Table 13.2 Electrode pair comparison

Electrode pairs	Voltage (V)	Current (mA)	Power (mW)
Copper–steel	0.46	0.35	0.161
Aluminium–cast iron	0.6	0.1	0.06
Copper–aluminium	0.5	0.4	0.2
Steel–zinc	0.9	0.6	0.54
Copper–zinc	0.9	1.07	0.965
Carbon–zinc	1.4	0.54	0.756

Source Maniyar et al. (2013)

Table 13.3 Soil properties

Soil type	Voltage (V)	Current (mA)
Dry black soil	0.75	0.01
Moist black soil	0.80	3.30
Red soil	0.78	2.88
Red soil with salt	0.785	8.60
Soil with Tulsi plant	0.787	12.50
Moist black soil with salt	0.80	30.00

Source Maniyar et al. (2013)

This system generates electricity from microbes in the soil by supplying them food regularly as well as using the electro-chemical approach (Tribune News Service Chandigarh 2017). This will bring a quantum jump in soil-based industries for production of electricity in India and elsewhere.

The world today is facing a serious problem of energy crisis. We are looking for new reliable clean renewable energy sources. Normally, micro-organisms utilize soil and other organic materials in the soil environment to produce free ions by their metabolic activity which are left uncollected (Maniyar et al. 2013). These free ions are collected using metal electrodes with soil acting as an electrolyte or medium for transfer of ions. Each electrode produces a certain potential voltage and few tens of million amperes of current. Thus a pair of positive and negative electrodes inserted in soil kept in an insulating container works as a 'soil cell'. The voltage can be increased by connecting multiple electrodes in series and current by connecting soil cells in parallel. Series and parallel combination of soil cells becomes a 'soil battery'. Maniyar et al. (2013) at Electronics and Telecommunication Engineering, Fr. C. Rodrigues Institute of Technology, Navi Mumbai established that the energy from the soil can be used as an uninterruptable power source to light a small cabinet along with a solar panel. The energy from Soil Battery can be used for lighting purpose when the solar panel is not functional or when its efficiency is low. The maximum power point tracking controller ensures optimum power output from both the soil battery and the solar panel, whatever be the battery voltage or the driving load. Thus a small grid comprising of soil battery, solar panel and a secondary battery can be used as an uninterruptable power supply with the optimum source in terms of power driving the load. They tested different combinations of electrodes (Table 13.2) and also the soil properties (Table 13.3). However, such emerging scopes of soils need more appreciation and encouragement on way to generate electricity to be utilized in future.

Further investigation made by Harish et al. (2016) indicates that the combinations of magnesium anode and coke cathode, zinc anode and graphite cathode, aluminium anode and carbon cathode, and zinc anode and copper cathode gave 2.05, 1.40, 1.10 and 0.9 V per cell, respectively. The power of a single Zn–Cu cell was measured to be few tens of microamperes. Small-scale power electronic devices such as calculators, electronic watches, baby toys and cell phones or even the white light LEDs were operated at site. Importantly, the voltage level was found to increase linearly by connecting multiple earth battery cells in series like commercial lead acid battery. The load current was found to increase by connecting earth cells in parallel. The source current capacities were also found to increase by increasing surface areas of the electrodes. However, single-cell voltage was found to remain constant irrespective of the electrode sizes. Thus, a pair of positive and negative electrodes inserted in soil and kept in an insulating container is a "soil cell". The voltage can be increased by connecting multiple electrodes in series and current by connecting soil cells in parallel. Series and parallel combinations of soil cells become a "soil battery". Thus, a small grid comprising of soil battery and a secondary battery can be used as a power supply (Harish et al. 2016). It is sure that in near future, the improved soil-based battery would be an aspiring energy source for both domestic as well as industrial purposes. Even the electric supplies as the energy source for industries and other purposes are necessarily connected with earth (electric earthing) that is controlled by type, texture, fineness, moisture-holding capacity and depth of soil. The success stories of soil battery and soil generator in days to come would significantly contribute to clean energy resource and will open a new avenue for employment.

13.9.4 Fossil Fuels, Radioactive Minerals and Gold

Fossil fuels (coal, crude oil, natural gas) are the major energy source for running the most of industries. The soils formed long periods ago at the earth's surface got subsequently buried (paleosols) with woody land plants like tree trunks, branches, leaves, twigs and roots turning into peat and coals (Source: Penn State University, USA, GEOSC 010—Geology of the National Parks). In fact, the organic matter so buried under sediments is slowly transformed into peat. If the peat is buried under more sediment, it becomes coal. As described in Wikipedia (2019), fossil fuel refers to buried combustible geologic deposits of organic origin from decayed plants and animals and converted to coal, crude oil, natural gas or heavy oils by exposure to heat and pressure in the earth's crust over hundreds of millions of years (Botkin and Keller 2003). Anaerobic digestion occurs naturally in some soils and in lake and oceanic basin sediments, where it is usually referred to as anaerobic activity (Koyama 1963). This is the source of marsh gas methane as discovered by Alessandro Volta in 1776 (Zehnder Alexander 1978; MacGregor and Keeney 1973). Even today in catenary sequence, soils in depressions or in topographic lows, indicate higher amount of carbon stock compared to topographic highs. Jharkhand has huge reserves of type-specific coals, while Northeast regions are known for crude oil and natural gas. However, contribution of soils (Paleosols) to the formation of fossil fuel deserves attention, although the available data are insufficient. Partially decomposed green plants, when buried under sediments for a long time, slowly transform into coal. There are many coal types, e.g. lignite, hard coal, sub-bituminous coal, etc. Obviously, the petroleum in NE region and coal fields in Jharkhand have linkage with paleosols. Coal mining, for example, could be looked into negatively so far as sustainability of soil productivity is concerned. In view of the fact that mining of coal is essential for energy source, let us confine to the visible impacts of mine spoil on the landscape and subsequently for their purposeful measure. Tripathy et al. (2016) systematically studied the impacts of mine spoil (excavated rock debris) being disposed in heaps, which destroyed the original soil habitat. However, Tripathy et al. (2016) observed that the average calculated annual carbon budget was 8.40 T/ha/year, of which 2.14 T/ha was allocated to above-ground biomass, 0.31 T/ha to belowground biomass, 2.88 T/ha to litter mass and 1.35 T/ha was sequestered into the soil. This work has shown that the development of biomass following the restoration of mine spoil was significant and that considerable quantities of carbon were stored in above and below-ground plant matter, and in the soil itself. It is of interest that appropriate restoration strategies can be used to establish a viable, healthy and sustainable ecosystem that imbibes carbon into former mine-impacted land and soil.

Monazite, a thorium bearing mineral in sands and soils, is appreciably found in Chilka lake of Odissa as well as in Kerala. Reports on monazite are also available in alluvial soils of Siwan district in Bihar (Mall and Mishra 1994). Besides, parts of Kerala and Tamil Nadu are high background radiation areas, because of the presence of large quantities of monazite in the soil. Thorium content in monazite ranges from 8 to 10.5% (Mall and Mishra 1994). These natural radioactive minerals have industrial values, though more emphasis is drawn on their roles in polluting food, water and environments (Narayana et al. 2001; Senthilkumar et al. 2012).

In some rivers in India, sands contain fine particles of gold as evident in Suvernarekha river flowing along Ranchi and in Sonu river in Jamui district of Bihar. Subarnarekha river flows through Jharkhand, West Bengal and Odisha. As per tradition, gold was mined near the origin of the river at a village named Piska around Ranchi. In Bihar, rivers like Balui, Kapan and Sonha around Ramnagar in Paschim Champaran often contain gold in sand particles and local people try to get gold particles through traditional way of filtration (http://www.magnifificentbihar.com).

13.9.5 Soil and Carbon Trading

Soils are well recognized for their ability to sequester carbon for mitigating climate change (Pathak et al. 2011, 2014; Srinivasarao et al. 2012). New policies are required to encourage acceptance and application of new technology related to soil carbon benefits or carbon trading. Friedrich (2008) stated that processes for carbon trading must be simple, transparent, consistent, comparable, complete, verifiable and efficient. Agreement for carbon gains under clearly defined management and climatic conditions must be sought and means of verification be established. Protocols defining field practices as well as verification and monitoring for carbon markets need to be developed in a harmonized and standardized way under local to specific agroecological conditions. The price for carbon should also reflect both the on-site and off-site value and societal gain in order to provide a real incentive for farmers (Friedrich 2008). India is also joining hands in managing soil carbon to mitigate climate change with the world. The warming in India in the course of past 100 years (1901–2007) was seen to be 0.51 °C with accelerated warming of 0.21 °C/10 years since 1970 (Krishna 2009). The mitigation of CO_2 emission from agriculture could also be achieved by increasing C sequestration in soil (Pathak et al. 2011; Srinivasarao et al. 2012). For C sequestration, the vital issues for considerations are (i) the storage of C should be for a longer term (30 years or more), (ii) all the GHGs (methane, N_2O) and all forms of C (organic and inorganic C) should be included for quantifying C sequestration, (iii) whole soil profile, rather than surface soil needs to be considered for assessing C stock, (iv) assessment should be at regional scale rather than at site or field scales, and (v) all the components of the agro-ecosystems and their life-cycle should be assessed. This is an interesting fact that during 1970–2013, the GHG emissions from Indian agriculture have increased by two-fold. The increasing utilization of fertilizers and other agri-inputs and the ascent in population of domesticated animals are the real drivers for this increasing GHGs emission. The relative contribution of Indian agriculture to the total GHGs emission has diminished from 33% in 1970 to 18% in 2013 (Pathak et al. 2014).

13.10 Conclusions

Important questions with the themes of the 21st World Congress of Soil Sciences were (i) how to feed a hungry planet? (ii) how to fuel an energy-hungry planet? (iii) how to quench a thirsty planet?, and (iv) how to clean up our polluted planet? But, another most emerging question is how to alleviate poverty and that too with soil resource itself. So, agriculture as a whole must be looked into as an industry wherein land use planning framework is linked with supply chain system and soil would be managed following the principles of conservation agriculture. Soil may be tested for generating electricity, though such efforts are at preliminary stage to develop soil lamp or even battery. More importantly, for a brighter soil science in future, industrial uses of type-specific soils may be encouraged and modernized but in a sustainable framework. Soil may thus be looked into as a unique resource for a number of industrial uses, but farming and associated communities need proper trainings, education and techniques to promote the framework of soil-based industries.

References

Adhikari T, Biswas AK, Kundu S (2010) Nano-fertilizer—a new dimension in agriculture. Indian J Fertil 6(8):22–24

Bhat AN, Sidheswaran P (1996) Industrial applications of clays and clay minerals. Clay Res 15A:2, 6–34

Bhatt J and Pandya VP (1998). Adsorption studies of textile dye by organo ~ clay complexes. Clay Res 17:64–71

Bhattacharya T, Pal DK, Srivastava P (1999) Role of zeolites in persistence of high altitude ferruginous Alfisols of the humid tropical Western Ghats, India. Geoderma 90:263–276

Bose HN (1948) Modern pottery manufacture. Bright and Singh Publishers, Bombay

Botkin DB, Keller EA (2003) Environmental science: earth as a living planet. Wiley, Hoboken, NJ, USA

Ceramics (1988) The random house college dictionary, revised edn

Chinnamuthu CR, Boopathi PM (2009) Nanotechnology and agroecosystem. Madras Agric J 96:17–31

Das DK (2003) Clay and clay minerals for soil health and human welfare. J Agric Phys 3(1–2):1–15

de Morais Tony (2009) The Medical Properties of Clay. Winter 2006 Magazine, October 16, 2009

Encyclopaedia (1947) Encyclopaedia Britannica, op. cit.

Encyclopaedia (1962) The American educator encyclopaedia, 1962 edn

FAO (2008) An international technical workshop, investing in sustainable crop intensification: the case for improving soil health. Integrated crop management, vol 6. FAO, Rome, p 149

Friedrich T (2008) Managing soil carbon to mitigate climate change: a sound investment in ecosystem services. A framework for action. FAO, Food and Agriculture Organization of the United Nations Conservation Technology Information Center, Dec 2008

Ghosh SK (1997) Clay research in India. In: 15th Prof. J.N. Mukheljee—ISSS foundation lecture. J Indian Soc Soil Sci 45:637–658

Ghosh SK, Bhattachmya T (1983) Clay minerals—their distribution and genesis in Indian soils. Advances in soil science. Books and Periodicals Agency, New Delhi, pp 216–277

Harish D, Dhanalakshmi TS, Gireesh HR (2016) Power supply using earth battery, vol 4, issue 2. ITSI Transactions on Electrical and Electronics Engineering (ITSI-TEEE). ISSN (PRINT): 2320-8945

Hartemink AE (ed) (2006) The future of soil science. International Union of Soil Sciences, Wageningen, The Netherlands

http://www.magnifificentbihar.com

https://draxe.com/10-bentonite-clay-benefits-uses/

http://www.toothwisdom.net/detox_mobilization.html

IIRR and ACT (2005) Conservation agriculture: a manual for farmers and extension workers in Africa. International Institute of Rural Reconstruction, Nairobi; African Conservation Tillage Network, Harare. ISBN 9966-9705-9-2

Indian Bureau of Mines (2013) Laterite. Indian minerals yearbook 2013 (part- III: mineral reviews), 52nd edn. Ministry of Mines, Government of India, Indira Bhavan, Civil Lines, Nagpur

Indian Bureau of Mines (2015) Indian minerals year book, part III. Govt. of India Ministry of Mines, Indian Bureau of Mines, Nagpur

Indian Minerals Yearbook (2013) Part- III: mineral reviews: kaolin, ball clay and other clays and shale (advance release), 52nd edn. Government of India, Ministry of Mines, Indian Bureau of Mines, Nagpur

IS 488 (1963) Bureau of Indian Standards http://www.indiamart.com/horizonenterprises-kutch

Kapoor BS, Singh HB, Goswami SC (1981) Analcime in a sodic profile. J Indian Soc Soil Sci 26:513–515

Koyama T (1963) Gaseous metabolism in lake sediments and paddy soils and the production of atmospheric methane and hydrogen. J Geophys Res 68(13):3971–3973. Bibcode: 1963JGR.68.3971K. https://doi.org/10.1029/jz068i013p03971

Krishna K (2009) Impact of climate change on India's monsoon climate and development of high resolution climate change scenarios for India. Presented at MoEF, New Delhi on 14 Oct 2009 (http://www.moef.nic.in)

Krishnamurthy K, Maheshwari C, Nithyavathy N, Mohanasundaram P, Deepasundar P, Thisanth G (2015) Design and development of roof tile production mechanism using advanced embedded processor control. Int J Innov Res Sci Eng Technol 4(4):104–110

Kulkarni RP (1973) Soil stabilization by early Indian methods. Maharastra Engineering Research Institute, Nasik, India

Kumaran KPN, Padmalal D, Nair KM, Limaye RB, Guleria JS, Srivastava R (2014) Vegetation response and landscape dynamics of Indian summer monsoon variations during holocene: an eco-geomorphological appraisal of tropical evergreen forest subfossil logs. PLoS ONE. https://doi.org/10.1371/journal

MacGregor AN, Keeney DR (1973) Methane formation by lake sediments during in vitro incubations. J Am Water Res Assoc 9 (6):1153–1158. Bibcode: 1973JAWRA ... 9.1153M. https://doi.org/10.1111/j.1752-1688.1973.tb05854.x

Majumdar RC (1957) The vedic age. George Allem & Unwin Limited, London

Mall J, Mishra BB (1994) Occurrence of monazite in some alluvial soils of North Bihar. Curr Sci 66:158–161

Maniyar M, Patil N, Saurabh Jadhav S, Dhokle S (2013) Uninterruptable power supply using earth battery and solar panel. Int J Emerg Technol Comput Appl Sci 3(2):107–112. ISSN (Print): 2279-0047, ISSN (Online): 2279-0055

Manjunatha SB, Biradar DP, Aladakatti YR (2016) Nanotechnology and its applications in agriculture: a review. J Farm Sci 29(1):1–13

Manoharan C, Sutharsan P, Dhanapandian S, Venkatachalapathy R (2012) Characteristics of some clay materials from Tamilnadu, India, and their possible ceramic uses. Cerâmica 58(347). http://dx.doi.org/10.1590/S0366-69132012000300021

Mishra BB (2017) Land economics vs land use planning. Agric Res Technol Open Access J 10(4):555791. https://doi.org/10.19080/artoaj.2017.10.555791

Mishra BB, Roy R (2015a) Soil science vs medical science: breakthrough in science for medicine. Bus Agric, 38–41

Mishra BB, Roy R (2015b) Soil science vs science for medicine. EC Agric 2(5):454–461

Mukherjee SK, Das SC, Raman KV (1971) Soil mineralogy. In: Review of soil research in India. Indian Society of Soil Science, pp 169–194

Narayan AC (2007) Peat deposits of west coast of India: implications for environmental and climate changes during late quaternary. J Coast Res 50:683–687

Narayana Y, Somashekarappa HM, Karunakara N, Avadhani DN, Mahesh HM, Siddappa K (2001) Natural radioactivity in the soil samples of coastal Karnataka of South India. Health Phys 80(1):24–33

Pathak H, Byjesh K, Chakrabarti B, Aggarwal PK (2011) Potential and cost of carbon sequestration in Indian agriculture: estimates from long-term field experiments. Field Crops Res 120:102–111

Pathak H, Bhatia A, Jain N (2014) Greenhouse gas emission from Indian agriculture: trends, mitigation and policy needs. Indian Agricultural Research Institute, New Delhi, xvi + 39 p

Poornam R (1966) Survey of ceramic industries in India. Souvenir, all India seminar on ceramic industries, Trichur, p 71

Rai V, Acharya S, Dey N (2012) Implications of nanobiosensors in agriculture. J Biomater Nanobiotechnol 3:315–324

Rajendran CP, Rajagopalan G, Narayanaswamy (1989) Quaternary geology of Kerala: evidence from radiocarbon dates. J Geol Soc India 33:218–222

Raman KV, Ghosh SK (1974) Identification and quantification of minerals in clays. Bull Indian Soc Soil Sci 9:117–140

Rao V (1966) Clays and their uses. Small Industries Service Institute, Trichur, p 2

Sankalia HD (1970) Some aspects of prehistorie technology in India, Indian National Science Academy, New Delhi

Sehgal JL (2006) Introductory pedology, revised edn. Kalyani Publisher, Ludhiana

Senthilkumar B, Manikandan S, Musthafa MS (2012) Natural radioactivity and dose rates for soil samples around Tiruchirapalli, South India using γ-ray spectrometry. Radiat Prot Environ (Serial Online) 35:43–51. Available from http://www.rpe.org.in/text.asp?2012/35/1/43/111409

Sivapullaiah PV, Savitha S (1996) Clay mixtures as liners for waste-disposal facilities. Clay Res 15A:35–43

Srinivasarao C, Venkateswarlu B, Lal R, Singh AK, Vittal KPR, Sumanta Kundu, Singh SR, Singh SP (2012) Long-term effects of soil fertility management on carbon sequestration in a rice-lentil cropping system of the Indo-Gangetic plains. Soil Sci Soc Am J 76(1):168–178

Tarafdar JC, Xiang Y, Wang WN, Dong Q, Biswas P (2012) Standardization of size, shape and concentration of nanoparticle for plant application. Appl Biol Res 14:138–144

Tiwary JR, Mishra BB (1990) Distribution of micronutrients in *tal* land soils (*Udic Chromustert*) of Bihar. J Indian Soc Soil Sci 38:319–321

Tribune News Service Chandigarh (2017) https://www.tribuneindia.com/news/chandigarh/soil-can-power-yourde

Tripathy N, Singh RS, Hills CD (2016) Soil carbon development in rejuvenated Indian coal mine spoil. Ecol Eng 90:482–490. https://doi.org/10.1016/j.ecoleng.2016.01.019

Wikipedia (2019) Fossil fuel, the free encyclopedia

Zehnder Alexander JB (1978) Ecology of methane formation. In: Mitchell R (ed) Water pollution microbiology, vol 2. Wiley, New York, pp 349–376. ISBN 978-0-471-01902-2

Soil and Future

14

Mariappan Velayutham and Bipin B. Mishra

Abstract

The future of India is the future of its soils since civilization itself rests upon the soil. Soil depends on diversity in multiple geological and geomorphological settings, relief features, landforms, climatic realms and natural vegetation types. As such, soil science education should address the emerging national as well as global issues in order to cover basic research related to pedogenesis, geomorphology, micromorphology, biodiversity, climate change, hydrological and ecological processes, mineral and chemical transformations, organic matter dynamics, photopedogenesis, water quality, biogeochemistry and elemental cycling including many other processes controlling ecosystem services as well as resilience. Pedology needs to be made comprehensive by linking its vital roles in soil functions and services. India must have its own system of soil classification and its linkage with land use options as well as land-use planning. Hyperspectral remote sensing integrated with satellite-based technologies has tremendous opportunities to generate reliable soil database. This chapter aims at looking Indian soils forward with the changes happening in differing magnitudes and domains, as ignoring such changes would be imprudent. Prediction of the future based on an extrapolation from the past may give a reliable road map following the governance to transform the soil resource on sustainable basis. Healthy soils can only give healthy food, fodder and ecosystem and make healthy people and healthy nation besides alleviating poverty and promoting the huge employment opportunities. India needs a Soil Health Revolution to keep the Indian soils "Ever Green".

Keywords

Soil and soil science • Present and future • Challenges • Opportunities • Governance • Land evaluation • Poverty alleviation • Employment provider

14.1 Introduction

Soil has its past, its present and its future, but there are well-defined steps that lead to a bright soil future. Human by nature is accustomed to encompass anticipation of the future, while future alone is what will happen in the time after the present even in the case of soil resource. Mother Earth (*Bhoomidevi*)–Mother Land–Mother Tongue–Mother—all comes in important succession in our value systems. Varma and Goswami (1982) have documented the ancient Indian Vedic views and praised towards the Mother Earth. Photosynthetic reaction in green plants and ion-exchange reaction in the soil solution are the nature's two fundamental reactions that sustain life on the earth. Any rational food-production system must harmonize the well-being of the soil–water–air–rhizosphere–biosphere including soil biodiversity in order to ensure full dependence of human beings on soil. Raychaudhuri (1984) critically documented that the ancient farming communities (3250 BC–1200 AD) were quite conscious about the nature of soil and its relation to the production of specific crops. They were intelligent and trained enough for the choice of a particular soil and crop

M. Velayutham (✉)
ICAR-National Bureau of Soil Survey and Land Use Planning, Amravati Road, University P.O., Nagpur, 440033, India
e-mail: velayutham42@yahoo.co.in

Indian Council of Agricultural Research, New Delhi, India

MS Swaminathan Research Foundation (MSSRF), Chennai, India

B. B. Mishra
Bihar Agricultural University, Sabour, Bhagalpur, 813210, India
e-mail: bbmsoil@rediffmail.com

Pedology and Land Use Planning, School of Natural Resources Management and Environmental Sciences, Haramaya University, P O Box 138 Dire Dawa, Ethiopia

International Union of Soil Sciences, Vienna, Austria

B. B. Mishra (ed.), *The Soils of India*, World Soils Book Series,
https://doi.org/10.1007/978-3-030-31082-0_14

and were quite conversant with the principles of the crop husbandry, suitable for that time. The exact chemical composition of different kinds of soils might not have been known to the ancient Indian farmers, but they made extensive experimental observations and obtained a masterly knowledge regarding the suitability of soils for the cultivation of different crops. The soil classification in those days was based on two grounds viz. medicinal and economic. The medical authorities like *Caraka* and *Susruta* had in view the efficiency of vegetable drugs, which depends on the nature of soils in which they grow and, on the other hand, ancient farmers could have experiential techniques to guess the productivity of different types of soils (Gangopadhyay 1932; Raychaudhuri 1984). But the scenario has been changed gradually with the state of Indian soils particularly during first-green revolution since 1960 onward.

India is on the verge of a looming soil crisis that can potentially impact its agriculture in the near future. One of the challenging issues lies with soil degradation, which is estimated to be 147 million hectares (M ha) of land, including 94 M ha from water erosion, 16 M ha from acidification, 14 M ha from flooding, 9 M ha from wind erosion, 6 M ha from salinity and 7 M ha from a combination of factors (Bhattacharyya et al. 2013). The current situation warrants immediate attention and urgent remedial measures. Many ICAR institutes viz. Indian Institute of Soil and Water Conservation Research, Dehradun and the Indian Institute of Soil Science (IISS), Bhopal are working on some critical aspects of problematic soils (Somasekhar 2017). It seems to be extremely serious because India supports 18% of the world's human population and 15% of the world's livestock population, but has only 2.4% of the world's land area. Despite its low-proportional land area, India ranks second worldwide in farm output. Agriculture, forestry and fisheries account for 17% of the gross domestic product and employ about 50% of the total workforce of the country (Bhattacharyya et al. 2015). In other estimates (IISS 2011), the human population increased from 683 million in 1981 to 1210 million in 2010 and projected to reach 1412 million in 2025 and to 1475 million in 2030. Now, to feed the projected population of around 1.48 billion by 2030, there is need to produce 350 million tonnes of food grains. The requirement of food demands of future could be met basically through intensive agriculture without little or no expansion in the arable land. The per capita arable land decreased from 0.34 ha in 1950–51 to 0.15 ha in 2000–01 and is expected to shrink to 0.08 ha in 2025 and to 0.07 ha in 2030 as projected in Vision 2030 of the Indian Institute of Soil Science (IISS 2011). In fact, the present is built upon the past while the future is predicted on the present and thus the future is rooted through the past being complimented solely by the present. Apparently, the future of soil and soil science is full of challenges, where we have to explore opportunities with available soil resources.

Soil being the lowest boundary of the earth's atmosphere varies f3.7 ha pe over the landscape and offers opportunities as to "what" can best be done efficiently, "how" could ensure the achievable goals and "where" with lowest risks or uncertainties could perform for bright future of soils. One of the most fundamental challenges locally and globally is how to convince the society that knowledge of soil science is seriously underutilized on way to overall sustainable development, poverty alleviation, employment generation, economic enrichment, ecosystem restoration including mitigation of climate change. The challenge for scientists is to assert the soil science as an independent discipline with interactive knowledge about natural bodies on the landscape.

Inayatullah (2008) described six pillars of future viz. mapping, anticipating, timing, deepening, creating alternatives and transforming. For a sound framework of future for soil and soil science in India, we have to examine each pillar in light of "present" of the soil, for example, (i) Mapping the present and the future could be done using methods and tools, (ii) Anticipating the future through evaluation and analysis of different emerging issues, (iii) Timing the future by establishing the pattern of major changes, (iv) Deepening the future through anticipating the emerging threats to soil resource, (v) Creating alternatives to present through evaluation of soil capability and subsequent improvement measures for production, and (vi) Transforming the present and creating the future through visioning the threats to soil in performing functions and restoring the ecosystem.

14.2 Soil Science Education and Research

Soil science education at undergraduate and graduate levels should address the emerging national as well as global issues in order to cover basic research related to pedogenesis, geomorphology, micromorphology, biodiversity, climate change, hydrological and ecological processes, mineral and chemical transformations, organic matter dynamics, water quality, biogeochemistry and elemental cycling including many other processes controlling ecosystem services as well as resilience. In soil science, there is a wide range of distinct scales the soil encompasses viz. spatial and temporal; spatial scales from the molecular to the landscape, while temporal scales from instantaneous processes to soil formation processes. However, the following questions are relevant to opt for soil science education;

(i) How could we enlist threats and limitations that restrict production potential of a type-specific soil?
(ii) Are available soil maps and databases sufficient to describe the diversity of soils and of their functions and services?
(iii) Can we deliver periodic, temporal and quantitative information on soil evolution from time zero of soil formation following the specific processes?
(iv) Can the type-specific soil be tested for different functions including human health hazard medical treatments and related health issues?
(v) How to develop awareness and creativity for different uses of type-specific soils in industries?

Soil is not merely a producer of crops, but more importantly as a natural body. Evidently, the man-made environments have caused imbalance of natural ecosystem functions considerably, wherein the soil has been deeply exploited for its production functions covering agriculture, horticulture, forestry and livestock. However, many other soil functions covering issues like (i) soil as an entity in itself (ii) soil as a part of ecosystem (iii) soil and human health (iv) soil and society (v) soil and space (vi) soil as a science and in art and (vii) soil in relation to other sciences and industries. Soils do possess attributes common to vital systems including (a) resistance, an ability to restore the existing conditions (b) residence time, capacity to store and release compounds (c) productivity, capability for plant growth and yield (d) resilience, quality to recover from disturbance; and (e) sustainability, a dynamic equilibrium of physical, chemical, mechanical and biological interactions within the soil environment for long enough time. Besides, some burning issues like frequent flood and drought, contamination of urban soils, global warming and availability of groundwater resources are of current and future concerns. All above issues need to be addressed in framing a smart education system in soil science in future and the ICAR Institutes and Agricultural Universities may review the inputs for inclusion in course curricula.

The Indian Council of Agricultural Research introduced a package in 2006 on Experiential Learning Programme aimed at equipping undergraduates with entrepreneurial skills. Such study may be undertaken to evaluate the performance of comprehensive Experiential Learning Programme curriculum in soil science too. Besides, interdisciplinary educational programs require broad training rooted in classical or traditional soil science including pedology complemented by geospatial modelling skills. The web-based distribution of 2D soil maps and data play a significant role to disseminate information, while scientific visualization and reconstruction techniques to create 3D and 4D soil-landscape models may communicate knowledge on soils. There is thus need of a revised concept of soil that will broaden the old conceptual barriers and open new horizons. The dissemination of science in soil from school to higher education levels has to be improved especially on interactions of soil with light, electricity, magnetism and even sound.

India has immense opportunities to expand the land productivity in order to meet the food requirements of growing population. The relevance of soil-less farming (hydroponics) is limited, since the cost of cultivation in hydroponics is very high as it requires high investment and technical management and no one would pay such high amount for the same thing which they can get cheaper in soil. Importantly, soil performs enormous known and hidden functions essential to meet the demands for livelihood. The future of soil and its science must address all such issues through education and research. Therefore, let us cultivate the future with a career in soil science, since civilization itself rests upon the soil and it deals with the very essence of our existence on the earth.

14.3 The State of Indian Soils

The earliest investigation on soils of India dates back to Leather (1898) who categorized the soils into four major groups viz. Indo-Gangetic alluvium, Black Cotton soil, Red soil and Laterite soil which are now in common knowledge with the Indian farmers and also easily recognizable by the people. Moreover, the immense variability and complexity of soil behaviour are also well perceived by practicing farmers who recognize the differential response of the land to soil management and production inputs among different soil types. The diverse rock formations and parent materials, climatic regimes, natural vegetations and anthropogenic actions, as soil-forming factors, have given to various kinds of soils, with field to field variations in soil properties even within a given landscape and village. Hence, the need of soil survey for soil resource inventory, classification and mapping at appropriate scales is inevitable. Remote sensing application with GIS tools has contributed significantly.

During the period 1986–1996, the National Bureau of Soil Survey and Land-Use Planning completed a Nationwide soil survey at 10 km grid interval and brought out the soil map of different states in 1:250,000 scale and soil map of India at 1:1 million scale. Based on this survey, Bhattacharyya et al. (2013) have classified the distribution of soils into 24 traditional major groups which according to the USDA Soil Taxonomy fall into seven orders, 22 Suborders, 78 great groups, 220 subgroups and 1247 soil families. The importance of exact soil grouping using soil taxonomy to decode the behaviour of climate and its change is worthwhile to appreciate the fact that soils have incredible memory and store the past episodes carefully as Paleosols (Pal et al. 2000a; Bhattacharyya 2014). However, there is need to correlate these nomenclatures with international classification system of the International Union of Soil Sciences (World reference base for soil resources). The proximate composition of soils and their physical, chemical and microbiological attributes have been studied and well documented by Murthy (1982), Lal et al. (1994), Pal et al. (2000b), Valayutham et al. (2003), Bhattacharyya et al. (2013), Bhattacharyya and Pal (2015) and Mandal (2011). However, efforts are being made to develop a national soil classification scheme in India (Mishra 2015, 2016).

14.3.1 Soil Governance

Swaminathan (2005–2006, 2010, 2012, 2017a, b), in the reports of the National Commission on Farmers entitled "Serving farmers and saving agriculture" have outlined the sustainable development of land and water resources for Evergreen Revolution in the farm sector and the pathways for attaining sustainable food security to all and National Policy for Farmers, which has led to the promulgation of the Food Security Act 2013 by the Parliament of India. India stands as the first and only country to have such an act in place. Soil governance refers to the strategies, policies and different decision-making processes by the state and central governments. It reflects a logical framework that discards any negative impact of agricultural land use to promote soil erosion or degradation, loss of soil water, soil contamination, soil pollution, soil sickness due to excess use of agrochemicals, loss of soil biodiversity, land shrinkage due to urbanization and/or non-farm activities, partial factor productivity and imbalanced fertilization. By and large, it is often encapsulated within the context of land governance. Besides, the soil governance often relates to atmospheric and anthropogenic issues, which as a consequence widens its scope of logistics. Obviously, the soil governance must work with research and development wing so that reliable database could be generated for appropriate measures. Besides, soil governance must be intended to promoting sustainable agriculture and ensuring food security as well as food safety. The Food and Agriculture Organization (FAO) has instituted the Global Soil Partnership to promote the governance of the global soil resources in order to ensure healthy and productive soils for a food-secure world besides supporting other need-based ecosystem services.

In addressing the soil-related issues, a policy statement must take note of the opportunities viz. (i) protection and conservation of soil on landscape as natural resources, untapped potential of soil resource, (ii) delineation of benchmark soils, (iii) strengthening of land ownership, (iv) alternative options against misuses or non-farming activities or shrinkage of productive land area, (v) technology revolution especially in the area of soil and land evaluation following the land-use planning, (vi) supply chain process to link point of produce (soil) to point of consumption (market) under the sole control of landowners and farmers in order to open tremendous employment opportunities and assuring poverty alleviation issues as measures in India, and (vii) revolution in informatics and communication and the opportunity of linking soil resource database with farmers, extension workers, policy makers, corporate sectors and scientists. The above issues could be presented in the form of a document and implemented as a procedure or protocol under policy that refers to the process of making decision preferably by the government or private sectors or the institutions involved for sustainable uses of soil resources across the country.

14.3.2 Functions and Services of Soils

There is often a lack of complete perspective on soils due to the fact that until mid-twentieth century, the soil resource was virtually not scarce, and farmers were neither looking for intensification in farming systems nor even thinking much for different uses of soils particularly for different industrial purposes. But with the changing scenario under pressing demand for land, multi-functionality for the soil has become appealing and accepting. Accordingly, soils are being put to many uses and it is generally exploited according to economic and social necessity depending upon its characteristics, properties, fertility and the socio-economic services. Soil function is somewhat vague concepts of "soil quality" and "soil health", in which soil evaluation makes sense in relation to specific soil functions (Baveye et al. 2016). The term "function" is not synonym to "service", which is merely an anthropocentric connotation, i.e. "Soil to serve the needs of humans". In the soil functions framework, no matter how large of a geographical area is considered, the soil is viewed as a complete system in itself, with a high degree of spatial heterogeneity, open to, and interacting in complex ways with, its surroundings (Baveye et al. 2016). The National Bureau of Soil Survey and Land-Use Planning accomplished a commendable work on agro-ecological delineation for promoting the sustainable agriculture (Velayutham et al. 1999a) and further to establish the agro-ecological subregions for planning and development (Velayutham et al. 1999b).

Can we save the India's ecosystems and millions of species, some of which may produce the foods and medicines of tomorrow in soils? Soil biodiversity loss in type-specific soils is of vital concern to decide the measures for its restoration. A soil microbiologist, Selman Waksman won a Nobel Prize in Physiology and Medicine in 1952 for his discovery of the antibiotic streptomycin, which has been a boon to medical treatment. More than 90% of the planet's genetic biodiversity is resident in soils but less than 1% of the microorganisms have been cultured and studied (Rao 2006). Soil evaluation in a landscape becomes the prerequisite for a quantitative framework within the boundary of soil governance so that land-use planning options could be grounded purposefully. Let us broaden the roles of the soils in the environment and there may be at least eleven broad soil functions as follow:

- Food, fibre, forest and biomass production
- Shelter for fauna, flora, microorganisms and biodiversity
- Algal–fungal–lichen interactions for biofuel energy source
- Physical and cultural environment (for living beings and mankind)
- Raw material for different construction works, pottery and ceramic
- Filteration, detoxification and transformation
- Gene pool
- Storage, adsorption, fixation and chelation
- Electricity and magnetism
- Protective human health care
- Interactions with light and electromagnetic radiation

Importantly, the differential roles of the clay fraction deserve quantification in different soil types following the corresponding response behaviours as documented by Velayutham and Bhattacharyya (2000), Bhattacharyya and Pal (2015), Pal (2003), Wilson (1999) and Sarma (1984). The occurrence of zeolites in some soils has been recorded as natural saviour for the soils to maintain its quality (Bhattacharyya et al. 2000). Velayutham and Pal (2016) have brought out the salient soil resilience features of the soil resources. However, soil resilience needs quantification in type-specific soils including benchmark soils in near future. Indian soils are by and large low to very low in organic matter and need careful efforts through scientific means for its enrichment steadily. Conservation agriculture with traditional experience in selecting the cover crops between two main crops in rotation would be a future task for success to build up organic matter in soils. Jha et al. (2010) observed that biochar, due to its unique aromatic structure and long mean residence time (MRT) in soil (>100 years), has the potential for long-term carbon sequestration in soil.

Mishra (2017) attempted to suggest the relationship between land economics and land-use planning in the current global context and emphasized the importance of such dynamic relationship for optimal land uses in the present scenario of World Trade Organization framework. This necessitates integration of supply chain system with land-use planning in order to ensure a profitable production process from the point of origin (soil) to the point of consumption (market). The future of soil science must be intended to harmonize such economically viable processes in order to alleviate poverty as well as provide employment to the youth. Such viable process needs to be applied in effective way to conservation agriculture as well as organic farming systems in future.

The future of soil and its science, by and large, rests on awareness, attitude, creativity, competitiveness and spirit of enthusiasm among students, researchers and soil scientists. This requires a set of conditions promoted in favour of soil science viz. (i) Critical assessment and review of course curricula covering different soil functions and services in different fields of accomplishment in order to get out of traditional soil science that must be demand driven, employment oriented and friendly with the emerging issues in smart soil science and research, (ii) Strengthening the governance in soil science covering all emerging issues on soil functions and services including production function under policies with full priority, (iii) Job or employment opportunities in soil science must be of high rating, since technological intervention has become highly sophisticated in soil science research, since conversion either with "soil conservation" or "soil reclamation" or "soil management" includes cumulative outcome of certain technological interventions—as traditional and scientific (iv) Support of institutional and regional soil science societies that can bring soil scientists together for enhancing knowledge and innovation and (v) Promoting donor support in the field of soil science research either through national or international donor agencies or through investment on way to promote industrial, environmental, medical or other uses.

14.3.3 Pedology to Establish Major Soil Functions

Pedologists have major concerns over how to pinpoint approaches for evaluating, characterizing, monitoring and predicting the changes in the state and development of soils. Additionally, what tools are required to make suitable predictions on soil and landscape conditions as well as sustainable land uses are vital. However, main focus in pedology lies on creation of taxonomic grouping for classifying a soil. But if one looks into the future, some questions are obvious viz. (i) what should we do to enlist all possible threats to soils within the scientific community and within society as a whole? (ii) How could we develop reliable strategies for evaluating both spatial and temporal soil changes in pedologic and edaphologic terms? (iii) How could we learn integrative science at landscape and watershed levels? and (iv) What can we expect to establish the visibility of pedology in particular and soil science at large? Soil science primarily focuses on pedogenetic processes and subsequently on soil functions and services in a dynamic soil–water–air–plant–aquifer–atmosphere system, which could be monitored through soil/land evaluation and soil classification. Soil scientists in the survey area study different relevant soil features in order to identify as well as quantify the existing limitations for subsequent improvement to suit the classified land use.

In fact, pedology provides a foundation for evaluating how existing land uses undergo interactions with the landscape. Hence, pedological investigation is prerequisite to edaphological interventions (Mishra 2015). Pedology is neither dead nor buried, but a critical necessity to discover the most relevant soil functions. There is need to sharpen the tools and techniques used to map, monitor and model the integrated processes in the pedosphere and to enrich the awareness about contributions of soil and its science to the society. Photopedology and hydropedology deserve more attention. The future of soil science in this sense seems to be both exciting as well as challenging. Soil plays roles from geosciences to agricultural, hydrological, ecological, medical, industrial, environmental, atmospheric and engineering sciences. The soil is by and large the basis of our living planet, and of everything we do on it, though it lies beneath our feet. Pal (2018) presented certain relevant facts on pedology and edaphology in India. The competeness of soil science in soil rests on pedology and its future.

14.3.4 Towards Land-Use Planning

Velayutham (1997) had reviewed the progress and achievement made by the National Soil Survey Programme since the 1950s. The way forward as outlined by Velayutham (2012, 2016a, b) for the National Action Plan on land resource inventory for integrated land-use planning of the Agricultural lands (142 M ha) deserves attention.

1. To undertake in a mission mode, the detailed soil survey at 1:12,000 scale at village level, of all revenue villages, with soil map superimposed on the revenue map (Cadastral map) as demonstrated through pilot project by Natarajan et al. (2002).
2. To delineate in all states the "Prime Agricultural lands", "Efficient Crop Zones" and "Special Agricultural Zones" using the "Benchmark soils concept" as representative examples.
3. To refine irrigation scheduling and water release in irrigation command areas based on the knowledge on soil–water holding capacity relationship and crop-water requirements during the crop period in the area. Promote community water harvesting of potential runoff water in stable farm ponds in rainfed agricultural lands to promote the raising of nurseries and to provide life-saving irrigation to get more per drop of water from rainfed crops.
4. To generate crop-weather forecasting and monitoring in rainfed areas based on soil-climatic zones (Krishnan 1988), choice of crops and varieties based on interpretation of soil map for soil available water capacity (AWC) in a given climatic zone (Murthy et al. 1978) and Normalized Differences Vegetation Index (NDVI) may form the core approach for climate-smart agriculture dovetailed with crop insurance scheme.
5. To work on "Land potential evaluation" based on climate, land characteristics, soil and crop productivity potential at benchmark sites (on-station and on-farm) in various "Agro-Ecological" and "Land utilization Regions" and further quantify periodically the "Land use change intensity Index" (LUCII) of the regions. Liu et al. (2014) have also worked and contributed to LUCII in China.
6. To promote theme and need-based soil survey and land-use planning and assessment research in the State Agricultural Universities and State Department of Agriculture/Agricultural Engineering through the Regional Centres of the ICAR-NBSS and LUP and disseminate the SWOT analysis of the local soils to the farmers' through web-based ICT at *Krishi Vigyan Kendras* (KVKs), Village Knowledge Centres and *Gram Sabhas*.
7. To dovetail land and soil-specific crop management recommendations and practices into the improved package of practices and agro-technology transfer modules of the research, training and extension agencies.
8. To provide the policy options and interventions by community and public and private sector agencies for judicious and sustainable management of land and water resources based on the collective expertise, experience and local socio-economic conditions.
9. To monitor at periodic intervals, in partnership with related R&D stakeholders, the extent of degraded lands, polluted lands, wastelands and fallow lands and the progress in their remediation, reclamation and restoration.
10. To generate data on the above lines and make their access to GIS framework at National Informatics District Centres, Village Knowledge Centres, Village

Resource Centres, e-Common Community Service Centres, *Krishi Vigyan Kendras* and *Gram Sabhas* to facilitate a web-based dynamic interaction among farmers, land managers and knowledge and technology providers. This will help in drawing up "Perspective Land use planning" at macro- and meso-levels and pragmatic "Farm level planning" at micro-level. Further, this will enable to take full advantage of the "Information Age" and the present "Digital India Paradigm" through e-land-use planning of natural resources—climate, land, water, biodiversity and human capital. The future strategy of land-use planning must be reliably adoptable and acceptable at farm level.

Mandal et al. (2001) have proposed a methodology for determination of crop-specific land quality index (LQI) for sorghum in semi-arid tropics of India. The method developed as LQI is a function of climatic quality index (CQI) and soil quality index (SQI). Naidu et al. (2017) have detailed, in the book "Land evaluation for sustainable agriculture", the role of soil survey and its translatory use for land-use planning and development at both macro- and micro-levels. Moreover, such approaches under LUP must be tested with ground truth approaches. The future for land-use planning will shape the future of soil and soil science and will be the key factor to develop a roadmap of Indian agriculture.

The Government of India is committed to double the farmer's income by 2022, wherein contribution of soils through scientific soil evaluation followed by effective land-use planning would alone be proved as dependable foundation base and will open a giant employment platform to the youth. Once the land-use planning is linked to supply chain process through corporate sector by connecting the soil to market, it will be the boon in the process of alleviating the rural poverty (Mishra 2017).

14.4 Advances in Soil Research

Soil science research in India has been advanced recently using sophisticated tools and technologies. Various uses of radioisotopes, X-ray diffractometer, electron microscope and applications of remote sensing, GIS and many others have particularly revolutionized the soil research momentum in different facets. Such technological advancement would lead to a strong foundation for future soil research in India as outlined below.

14.4.1 Geo-informatics in Soil Science

With the availability of geo-informatics, sensors and nanotechnology for the first time in history, the technologies for global data collection at multiple scales are now available. Combining these technologies with digital databases along with their incorporation into geospatial models should afford many opportunities for better understanding of soil ecosystems and associated problems in the areas of land degradation; crop cover, forestry and assessment of organic carbon and nutrient status of soils, etc. Soil information systems (SIS) research works within geographic information systems (GIS), which encompasses spatial data collection, spatial statistics, spatial modelling, data display, as well as management and ethical issues, including data integrity and usage (IISS 2011, 2015).

14.4.2 Exploring Soil Biodiversity and Genomics

Microbial identification usually requires isolation from soil through growth on standard laboratory media; however, lesser than 1% of the diversity can be recovered via DNA or RNA sequences (Rao and Patra 2009). There is a need to characterize the vast amount of biodiversity of fauna and flora present in Indian soils (IISS 2011, 2015). Exploring new bio-inoculants/bio-fertilizers, liquid biofertilizers and micro-consortium of organisms for sustaining soil health is the need of the hour.

14.4.3 Use of Biochar for Carbon Sequestration

The application of biochar to soil is proposed as an innovative approach to start a significant and long-term sink for atmospheric carbon dioxide in terrestrial ecosystems (Lehman et al. 2006). Biochar, due to its unique aromatic structure and long mean residence time (MRT) in soil (>100 years), has the potential for long-term carbon sequestration in soil (Jha et al. 2010).

14.4.4 Application of Nanotechnology, NIR and MIR Spectroscopy

Preliminary investigations at IISS clearly indicated that low-grade rock phosphates can easily be made as a source of P to the plant when they are converted to nano-size (<100 nm). Utilization of nanoparticles of indigenous rock phosphate, glauconite/waste mica, pyrite, calcite/dolomite and sphalerite/smithsonite as potential source of P, K, S, Fe, Ca, Mg and Zn to the plant may solve much of our plant nutrient problems (Patra et al. 2014).

Visible–NIR spectroscopy is a non-destructive analytical technique that can be used for measurement of many soil properties simultaneously. These properties can be utilized for development of instruments or sensors for rapid measurement of soil properties on the go and can be utilized in precision farming. Satellite imagery and spectro-radiometers can help immensely in achieving the desired objectives. Besides, mid-infrared (MIR) spectroscopy has the potential to provide a good alternative that may be used to enhance or replace conventional methods of soil analysis (IISS 2011, 2015).

14.4.5 Promoting Conservation Agriculture

It is estimated that over the past few years, adoption of no-tillage has been gaining momentum in India and spread about <5 M ha area. The potential of C sequestration in C depleted soils of India is high with adoption of conservation agriculture. It is also estimated that most parts of the country will receive higher rainfall in 2020, 2050 and 2080 than the current value, so this changing scenario can be converted to suitable opportunities in conserving and sequestering C in Indian soils bringing with it assured benefits of improved soil structure, improved soil physical, chemical and biological as well as reduced soil erosion/degradation. It is estimated that over the past few years, adoption of no-tillage has been gaining momentum in India and spread about <5 M ha (IISS 2011, 2015). Conservation agriculture is truly a soil-based intervention with complete prescription towards "Ideal sustainability", but still needs experimental as well as experiential database to write its success story for future.

14.5 Challenges and Opportunities for Future

International Union of Soil Sciences (IUSS) in 2006 articulated the future of soil science in the new millennium. The science journal published a special issue in 2004 (Vol. 304, Issue 5677) on "Soils—The Final Frontier" suggesting that the ground beneath our feet is still as alien as a distant planet (Science 2004). The present "Decade" of soils (2015–2024) is opportune for making a "soil health care revolution" to sustain "Evergreen Revolution" (Velayutham 2017).

While highlighting the future of soil science, Samra (2006) stressed on topmost priority in developing appropriate human resources to compete in the market-driven R&D portfolio. At the same time, the nanotechnologies are going to obliterate rigid boundaries between physical, chemical and biological sciences. Silicon chip of the fastest computers processors may be replaced by the biochips with billion time's higher speed of conducting messages and telecommunication services matching the speed of human imagination and soil scientists may watch their partnership possibilities. Soil is being looked upon as unexplored great gene pool especially after the successful deployment of Bt genes in various crops and commodities. So far, there was a very subdued interest and investments on the explorations of soil microbes, their characterization, DNA fingerprinting, quantifying structural/functional genomics and documentation. This new phase of interests, of course, will be destined by the rapidly growing intellectual property and geographical indicator rights. Most of the energy production and consumption are environmentally degradable. High urbanization is likely to produce large quantities of solid wastes, domestic and industrial effluents which will be recycled in peri-urban agriculture. Chemical compositions of these bio and industrial wastes are changing very rapidly due to the production and consumption of a large variety of pharmaceuticals, soaps, sanitation products, industrial processes, etc. (Samra 2006).

Importantly, the urban expansion frequently seals soil, and in most cases removes it from any potential future use. It is essential that soil scientists are involved in sustaining the use of soils within these urbanization processes. Restoration of the productivity of such degraded soils is not only an ecological necessity but a socio-economic imperative to ensure the livelihood security of the inhabitants (NBSS and LUP 1994). Soil scientists play major role to address these complex nature of environmental, biodiversity and land use challenges being faced by mankind now and in the future. Amongst the soil degradative processes, water erosion, globally affecting more than half the area, has been recognized as the greatest threat to landscapes and ecosystems (Minhas 2006).

India's energy demand is increasing day by day. Alternatives to fossil fuel must take effect in near future, wherein biofuels, H_2 produced from biomass, soil battery and even soil generator may be proved as the alternative energy sources. Carbon sequestration in soil ecosystems may be another option of off-setting fossil fuels. Generating technologies for growing biofuel plantations on degraded and marginal soils are a win-win strategy in India. Soil scientists should have a commitment to achieving energy security and we have to work for bright future of soil science. Biofuel production from jatropha plantation on wastelands is considered a means of addressing concerns about climate change and improving energy security while at the same time providing an additional source of income for the land users. The assessment was made for a wasteland located in the Velchal watershed, Andhra Pradesh, which recently was converted to a biofuel plantation with Jatropha. A win-win situation between improved land productivity and soil carbon content was observed for the Jatropha plantations (Garg et al. 2011).

Harish et al. (2016) in India demonstrated soil battery and soil cell, but more efforts in future may substantiate the global mission of generating the clean energy using soil resource. Recently, soil has been used in China also to generate electrical power in microbial fuel cells (MFCs) and exhibited several potential applications. Jiang et al. (2016) studied the effect of soil properties on the generated electricity and the diversity of soil source exoelectrogenic bacteria and the results showed that soils with higher organic carbon (OC) content but lower soil pH generated higher peak voltage and charge. On the one hand, soil is capable of sequestering carbon to mitigate the challenges of climate change; while on the other hand, there may be possibility to generate electric power with continued efforts.

Soil, as a part of natural heritage, needs to be preserved in its diversity. The full-length information on soil biodiversity, soil micromorphology, soil biogeochemistry and soil biotechnology is scanty. Besides, the linkage between human health and soil or clay has established another promising field of opportunities, when we can trust in soil-based medical treatment or pedomedicine or protective medical therapy.

Besides, the in situ assessment of soil properties remains a formidable task despite decades of investigation and basic research in soil science. However, over the past few decades, remote sensing application provides some solution for rapid soil assessment (Das et al. 2015). Such approaches are reliable, fast, temporal, non-destructive and synoptic with large spatial coverage. Moreover, remote sensing data are often used for soil classification, soil resources inventory and mapping (Ray et al. 2002, 2004), soil moisture assessment (Engman and Chauhan 1995) and soil degradation (salinity) mapping. Particularly, hyperspectral remote sensing (HRS) is emerging as a promising tool for its capability to measure the reflectance of earth surface features including soil resource at hundreds of contiguous and narrow wavelength bands (Das et al. 2015).

In India, limited efforts have been made for spectral library generation. The National Bureau of Soil Survey and Land-Use Planning (NBSS and LUP), Nagpur, developed a spectral database of 128 surface soils collected from different physiographic/climatic regions of India (NBSS and LUP 2005). Spectral reflectance properties (350–1800 nm) of some dominant soils occurring at different altitudinal zones in Uttaranchal Himalayas have also been examined (Saxena et al. 2003). Besides, DRONE (Dynamic Remotely Operated Navigation Equipment) could be considered as an option for research and development. It is suitable under Indian condition since the land holdings are comparatively small (<1 ha). The Indian satellite with hyperspectral payload, IMS-1, launched in April 2008 has a Hyperspectral Imager (HySI). The major objective is to develop prototypes for drone-based soil health monitoring system using hyperspectral remote sensing (HRS) sensors besides other useful applications (Das et al. 2015). However, joint mission of ISRO–NASA has made an execution plan on 04 February 2016 to capture the surficial mineral deposits in some part of Rajasthan (SAC 2016).

Soil science must address environmental issues effectively. There is need to broaden the scope of soil not only as a medium for plant growth but equally for other soil functions viz. construction, pottery, brick–tile making, nanotechnology, human health, etc. covering industries. Soil science must be a complete discipline with direct linkage to hydrology, climatology, geology, ecology, biology, chemistry, physics, economics social, and political science in order to understand the soil's ecosystem services besides decision-making process either in land-use planning or supply chain system from point of origin (soil or land) to the point of consumption (market).

The processes, driving interactions of the pedosphere with biosphere, need to be critically assessed for sustaining the productivity and promoting the biodiversity, improving the air quality and mitigating the greenhouse effect, disposing the waste materials and sequestrating the CO_2 in soil mass besides improving the quality of water resources. Importantly, overexploitation of soil is causing the second-generation problems of nutritional disorders, heavy metal toxicity, global warming, changing trend of climate, contamination and pollution. The speedy growth in population, urbanization and industrialization followed by shrinking of productive land areas as well as intensification in land use option has accelerated the yield of bio-waste materials viz. farm garbage, crop residues, animal waste, municipal liquid and solid waste including polluted river water. There is thus need of recycling of organics in order to minimizing the environmental pollution as well as improving the soil carbon stock, soil productivity, soil health, and crop productivity (Manna et al. 2018).

The tree is known by its fruits. The discipline and its institutions are known by their impact and contribution to the well-being of the society. Soil science is the mother of all-natural sciences and in future, it needs broad integration with trans-disciplinary earth science, social science and humanities in the network society. As the famous saying goes, "We have not inherited the land from our forefathers, but have borrowed it from our grandchildren. Hence it is important that we hand over to them the quality of the land resource undiminished and improved". Present and future soil scientists and soil science societies are the custodians of this Natural Resource Trust and Treasure of the society, for whom without exception, soil is the bottom line, that needs handling with care and reverence, for Mother-Earth never betrays. Our agricultural future and in turn our own future lie on the attitude, care and quality that we give to our land, water and soil resources. As Franklin D. Roosevelt rightly said, "The Nation that destroys its soils destroys itself" is a reminder forever. This quotation is not merely one of quotable quotes, it is existential. As Hartemink (2006) remarked, "the future of soil science is brighter than ever" and this is more relevant to Indian soils, soil scientists, soil science societies and all who are responsible for restoration of soil health and sustainability.

The fundamental challenges include type-specific degradation and pollution, environmental and climatic imbalance, shrinkage of land areas and intensification in agriculture, urbanization and industrial uses of soil. The immediate question is how to convince society that knowledge of soil science has to be broadened and utilized in order to promote national and local economy. Thus, soil is the engine to the future agricultural prosperity of India.

Nowadays, there is often a question, "Will soil play a significant role in food production even after 100 years as it does today?" It is experienced that soil, in general, is not yet sustainable, although soil care has improved. Another challenge in future may be to move away from our reliance on good soil and to allow large tracts of the most vulnerable soils to repair. It is presumed that healing the damaged or wounded soils may be an important step towards sustainability. Besides, soil scientists would also work in a very small land area for food production following the basic principles already advancing under vertical farming. This will be the true future of soil and soil science, since soil holds not only our past and present, but our future too. Such situation calls for transformation in uses of soil and application of soil science in different soil functions.

14.6 Conclusions

By and large, we are looking for function-based soil quality that is how well soil does what we want it to do. Stability of soil properties and functions is often driven by resistance (capacity to resist change) and resilience (capacity to return to prestressed condition). Thus, the future of soil and its science should not depend only on data and facts but requires sincere motivation, enthusiasm, creativity and reliability for sound interpretation and application. Whether we are producer or consumer or simply landowner, the health of soils may impact the future health of our food system and so our community or society and nation as a whole. In India, the future of soil-less farming (hydroponics) is limited, since its cost of cultivation is very high and it requires high technical management. However, the future of soil is not only challenging but full of opportunities.

References

Baveye PC, Baveye J, Gowdy J (2016) Soil "Ecosystem" services and natural capital: critical appraisal of research on uncertain ground. Front Environ Sci. https://doi.org/10.3389/fenvs.2016.00041

Bhattacharyya T (2014) Pedology: the grammar of soil science. In: 79th annual convention of ISSS. The 32nd Prof. J. N. Mukherjee foundation lecture

Bhattacharyya T, Pal DK (2015) The state of Indian soils. State of Indian agriculture. NAAS Publication, New Delhi, pp 6–35

Bhattacharyya T, Pal DK, Srivastava P, Velayutham M (2000) Natural zeolites as saviour against soil degradation. Gondwana Geol Soc Mag 16:27–29

Bhattacharyya T, Pal DK, Mandal C, Chandran P, Ray SK, Sarkar D, Velmourougane K, Srivastava A, Sidhu GS, Singh RS, Sahoo AK, Dutta D, Nair KM, Srivastava R, Tiwary P, Nagar AP, Nimkhedkar SS (2013) Soils of India: historical perspective, classification and recent advances in knowledge: a review. Curr Sci 104:1308–1323

Bhattacharyya R, Ghosh BN, Mishra PK, Mandal B (2015) Soil degradation in India: challenges and potential solutions. Sustainability 7(4):3528–3570. https://doi.org/10.3390/su7043528

Das BS, Sarathjith MC, Santra P, Sahoo RN, Srivastava R, Routray A, Ray SS (2015) Hyperspectral remote sensing: opportunities, status and challenges for rapid soil assessment in India. Curr Sci 108 (5):860–868

Engman ET, Chauhan N (1995) Status of microwave soil moisture measurements with remote sensing. Remote Sens Environ 51 (1):189–198

Gangopadhyay R (1932) Some materials for the study of agriculture and agriculturists in ancient India. N.C. Mukherjee & Co., Serampore

Garg KK, Karlberg L, Wani SP, Berndes G (2011) Biofuel production on wastelands in India: opportunities and trade-offs for soil and water management at the watershed scale. Biofuels Bioprod Biorefin Biofpr 5:410–430. ISSN 1932-104X

Harish D, Dhanalakshmi TS, Gireesh HR (2016) Power supply using earth battery, vol 4, issue 2. ITSI Transactions on Electrical and Electronics Engineering (ITSI-TEEE). ISSN (PRINT): 2320-8945

Hartemink AE (ed) (2006) The future of soil science. International Union of Soil Science, Wageningen, The Netherlands

IISS (2011) Vision 2030. Indian Institute of Soil Science, Indian Council of Agricultural Research, Nabi Bagh, Berasia Road, Bhopal. www.iiss.nic.in

IISS (2015) Vision 2050. Indian Institute of Soil Science, Indian Council of Agricultural Research, Nabi Bagh, Berasia Road, Bhopal. www.iiss.nic.in

Inayatullah S (2008) Six pillars: futures thinking for transforming. Foresight 10(1):4–21. https://doi.org/10.1108/14636680810855991

Jha P, Biswas AK, Lakaria BL, SubbaRao A (2010) Biochar in agriculture—prospects and related implications. Curr Sci 99 (9):1218–1225

Jiang Y-B, Zhong W-H, Han C, Deng H (2016) Characterization of electricity generated by soil in microbial fuel cells and the isolation of soil source exoelectrogenic bacteria. Front Microbiol 7:1776. https://doi.org/10.3389/fmicb.2016.01776. PMCID: PMC5099896, PMID: 27877168

Krishnan A (1988) Delineation of soil climatic zones of India and its practical applications in agriculture. Fertil News 33(4)

Lal S, Deshpande SB, Sehgal J (1994) Soil series of India. NBSS publication no 40, p 684

Leather JW (1898) The composition of Indian soils. In the agriculture ledger, pp 399–400

Lehmann J, Gaunt J, Rondon M (2006) Bio-char sequestration in terrestrial ecosystems—a review. Mitig Adapt Strat Glob Change 11:403–427. https://doi.org/10.1007/s11027-005-9006-5

Liu JW, Jin TT, Liu GH, Li ZS, Yang RJ (2014) Changes in land use and soil and water conservation of the upper and middle reaches of Heihe river basin during 2000–2010. Acta Geogr Sin. https://doi.org/10.5846/stxb291303010328

Mandal B (2011) Soil organic carbon research in India: a way forward. In: 29th JN Mukherjee—ISSS foundation lecture. UAS, Dharwad

Mandal DK, Mandal C, Velayutham M (2001) Development of a land quality index for sorghum in Indian semi-arid tropics. Agric Syst 70:333–350

Manna M, Rahman MM, Naidu R, Khanna SS (2018) Bio-waste management in subtropical soils of India: future challenges and opportunities in agriculture. Adv Agron. https://doi.org/10.1016/bs.agron.2018.07.002

Minhas PS (2006) Future of soil science: fostering multidisciplinary linkages. In: Hartemink A (ed) The future of soil science. International Union of Soil Sciences, Wageningen, pp 95–96

Mishra BB (2015) Soil classification: issues and opportunities for Indian system. In: The 26th Dr. S. P. Raychaudhury memorial lecture. J Indian Soc Soil Sci 63(Supplement):S41–S52

Mishra BB (2016) Indian system of soil classification: a way forward. Agric Res Technol Open Access J 3(2):555606. https://doi.org/10.19080/artoaj.2016.03.555606

Mishra BB (2017) Land economics vs land use planning. In: Proceedings of national conference on climate change and agricultural production. BAU, Sabour, pp 42–51

Murthy RS (ed) (1982) Benchmark soils of India. NBSS Publication, NBSS&LUP, Nagpur, p 374

Murthy RS, Sankarnarayana HS, Hirekerur LR (1978) Interpretation of soil map for available moisture in a given climatic type. J Indian Soc Soil Sci 26:63–70

Naidu LGK, Ramamurthy V, Challa O, Velayutham M (2017) Land evaluation for sustainable agriculture. Brillion Publishing Co., New Delhi, pp 356

Natarajan A, Krishnan P, Velayutham M, Gajbhiye KS (2002) Land resources of Kudankulam, Vijayapatti and Erukkandurai villages, Radhapuram taluk, Tirunelveli district, Tamil Nadu. NBSS&LUP (ICAR) publication no 36, p 277

NBSS&LUP (1994) Global assessment of soil degradation (GLASOD) guidelines. National Bureau of Soil Survey and Land Use Planning, Nagpur, India

NBSS&LUP (2005) Annual report-2005. ICAR-National Bureau of Soil Survey and Land Use Planning, Nagpur, India

Pal DK (2003) Significance of clays, clay and other minerals in the formation and management of Indian soils. J Indian Soc Sci 51:338–364

Pal DK (2018) Simple methods to study pedology and edaphology of Indian tropical soils. Springer International Publishing AG. Part of Springer Nature. https://doi.org/10.1007/978-3-319-89599-4. ISBN 978-3-319-89598-7

Pal DK, Bhattacharyya T, Deshpande SB, Sarma VAK, Velayutham M (2000a) Significance of minerals in soil environments of India. NBSS review series 1, 68 pp

Pal DK, Velayutham M, Deshpande SB, Srivastava P, Durge SL (2000b) Climate change, degradation, polygenesis in vertisols of the Purna valley (Maharashtra) and their resilience. NBSS publication no 83, p 35

Patra AK, Lakaria BL, Elanchezhian R, Neenu S, Mandal A, Sinha NK, Meena VD, Subba Rao A (2014) Annual report 2013–14. ICAR-Indian Institute of Soil Science, Bhopal, Madhya Pradesh, 132 pp

Rao DLN (2006) Maintaining the soil ecosystems of the future. In: Hartemink A (ed) The future of soil science. International Union of Soil Sciences, Wageningen, pp 116–118

Rao DLN, Patra AK (2009) Soil microbial diversity and sustainable agriculture. J Indian Soc Soil Sci 57(4):513–530

Ray SS, Singh JP, Dutta S, Panigrahy S (2002) Analysis of within field variability of crop and soil using field data and spectral information as a pre-cursor to precision crop management. Int Arch Photogram Remote Sens Spat Inf Syst C 34(7):302–307

Ray SS, Singh JP, Das G, Panigrahy S (2004) Use of high resolution remote sensing data for generating site-specific soil management plan. Int Arch Photogram Remote Sens Spat Inf Syst B 35(7):127–131

Raychaudhuri SP (1984) Development of soil productivity in ancient and mediaeval India. Indian Soc Soil Sci Bull 14

SAC (2016) Space Application Center.pdf, India. SAC Courier 41(3). Available at http://www.sac.gov.in/SACSITE/SACCourier/July2016.pdf

Samra JS (2006) Future of soil science. In: Hartemink A (ed) The future of soil science. International Union of Soil Sciences, Wageningen, pp 125–127

Sarma VAK (1984). Mechanisms and rate of release of potassium form potassium-bearing minerals in soils. In: Mineralogy of soil potassium. PRII review series I. Potash Research Institute of India, Gurgaon, Haryana, pp 55–61

Saxena RK, Verma KS, Srivastava R, Yadav J, Patel NK, Nasre RA, Barthwal AK, Shiwalkar AA, Londhe SL (2003) Spectral reflectance properties of some dominant soils occurring on different altitudinal zones in Uttaranchal Himalayayas. Agropedology 13 (2):35–43

Science (2004) Soils—the final. Front Sci 304(5677):1549–1700

Somasekhar M (2017) India on verge of looming soil crisis: report. Hindu Bus Line

Swaminathan MS (2005–2006) Reports of the national commission on farmers. Ministry of Agriculture and Cooperation and Farmers Welfare, GoI

Swaminathan MS (2010) Science and sustainable food security. World Scientific Publishers, p 436

Swaminathan MS (2012) Remember your humanity: pathways to sustainable food security. New India Publishing Agency, New Delhi, p 208

Swaminathan MS (2017a) In search of biohappiness, 2nd edn. World Scientific Publishers, p 232

Swaminathan MS (2017b) The quest for a world without hunger, vol 1. World Scientific Publishers, p 420

Varma SK, Goswami NN (1982) The Vedic views on soil as the soul of infinite life. In: Souvenir 12th international congress of soil science, New Delhi

Velayutham M (1997) Mapping soil resources for optimal land use planning. In: Singh GB, Sharma BR (eds) 50 years of natural resource management research. Division of NRM, ICAR, New Delhi

Velayutham M (2012) National soil information system (NASIS) and land resource mapping for perspective land use planning and pragmatic farm level planning. In: 4th Dr. S. V. Govindarajan memorial lecture. Madras Agric J 99(4–6):147–154. (https://sites.google.com/site/majmasu/archive/99-4-6)

Velayutham M (2016a) Soil and land use survey for state level perspective land use planning and pragmatic farm level planning. In: Rao MV et al (eds) Proceeding conference integrated land use planning for sustainable agriculture and rural development. Apple Academic Press, pp 16–19

Velayutham M (2016b) Land resource inventory for integrated land use planning. In: Souvenir international conference on "Integrated Land Use Planning for Smart Agriculture", pp 1–3

Velayutham M (2017) Soil health revolution—a must for sustaining sustainable agriculture. Lecture delivered at the 104th Indian science congress, SV University, Tirupathi, 4 Jan 2017

Velayutham M, Bhattacharyya T (2000) Soil resource management. In: Special publication for the international conference on management of natural resources for sustainable agriculture towards 21st century, New Delhi, 14–18th Feb. Soils, pp 1–133

Velayutham M, Pal DK (2016) Soil resilience and sustainability of semi-arid and humid tropical soils of India: a commentary. Agropedology 26(1):1–9

Velayutham M, Mandal DK, Mandal C (1999a) Agro-ecological delineation for promoting sustainable agriculture. J Soil Water Conserv India 4:1–26

Velayutham M, Mandal DK, Mandal C, Sehgal J (1999b) Agro-ecological sub-regions of India for planning and development. NBSS&LUP publication 35, 452 pp

Velayutham M, Bhattacharrya T, Ramamurthy K, Venugopalan M (2003) Natural resources in terms of soil, water and environment in India and their management in reshaping our earth. Bhoovigyan Vikash Foundation Publication. Tata_McG Raw Hill Publishing Co., New Delhi

Wilson MI (1999) The origin and formation of clay minerals in soils: past, present and future perspective. Clay Miner 34:7–25

Index

A
Ability of soil to restore its quality after disturbance, 234
Acidification and aluminium toxicity in acid soils, 239
Acid soils, 228
Acid Soils of India: Their Genesis, Characteristics and Management, 26
Acid sulphate soils, 112
Actual and potential productivity, 198
Actual productivity, 10
Adsorption-desorption dynamics, 155
Adsorption isotherms, 119
Aerobic decomposition, 152
Aerobic microbial oxidation, 52
Aggradational plain, 65
Aggregate formation and aggregate stability, 225
Agricultural colleges, 21
Agricultural School at Saidapet, Chennai, 21
Agricultural sustainability, 216
Agriculture, 11
Agriculture road map, 189
Agroclimatic as well as agro-ecological zones, 6
Agro-climatic region, 160, 225
Agro-climatic zones, 41
Agro-ecological sub-regions, 7
Agroforestry interventions for rehabilitation of lands, 228
Agro-technology transfer modules, 266
Albite, 111
Alfisols, 82, 239
All India Coordinated Research Projects, 228
Allophane-humic acid complexes, 120
Allophone, 114
Alluvial soils, 110
Alluvial soils of the Indo-Gangetic plains, 19
Alluvium, 85
Almora-Dudatoli Crystalline thrust, 65
Alteration by anthropogenic perturbations, 152
Alternate wetting and drying, 225
Alternating dissolution and re-precipitation, 120
Ammonification, 152
Amorphous ferri-aluminosilicates, 112
Amorphous minerals, 112
Amphibole, 112
Anaerobic continuous submergence, 225
Anaerobic decomposition, 152
Anaerobic microbial reduction, 52
Annual mean temperature, 48
Anorthite, 111
Antibiotics and anticancer drugs, 253
Anti-diarrheal drugs, 253
Application of gypsum, 33
Applications of remote sensing, GIS, 267
Arabian Sea, 46
Arable and non arable lands, 213
Aravalli, The, 68
Archaean Era, 60
Archaeological records and literatures, 18
Argillic horizon, 96
Aridisols, 82
"A Soil' or "Pedon', 4
Assam-Burma ranges, 65
Atmospheric circulation patterns, 42

B
Backbone of construction industry, 244
Back water, 91
Ball clay, 245, 246
Basic principles of soils, 10
Beidillite-nontronite type of minerals, 118
Below-ground biomass-C inputs, 235
Benchmark series, 105
Benchmark soil as a reference point, 160
Benchmark soils, 7, 224
Benchmark soils in agro-ecological regions, 160
Benchmark soils of India, 35
Benchmark soils with corresponding soil series and classification, 183
Beneficial soil fauna, 151
Benefit cost ratio, 229
Bengal famine in 1943, 23
Bengal gneiss, 245
Bentonite, 119
Bentonite as a detoxifier, 253
Bentonite clay, 119
Bentonite linings, 245
Bidirectional reflectance distribution function, 105
Biochar, 267
Biochar application, 239
Biochar in agricultural soils, 153
Bio-climatic systems, 82
Bio-engineering measures, 228
Bio-enriched compost, 224
Biofertilizer, 9, 224, 267
Biofuel production, 268
Biogas Technology Development, 229
Biogeochemical cycles of soil nutrients, 143
Bio-inoculants, 267
Biological buffering capacity, 235

B. B. Mishra (ed.), *The Soils of India*, World Soils Book Series,
https://doi.org/10.1007/978-3-030-31082-0

Biological resilience in soil, 236
Biotite, 116
Black cotton or regur soils of the Deccan plateau, 19
Black cotton soil or regur, 82, 263
Black (*regur*) soils , 17, 111, 217
Brahmaputra basin, 65
Brick and tile clays, 246
Bricks and clay roof tiles, 248
Bright soil future, 261
Bronze Age Harappan civilization, 139
Brown forest soils, 96
Buffering capacity of soil, 235
Buffering features and resilience, 243
Buildings, schools, colleges, 244
Bulk density, 236
Bundelkhand gneiss, 68
Buried combustible geologic deposits, 257
Burning of crop residues, 52

C
Ca-feldspars, 115
Calcic and petrocalcic horizons, 138
Calcium carbonate, 118
Cambic, 96
Capacity of a soil to function within ecosystem boundaries, 215
Capacity of a soil to recover its functional and structural integrity, 234
Carbon dioxide, 52
Carbon dioxide budget, 52
Carbon fixation mechanism/photosynthesis, 151
Carbon sequestration, 32, 268
Carbon sequestration potential, 225
Carbon, the basic building block of life, 151
Carrier of ancient plant parts for synthesis of fossil fuels, 243
Catena, 117
Catenary relationships, 132
Cation exchange capacity, 195, 217
Centuries-old experience of the Indian farmers, 23
Ceramic pottery, 252
Charge characteristics of Vertisols, 118
Chemical buffering capacity, 235
Chemical degradation, 217
Chemihydropedoturbation, 135
Chlorite, 110
Chloritized vermiculite, 110
Classifying the Indian soils, 7
Clay based bricks, 248
Clay content, 239
Clay enriched B-horizon, 92
Clay-humus complex, 120
Clay illuviation, 130
Clay Minerals Society of India, 22
Clay mineral structure, size and crystallinity, 124
Clay-organic interactions, 120
Clay or soil eating, 255
Clay pedofeatures and pedogenic $CaCO_3$, 134
Clay roofing tiles, 248
Clay tablets, 255
Climate change, 42, 269
Climate change abatement, 145
Climate mitigation and Bioremediation, 32
Climatic database, 194
Climatic zone concept, 6
Climatic zones of India, 82
Clinical and medical relevance of type specific soils, 253
Cluster Approach and Participatory Guarantee System, 229
C:N ratio, 48
Coal mining, 257
Coastal plains, 77
Coated calcium carbide and dicyandiamide, 225
Coefficient of improvement, 10, 213
Coefficient of linear expansion, 119
Cold arid shallow skeletal soils, 160
Cold waves, 42
Compact and ferruginous variety of laterite, 247
Compacted soil, 239
Comparative assessment technique, 223
Complex computer simulation, 155
Complexity of subsoils, 134
Concept of land economics, 192
Conservation agriculture, 10, 33, 234, 255, 268
Conservation and improvement measures, 234
Construction, conservation and land fillings, 243
Constructions of houses, 244
Contamination of urban soils, 263
Coordinated research projects on soil productivity and fertilizer use, 27
Corporate sector, 255
Correctable limitations, 213
Correlation of classification system, 103
Cost effective amelioration techniques, 228
Creation of taxonomic grouping for classifying a soil, 265
Cretaceous period, 61
Critical aspects of problematic soils, 262
Critical limits of nutrients, 28
Crop growth models for land evaluation, 210
Crop performance, 197, 198
Crop simulation models, 212
Crystalline minerals, 112
C-sequestration, 114
Cuddapah, The, 68
Cultivation in hydroponics, 263
Current land use, 195
Cycling of nutrients by diversification of land-use types, 145
Cyclone Phailin, 47
Cyclone prone, 41
Cyclonic storms, 43

D
Dagshai formation of the Himalayan Forestland, 138
3D and 4D soil-landscape models, 263
Decalcification of carbonates, 136
Deccan traps, 59
Decline in soil quality, 6
Decline in soil quality and risk of soil degradation, 191
Decorative ceramics, 250, 252
Degradation, 234
Degree of profile development, 85
Dehydrogenase activity, 223
Delineation of benchmark soils, 264
Demand for land, 233
Depolymerization of polymeric soil organic carbon, 151
Desertification and to reduce erosion hazards, 32
Detoxifying potential of clays, 253
Devonian fossils and corals, 63
Dharwar system, 62
Diara and Tal land, 7
Diara land, 67
Diffuse double layer, 119
Dioctahedral low-charge smectites, 116
Direct-seeding of rice, 225
Distribution of soil health cards, 228

Diversion of land, 6
Diversity of soils, 263
Division of Soil Science and Agricultural Chemistry, 20
Dominance of kaolinite in pre-monsoon Gangetic silt, 110
Double of the farmer's economic growth, 213
Drought-prone zones, 7
Dynamic soil-water-air-plant-aquifer-atmosphere system, 266

E
Earthen pots, 249, 250
Earthenwares, 249
Earth's electrons, 4
Eastern Ghats and the Tamil Nadu Upland, 75
Eco-friendly source of soil organic carbon, 222
Economical and ecological viability assessment, 196
Economic enrichment, 262
Eco-physiological index, 235
Ecosystem and climate change mitigation, 243
Ecosystem requirements, 222
Ecosystem restoration, 262
Ecosystems, 41
Ecosystem services, 222, 262
Effective biofertilizer strains of Rhizobium, 224
Effective detoxifier, 253
Effective microorganisms, 154
Effective microorganisms-based composts, 154
Elastic attributes, 235
Electron microscope, 267
El-Nino events, 44
Embankments, dams including land fillings, 244
Emerging deficiencies of secondary and micronutrients, 29
Emerging threat, 233
Employment generation, 262
Employment provider, 13, 255
Endogenous factors, 235
Energy crisis, 256
Energy from soil battery, 256
Energy from the soil, 256
Energy-hungry planet, 258
Engineering, 11
ENSO events, 50
Entisols, 82
Entisols and Aridisols, 91, 161
Entrepreneurial skills, 263
Environmental consequences of anthropogenic perturbations, 155
Erosion induced loss in crop production, 222
Establishment of agricultural universities, 27
Eurasian plate, 60
Evaluation of soil structure directly in the field, 130
Evergreen revolution, 213
Exchangeable sodium percentage, 101, 119
Exogenous factors, 235
Experiential Learning Programme, 263
Extent and distribution of different soil types, 91
Extreme climatic events, 41

F
Factor ratings, 196
Fallow periods to restore fertility, 17
Farm level planning, 267
Farmyard manure, 236, 239
Feldspars, 111
Fe-rich clay pedo features, 118
Ferrihydrite, 114
Ferruginous Alfisols, 116
Fertigation, 228
Fertile floodplains, 17
Fertility capability classification, 198
Fertilizer Control Order, 228
Fertilizer response ratio, 217
Fertilizers encapsulated in nanoparticles, 253
Filtration, storage and detoxification of water and nutrients, 243
Fireclays, 246
First green revolution, 262
First large ammonium sulphate plant at Sindri, 23
Fixation of FA and HA by allophone, 120
Flash flood damage, 44
Floodplain soils, 4, 7
Fluvisols in USDA soil taxonomy, 85
Fly ash, 239
Food-secure world, 264
Food Security Act, 264
Forest and hill soils, 94
Fossil fuels, 257
Foundation base for plant growth, 243
Four major soil regions, 19
Free-swell index, 119
Frequent flood and drought, 263
Functions of a soil, 243
Fundamental soil policies, 5

G
Ganga plains, 67
Genesis and development of soils, 130
Geo-informatics, sensors, 267
Geo-referenced soil fertility maps, 223
Ghaggar plain, 66
GHG mitigation, 52, 53
Gibbsite, 111
GIS tools, 189
Gleyic, 133, 136
Globally-averaged atmospheric levels, 48
Global Soil Partnership, 264
Global warming, 52, 263
Global warming potential, 154, 225
Gold, 257
Gondwana, 59
Gondwana clay deposits, 119
Gondwanas series, 60, 69
Government of India, 228
Great Bihar Earthquake of 1934, The, 20
Greater Himalayas, 64
Great famine in Bengal and Orissa, 19
Great plains, The, 64
Greenhouse effect, 52
Greenhouse gas emission, 145, 222
Green manuring with *sesbania*, 228
Green Revolution, 21
Groundwater resources, 263
Grow more food, 23

H
Hailstorm, 43
Halloysite, 112

Harappa civilization, 250
Healing the wounded soils, 270
Heat and cold wave, 50
Heavy metals, 9
Higher resolution mapping of soils using GIS, 133
Hill agro-ecosystem, 234
Hilly soils, 111
Himalayan orogeny, 61, 63
Himalayas, The, 63
Histosols, 82
Hornblende, 121
Hot-arid desert and saline soils, 161
Hot-arid red and black soils, 164
Hot humid/per-humid red and lateritic and alluvial soils, 182
Hot semi-arid coarse-loamy alluvial soils, 164
Hot semi-arid moderately deep black soils, 167
Hot semi-arid old alluvial soils, 165
Hot semi-arid to dry sub-humid alluvial and Tarai soils, 165
Hot semi-arid with mixed red and black soils, 168
Hot semi-arid with red loamy soils, 169
Hot sub-humid alluvial soils, 173
Hot sub-humid coastal and deltaic alluvial soils, 181
Hot sub-humid loamy to clayey alluvial soils, 176
Hot sub-humid red and lateritic soils, 171
Hot sub-humid red and yellow soils, 171
Hot sub-humid with moderately deep black soils, 170
Household materials and utensils, 249
Human health and clinical therapy, 243
Human shelter, 244
Humification rate, 222
Hungry planet, 258
Hydrobiotite, 110
Hydropedology, 266
Hydroponics, 5
Hydroxy-interlayered smectite, 115
Hydroxy-interlayered vermiculite, 115
Hydroxyl-aluminium polymeric component, 112
Hyper-spectral remote sensing, 223, 269

I
Idols of gods or goddesses, 252
Illite, 109
Illuvial clay-rich B horizons, 136
Illuvial horizons, 134
Imbalanced fertilizer use, 9
Imogolite, 114
Imperial Agricultural Chemist, 20
Imperial Agricultural Research Institute at Pusa in Bihar, 20
Improper land use/land cover, 234
Inceptisols, 82, 237, 239
Incomplete reduction process, 152
Increasing magnification with polarized and reflected light, 129
Indian Council of Agricultural Research, 21
Indian craton, 59
Indian Ocean, 45
Indian Society of Soil Science, 22
Indian Society of Soil Survey and Land Use Planning, 22
Indian soil classification scheme, 98
Indian system of soil classification, 25, 132
Indian vedic views, 261
India's energy demand, 268
Indo-Australian plate, 60
Indo-Gangetic alluvium, 82, 263
Indo-Gangetic plain, 63, 217
Indus civilization, 250
Indus plains, 66
Industrial uses of soils, 244
Industry, 246, 247
Indus Valley civilization, 18
Information Age, 267
Integrated land use planning, 266
Integrated mitigation options, 42
Integrated nutrient management, 153, 228, 236
Integrated Watershed Management Programme, 229
Interaction of clay minerals with pollutants, 121
Interactions between soil microorganisms and its physical environment, 149
Interglacial soil, 136
Intermittent wetting and drying, 154
12th International Congress of Soil Science, New Delhi, 1982, 22
Interstratification of Sm/Kao minerals, 118
Intra zonal soils, 94
Iron compounds responsible for brunified B horizons, 134
Iron manganese features, 136
Irreversible degradation, 234
Islands, 64

J
Jhelum and Ravi, 64
Jurassic period, 60

K
Kaolin and attapulgite, 245
Kaolinite, 109
Kaolin or china clay, 245
Karewa Group in Kashmir, 136
Kerala plain, 77
Kharif season, 8
Konkan plain, 75
Kosi and Brahmaputra rivers, 64
Krishi Vigyan Kendras, 266, 267
Kumhar or kumbhkar, 250, 251
Kumhar (potter), 2

L
Land area shrinkage, 6
Land capability classification, 198, 213
Land characters and land qualities, 194
Land economics and land use planning, 265
Land evaluation, 192
Land evaluation activities, 193
Land evaluation and soil testing, 255
Land evaluation approaches, 197
Land evaluation for sustainable agriculture, 267
Land irrigability classification, 198
Land potential evaluation, 266
Landscape and watershed levels, 266
Land shrinkage, 4
Land suitability assessment, 193
Land suitability assessment by parametric approach, 198
Land suitability by fuzzy AHP and TOPSIS methods, 198
Land unit, 193, 194
Land use change intensity Index, 266
Land use changes, 192
Land use intensification, 2
Land use planning, 10, 192, 193, 264, 265
Land use planning and analysis system, 198
Land use planning at grass root level, 229

Land use plans, 189
Land use requirements, 195
Land utilization regions, 266
Land utilization type, 195
Langmuir and Freundlich adsorption equations, 120
Large scale land resources database, 106
Late Pleistocene-Holocene sedimentary successions within the Ganga plains, 137
Late Proterozoic, 60
Late Quaternary palaeosols, 136
Laterite, 19, 247
Laterite as a building stone, 247
Laterites and lateritic soils, 93
Laterite soil, 82, 111, 263
Leaching and denitrification or volatilization, 217
Leaf colour chart, 225
Lesser Himalayas, 64
Li-clays, 118
Life-supporting enterprises, 6
Life-supporting functions, 4
Lightening candle or *deep*, 249
'Light festival' of India, 249
Light texture soils, 23
Lime requirement, 34
Liquid biofertilizers, 267
Lithology and Soil, 61
Living and non-living components of an ecosystem, 143
Long term fertilizer experiments, 228
Lorentz-Polarization factor correction, 124
Loss of biodiversity, 4
Loss of resilience, 234
Lower Paleozoic, 63
Lowest boundary, 4
Low organic matter, 217

M
Macro Management of Agriculture, 229
Madagascar and southern Africa, 60
Maharashtra plateau, 75
Major land resource areas, 159
Major soil groups, 82
Management measures to overcome the limitation, 205
Management of Salt-Affected Soils and Use of Saline Water, 33
Mangalore tiles, 249
Man-made environments, 263
Matching of crop requirements, 201
Mean surface air temperatures, 48
Measures to enhance soil resilience, 235
Medical science, 12
Medicinal values of bentonite or montmorillonite clays, 253
Mesozoic, 63
Metabolic quotient, 223
Metahalloysite, 110
Metal contamination, 9
Methane, 54
Mica-montmorillonite and vermiculite-chlorite interstratified minerals, 112
Mica-rich parent materials, 112
Mica-smectite or mica-vermiculite mixed-layer minerals, 110
Micro- and Secondary Nutrients and Pollutant Elements, 228
Microbial degradation of pesticide, 154
Microbisols, 154
Microwave remote sensing, 47
Mineralization of organic nitrogen, 152
Minerals Act 1957 of India, 244
Mining activities, 234
Mining zeolites from shrink-swell soils, 247
Mission Organic Value Chain Development for North Eastern Region, 229
Mitigating climate change, 235
Mitigation of climate change, 262
Mitigation of CO_2emission, 258
Mixed layer minerals, 110
Modification for different soil orders of India, 98
Mollic epipedon, 96
Mollisols, 82
Mollisols in the Western Ghats, 119
Monazite, 257
Monsoonic climate, 6
Montmorillonite, 109
Montmorillonite, vermiculite, illite and a few unrecognized minerals, 110
Mother Earth, 2
Mother Earth (*Bhoomidevi*) - Mother Land - Mother Tongue – Mother, 261
Mottles, 136
Mud therapies, 253
Mudwall, 247
Multani mitti, Fuller's earth, 254
Multani mitti in India, 254
Multi-nutrient deficiencies, 8, 217
Multiple electrodes in series, 256
Multiple types of disturbances, 234
Municipal solid waste compost, 224
Mycorrhiza, 9
Mycorrhizal fungi, 153

N
Nanotechnology, 267
Nano-technology and nano-clay, 252
National Centre of Organic Farming, 229
National Commission on Farmers, 264
National food and nutritional security, 27
National Food Security Mission, 229
National Horticulture Mission, 229
National Mission on Sustainable Agriculture, 228
National Mission of Sustainable Agriculture, National Mission on Soil Health Cards, 228
National Network on carbon credit and carbon sequestration, 235
National Policy for Farmers, 6
National Rehabilitation and Resettlement Policy, 6
National roadmap for assessment of soil quality, 233
National soil classification scheme in India, 264
Nationwide soil databases and soil interpretations, 159
Nation-wide soil survey, 263
Natural disasters, 41
Natural zeolites as building stone, 247
Negative outcomes of land-management practices, 145
Net-C balance, 153
Network Project on Soil Biodiversity-Biofertilizers, 228
Nilgiri gneiss, 57, 73, 169
Nitrification inhibitors, 225
Nitrogen and carbon dynamics, 24
Nitrogen in soils, crops and fertilizers, 26
Nitrous oxide, 54
Nitrous oxide emission and nitrate pollution, 225
Nodulation and nitrogen fixation, 224
N_2O emission, 154

Non-renewable resource, 3, 234
Northeast Monsoons/Post-Monsoon season, 42
North Indian Ocean, 45
No-till agriculture, 225
Nutrient balance, 216
Nutrient based fertiliser subsidy to improve soil health, 228
Nutrient Based Subsidy, 228
Nutrient Based Subsidy scheme, 228
Nutrient demand and supply, 145
Nutrient index, 29
Nutrient mining, 1, 4, 7, 28
Nutrient retention, 217
Nutrient stripping by crop off-take, 145
Nutrient turn over, 9
Nutrient use efficiency, 8, 9, 32, 33, 217, 224

O

Ochric epipedon, 96
Olivine, 120
Open system, 4
Optical properties, 129
Organic amendments and balanced fertilization, 153
Organic matter, 265
Organic Matter and Organic Residue Management for Sustainable Productivity, 26
Organo-clay complex, 245
Organo-clays, 253
Organo-metallic complexes, 115
Organophosphorus compounds, 152
Origin of pottery, 250
Outer Himalayas, 64
Overall sustainable development, 262
Over-reliance on chemical fertilizers, 222
Oxisols, 82

P

Palaeosols dating to the middle and late Pleistocene, 136
Palaeosols in the Thar desert, 137
Palaeosols of the Middle and Upper Siwalik Groups, 137
Paleosols, 118, 264
Palygorskite mineral, 117
Paramparagat Krishi Vikas Yojana, 229
Parent material, 57
Partial factor productivity, 4, 8, 233
P-deficient tropical soils, 153
Peat deposits, 252
Peaty and marshy soils, 96
Pedal microstructure and blocky aggregates, 136
Pedogenesis, 77, 234
Pedogenesis on moon, 266
Pedogenetic processes, 266
Pedogenic and diagenetic features of lithified paleosols, 138
Pedogenic carbonates, 118
Pedogenic carbonates in Holocene soils of the Gangetic Plains, 137
Pedogenic clay coatings, 129
Pedologic as well as edaphologic limitations, 10
Pedology and edaphology, 140, 143
Pedomedicine, 269
Peds and voids, 139
Peninsular India, 45
Peninsula, The, 64
Per capita arable land, 262
Permanent manure trials, 22
Permanent plot experiment, 19
Permo-Carboniferous glaciations, 60
Permo-Triassic, 63
Pesticide molecules, 124
Petrographic microscope, 129
Phipps Laboratory, 20
Phosphatase activity, 223
Phosphate solubilising bacteria, 224
Phosphate solubilizing microorganisms, 32
Phosphocompost, 224
Phospho compost technology, 32
Phosphorus in soils, crops and fertilizers, 26
Photopedogenesis, 261
Photopedology, 266
Phyllosilicate, 121
Physical buffering capacity, 235
Physical environment of soil, 221
Physical features on landscapes, 59
Picnic spots, 244
Plagioclase feldspars, 118
Plant Growth Promoting Rhizobacteria, 9, 224
Platy microstructure, 134
Playgrounds, 244
Plinthite, 93
Ploughing of soils, 17
Podzolization, in the Himalayas, 96
Policies and governance, 5
Polluted planet, 258
Polygenetic features, 118
Population explosion, 234
Pore size distribution, 119
Potassium in Soils, Crops and Fertilizers, 25
Potentially mineralizable N, 223
Potentially viable soils, 6
Potential productivity, 10
Pottery and Ceramic, 250
Pottery, porcelain and refractory materials, 246
Pottery production, 251
Poultry manure, 239
Poverty alleviation, 255, 262
P-poor soil, 152
P-poor tropical soils, 145
Pradhan Mantri Krishi Sinchayee Yojana, 229
Precambrian, 62
Preferred land-use options, 41
Pre-monsoon season, 43
Pre-monsoon season and post-monsoon season, 45
Prime agricultural land, 106
Prime and marginal lands, 213
Prime land classification, 198
Principles of land evaluation, 193
Production function of soil, 233
Production of NP complex fertilizers, 29
Production of organic fertilizers, 228
Production potential of a type specific soil, 263
Productivity, 263
Protection and conservation of soil, 264
Protection of soil resource, 5
Protective medical therapy, 269
Protective medical treatment, 255
Proterozoic orogen, 60

Pseudo-chlorite, 115
P-solubilizing microorganisms, 152
Public awareness and policy, 5

Q
Qualitative and quantitative approaches for land evaluation, 198
Quartz, 110
Quartzite, 115
Quaternary period, 63

R
Radioisotopes, 267
Rainfed conditions and waterlogging, 217
Rajasthan bentonite, 245
Rajmahal hills, 60
Rajmahal Trap, 111
Raw materials, 243
Reclamation and Development of Alkali and Acid Soils, 229
Reclamation of sodic soils, 23
Recognition of a need for change, 193
Recycling of organic residues, nutrient elements and waste materials, 243
Red and laterite soils of the peninsular and east India, 19
Red and lateritic soils, 93, 217
Red and yellow soils, 111
Red sandy soils, 111
Red soil, 82, 263
Reflectance of earth surface features, 269
Regular interstratification of mica-vermiculite, 112
Remediation of degraded, problematic and polluted soils, 32
Remote sensing application with GIS tools, 263
Replenishment by mineral weathering of soils, 145
Requirement of food demands, 262
Residence time, 263
Resilience, 263
Resilience index, 239
Resilience of degraded Vertisols, 234
Resistance, 263
Resource-conservation technologies, 155
Resource-conserving technologies, 155
Resource-efficient evaluation measures, 229
Resource use efficiency, 213
Restoration and enrichment of biodiversity, 243
Re-union hotspot, 59
Revolution in informatics and communication, 264
Rheological properties of clays, 119
Rhizoliths, 118
Rice root intensification, 154
Rigvedic Indian civilization, 18
Riverbank erosion, 5
Roads and highways, 244
Rooting depth of crops, 235
Rothamsted Classical Experiments, 22
Rothamsted model, 19
Royal Agricultural Society, 19
Royal Commission on Agriculture, The, 20
Rural universities, 27
Russian soil classification schemes, 98

S
Saccharum spontaneum, 4
Saline alkali soils, 112
Salinity and alkalinity, 217
Salinity and alkalinity in soils, 233
Salinity and sodicity, 228
Salt affected soils, 94
Sandstone, 115
Scarce land resource, 4
Second green revolution, 13, 213
Second World War, 23
Sediment chemistry, 152
Self-terminating process, 118
Separation of Madagascar and India, 59
Sequestration of inorganic C, 119
Serving farmers and saving agriculture, 264
Sesquioxides, 92
Shale, 115, 246
Shape and volume of pores, 134
Shifting agriculture, 234
Shrinkage of productive land area, 264
Shrink-swell soils, 117
Silica: alumina and SiO2: sesquioxide molar ratio, 112
Silica sand, 244
Silt as a soil conditioner, 245
Siltation, 244, 245
Simulation models, 212, 213
Si-poor heulandites, 119
Siwaliks, 66
Slow permeability, 164, 166, 167, 169, 181, 199
Smart Soil Science Education, 11
Smectite, 110
Socio-economic data, 197
Soil acidification or alkalization, 235
Soil amendments, 239
Soil and carbon trading, 257
Soil and human health, 263
Soil and land evaluation, 264
Soil and reflectance data for spectral library, 105
Soil and society, 263
Soil and space, 263
Soil as an entity in itself, 263
Soil as a part of ecosystem, 263
Soil as a science and in art, 263
Soil as a system, 8
Soil-associated crop requirements, 195
Soil available water capacity, 266
Soil based Agro-technology Transfer, 106
Soil based medical treatment, 269
Soil battery and soil cell, 268
Soil battery and the solar panel, 256
Soil being the lowest boundary of the earth's atmosphere, 262
Soil biodiversity, 9, 222, 235, 269
Soil biodiversity loss, 264
Soil biogeochemistry, 269
Soil biogeochemistry modeling, 155
Soil biotechnology, 269
Soil-borne pathogens, 13
Soil carbon sequestration, 54, 222
Soil classification, 25, 98
Soil classification system, 159
Soil colloids and their relationship to the soil-fertility problem, 24
Soil conservation research, 24, 32
Soil contamination/pollution, 35
Soil C sequestration, 145
Soil degradation, 4, 262
Soil degradation by mismanagement, 7
Soil enzymes, 236
Soil evaluation, 9
Soil evaluation and land use planning, 267

Soil evaluation procedure, 233
Soil evolution from time zero, 263
Soil fertility, 216
Soil fertility maintenance programme, 28
Soil fertility maps for major and micro-nutrients, 27
Soil forming processes, 131
Soil generator, 257
Soil health, 4, 154
Soil health' and soil quality, 216
Soil health assessment, 223
Soil health cards, 223
Soil health care revolution, 268
Soil-health enhancement, 222
Soil health (or quality) management, 145
Soil index rating, 198
Soil in relation to other sciences and industries, 263
Soil less farming, 263
Soil Macro-organisms, 149
Soil-made *kullhar*, 2
Soil management for sustainable agriculture in dryland areas, 26, 33
Soil management in relation to land degradation and environment, 26
Soil map based on the Russian concept, 25
Soil map by placing the soils in different climatic zones, 25
Soil map of India, 82, 263
Soil map on geological formations, 25
Soil microbial biomass C and N, 154
Soil microbial diversity, 32
Soil micromorphology, 269
Soil museum, 1
Soil organic matter quality, 34
Soil physical quality, 222
Soil pollution and remediation, 32
Soil quality and carrying capacity of the soil resources, 229
Soil quality index formulation, 225
Soil related constraints in crop production, 26
Soil resilience and resistance, 239
Soil resilience in rainfed soils, 234, 239
Soil resource inventory, 189, 239
Soil resources map, 7
Soil Resources Mapping programme, 82
Soil respiration, 223, 239
Soil science-based medical education, 11
Soil science education, 11, 262
Soil Science in India, 26
Soil science research, 11
Soil series, 160
Soil series qualifying for Benchmark, 160
Soil series, soil associations and other soil mapping units, 194
Soil's function, 41
Soils in different climatic zones, 82
Soils in industries, 263
Soil-site parameters, 197
Soils of different agro-ecoregions, 85
Soil stabilization, 250
Soil suitability classification, 198
Soil suitability classification-statistical approach, 198
Soil survey, 159
Soil survey and classification aid in transfer of technology, 201
Soil Taxonomy classification, 216
Soil Test and Fertilizer Recommendation, 223
Soil test based balanced and integrated nutrient management, 224
Soil test based fertilizer recommendations for targeted yields of crops, 27
Soil Test Crop Response, 228
Soil test crop response correlation, 27
Soil testing programme, 35, 255
Soil testing service, 223
Soil to recover to its original potential, 234
Soil to serve the needs of humans, 264
Soil water balance, 8
Solid and liquid wastes, 9
SOM turnover, 145
Sorption-desorption studies of 134Cs on clay fractions, 122
Source and sink of carbon, 54
Southwest monsoons, 42
Spatial and temporal soil changes, 266
Spatial pattern of mean annual temperature, 50
Spectral library for Indian soils, 105
Spodosols, 96
Stabilization of SOC, 153
Stabilization of soil structure, 250
Straight nitrogenous fertilizers, 29
Strengthening of land ownership, 264
Streptomycin, 253
Studies with 15N and 32P for the balance sheet of nitrogen and phosphorus, 27
Subsidy for fertilizers fortified with zinc and boron, 228
Subsidy on city compost, 228
Subsoil hard pan, 23
Subsoils B-horizon, 96
Sub-surface drainage and bio-drainage technologies, 228
Sub-surface drainage technology, 229
Suitable crop rotations, 34
Summers and winter monsoons, 44
Supercontinent of Pangaea, 60
Supply chain process, 192, 255, 264
Surface crusting and hardening, 33
Sustainability, 235, 263
Sustainable agricultural production, 191
Sustainable agriculture, 32
Sustainable soil health, 8
Sustainable Soil Productivity under Rice-Wheat System, 26
Sustainable uses of soil resource, 264
Swachh Bharat Abhiyan, 229
Swelling for moisture retention, 217
System of Rice Intensification, 225

T

Tal land, 67
Task Force on balanced and integrated use of fertilisers, 228
Technology revolution, 264
Tectonic evolution, 60
Tertiary period, 63
Tethys Ocean, 60
Tetrahedral and octahedral charge, 118
Thar desert, 42
Thematic maps on N, P, K index values using ARC GIS platform, 28
Theory of soil acidity, 24
Thermal and photo-chemical fixation of atmospheric nitrogen, 25
Thermodynamic models, 120
Thin sections of a soil, 129
Thirsty planet, 258
Thixotropic critical stresses, 119
Threats to soils, 233, 265
Threshold limits of heavy metals, 9
Tile industry, 250
Total food basket, 217
Total food security, 229
Tourmaline, 120
Traditional Indian composting system into the Indore method, 23
Transformation in soil and soil science, 270

Transformation of smectite to kaolin, 116
Translocation and accumulation of soil components down the profile, 134
Trioctahedral high-charge smectites, 116

U
Ultisols, 82
Ultisols and Oxisols, 34, 82, 101, 217
Uncertain weather, 41
United Nations Millennium Development Task Force, 216
United Planters' Association of Southern India, 21
Universal soil classifications, 82, 98
Universal soil loss equation, 59
Untapped potential of soil resource, 264
Upper Paleozoic, 59
Urbanization, 5
Urvara (fertile) and *Anurvara* or *Usara* (sterile), 18
USDA 7th approximation in classifying the Indian soils, 25
USDA soil taxonomy, 98, 263
Use of radio tracers in soil fertility research, 27
US soil taxonomy, 98

V
Vale of Kashmir, 64
Value added fortified/customised fertilisers, 228
Vedas, the *Upanishads*, 17
Vermiculite, 110
Vermincompost, 224
Vertic features, 3
Vertisols, 82
Vindhyan, 59
Vindhyayan, 68
Visible near infrared diffuse reflectance spectroscopy, 223
Visible–NIR spectroscopy, 267
V V Dokuchaev, 4

W
Warm humid to per-humid loamy to clayey alluvial soils, 177
Warm per-humid red and yellow soils, 179
Warm per-humid shallow and skeletal soils, 177
Warm sub-humid to humid sub-montane shallow and skeletal soils, 174
Water logged saline soils, 23, 229, 230
Waterlogging, 233
Watershed Development Component, 229
Water trading, 8
Weathering rind of the earth's surface, 57
Wide nutrient gap, 9
Wind erosion, 228
Winter season, 43
12th World Congress of Soil Science, The, 22
World reference base for soil resources, 103, 132, 264

X
Xenobiotic compounds, 154
Xenobiotics application in agricultural fields, 145
X-ray diffraction, 109
X-ray diffractometer, 267

Z
Zeolite in nanoscale acts as a reservoir for nutrients, 247
Zeolite in sand fraction of calcareous Vertisols, 118
Zeolites, 247

MIX
Papier aus verantwortungsvollen Quellen
Paper from responsible sources
FSC® C105338

If you have any concerns about our products, you can contact us on
ProductSafety@springernature.com

In case Publisher is established outside the EU, the EU authorized representative is:
Springer Nature Customer Service Center GmbH
Europaplatz 3, 69115 Heidelberg, Germany

Printed by Libri Plureos GmbH
in Hamburg, Germany